CIVIL AND ENVIRONMENTAL SYSTEMS ENGINEERING

Charles S. ReVelle
The Johns Hopkins University

E. Earl Whitlatch, Jr.
The Ohio State University

Jeff R. Wright
Purdue University

PRENTICE HALL, Upper Saddle River, New Jersey 07458

Library of Congress Cataloging-in-Publication Data

ReVelle, Charles.
Civil and environmental systems engineering / Charles S. ReVelle, E. Earl Whitlatch, Jr., Jeff R. Wright.
p. cm.
Includes bibliographical references and index.
ISBN 0-13-138678-6
1. Civil engineering—Linear programming. 2. Environmental engineering—Linear programming. 3. Engineering economy—Linear programming. 4. System analysis—Data processing. I. Whitlatch, Elbert E. (Elbert Earl), 1942- . II. Wright, J. R. (Jeff R.)
TA153.R38 1997
624'.01'51972—dc20 96-220
CIP

Acquisitions editor: ***Bill Stenquist***
Editorial/production supervision and interior design: ***Sharyn Vitrano***
Editor-in-chief: ***Marcia Horton***
Managing editor: ***Bayani Mendoza DeLeon***
Copyeditor: ***Sally Ann Bailey***
Cover designer: ***Karen Salzbach***
Director of production and manufacturing: ***David W. Riccardi***
Manufacturing buyer: ***Julia Meehan***
Editorial assistant: ***Meg Weist***

Printed in the United States of America

10 9 8 7 6 5 4 3 2 1

ISBN 0-13-138678-6

Prentice-Hall International (UK) Limited, *London*
Prentice-Hall of Australia Pty. Limited, *Sydney*
Prentice-Hall Canada Inc., *Toronto*
Prentice-Hall Hispanoamericana, S.A., *Mexico*
Prentice-Hall of India Private Limited, *New Delhi*
Prentice-Hall of Japan, *Tokyo*
Simon & Schuster Asia Pte. Ltd., *Singapore*
Editora Prentice-Hall do Brasil, Ltda., *Rio de Janeiro*

To Penny, Janet, and
Delores

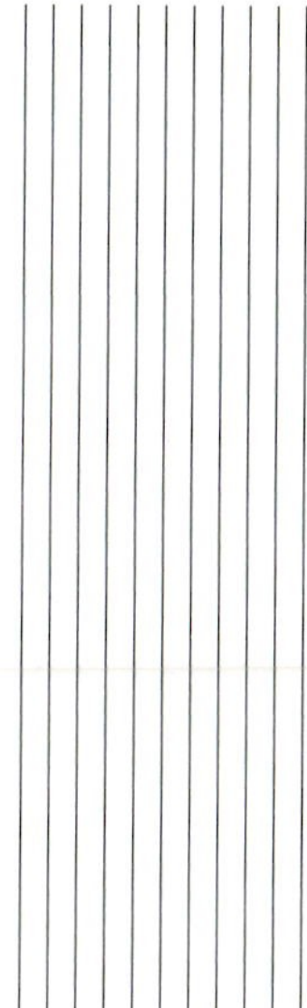

Contents

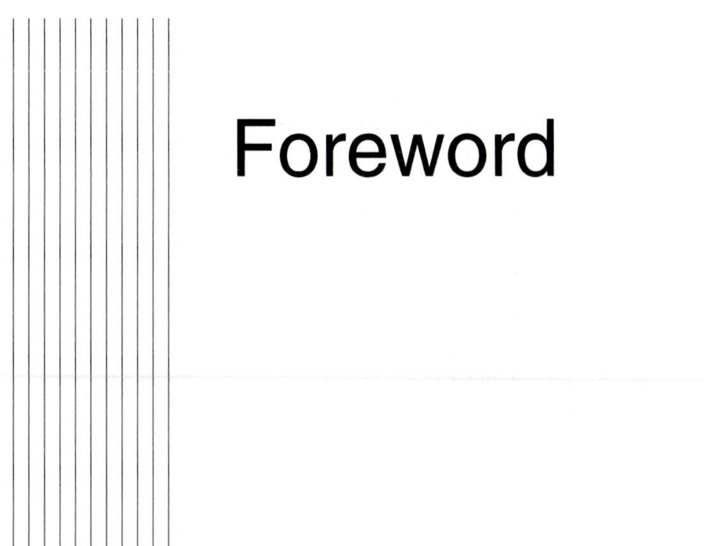

Foreword

Decisions, decisions, decisions! Everyone's lifetime is filled with an endless string of opportunities and challenges to decide or choose among alternative courses of action, and for each of these the consequences are real. Some are good, some are bad, and most are somewhere in between. Not everyone relishes the idea of making decisions, preferring the role of an interested bystander content with just letting things happen. Obviously there are consequences even if one decides to watch and not choose.

Civil and environmental engineers rarely have the luxury of being spectators since invariably they are retained primarily to present alternatives (choices) that serve the interests of their clients. Every client has a unique set of interests and perspectives. While such interests or concerns ofttimes have quite personal underpinnings, the overall goals and objectives of the decision maker may reflect very different concerns depending upon whether they see themselves as being in the private or public sector.

For example, a client who is a private developer of housing may recognize the need to build a water supply system in order that her houses are saleable. In doing so she is motivated to create a system that will meet acceptable design standards, but wants to spend the least amount possible on these facilities. (After all, she wants to sell houses—not water.) In contrast, a municipal engineer faced with a similar challenge, providing water supply for a new subdivision within the community, is surely concerned about creating a cost effective system, but would also have other concerns in mind. Such things as design and useful life of the facilities, sources of supply, future growth, cost of money, choice of technology, etc. are factors that are likely to produce different alternatives and consequences. The engineering science might be the same in each instance, but the engineering decision making environments are very different.

Just as there are powerful engineering technologies for designing water supply systems, there are powerful engineering management technologies for designing engineering decision strategies. This book presents these technologies in the context of civil and environmental engineering management. The authors bring to this volume the expertise and experience of superb model builders and extraordinary skills as teachers. Building models and finding ways to solve them are highly developed art forms that are essential elements of engineering. All engineering designs depend upon building and applying models that represent reality in a way that is manageable and sufficiently accurate. The test, of course, is whether they produce results that are meaningful and useful.

What the authors present in this excellent volume are the powerful tools that enhance and facilitate the decision-making processes, which is the very substance of what engineers have been and will continue to be asked to do. That's what systems engineering is about. And because these tools and skills can be applied in an endless variety of problems, issues, etc. they provide civil and environmental engineers the ability to address a broad range of management, policy, and design challenges. Armed with these tools, civil and environmental engineers will be better prepared not only to find better engineering alternatives, but also to extend the influence of our profession into spheres of planning and policy making, thereby improving the quality of life for us all from a more technical perspective.

WALTER LYNN
Director, Center for the Environment
Cornell University

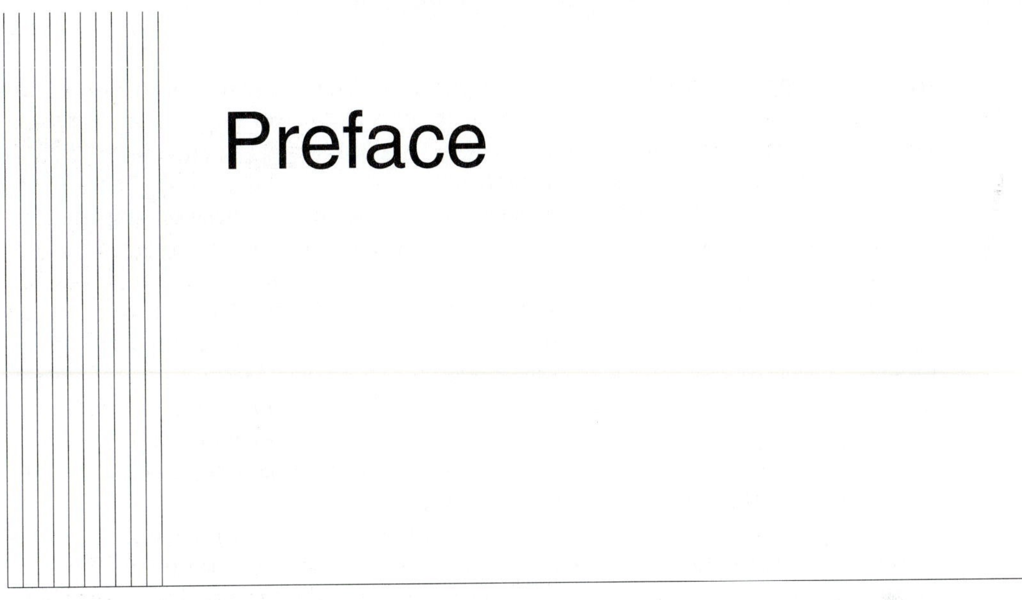

Preface

This text is designed for a junior or senior year course on systems analysis or systems analysis and economics as applied to civil engineering. This civil system/engineering economics course has evolved over roughly the last 25 years and draws on the fields of operations research and economics to create skills in problem solving. Because of the presence of several more advanced sections and sections focusing on applications, the book may also find use as a text for first-year graduate courses that introduces students to civil systems.

As the field of operations research evolved from its origins during World War II, one area in particular grew in popularity. That area, known as mathematical programming, found wide application as a means to optimize not only the design of chemical and mechanical systems in industry but also as a means to find promising alternatives in civil and environmental engineering decision problems. Most popular among the computer-based optimization techniques has been and continues to be the method known as linear programming, a procedure which operates on one or more objectives subject to economic, resource or logic constraints.

Mathematical programming and linear programming, in particular, have found wide application in civil engineering problem solving. These techniques have been used in structural design, in highway alignment, in intersection light timing, in subway and rail route design, in traffic prediction, in terminal location, in the routing of collection vehicles, in the routing of hazardous wastes, in equipment selection, in landfill location, in the siting of transfer stations, in crew scheduling and allocation, in waste treatment plant design and location, in waste load allocations on a river, in the design of hydrologic models, in the selection of projects to bid on, in the design of water distribution systems and sewer systems, in cost sharing, in reservoir design and operation, in fire station siting, in ambulance deployment, and in many other civil and environmental engineering areas. The power of these tools to develop efficient alternatives is enormous. So many applications have been created that a number of journals have been established with principal emphases on civil systems optimization problems; these include Water Resources Planning and Management, Transportation Science, Civil Engineering Systems, and Water Resources Research, and The Journal of Infrastructure Systems, among others.

Our treatment of linear programming and other forms of optimization is pragmatic. We prove no theorems but do, however, provide a description of how and why linear programming works. If we did not, we would be handing the student a "black box" and telling the student to "believe." Instead of theory, we offer application in large quantities to motivate the student to learn the methodologies. We first offer problems that are not terribly difficult to formulate, and then problems which demand greater skill to put in solvable forms. Our thrust is to build up skills in an orderly fashion as there are greater and lesser challenges in formulation and greater and lesser challenges in solution method. Later chapters are, of course, the most demanding. These later chapters are a unique feature of this text. Titled as "Lessons in Context" and the name of an application field, these chapter offer new techniques within the framework of a problem setting, a problem setting that demands the new methodologies that are then introduced. Our experience suggests that the "need" for the methodologies helps to motivate students to learn them.

A second focus of this book, in addition to linear programming and associated tools optimization, is the closely allied field of engineering economics. At first glance, our treatment of engineering economics would appear to be guided by the need to cover all topics necessary to prepare undergraduate engineers for the professional engineers examination on this important topic. These topics include the time value of money, cash flow analysis, and selection of economic alternatives. Because cost analysis and economic analysis over time is an important consideration in the development of models that help identify optimal management decisions—virtually all engineering management decisions in the public sector involve significant, and often enormous, cost considerations over potentially long periods of time—our presentation of engineering economics provides students a solid foundation upon which to compute important model parameters. Indeed, the modeling context provided for this important topic gives added relevance to this all too often suppositious subject.

These two related topics, optimization/systems analysis and engineering economics, are the core of this book. When the student has completed a course in these topics using this text, or has read this book independently, as it is quite possible to do, he or she will have learned the most modern skills available for the design, operation and evaluation of civil and environmental engineering systems.

The team of authors, ReVelle, Whitlatch and Wright, is well credentialed to provide a text that delivers both solid technical content and quality communication. ReVelle, a professor at Johns Hopkins for more than 25 years, studied with one of the originators of systems analysis in water management and teaches a course in civil systems regularly. ReVelle is also the author, with his wife Penelope, of The Environment, a basic college text that has appeared in three editions and more recently of The Global Environment. Whitlatch, a professor in civil engineering at Ohio State, has been teaching a popular and well-received civil systems course for over 20 years. Wright, a professor of civil engineering at Purdue and the editor-in-chief of the Journal of Infrastructure Systems, has been teaching courses on civil systems and engineering economics for more than 15 years. The authors have collaborated on research for many years. All three authors have distinguished records of research and application. They enjoyed writing the text together and will be interested in your comments.

CHARLES S. REVELLE
The Johns Hopkins University

E. EARL WHITLATCH, JR.
The Ohio State University

JEFF R. WRIGHT
Purdue University

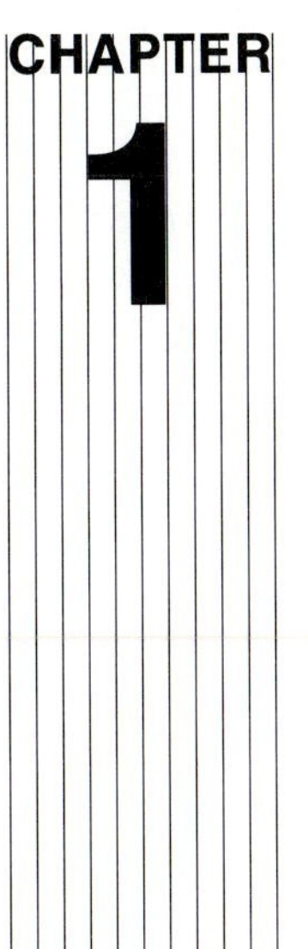

Explaining Systems Analysis

1.A INTRODUCTION: BUILDING MODELS

This book is about building mathematical models that assist in the design, management, or development of policy relative to natural or constructed systems. "Mathematical models" may not be a familiar term, but virtually all people have experience with models of other types. As children, many will have built and flown paper airplanes. Some will have built model airplanes of balsa or plastic. Others may have attached motors to these planes and, as the writer did, attempted to fly them. Such experiences led the writer to become a civil engineer as opposed to an aeronautical engineer and hence directly to this book.

When we were younger, most of us participated in models of constructed human systems. These models or simulations gave us the opportunity to experiment with other roles. You probably played cops and robbers, cowboys and Indians, space invaders, "house," doctor/hospital, circus, or any other games that allowed you to simulate other environments. Monopoly,™ while entertainment, tested financial acumen in an environment with much randomness. The military regularly conducts "war games" to test the readiness of its forces. Fire drills are used to simulate the conditions and situations that may occur during possible fire episodes. Physical and role-playing models pervade society and enrich it with the insights they provide.

Since the early 1900s, physicists, engineers, chemists, biologists, mathematicians, and, more recently, economists have been building mathematical models to assist them in understanding atomic, mechanical, chemical, biological, and economic systems. For the most part, these models were built using differential and difference equations. Such models are still being built throughout the sciences, social sciences, and engineering as a means to explain natural phenomena. These mathematical representations are called collectively *descriptive models* because they offer, for a given set of inputs and initial conditions, a description of the phenomena under study.

Since World War II, two important developments have transformed the world of mathematical modeling: (1) the creation of a new mathematics, a mathematics of decision making, and (2) the invention and continual improvement of the digital computer that has been made possible by the silicon chip. The invention and subsequent development of the computer have made descriptive models far more powerful, far more capable of mimicking natural phenomena—even on a global scale. The invention of a mathematics of decision making, on the other hand, has opened avenues of research not previously thought possible. Although this new mathematics was originally applied to only small problems, much as descriptive models were, the rapid evolution of the computer has now made possible consideration of problems far beyond the ken of those who originated the mathematics of decision making. Thus, the computer facilitated enormously the application and development of both kinds of mathematics. We called the first type of mathematical representation a *descriptive model* because it describes.

In contrast, the representation that uses the mathematics of decision making is called a *prescriptive model* because it *prescribes* a course of action, a design, or a policy. The *descriptive model* is said to answer the question, "*If I follow this course of action, what will happen?*" In contrast, the prescriptive model may be said to answer the question, "*What should I do?*" Implied in this question of what to do may be some notion of cost or of effectiveness, so that the course of action derived is the least costly or the most effective. Another term for a prescriptive model is an *optimizing* or *optimization model* in the sense that the policy or design that is found achieves the *best* value of some objective. This book will focus on prescriptive models, but it is often true that descriptive models may be contained within prescriptive models. These descriptive models are sometimes so simple that a reader may not even realize that a descriptive model is being used. At an early point in the text, we will pause in the building of a prescriptive model—a model on how to operate a furniture factory—to take note of the descriptive model that underlays a prescriptive formulation.

The descriptive/prescriptive classifications are just one way we can divide types of models. Another way we can divide model types is by the kind of data they utilize. Some models utilize data that are considered to be known with relative certainty. An example might be the number of table tops of a given size that can be cut from a 4- by 8-foot sheet of plywood. Except for occasional cutting errors or flaws in the plywood, that number is fixed. Another example is the oper-

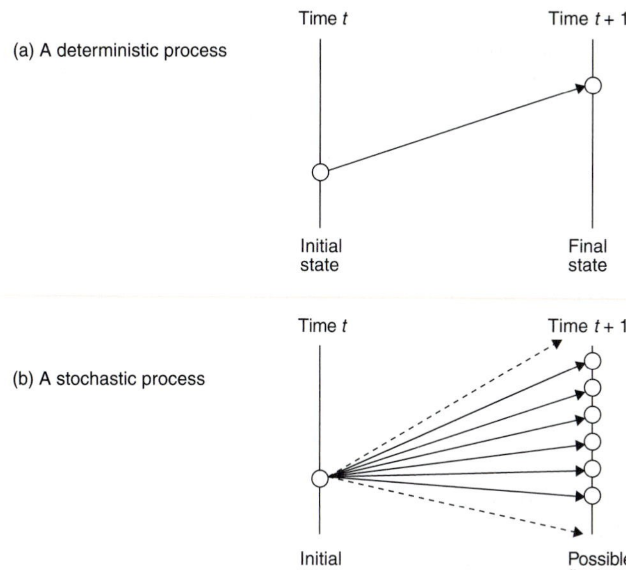

Figure 1.1 Only a single outcome occurs when a process is driven by a deterministic force (a). Multiple outcomes are possible when a process is driven by a stochastic phenomenon (b).

ation through time of a materials stockpile in which a given month's demand is fairly predictable year to year. Models of this type in which data elements are not variable but are relatively fixed quantities are referred to as *deterministic models*. Given an initial contents of the stockpile and a given release of materials and purchase of new materials during a unit of time, a deterministic model suggests that there is just one possibility for the final, end-of-period condition of the stockpile. That is, only a single outcome can occur from the month's events given the choice of action (see Figure 1.1a).

In contrast to deterministic models, other models might utilize data elements that are not precisely known but can be characterized by a mean and/or a variance. The August evaporation depth from a reservoir might fit in this category. A very warm August will result in a greater depth of evaporation than will a cooler August, and no one really knows in advance what temperature will prevail in August. Models in which the data elements are variable—capable of taking on any value from a range of values—are called *stochastic models*. Given an initial value of the storage in the reservoir, and a known amount of release and of inflow, the stochastic model suggests that the contents of the reservoir cannot be stated with certainty because of such things as the variable rate of evaporation from the surface of the storage pool. Many end-of-period values of storage are possible. These two additional model classifications allow the characterization of four basic types of models as summarized in Table 1.1.

The intersections of these two sets of categories—deterministic/stochastic and prescriptive/descriptive—give rise to a four-way classification table that contains nearly all major model types that are utilized today. For instance, models that are

TABLE 1.1 TYPES OF MODELS BASED ON TWO-WAY CLASSIFICATION

	Deterministic	Stochastic
Prescriptive	Linear programming Integer programming Multiobjective programming Dynamic programming	Stochastic programming
Descriptive	Difference equations Differential equations	Stochastic differential equations Queueing theory Monte Carlo simulation

deterministic and descriptive are known as differential equations or models that use difference equations. These are models you probably encountered in your calculus or applied mathematics courses—they typically incorporate empirically derived parameters, rate constants, and known (or at least assumed) initial conditions. These models may be linear or nonlinear, depending on the nature of the system or on how realistic the model structure or function needs to be for a particular application.

The intersection of descriptive and stochastic models contains several types of models. One is the differential equation/difference equation model coupled with parameters that are random variables. Such equations are called stochastic differential equations—a form of mathematics that becomes exceedingly complex when the equation(s) contain more than one random parameter. As soon as two random parameters are introduced, the structure of the correlation between them is also necessary in order to create the model. Another model form at the intersection of these two types of models is embodied in the mathematics of queueing theory or, more generally, stochastic processes. These mathematics presume known parameters that describe arrivals and departures in a random environment. A third model type at this intersection is known as "simulation," a computer-intensive form of modeling that generates realistic events and system responses through time. Here, the statistics of the events and the responses are designed to correspond to the actual statistics of parameters in the system being studied. All three model types allow the modeler to observe a range of possible outputs that evolve through time from a set of initial conditions.

Another intersection is that of prescriptive models with deterministic models. These deterministic optimization models are known by various names depending on whether the mathematical descriptions are linear or nonlinear, are static or evolve through time, or have particular forms. One of the model names is linear programming, the form of optimization that is emphasized in this text. Others include quadratic programming, gradient methods, optimal control theory, dynamic programming, multiobjective programming, and integer programming. The last three of these items will be treated in this text to a modest depth. Quadratic programming deals with problems having a quadratic objective function. Gradient methods follow slopes of objective functions to optimal solutions. Optimal control theory finds an optimal control or decision function in time or an optimal trajectory. Dynamic programming typically considers problems with a number of time stages. Multiobjective programming operates on more than one objective and derives trade-offs between objectives. Integer programming considers only integer valued decisions as practical or desirable.

The reader will be interested in the meaning of the word programming in this context. The body of all optimization methods is known as *mathematical programming*, and its subspecialties are known as linear programming, dynamic programming, and so on. The term "programming" is commonly confused by the uninitiated with computer programming. This programming is a term that means "scheduling," "the setting of an agenda," or "creating a plan of activities." Some confusion seems inevitable, however, because virtually all optimization, except that done as learning exercises for homework, classes, or labs, requires extensive use of digital computation.

The remaining intersection of model categories is that of prescriptive and stochastic models. There is a form of optimization that deals with models whose parameters are random variables. It is known as stochastic programming, and it requires of the student a reasonable background in probability theory. The importance of stochastic programming to applied studies is growing, and is also probably one of the most challenging forms of mathematical programming. We do not discuss this modeling methodology at the depth that we treat other forms of programming.

Of the various forms of programming or optimization, the most widely used in practice is linear programming, a principal subject of this book. The wide application of linear programming is not simply a matter of having an efficient method of solution available, although an efficient solution method is available. No matter how efficient a solution procedure is, a solution procedure for a form of optimization that does not closely match the needs of real problem settings would not be expected to be utilized widely. The wide use of linear programming then is not driven as much by an available solution method as by the nearly universal form of the linear programming problem statement. That form places an *objective* or goal alongside *constraints*—conditions that any solution must satisfy.

Many objectives are possible for a linear programming problem. These include, but are not limited to, minimum cost, maximum production, maximum equity, and maximum profit. Constraints are natural limits on achievement. The most commonly constrained quantities are resources such as personnel, vehicles, level of investment, time, and materials. Other constraints that will be taken up in the chapter on integer programming enforce the logic of system development. For instance, a particular link of a road network must be built before some other more remote link can be built. Finally, some constraints provide only definitions.

The language of objectives and constraints of the linear programming problem turns out to be the language of real problem statements. Problems in the field are often stated in precisely this format of objective and constraints even by people who have absolutely no training in systems engineering or optimization. It is an absolutely striking and unforgettable phenomenon to find people without any systems training describing their problems in the language of the linear programming problem. It is the naturalness of the linear programming problem statement that accounts for the remarkable appeal of this form of optimization. Some other forms of optimization also have this structure, but no methodology that uses this structure is more versatile and more available than is linear programming.

1.B A HISTORY OF SYSTEMS AND OPTIMIZATION

Two great events punctuate the history of applied mathematics. The first step, the invention of calculus, occurred in the seventeenth century. Sir Isaac Newton was one of two inventors of the calculus. Although Newton in Britain was the first to create the calculus in 1665–1666, Baron von Gottfried Wilhelm Liebniz independently invented the calculus in 1675. In those days, publication of ideas was often long delayed. Liebniz published his calculus in 1684, nine years after he conceived the idea. Newton, in order to retain claim to his invention, rushed into print by 1687. The world of mathematics was never the same, and college students were thereafter punished for Newton's and Liebniz' original sin. Newton's invention was stimulated not by a consideration of abstract issues, but by his interest in explaining the effects of planets on one another. That is, the calculus was invented to solve a general problem that Newton was considering. In the same way, the invention of linear programming was to spring from the necessity of solving real problems.

Almost three centuries later, another event shook and reoriented not only the world of mathematics, but also the field of economics. The invention of linear programming was to influence not only economics but would form the core of an entirely new discipline, operations research or systems engineering. In the same way that the calculus can be traced to two central figures, the development of linear programming is attributed to several towering players.

Koopmans in the United Kingdom and Kantorovich in the former U.S.S.R. at about the same time independently attacked the problem of least cost distribution of items. Kantorovich's work (1939) was suppressed for more than 20 years by Soviet authorities, but Koopmans came to the United States where he imbued his student Dantzig with the importance of providing a practical method of solution to the problem forms he was proposing. In 1947, in conjunction with a U.S. Air Force research project, Dantzig invented the simplex procedure for solving linear programming. His procedure, with modifications to take advantage of modern computers, is in wide use today, and is the procedure we will teach. In the period 1948–1952, Charnes and his coworkers pioneered industrial applications of linear programming and created the simplex tableau—the data storage methodology used in the repeated calculations of the simplex procedure. Charnes and coworkers went on to adapt linear programming to deal with convex rather than linear functions, to invent goal programming, and to create new forms of optimization to deal with problems characterized by random parameters. Dantzig went on to make major contributions to the solution of network problems. Koopmans and Kantorovich received the 1975 Nobel Prize in Economics. The magnificent achievements of Dantzig and Charnes are yet be honored in such a way.

Charnes did not stop at industrial applications of linear programming. With students and coworkers, he pushed on to the first applications of linear programming to civil and environmental engineering. Charnes thus occupies a central role in the initiation of civil and environmental systems engineering. One of his students, Lynn, has been particularly influential both in research in environmental systems and in training new students. The authors of this text fall from the tree of Charnes and Lynn. They share with these two people an enormous enthusiasm for the application of systems methodology. That enthusiasm is the wellspring that nourishes this book.

There is more to the history of systems, though, than the development of the mathematics and its application in new settings. It is said that the calculations and data manipulations that were needed for the first application of the simplex procedure were so extensive and voluminous that the application was carried out on a large tablecloth. This calculation procedure was carried out for a linear problem of a size that the reader can do by hand and minute by today's standard of what can be solved. Problems solved today may have dimensions that are more than three orders of magnitude (1000 times) larger than those first linear programming problems. The difference, of course, that makes the solution of such large problems possible, is the appearance and explosive evolution of the computer. From the 1950s onward, computers have advanced in speed and power, making possible the solution of larger and larger linear programming problems.

Whereas a careful person may solve by hand calculations alone a problem with perhaps ten variables and five constraints, modern codes on up-to-date mainframe computers can handle problems with 40,000 variables and 20,000 constraints. With continual advances in computer technology, even these limits will soon be far surpassed. Arguably, we can say that linear programming would today be a fascinating but small branch of applied mathematics and economics if it were not for the codevelopment in time of the electronic computer. With the development and evolution of the computer, linear programming has become the foremost mathematical tool of management science and of industrial and engineering management and design.

1.C APPLICATIONS OF LINEAR PROGRAMMING

To provide you with an idea of how widely utilized linear programming and its derivative types of optimization are, we describe a number of settings in the public sector, in industry, and in business where linear programming and allied methods have been put to use.

1.C.1 Distribution, Warehousing, and Industrial Siting

From the beginning, distribution has played a key role in the development of linear programming. It was the problem of least cost distribution of goods from multiple sources to multiple destinations that motivated both Koopmans and Kantorovich to structure their linear programs; these problems are known today as "transportation problems," and very large problems are solved routinely, even on desktop computers. Transportation problems assume direct and separate shipments that are made over known routes. Another class of problems, known as "delivery" or "routing" problems, also move goods from multiple origins to multiple destinations, but the challenge of these problems is to find the tours or routes for vehicles that will drop off the needed amounts at multiple destinations as the vehicles traverse the prescribed route. These problems can be solved as linear integer programs.

Warehouse problems, another class of problem that has been solved by linear programming, take up the issue of optimal stocking of goods and their release through time to the distribution system. The siting of warehouses between factories and markets has also been studied with linear integer programming, as has the siting of manufacturing plants, which may supply either warehouses or customers.

1.C.2 Solid Wastes Management

Within the environmental area, linear programming and allied methods have been used to site landfills and transfer stations and to create solid waste districts. It has also been used to route solid waste collection vehicles through street networks and, within the framework of hazardous waste management, to route spent nuclear fuel from power plants to storage sites.

1.C.3 Manufacturing, Refining, and Processing

Some of earliest applications of linear programming took place in these areas. The problem Koopmans called "activity analysis" consists of choosing which items to manufacture to achieve either least cost or maximum profit given constraints on the total amount of each of various resources that are consumed in the manufacturing process. We will describe the activity analysis problem in a more formal way later in this chapter. Another manufacturing area to which linear programming and allied methods have been applied is the design of factory floors. Known as the "facility layout problem," this model sites the various activities on the factory floor to minimize interaction costs.

The operation of a refinery, especially the blending of aviation fuel, was also a problem studied early in the history of linear programming—in this case by Charnes and coworkers. Chemical process design remains a fertile area to this day for the application of linear programming.

1.C.4 Education Systems

Educational systems are a rich setting for the application of systems methodology. Linear programming (LP) and allied procedures have been applied to class and room scheduling. These methods have also been used in school bus routing and to draw school district boundaries for efficient transportation. LP has also been utilized to allocate pupils to schools to achieve mandated desegregation plans. LP models have been used for enrollment planning at colleges.

1.C.5 Personnel Scheduling and Assignment

Systems techniques have found application in the scheduling of personnel through shift rotations. They have also been used to assign people to jobs or tasks in large organizations. The scheduling and assignment of airline crews to flight legs is an ongoing and important use of linear integer programming. Manpower planning models have also been developed using linear programming to project needs and policies relative to such areas as physician and nurse availability. The application to lawyer availability apparently has not been necessary. The efficient assignment of crews to snow plows for winter highway maintenance has been structured as a linear programming problem as well.

1.C.6 Emergency Systems

Since the late 1960s, linear programming and allied methods have been applied to the siting of fire engines, fire trucks, fire stations, and ambulances. Early problem statements focused on minimizing average travel time given budget constraints. Later formulations required or sought "coverage," the stationing of at least one vehicle within a travel time or distance standard of every point of demand. Cost was either a constraint or an objective in these models. Most recently congestion in emergency systems has been investigated with the emphasis on ensuring the actual availability of a server within the time standard at the moment of a call. Dozens of linear programming models have been built in the area of emergency facility siting. Other areas of siting have also been investigated, and these are referred to in the section on "Distribution, Warehousing, and Industrial Siting." In fact, the whole area of siting decisions is receiving such wide attention that a new journal, *Location Science*, debuted in 1993.

1.C.7 The Transportation Sector

Linear programming or a variant has been used extensively in the design of transportation networks including highway networks, rail networks, and airline networks. Efficiency and cost objectives have been utilized in such formulations with constraints on connectivity (or continuity) of the network or on population proximity. LP has also been used to design bus routes, assign drivers to buses, and schedule buses. Traffic light timing at intersections and at freeway entrance ramps have also been approached as linear programming problems. Goods movement, as mentioned in the "Distribution, Warehousing, and Industrial Siting" entry, is a classic application of linear programming. Empty railcar movement has also been structured as a linear programming problem as has the selection of freight terminals to open or close and the specification of hubs in an airline network. The development of pipeline networks for oil and natural gas can also be structured as a linear programming problem. Military applications of linear programming are often of a goods movement/logistics nature. The vertical alignment or grade design of highways, as well as the determination of optimal cut-and-fill strategies, can be cast as a linear programming problem.

1.C.8 Sales

The area of sales has proved an important area for the application of linear programming. The famous traveling salesman problem, whose first solution was provided by Dantzig and coworkers, seeks the routing of a salesperson through a set of cities with the least total route length. The design of sales territories considers the creation of compact (for efficient travel) districts of roughly equal sale potential for the sales representatives to cover. Media selection, that is, selection of the mix of magazines, broadcast networks, and so on, for the display of advertisements has been approached as a linear programming problem. Executive compensation is a famous application of linear programming due to Charnes and coworkers.

1.C.9 Electric Utility Applications and Air Quality Management

Linear programming has found application in the electric utility industry. Applications include the design of transmission networks, power dispatching between and among utilities, and power plant siting, to name a few areas of focus. Related to power plant siting is the environmental area of air quality management from stationary sources (such as power plants). Here the problem is to select the level of pollutant removal (say, sulfur dioxide) from stack gases at power plants in the region so that downwind concentrations of the pollutant are less than some critical level.

1.C.10 Telecommunications

A relative newcomer to optimization applications is the area of telecommunications. Although the allocation of radio frequencies has been approached as a linear programming problem, most efforts have been directed toward the creation of efficient telecommunications networks. This latter area has very strong relations to transportation network design but with certain added features. For instance, computer networks can be structured with a central computer having connections to smaller computers, which themselves have dispersed users. Both the smaller computers and central computer need to be sited. Telephone networks begin with customers connected to local exchanges, but the local exchanges are connected to a central exchange, which is linked to long-distance lines. The customers may require a backup connection, so that if their local exchange goes down, communication is not disrupted.

1.C.11 Water Resources and Water Quality Management

Of all the environmental areas to adopt systems methodology, probably the most active applications have taken place in water resources and water quality management. Initiated by Charnes and by Lynn and by researchers at the Harvard Water Program in the late 1950s and early 1960s, linear programming has been widely applied to the design and operation of reservoirs. Optimization models have been built to operate reservoirs in both deterministic and stochastic environments. Models have dealt not only with single-purpose reservoirs, such as those devoted only to water supply, but also to multipurpose reservoirs, which may have a variety of conflicting uses: flood control, recreation, habitat preservation, irrigated agriculture, and the like. In addition, the models have been extended to a number of reservoirs operating on a number of rivers to be used jointly toward one or more common purposes, such as drinking water supply, irrigation, and hydropower. Water distribution networks have also been designed by means of linear programming. Water quality management models also use linear programming: they seek the least cost removal of organic wastes from many wastewater treatment plants on a river. Least total cost of removal is the goal, with levels of dissolved oxygen in the river constrained to be greater than some needed level for fish survival, for example.

1.C.12 Agriculture and Forestry

Linear programming has been used to choose the set of activities on a farm that maximize profit. LP has also been applied to determining optimal volumes for grain reserves and to site grain reserves in developing nations. In forestry, linear programming has seen wide adoption for the management of the National Forests.

The timing of cutting has been determined that maximizes the value of the harvest while providing aesthetic values through spatial limitations on cutting activities.

1.C.13 Civil Infrastructure and Construction

Civil works planning provides a number of areas where linear programming has found important application. Linear programming and allied methods have also been used to site regional wastewater treatment plants and to allocate communities to those plants. The share of costs that each community should bear in such a system has also been structured as a linear programming problem with the goals of achieving both full cooperation of potential partners as well as equity of treatment of the partners. The sequence of repair of road networks, including determining which bridges to repair, and the siting of repair and maintenance garages along the network are other applications in infrastructure planning. The timing of the building of new reservoirs is also a linear programming problem. The building process itself, the ordering of activities of construction, is a linear program of some fame. Excavation in highway construction can also be approached by linear programming; the movement of earth from the site of cutting to the borrow pits is determined by a linear programming model. The vertical alignment of a highway is likewise a problem in which least cost levels of cut and fill need to be determined subject to grade and curvature constraints. Increasing opportunities for applications of systems science and engineering to problems of civil infrastructure management and planning has resulted in establishment of a new journal published by the American Society of Civil Engineers (ASCE) called the *ASCE Journal of Infrastructure Systems*, which began in early 1995.

1.D INTENT OF THE TEXT

The diversity of applications already studied using this technology suggests that numerous linear programming and related optimization problems remain to be built. This text is aimed first at providing you the background to understand why and how linear programming and related problems can be solved, and to enable you to build new models. The why and how of linear programming solutions are basically a set of mathematical issues, but mathematics with a decided purpose. The reader needs to understand the mathematics to know what kinds of models are soluable and what kinds are not; the mathematics of optimization is not yet so advanced that all forms can be solved.

The second and central purpose of this text is to teach you—by example—how to build linear programming and other decision models for real situations. Most of the text, with the principal exception being the single chapter on the simplex procedure, will be aimed at achieving that purpose—learning to build decision models. There is, unfortunately, no formula for building these models. These are, however, approaches that generally work (although they do not work in all situations). It is not pretentious to say that building linear programming problems is an art, an art that is built up through experience—in the same way that a fine cabinetmaker learns the art of carving and creating from the building of many pieces of furniture. We can provide a few thoughts on model building, but then the art must be pursued, example by example. You will encounter times and cases outside this book when no amount of art or prior experience will be sufficient to build a solvable model, although in this book all exercises will have a solution. Does the fact that sometimes no solvable model can be built mean that the reader's art is simply not developed enough? Not necessarily; some problem settings may not yield to the modeler's art. We hope, however, that you will find and pursue such problems, if only for a little while, for much that is useful is to be learned in attempting to structure a problem.

1.E A FEW RULES FOR MODEL BUILDING

The famous French artist Paul Gauguin once said, "Art is either plagiarism or revolution." Either is acceptable in the quest for efficient solutions to difficult public sector engineering planning and management problems. The novice model builder is well advised to borrow as much experience as possible from those who have tread before:

1. Keep the model as absolutely simple as possible while still answering the question at hand. Try to use only those parameters—such as size, temperature, elevation, composition, cost, distance, and so on—whose values can be obtained. The level of uncertainty about the values themselves should be relatively small.
2. Let x equal the unknown level of a decision or, where many decisions are involved—as is most commonly the case—let x_j equal the unknown level of the jth decision. Determining these problem features that can be controlled is crucial to creating decision variables.
3. Try to list all possible constraints and to enumerate the objective or objectives. A constraint is what you must achieve without fail. An objective is a goal you would like to achieve, the way you evaluate the goodness of a solution, and a yardstick against which to measure alternative solutions.

   A constraint might be a composition of gravel with at least a certain percentage of coarse material or a road with a slope at no more than a certain grade. A goal might be the least cost gravel mixture or the least cost road profile—attainment of these goals is limited by the constraints that must be adhered to.
4. Solving models will probably be easiest when processes are additive and proportional; by this we mean linear. The concept of linearity will be defined shortly.

Although the simplex algorithm produces an optimal solution to the mathematical problem (model) you will structure, the model is not intended to provide an optimal solution for a decision maker. Distrust people who claim that a model solution should be implemented because it is "optimal." Models are used as a tool for evaluation, as a provider of insights. The real world is far too complex to claim that a model can reflect it so accurately that the optimal mathematical solution is the optimal "real-world" solution. Instead, we hope that model solutions are good solutions, reflecting possible desirable answers, and, further, that an analysis of tradeoffs will provide deeper insight for decision making. Perhaps such warnings temper the reader's enthusiasm. They should not. Despite the shortcomings of the models we build, they can provide insights and understanding that intuition and human judgment cannot.

Probably someone could make up more rules than this. They might or might not be helpful. In a certain sense, we have just handed you a chisel, told you the angle at which to hold it and said you are ready to carve, well almost ready. First, we will take a look at a number of models that others have carved and examine the steps they took and the angle at which they held the chisel.

1.F MODELS IN CIVIL AND ENVIRONMENTAL ENGINEERING

By civil and environmental systems engineering, we mean that branch of engineering that models decisions and designs in these areas. Examples of such decisions include which landfill sites to operate, what road segment to repair next, the time staging of water treatment facilities, the allocation of pollution control burdens, the sharing of costs in regional facilities, the size of a reservoir, and many more similar issues. Each of the models presented in this first chapter is simplified somewhat to make it intelligible, but underlying these models are more complex and realistic representatives of problems in decision making that have been used successfully in practice. While the appearance of these models will seem quite different from one another, they actually share a common structure. It is that structure which allows even the largest of these models to be solved quickly and efficiently, as will be discussed in subsequent chapters.

1.F.1 The Form of a Linear Program

The linear programming model has, as indicated earlier, a linear objective and linear constraints. To illustrate the model form, we will present examples, but it is useful to examine first a picture of how the examples ought to look. First, we pose a model: in this case it is the activity analysis problem—the problem that Koopmans and Dantzig suggested when linear programming was still just an idea and the simplex algorithm was still being born.

The activity analysis problem is one of choosing the extent to which each of a number of activities (indexed by j) is to be undertaken, for example, how many items of a given type to manufacture. Each activity brings a constant degree of profit as it is increased in extent. Each activity utilizes one or more resources, also at a constant rate; each of the resources is available in only limited quantity.

We define the following parameters (constants) and variables for an n-variable m-constraint problem:

j, n = index and total number of activities or decisions;

i, m = the index and total number of resources;

x_j = the extent of the jth activity or decision;

c_j = cost or profit or payoff for each unit of the jth activity that is undertaken;

a_{ij} = the coefficient of the jth activity in the ith constraint. The coefficient represents the amount of resource i that is consumed for each unit of activity j that is undertaken; and

b_i = the amount of each resource i available to the manufacturing process.

The objective of profit maximization is written.

$$\text{Maximize } Z = c_1x_1 + c_2x_2 + c_3x_3 + \cdots + c_nx_n.$$

The constraint that limits the use of resource i to b_i units in total is

$$a_{i1}x_1 + a_{i2}x_2 + a_{i3}x_3 + \cdots + a_{in}x_n \leq b_i.$$

This form of constraint must be written for each kind of resource that is consumed in the group of activities. Each activity must be undertaken to either some positive extent or not at all. That is,

$$x_1 \geq 0, x_2 \geq 0, x_3 \geq 0, \ldots, x_n \geq 0.$$

Note that the objective is the sum of linear terms and only linear terms; for example, c_1x_1 and c_2x_2 are linear in x_1 and x_2, respectively. Further, the constraint for the ith resource and hence for any resource has only linear terms.

The complete linear program is summarized as

$$\text{Maximize } Z = c_1x_1 + c_2x_2 + c_3x_3 + \cdots + c_nx_n$$

$$\text{Subject to:} \quad a_{i1}x_1 + a_{i2}x_2 + a_{i3}x_3 + \cdots + a_{in}x_n \leq b_i, \quad i = 1, 2, \ldots, m$$

$$x_1 \geq 0, x_2 \geq 0, x_3 \geq 0, \ldots, x_n \geq 0.$$

A general linear programming problem allows either *maximization* or *minimization* of the objective and permits the relationship between the left-hand and right-hand sides of the constraints to be of any form, *less than-or-equal to*, *greater than-or-equal to*, or simply *equal to*. The general statement of the linear programming problem also places all linear terms on the left-hand side and all constant (unmultiplied) terms on the right-hand side of the relational indicator. It is this form that you will see in the examples that follow and throughout this text. It is this form that you will seek to produce when you structure problems—because this form can be solved. Several example linear programs follow.

1.F.2 Example 1-1: Shared Recycling Program(s) in Neighborhood Communities

Three communities A, B, and C, not too distant from one another, are individually considering recycling programs/projects. Thoughtful citizens in the three communities have suggested that cost savings may result if the towns cooperate in a joint project to be located somewhere between the three. Others have suggested that some pair of towns may be suited to cooperate but not all three. The towns have a tradition of independence suggesting that cooperation may not be in the cards, even if some coalition appears cost effective. The least excuse could sink a cooperative project.

The planning boards of each of the towns have voted jointly to engage an engineering consultant to evaluate and compare five alternatives. These five alternatives, shown in Figure 1.2 are (1) A, B, and C operating independently of another; (2) A and B cooperating in a joint effort called AB while C acts independently; (3) B and C cooperating in a joint effort called BC, while A acts independently; (4) A and C cooperating in a joint program while B acts independently; and (5) A, B, and C cooperating in a unified project. As the situation turns out, the joint project consisting of A, B, and C in a common venture is the most cost-effective alternative. Costs in millions of dollars of all the alternatives, including go-it-alone costs, are shown in the figure.

Although the grand coalition is clearly to be preferred from a cost standpoint, it is possible for the cost burden to be allocated in ways that make the potential

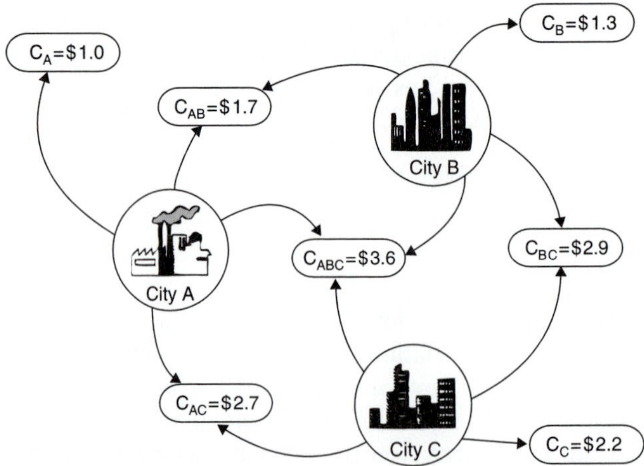

Figure 1.2 Each of three cities has four options and associated costs for community recycling programs. The least cost regional solution is the development of a facility that would be shared by all cities: total cost = \$3.6 million.

participants decline to join. For instance, because the cost of the grand coalition is 20% less than the sum of the go-it-alone costs, we could try to distribute the 20% savings to each of the potential participants. This distribution of savings ensures that the money collected from the three communities will be enough to pay for the grand coalition. This efficient regional solution would allocate a cost of $0.8 million to A, $1.04 million to B, and $1.76 million to C, for a total of 3.6 million, the cost of the grand coalition. Unfortunately, cities A and B also have an option of developing a joint facility, and would exclude city C. This total cost to cities A and B would be $1.7 million, which is $0.14 million less in total than they could spend if they each accepted a 20% discount to join the grand coalition. With city C going it alone, the total regional cost would be $3.9 million. The problem is that the total least cost solution will not be feasible from a practical perspective, because cities A and B will be reluctant to participate in the grand coalition, which from their perspective would be a partial subsidy to city C.

Some method is needed to allocate costs that will not favor this inferior subcoalition. We will try linear programming. Define the following notation:

$$Y_A = \text{cost allocated to community A;}$$

$$Y_B = \text{cost allocated to community B; and}$$

$$Y_C = \text{cost allocated to community C.}$$

A constraint can now be written that ensures that the sum of the costs allocated to A and B must be no greater than the cost they would face if they formed a partnership AB ($1.7 million). That is

$$Y_A + Y_B \leq \$1.7.$$

Similarly, the sum of the costs allocated to A and C must be no greater than the cost they would face if they formed a partnership AC. That is

$$Y_A + Y_C \leq \$2.7.$$

In addition, the sum of the cost allocated to B and C must be no greater than the cost they would face if they formed a partnership BC. That is,

$$Y_B + Y_C \leq \$2.9.$$

Finally, the total of costs allocated to all three participants should be as large as possible so that sufficient monies are available to undertake the project associated with the grand coalition. That is

$$\text{Maximize } Z = Y_A + Y_B + Y_C.$$

The problem is summarized as

$$\begin{aligned} \text{Maximize } Z &= Y_A + Y_B + Y_C \\ \text{Subject to:} \quad & Y_A + Y_B \leq \$1.7 \\ & Y_A + Y_C \leq \$2.7 \\ & Y_B + Y_C \leq \$2.9 \end{aligned}$$

and all allocations must be positive

$$Y_A, Y_B, Y_C \geq 0.$$

The solution that maximizes the objective while satisfying all three constraints is $Y_A^* = 0.75$, $Y_B^* = 0.95$, and $Y_C^* = 1.95$ (the asterisk indicates that this particular value of the variable is the *optimal* value for that variable). The total of costs allocated is $Z^* = \$3.65$, enough to build the project associated with the grand coalition, and with money left over that perhaps could be used for administrative costs.

1.F.3 Example 1-2: Blending Water Supplies

A city is growing in population and the public works department projects the need for an increased water supply. The current water supply is drawn from a reservoir on a local stream and is of good quality, but future sources—the ones to be tapped next—have various problems. Water from a nearby aquifer is available in adequate supply, but its hardness level is too high unless it is blended with a lower hardness source. The total pounds of hardness per million gallons is limited to 1200. Water from a distant stream is of sufficient quality, but the cost to pump the water to the treatment plant is quite high and a pipeline would have to be built.

The city is conducting its planning in stages. The first stage is to plan for ten years from the present. The three sources are source 1–the current supply, source 2–the aquifer, and source 3–the distant stream. The costs to obtain water in dollars per million gallons, the supply limits in millions of gallons per day, and the hardness in pounds per million gallons are given in Table 1.2. For example, if the present water supply is developed further, up to 25 million gallons per day (mgd) could be made available from source 1 at a cost of \$500 per million gallons, and each additional million gallons from this source would contribute 200 pounds of hardness to the total water supply. A total of 150 additional million gallons per day is needed by the end of ten years. The city council members are interested in a least cost strategy for expanding the water supply, while ensuring that the water supply remains of sufficient quality.

TABLE 1.2 COST, SUPPLY, AND QUALITY OF AVAILABLE WATER SOURCES.

	Source 1	Source 2	Source 3
Cost (\$/mgd)	\$500	\$1,000	\$2,000
Supply limit (mgd)	25	120	100
Hardness (lb/mg)	200	2,300	700

Let: x_1 = millions of gallons per day to be drawn from source 1;

x_2 = millions of gallons per day to be drawn from source 2; and

x_3 = millions of gallons per day to be drawn from source 3.

The linear program formulation for the problem that the city wants to solve is

$$\text{Minimize } Z = 500x_1 + 1000x_2 + 2000x_3$$

$$\text{Subject to:} \quad x_1 + x_2 + x_3 \geq 150$$

$$200x_1 + 2300x_2 + 700x_3 \leq (1200)(150)$$

$$x_1 \leq 25,\ x_2 \leq 120,\ x_3 \leq 100$$

$$x_1, x_2, x_3, \geq 0.$$

The objective is a minimum cost choice of water sources. The first constraint indicates the requirement for 150 million gallons per day, somehow divided among the three sources. The second constraint limits the pounds of hardness in the blended water. If the hardness concentration is limited to 1200 pounds per million gallons and 150 million gallons are required, the total pounds of hardness is limited to 1200 × 150 = 180,000 pounds. The solution to this problem is $x_1^* = 25.0$, $x_2^* = 54.7$, and $x_3^* = 70.3$ in million gallons per day (rounded to the nearest one-tenth) with an objective function value of $Z^* = \$207{,}800$ at optimality.

1.F.4 Example 1-3: A Furniture Factory—An Activity Analysis

A furniture factory specializes in dormitory furniture, making its product of substantial weight and very durable. It makes only student desks, desk chairs, dressers, and tables that are used in lounge space. The factory uses maple for exterior surfaces where resistance to direct impact is important, and pine for interior applications. Limited amounts of maple and pine are available each month to the factory. Different amounts of maple and pine are required in each of the three pieces of furniture and costs are different as well. Table 1.3 indicates costs and

TABLE 1.3 PRICES, COSTS, UTILIZATION FACTORS, AND SUPPLIES FOR EXAMPLE 1-3.

Item	Desks (1)	Desk Chairs (2)	Dressers (3)	Tables (4)	Amount Available (board-ft)	Cost ($/board-ft)
Pine	7	1.5	10	4	4000	2.00
Maple	15	4	22	10	2000	4.00
Revenue	$150	$100	$250	$170		

availabilities of the resources as well as the board-feet of wood used in each of the four products. The selling prices of the four products are also included. For example, each desk chair that the company produces requires 1.5 board-feet of pine and 4 board-feet of maple and sells for $100. A total of 4000 board-feet of pine is available each month at a cost of $2.00 per board-foot.

Assuming that labor and other resources such as shop and inventory space, and machines and tools are adequate, what items should the factory produce each month and in what quantities to maximize its profit?

Let: x_1 = the number of desks to make;

x_2 = the number of desk chairs to make;

x_3 = the number of dressers to make; and

x_4 = the number of tables to make.

The constraint on the pine resource ensures that the amount of pine utilized is not greater than the amount available:

$$7x_1 + 1.5x_2 + 10x_3 + 4x_4 \leq 4000.$$

It is interesting here to note that we are using a set of four linear models to predict the amount of pine that will be consumed in the four activities. That is, we are assuming that pine consumption is directly proportional by the factor of seven to the number of desks that will be made. The first desk, the last desk, and all desks in between will each consume 7 board-feet of pine. This is a predictive model of the form we discussed earlier and it is being embedded in the constraints of a prescriptive model. The constraint on maple limits the amount used in all four products to not more than the amount of maple available:

$$15x_1 + 4x_2 + 22x_3 + 10x_4 \leq 2000.$$

The cost of making these items in terms of the value of wood resources is the cost of pine times the amount of pine utilized plus the cost of maple times the amount of maple utilized. The appropriate cost function is computed as

$$\text{cost} = 2(7x_1 + 1.5x_2 + 10x_3 + 4x_4) + 4(15x_1 + 4x_2 + 22x_3 + 10x_4)$$

or

$$\text{cost} = 74x_1 + 19x_2 + 108x_3 + 48x_4.$$

Similarly, total revenue from sale of the items can be computed using a linear function of these same decision variables:

$$\text{revenue} = 150x_1 + 100x_2 + 250x_3 + 170x_4$$

so that profit Z, the difference between revenue and cost, is

$$Z = 76x_1 + 81x_2 + 142x_3 + 122x_4.$$

The complete linear program to be solved can be summarized as

$$\begin{aligned}
\text{Maximize } Z &= 76x_1 + 81x_2 + 142x_3 + 122x_4 \\
\text{Subject to:} \quad & 7x_1 + 1.5x_2 + 10x_3 + 4x_4 \le 4000 \\
& 15x_1 + 4x_2 + 22x_3 + 10x_4 \le 2000 \\
& x_1, x_2, x_3, x_4 \ge 0.
\end{aligned}$$

Even though this is a very small linear program—one having only four decision variables and two simple constraints—the optimal solution is not at all obvious. The maximum monthly profit for the furniture factory ($Z^* = \$40{,}500$) would result from the manufacture of 500 desk chairs ($x_2^* = 500$) and *no other products*! Yet desk chairs are clearly not the most profitable product line for the company. Examine the formulation and convince yourself that this production strategy would indeed be optimal given the constraint equations and parameters given. Develop in your mind an explanation that you might use in making a production recommendation to the factory manager. Are there other factors that might be important in determining a final production strategy that have not been considered explicitly in your model? An important aspect of modeling is gaining an appreciation for both the strengths and limitations of models and their solutions. The evaluation of prescriptive models is a topic of Chapter 3.

1.F.5 Example 1-4: Grading a Portion of a Highway

A portion of a highway through hilly terrain is to be graded to elevations that meet acceptable standards of vertical curvature. That is, the slope and rate of change of slope of the road must be within preset bounds. To achieve this vertical alignment, an acceptably smooth curve is superimposed conceptually on the surface, resulting in two areas to be cut and three areas to be filled. The areas to be cut and the volume to be cut from them are listed in Table 1.4 as a_1 and a_2. In a similar fashion,

TABLE 1.4 REQUIREMENTS AND DECISIONS IN THE CUT-AND-FILL PROBLEM

Fill areas (j)⇒ ⇓ cut areas (i)	$j = 1$	$j = 2$	$j = 3$	Relationship of row sum	Amount available from cut area i
$i = 1$	c_{11} x_{11}	c_{12} x_{12}	c_{13} x_{13}	$\ge$	a_1
$i = 2$	c_{21} x_{21}	c_{22} x_{22}	c_{23} x_{23}	$\ge$	a_2
Relationship of column sum	$\ge$	$\ge$	$\ge$		
Amount required for fill area j	b_1	b_2	b_3		

the areas to be filled and the volume of fill required, b_1, b_2 and b_3, are also shown in Table 1.4.

The remainder of Table 1.4 consists of blocks whose elements are the amounts to be transported from a particular cut area to a specific fill area. Since these amounts are not known in advance, each value is represented by a decision variable x_{ij} whose subscripts (ij) indicate the amount transported from cut area i to fill area j. Thus, for example, x_{12} is the amount transported from cut area 1 to fill area 2. The relations of the decision variables in the constraints can be inferred from their arrangement in the table. The two kinds of constraints are cut constraints and fill constraints. The cut constraints say that no more material can be transported to the three fill areas than is available for cutting at area i, namely, a_i. The fill constraints ensure that a certain amount of material, b_j, is required to be supplied at j from the two cut areas.

Within each cell is a quarter moon in the northwest corner of the cell. The number in the quarter moon, c_{ij}, is the cost to transport one unit of material from cut area i to fill area j. The specific value reflects mainly the distance separation between the two sites. Each value or cost is indicated and this allows us to construct an objective function of minimum cost. Again, all variables must be positive or zero; none can be negative. More total volume of earth is available from cuts than is needed for fills.

The problem can now be stated, as read from Table 1.4, as follows:

$$\text{Minimize } Z = c_{11}x_{11} + c_{12}x_{12} + c_{13}x_{13} + c_{21}x_{21} + c_{22}x_{22} + c_{23}x_{23}$$

$$\begin{array}{llllllll}
\text{Subject to:} & x_{11} + & x_{12} + & x_{13} & & & & \le a_1 \\
 & & & & x_{21} + & x_{22} + & x_{23} & \le a_2 \\
 & x_{11} + & & & x_{21} & & & \ge b_1 \\
 & & x_{12} + & & & x_{22} & & \ge b_2 \\
 & & & x_{13} + & & & x_{23} & \ge b_3 \\
 & & & & & & x_{11}, x_{12}, x_{13}, x_{21}, x_{22}, x_{23} & \ge 0.
\end{array}$$

The solution to this linear program would be an optimal strategy for transporting cut material from each roadway section requiring excavation, to sections requiring fill so as to achieve the least cost earthwork management plan. This example differs from the previous ones in that the model constants—transport costs, section cut availability, and section fill requirements—are represented symbolically rather than numerically. While for this small example problem a symbolic representation is probably not necessary, an actual application of this model may contain hundreds or even thousands or tens of thousands of sections, each with its own set of parameters. Symbolic notation allows even the largest of models to be formulated and specified in an efficient manner.

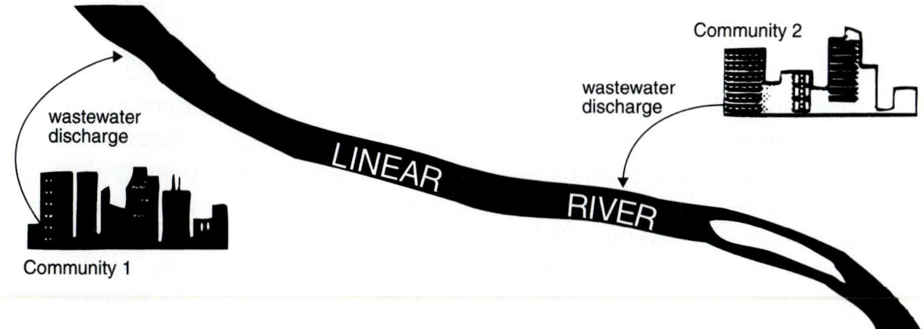

Figure 1.3 The hypothetical Linear River receives wastewater discharges from two municipalities.

1.F.6 Example 1-5: Cleaning Up the Linear River

Two communities are on the Linear River shown in Figure 1.3. Each has been discharging its untreated water into the Linear River, resulting in low dissolved oxygen levels, foul odors, and a lack of desirable fish in the river. Dissolved oxygen levels above 5 milligrams per liter generally reflect good stream health, but 6 milligrams per liter is better and 4 may be acceptable. The two communities have both agreed to clean up their wastewater by installing water pollution control plants. Their goal is to increase the dissolved oxygen levels in the river by removing organic wastes from the wastewater, because the organic wastes, when degraded by microbes, are the cause of low dissolved oxygen levels.

The two communities have decided to share the costs of cleanup, so they are seeking the fractional removal levels that will achieve dissolved oxygen above 5 milligrams per liter at the least total cost. By suitable calculation,[1] the lowest dissolved oxygen level in the reach between communities 1 and 2 can be modeled as a linear function of the fractional organic removal (or efficiency) at the treatment plant at community 1. In a similar fashion, the lowest dissolved oxygen in the reach below community 2 can be written as a function of the fractional organic removals (efficiencies) of wastewater treatment at both communities 1 and 2. These two functions are

$$1.0 + 0.1\varepsilon_1 = \text{lowest dissolved oxygen in the first reach}$$

and

$$2.0 + 0.02\varepsilon_1 + 0.02\varepsilon_2 = \text{lowest dissolved oxygen in the second reach,}$$

where

[1]See ReVelle, Loucks, and Lynn, "A Management Model for Water Quality Control," *Journal of the Water, Pollution Control Federation*, July, 1967.

ε_1 = the organic removal efficiency of community 1, and

ε_2 = the organic removal efficiency of community 2.

For physical reasons, both efficiency levels must be greater than 35% (reflecting the removal of settleable solids). The communities have decided that the dissolved oxygen in the upper reach must be kept above 6 milligrams per liter, and above 4 milligrams per liter in the lower reach. The cost per percent of efficiency at the first treatment plant is $20,000. The cost per percent efficiency of the second treatment plant is $10,000. The optimization problem is to minimize cost subject to dissolved oxygen constraints. All efficiencies must be zero or positive. That is,

$$\text{Minimize } Z = 20{,}000\varepsilon_1 + 10{,}000\varepsilon_2$$

$$\text{Subject to:} \quad 1.0 + 0.1\varepsilon_1 \geq 6$$

$$2.0 + 0.02\varepsilon_1 + 0.02\varepsilon_2 \geq 4$$

$$\varepsilon_1, \varepsilon \geq 0.$$

Modifying both constraints such that all terms containing decision variables are on the left of the relation and all constant terms are on the right of the relation, the constraint set becomes

$$\varepsilon_1 \geq 50$$

$$\varepsilon_1 + \varepsilon_2 \geq 100.$$

The solution for this problem is $\varepsilon_1^* = 50$, $\varepsilon_2^* = 50$, and $Z^* = \$150{,}000$.

CHAPTER SUMMARY

Decision support models provide a mechanism by which engineers can explore design, operations, management, and planning alternatives in a systematic manner, allowing the explicit evaluation of different options. Different modeling frameworks support the formulation and analysis of problems displaying different characteristics. Some modeling methodologies provide descriptions of how systems operate and thus how they might respond to changing conditions. Others provide recommendations about how a system should be managed to best achieve a desired outcome. Some models are designed to acknowledge the explicit stochastic nature of a system, while others assume that the system of interest is deterministic, or may safely be assumed so. The challenge for students of engineering systems analysis is to understand the strengths and weaknesses of available modeling methodologies, and to be able to adopt and adapt the most appropriate one for a given problem application.

The focus of this book is the design and use of optimization models to study engineering decision making. Models of this type (1) specify a solution space consisting of all decision alternatives that satisfy a set of precisely defined constraint equations, and (2) evaluate all of these alternatives using an objective function that measures how well a particular alternative within this solution space performs. These models are able to determine which alternative is "best" within the context of the assumptions being made about system structure and function. When the system being studied can be adequately represented by a system of linear constraint equations and a linear objective function, an efficient procedure is available for finding this optimal solution. A wide range of important engineering problems can be formulated and, therefore, solved in this manner.

EXERCISES

1.1. Consider a typical summer day in your hometown. List the public sector services that you encounter and try to envision the decision maker who is responsible for providing those services. For each service you are able to identify, list the major objectives in the mind of that decision maker while (s)he is developing management strategies.

1.2. The administration of your college or university includes an office that is responsible for scheduling classes and making classroom assignments. What are the objectives behind the development and evaluation of different assignments? What are the constraints within which these assignments are made?

1.3. List five personal decisions you have made in the last 24 hours that could be specified within the framework of the mathematical program (for example, selecting what to wear today), and list the objectives and constraints you considered in identifying alternatives and making your decision.

1.4. Your city council has just approved the use of municipal revenues for building a new fire station that will also house a rapid-response paramedic unit and a rescue vehicle. You have been appointed as the citizens' representative to a committee that will select the location for this new facility. List, as specifically as you can, the objectives to be optimized in selecting the location. Prioritize these objectives from the perspectives of each of the following groups:

(a) the mayor of your city
(b) the chamber of commerce for your city
(c) the residents of your city
(d) the association of relators in your city
(e) the merchants association in your city

1.5. Reconsider the problem of allocating costs to neighboring communities for implementation of a shared recycling program (presented as Example 1-1). Consider how the optimal cost allocation strategy of $Y_a^* = \$0.75$ million, $Y_b^* = \$0.95$ million, and $Y_c^* = 1.95$ million might be different if the cost of a cooperative facility between City B and City C were lower. Would a three-way partnership be the optimal solution to this problem regardless of the cost of this two-city coalition?

1.6. The local transit authority has received a grant to expand its bus fleet by 30%. This expansion will provide the opportunity for approximately ten new service routes. Suggest how an optimization model might be structured to assist with the design of these new routes and the modification of existing routes.

1.7. Due to demographic changes and the resulting reduction in tax revenues, your community school corporation is faced with the problem of closing one or more elementary schools, and redrawing service areas for those scheduled to remain open. Describe in general terms the structure of an optimization model that could be developed to help suggest an optimal strategy for school consolidation. Include in your description: (a) what constraints should be considered in your model, (b) what would be the most appropriate objective function for your model, and (c) what data would need to be collected in order to build your model. If you had such a model, suggest how it might be used to address specific complaints that might arise regarding any one specific consolidation alternative.

1.8. The solid waste landfill that serves your community is nearing capacity. The city council is starting to explore alternatives to avoid disruption in service. Options that have been proposed include (a) expanding the existing landfill, (b) buying land and building a new city-owned land-fill, (c) contracting with a company that will take waste out of the region, and (d) building an incineration facility and, at the same time, expanding the existing recycling program. How would you structure an optimization model to help evaluate these alternatives?

REFERENCES

BADIRU, A. B. and P. S. PULAT, 1995. *Comprehensive Project Management: Integrating Optimization Models, Management Principles, and Computers*, Prentice Hall, 576 pages.

BELTRAMI, E. J., 1977. *Models for Public Systems Analysis*, Academic Press, 218 pages.

CASTI, J. L., 1995. *Five Golden Rules: Great Theories of 20th-Century Mathematics—and Why They Matter*, Wiley, 235 pages.

CHURCHMAN, C. W., 1979. *The Systems Approach*, New York: Laurel Press, 243 pages.

GREENBERG, M., 1978. *Applied Linear Programming for the Socioeconomic and Environmental Sciences*, New York: Academic Press, 327 pages.

LEE ROY BEACH, L. R., 1993. *Making The Right Decision: Organizational Culture, Vision and Planning*, Prentice Hall, 224 pages.

LEVIN, R. I., RUBIN, D. S., STINSON, J. P., and E. S. GARDNER Jr., 1992. *Quantitative Approaches to Management, Eighth Edition*, McGraw-Hill, 920 pages.

MATHUR, K. and D. SOLOW, 1994. *Management Science: The Art of Decision Making*, Prentice Hall, 976 pages.

MEREDITH, D. D., WONG, K. W., WOODHEAD, R. W., and R. H. WORTMAN, 1985. *Design and Planning of Engineering Systems*, Prentice Hall, 384 pages.

TOCHER, K., 1974. *The Art of Simulation*, English University Press, 184 pages.

CHAPTER 2

A Graphical Solution Procedure and Further Examples

2.A INTRODUCTION

In this chapter we continue to explore the linear programming problem and its variations, but we also provide a first method of solution. The solution procedure we describe is easy to understand, but to use it, we need to restrict our attention to problems with two, or at most three decision variables. Such problems do occur, but rarely, and may be contrived. Nonetheless, the procedure demonstrates that these problems can be solved and, more importantly, provides extremely valuable insights about the general nature of solutions to linear programming problems. With this as a foundation, Chapter 3 presents a general algorithmic procedure that can be used to solve any linear program.

The method of solution demonstrated in this chapter is graphical and we apply it to two different problems: the first (Example 2-1), a problem of managing a materials production operation, and the second (Example 2-2), a problem that seeks to divide work effort efficiently between two mines. While the models formulated to solve these problems are quite different, their basic structure and the solution procedure used to solve them are identical. This second example is then extended (as Example 2-3) and the formulation modified to incorporate a nonlinear objective function.

In addition to solving these small linear programs, we also present several additional problems and model formulations in this chapter (Examples 2-4 through

2-7). Although these additional problems are too large to be solved graphically, they represent additional examples of the flexibility and extensibility of the linear program for use in engineering management. Later chapters will present systematic procedures for solving these increasingly sophisticated management models.

2.B SOLVING LINEAR PROGRAMS GRAPHICALLY

2.B.1 Example 2-1: Homewood Masonry—A Materials Production Problem

Problem Statement. Homewood Masonry is a small owner-operated firm that produces construction materials for the residential and commercial construction industry in the region. The company specializes in the manufacture of two widely used building products: (1) a universal concrete patching product called HYDIT and (2) a decorative brick mortar called FILIT. These products are in great demand, and Homewood can sell all of the HYDIT it produces for a profit of $140 per ton, and all the FILIT it can produce for a profit of $160 per ton.

Unfortunately, some of the resources needed to manufacture these products are in limited supply. First, the demand for both HYDIT and FILIT is due in large part to their special adhesive characteristics, which result from the use of a special ingredient in the blending process—Wabash red clay (from the banks of the Wabash River in Indiana). Each ton of HYDIT produced requires 2 cubic meters of this red clay, and each ton of FILIT produced requires 4 cubic meters. This material is in limited supply; a maximum of 28 cubic meters of red clay is available each week. Second, the operator of the machine used to blend these products can work only a maximum of 50 hours per week. This machine blends a ton of either product at a time, and the blending process requires 5 hours to complete. Last, each material must be stored in a separate curing vat further limiting the overall production volume of each product; the curing vats for HYDIT and FILIT have capacities of 8 and 6 tons, respectively. These resource limitations are summarized in Table 2.1.

TABLE 2.1 RESOURCE REQUIREMENTS AND AVAILABILITY FOR THE HOMEWOOD MASONRY PROBLEM

Resource	HYDIT	FILIT	Available
Wabash red clay	2 m^3/ton	4 m^3/ton	28 m^3/wk
Blending time	5 hr/ton	5 hr/ton	50 hr/wk
Curing vat capacity	8 tons	6 tons	
Profit	$140/ton	$160/ton	

You have been hired to help with the development of a strategy for operating the material production process. What is the optimal production strategy for Homewood Masonry given these data?

Model Formulation. The formulation of a production model for the Homewood Masonry problem begins by translating the major elements of the problem as presented above into a *pseudomodel*; a word description of the equations that we will need to capture the important elements of the problem. In this case, the objective of our model is to determine the amount of each product to make each week so as to maximize the overall profits resulting from sales:

Maximize total weekly profit

and four production conditions must be satisfied:

- The total available supply of Wabash Red Clay cannot be exceeded each week;
- The blending machine cannot be used for more than 50 hours per week;
- The storage capacity for HYDIT may not be exceeded each week; and
- The storage capacity for FILIT may not be exceeded each week.

For a given solution to be *feasible*, it must satisfy all three of these constraint conditions simultaneously. For a given solution to be *optimal*, it must, in addition, provide the largest value for the objective function (total profit). For any linear programming formulation, the constraint relationships determine feasibility, while the objective function is used to determine optimality.

To translate the pseudomodel into a linear program, it is necessary to define the decision variables that will be used in the model. In this case we might first ask ourselves "in determining a production policy for the manufacture of HYDIT and FILIT, what is the set of decision variables that the production engineer controls?" Clearly, the engineer does not control profit per volume of product sold (at least not directly), nor does he/she control the availability of resources. As the problem is stated, the engineer is faced with the decision of how much of each product to manufacture. Indeed, we assume that the engineer has total control over, and responsibility for, this decision.

Let: x_1 = the number of tons of HYDIT to produce each week.

x_2 = the number of tons of FILIT to produce each week.

We may now write an expression for total profit as a function of weekly sales as follows:

$$\text{total weekly profit} = 140x_1 + 160x_2$$

or, written as the objective function for our model,

$$\text{Maximize } Z = 140x_1 + 160x_2.$$

This expression will be used to evaluate a (generally, very large) number of problem solutions.

We can use these decision variables to write general expressions that will relate any possible level of production to the resulting level of resource usage. For example, the total volume of the red clay resource that will be used for a given production strategy is a linear function of production volumes of our two materials as presented in Table 2.1:

$$\text{total red clay used} = 2x_1 + 4x_2.$$

Because our production strategy may be constrained by the availability of this material, our model must include an explicit constraint on the consumption of red clay:

$$2x_1 + 4x_2 \leq 28.$$

Similarly, we can write expressions for the usage of the blending machine as a function of production:

$$\text{total blending machine time} = 5x_1 + 5x_2$$

or, written as a linear constraint:

$$5x_1 + 5x_2 \leq 50.$$

Finally, two additional model constraints will ensure that storage for the materials will not be exceeded:

$$\text{HYDIT produced each week} \leq 8 \text{ tons},$$

$$\text{FILIT produced each week} \leq 6 \text{ tons},$$

or

$$x_1 \leq 8,$$

$$x_2 \leq 6.$$

The complete linear program for the Homewood Masonry production model may now be presented:

$$\begin{aligned} \text{Maximize } Z &= 140x_1 + 160x_2 \\ \text{Subject to: } & 2x_1 + 4x_2 \leq 28 \\ & 5x_1 + 5x_2 \leq 50 \\ & x_1 \leq 8 \\ & x_2 \leq 6 \\ & x_1, x_2 \geq 0. \end{aligned}$$

The last term in our formulation restricts our decision variables from taking on negative values. Negative production of one type of material is not feasible; even though doing so in a theoretical sense might result in more resources being available to manufacture an additional amount of the other material. Nonnegativity of decision variables is an important assumption as will be more fully appreciated when we discuss the algorithm used to solve linear programs having large numbers of decision variables.

Because this model formulated to solve the Homewood Masonry production problem uses only two decision variables, we can solve this problem graphically as presented later.

2.B.2 A Graphical Solution for the Homewood Masonry Problem

Problems containing fewer than four decision variables can be solved graphically because the solution space for all possible combinations of these variables can be mapped into 2-space in the case of two variables, and 3-space in the case of three variables. In addition, that solution space can be further partitioned by plotting the constraint equations. Finally, the objective function can be plotted and used to find the optimal solution.

Consider a solution space for the Homewood Masonry problem as presented in Figure 2.1 where the total volume of production of HYDIT is plotted on the horizontal axis (x_1), and the total volume of production of FILIT on the vertical axis (x_2). All possible combinations of production levels can be represented in this half-space. Next, each constraint equation from our model can be plotted as a line containing all points that satisfy that equation with strict equality. All points on one side of these lines—the side indicated by the open arrow—satisfy the original inequality, while all points on the other side of these lines violate that constraint. In the case of an equality constraint, only points falling directly on the corre-

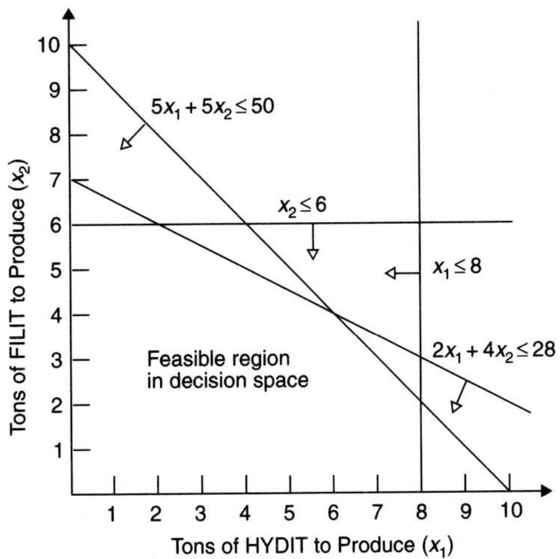

Figure 2.1 Feasible solutions for a linear programming problem are those points that satisfy all constraints simultaneously; the shaded region in the graph above. Note that the boundary of the feasible region is feasible.

sponding line would satisfy that condition. The feasible region for a linear program is defined as the set of solutions (values for the decision variables) that satisfy all constraint equations simultaneously. The feasible region for the Homewood Masonry problem is thus the shaded area of Figure 2.1. Convince yourself that the solution space for a linear program will always consist of (1) an infinite number of feasible solutions, (2) a single feasible solution, or (3) no solutions (it is possible that no point satisfies all constraints).

After the feasible region has been identified (graphed), the objective function can be used to evaluate all feasible solutions. Figure 2.2 shows the feasible region for the Homewood Masonry problem with the objective function—$Z = 140x_1 + 160x_2$—plotted at four different locations in decision space: $Z = 0$, $Z = 560$, $Z = 1120$, and $Z = 1480$. Note that each of these lines is parallel to each other. This is because regardless of the value of the objective function at a particular solution, its gradient is constant, as determined by the coefficients that multiply the decision variables in the objective function. At the lower left-hand portion of Figure 2.2, the objective function gradient is plotted passing through the origin ($x_1 = 0$, $x_2 = 0$) such that each point on this line will produce a value of zero for the objective function Z. Note that the only point on this line segment that is feasible (intersects the feasible region in decision space) is the origin.

Next, we plot the objective function gradient passing through the points (4, 0) and (0, 3.5). These points, as well as all points on that line, result in a value for the objective function of 560, which, because we are maximizing the objective function, is an improvement over the first objective function. Furthermore there are an infinite number of feasible solutions that give this same value for the objective function–all points on the line segment (4, 0), (0, 3.5) are feasible and give an objective function value of 560. By continuing to "move" the objective function gradient in the direction of improvement (upwards and to the right in this case, as in-

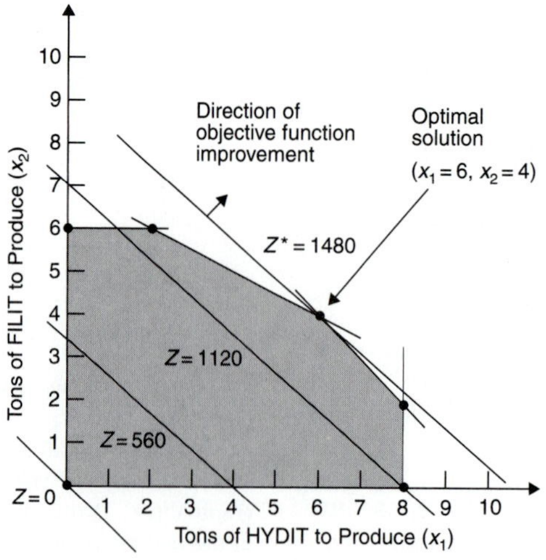

Figure 2.2 As the objective function is passed through the feasible region from the origin, the value of the objective function increases. Because the objective function is being maximized, the point at which it last intersects the feasible region is the optimal solution.

dicated by the solid arrow near the Direction of Objective Function Improvement), we can graphically evaluate all points in the feasible region. That point where the objective function last intersects the feasible region is the feasible solution that provides the best (optimal) value of Z while still satisfying all problem constraints. This point is thus the *optimal solution* to this problem; in this example, that point is $x_1 = 6$, $x_2 = 4$ with an objective function value $Z = 1480$. It should be obvious that the point $x_1 = 0$, $x_2 = 0$ would be the optimal solution if the problem were a minimization problem, though the feasible region and the objective function gradient would be the same.

While the models that can be solved graphically are generally too small to be of practical value, the visual solution procedure just described is enormously important in understanding how analytical solution methodologies work. The concepts of feasibility and optimality are the same regardless of problem size.

2.B.3 Types of Linear Program Solutions

Whenever a linear programming model is formulated and solved, the result will be one of four characteristic solution types. The graphical framework just developed while solving the Homewood Masonry problem is useful for visualizing these solution types.

Problems Having Unique Optima. The solution to the Homewood Masonry problem was achieved by first graphing the feasible region in decision space, plotting the gradient of the objective function on the same graph, and then shifting the objective function gradient in the direction of improvement until it last intersected the feasible region (see Figure 2.3a). In this case the intersection between the feasible region and the set of points satisfying the equation

$$140x_1 + 160x_2 = 1480$$

consisted of a single point $x_1 = 6$ and $x_2 = 4$. This point is the only point on this line that satisfies all constraint equations simultaneously. Consequently, the optimal solution to the linear program is a unique one; the solution is said to have a *unique optima* or *unique optimal solution.* It is possible, however, that more than one (perhaps an infinite number) solution would be optimal.

Problems Having Alternate Optima. As demonstrated in the previous section, the orientation of the objective function in decision space is determined by the coefficients that multiply the decision variables. For example, if the coefficient on x_2 in the original objective function is decreased relative to the coefficient on x_1, the gradient becomes steeper (more negative). If the original objective function is replaced by

$$\text{Maximize } 140x_1 + 140x_2$$

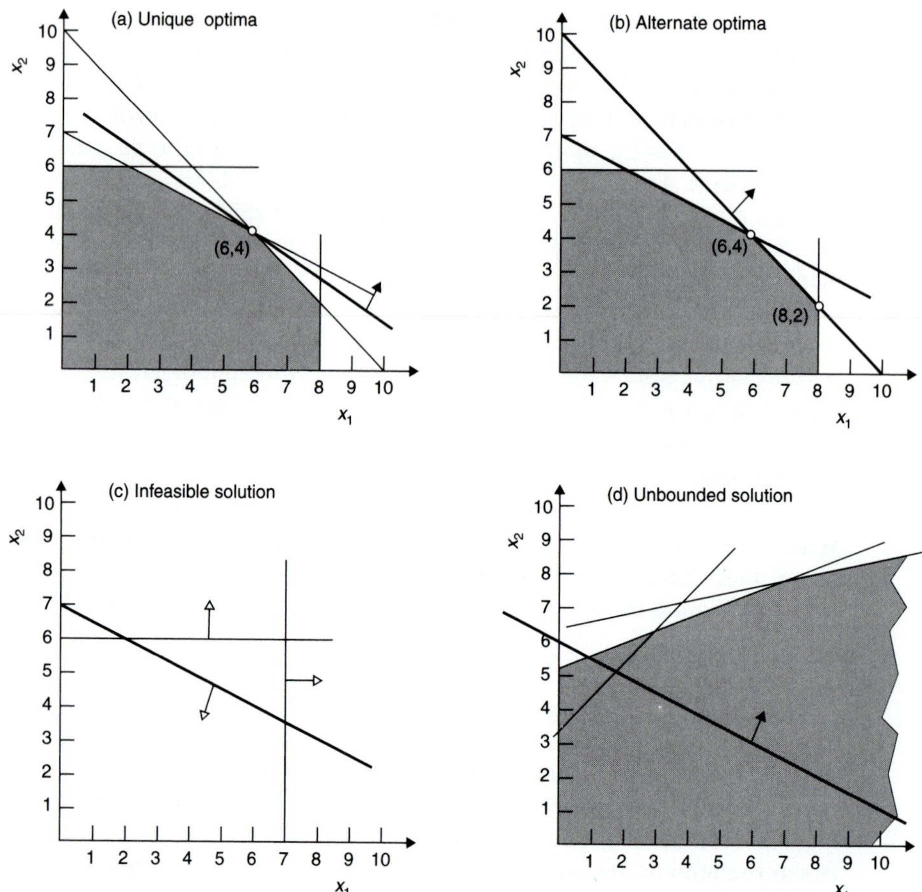

Figure 2.3 Solving a linear program results in one of four possible outcomes: (a) problems having unique optima (a single optimal solution), (b) problems having alternate optima, (c) problems having no feasible solution, and (d) problems that are unbounded.

and the problem is resolved graphically, the intersection of the objective function gradient and the feasible region at optimality becomes a line segment as shown in Figure 2.3b; all points on the line segment connecting the points (6,4) and (8,2) yield the same value for the objective function and satisfy the equation

$$140x_1 + 140x_2 = 1400.$$

This problem thus has an infinite number of optimal solutions, or is said to have *alternate optima.* Alternate optima are actually more common as the size of the linear programming problem (number of decision variables and constraints) increases, and it is important to be able to recognize their presence. An important consideration in our discussion of a general solution procedure for linear programs—the topic of Chapter 3—will be the identification of conditions that indicate the presence of alternate optimal solutions to linear programming problems.

Problems Having No Feasible Solution. It is possible that there are no feasible solutions for a given problem formulation or, due to errors in formulating logical constraints or mistakes in inputting a problem formulation to a model solver (computer program), the problem is overconstrained to the extent that there is no solution satisfying all constraint equations simultaneously. Figure 2.3c shows three constraints of a hypothetical model plotted in decision space in such a way that there is no feasible solution; the problem is said to be *infeasible.* When solving linear programs graphically, infeasible solutions are easy to detect and avoid. In larger problems, it is sometimes difficult to identify if the model is incorrectly formulated, or if an input data coding error has been made—either obvious errors such as an incorrect relation, or a more subtle error such as a typographical error in a variable name. We will discuss in Chapter 3 how infeasible solutions are detected by generalized solution procedures.

Problems That Are Unbounded. As infeasible problems result from problems that are overconstrained, we may also encounter problems that are underconstrained. For example, consider the feasible region of a hypothetical problem presented in Figure 2.3d. As before, the feasible region is shown shaded and constrained by three constraints. The objective function gradient and its direction of improvement are also shown. Note that for any feasible solution in this decision space, we can always find another solution that gives a better value of the objective function such that the objective function for this problem could be moved upwards and to the right without limit. Such a problem is said to be *unbounded*; a situation that can also be identified by the solution procedure to be presented in the next chapter. Like infeasible solutions, unbounded solutions generally indicate that logical or typographical errors have been made in model formulation or input. Detection of unbounded solutions when using analytical solution procedures will also be described more fully in Chapter 3.

2.B.4 Example 2-2: Allocating Work Effort at Two Mines

Problem Statement. A company that produces both copper and nickel owns two mines whose ores contain both minerals. The two mines have different costs and different production rates. Furthermore, their products are different as well. The following data describe the mines and their outputs:

Fraction of ore of mine 1 that is copper = 0.25

Fraction of ore of mine 2 that is copper = 0.15

Fraction of ore of mine 1 that is nickel = 0.00875

Fraction of ore of mine 2 that is nickel = 0.035

Tons of ore per day produced by mine 1 = 1600

Tons of ore per day produced by mine 2 = 1000

Cost in thousands of dollars to operate mine 1 each day = 3

Cost in thousands of dollars to operate mine 2 each day = 9

From these data we can calculate the tons of ores that can be produced for each day of operation of each of the mines. For example, the tons of copper produced per day from mine 1 is

$$\text{fraction copper from mine 1} \times \text{daily tons of ore from mine 1}$$

or

$$0.025 \times 1600 = 40$$

Tons per day of copper from mine 2 and tons of nickel per day from each mine are calculated in a similar fashion.

The company has contracted to provide 100 tons per week of copper and 140 tons per week of nickel. Daily operating costs have been estimated at $7000 and $5000 for mine 1 and mine 2, respectively. We will assume that neither mine can be in operation for more than 5 days per week. A summary of these problem data is tabulated in Table 2.2. Our goal is to set up a linear program that will show the company how to meet its contract obligations at least total cost.

Model Formulation. To formulate a production model for the mining operation problem, we again begin with a pseudomodel. The objective of our model is to determine a strategy for operating both mines so as to minimize total operating costs:

$$\text{Minimize total operating cost}$$

while adhering to all production constraints:

TABLE 2.2 PARAMETERS FOR THE MINING PROBLEM

Resource	Mine 1	Mine 2	Weekly Requirements (tons)
Daily tons of copper	40	15	100
Daily tons of nickel	14	35	150
Daily operating costs ($)	$7000	$5000	

- At least 100 tons of copper must be mined each week;
- At least 140 tons of nickel must be mined each week;
- Mine 1 may not be operated more than 5 days each week; and
- Mine 2 may not be operated more than 5 days each week.

As the problem is specified, mining cost is a function of the number of days that each mine is operated. Indeed, the manager of the operation is responsible for deciding how many days each mine will be in operation during a given week. Our model decision variables are defined accordingly:

Let: x_1 = number of days per week that mine 1 is operated, and

x_2 = number of days per week that mine 2 is operated.

Total weekly production cost may be expressed as a linear function of these decision variables:

$$\text{total cost} = 3x_1 + 9x_2.$$

The tons of copper per week from mines 1 and 2 may then be computed as

$$\text{weekly copper production} = 40x_1 + 15x_2$$

and our model must ensure that at least 100 tons are produced each week. The corresponding constraint equation is thus

$$40x_1 + 15x_2 \geq 100.$$

Similarly for total nickel production:

$$\text{weekly nickel production} = 14x_1 + 35x_2$$

which must be at least 140 tons per week:

$$14x_1 + 35x_2 \geq 140.$$

Two additional constraints must be included in our model to restrict the operation of each mine to no more than 5 days each week, and the linear program formulation for the mining operation problem is complete:

$$\text{Minimize } Z = 3x_1 + 9x_2$$

$$\begin{aligned} \text{Subject to:} \quad 40x_1 + 15x_2 &\geq 100 \\ 14x_1 + 35x_2 &\geq 140 \\ x_1 \qquad\qquad &\leq 5 \\ x_2 &\leq 5 \\ x_1, x_2 &\geq 0. \end{aligned}$$

Problem Solution. Because the mining problem has two decision variables, a graphical solution is again possible by first plotting the model constraints

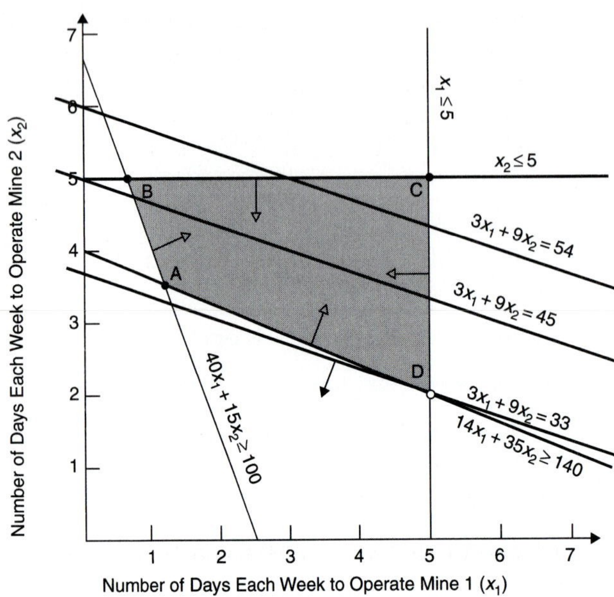

Figure 2.4 Graphical solution to the problem of allocating effort between two mines.

TABLE 2.3 PARAMETERS FOR THE MINING PROBLEM

Corner Point	x_1	x_2	Objective Function Value
A	1 3/17	3 9/17	35.29
B	5/8	5	46.88
C	5	5	60.00
D	5	2	33.00

to delimit the feasible region, and then evaluating all feasible solutions using the objective function. The graphical solution for this example problem is presented as Figure 2.4.

As with the previous example, the feasible region for the mining operation problem is shown as the shaded portion of the solution space. Note, in particular, that the feasible region includes four *corner points*—points at which pairs of constraint equations are satisfied as strict equalities. These points are labeled A, B, C, and D in Figure 2.4 and in Table 2.3. In this example, the solution that satisfies all constraints and that provides the smallest value of the objective function is the optimal solution–point D. Perhaps you have figured out that an alternate solution method might be to solve all pairs of constraint equations for x_1 and x_2 and simply select that solution having the lowest value. Indeed, as can be seen from the graphical solutions for both example problems, and because of the special properties of the feasible region for a linear program to be discussed later, the optimal solution for a linear program will *always* be one of these corner point solutions.

Unfortunately, for all but very small linear programs, there are so many of these *extreme point* solutions that it is impossible to enumerate them all, even with the most sophisticated computer technology. Consequently, we must rely on efficient analytical solution methods that don't require the identification of all corner point solutions. Fortunately, there exist very powerful procedures that can solve linear programs having tens of thousands of variables and constraints, with very little computational effort. The most widely used of those methodologies will be presented in detail in Chapter 3.

2.C MORE EXAMPLE PROBLEMS

2.C.1 Example 2-3: The Mining Company Problem with a Nonlinear Objective Function

The mining company in Example 2-2 is negotiating contracts with several new customers that would increase its weekly requirement for producing copper to 330 tons and nickel to 225 tons. To do so, management might have to operate one or both mines on the weekends, which would mean increased labor costs. It is estimated that these additional costs would be \$6,000 per weekend day for mine 1 and \$12,000 per weekend day for mine 2. How can we modify our model to incorporate this new information?

As a first step, we might draw the cumulative cost curve for the operation of mine 1 as a function of days worked and then the cumulative cost curve for the operation of mine 2 as in Figure 2.5. Next, we define two new decision variables that will allow this additional cost information to be incorporated explicitly into our model.

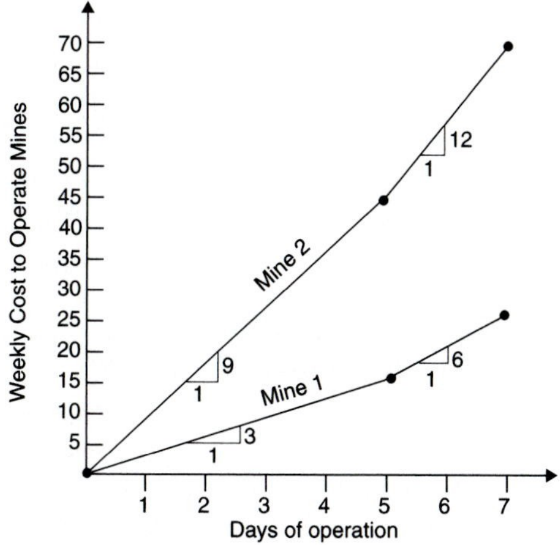

Figure 2.5 Modified cost function for the mine operation problem.

Let: y_1 = additional (weekend) days that mine 1 is operated, and

y_2 = additional (weekend) days that mine 2 is operated

where both y_1 and y_2 are limited to two days or less. The objective function and constraints for this problem are now

$$\text{Minimize } Z = 3x_1 + 6y_1 + 9x_2 + 12y_2$$

$$\begin{aligned} \text{Subject to:} \quad & 40x_1 + 40y_1 + 15x_2 + 15y_2 \geq 330 \\ & 14x_1 + 14y_1 + 35x_2 + 35y_2 \geq 225 \\ & x_1 \leq 5 \\ & y_1 \leq 2 \\ & x_2 \leq 5 \\ & y_2 \leq 2 \\ & x_1, x_2, y_1, y_2 \leq 0. \end{aligned}$$

Because this model has more than three decision variables, it is not possible to plot the feasible solution nor solve the problem graphically. However, computer programs for solving such a problem are widely available for virtually any computer. The computer generated solution for this problem is

$$x_1 = 5, x_2 = 3.7, y_1 = 1.9, y_2 = 0, \text{ and } Z^* \cong \$60{,}000.$$

That is, to fulfill contractual obligations at least total cost, mine 1 will operate 5 days during the week and 1.9 days on the weekend, while mine 2 will only operate 3.7 days during the week and will not operate on weekends.

This model is slightly larger than that presented in Example 2-2, but its overall structure is identical. Yet the problem that the model addresses is more realistic because the cost function—which is actually the sum of two nonlinear functions—more accurately reflects the true costs faced by company management. Does this solution make sense? How difficult would it be to identify this solution using an exhaustive search procedure? How difficult would it be to modify this model to incorporate changes in parameters such as ore production rates, operating costs, changing demand requirements, and so on? Why can an optimal solution still be obtained by linear programming even though the cost function is nonlinear?

2.C.2 Example 2-4: The Thumbsmasher Lumber and Home Center—Producing Paints of Various Qualities Using Ratio Constraints

Problem Statement. The Thumbsmasher Lumber and Home Center is so large that it produces its own brand of paint. This problem is concerned with

TABLE 2.4 PAINT SPECIFICATIONS FOR THE TWO PAINTS

	Latex A	Latex B	Latex C	Price/Gallon
Cost ($/gallon)	2.50	2.00	1.00	
Inferior	≥ 20%(1)*	None	≤ 60%(4)	$9.00
Superior	≥ 50%(2)	≥ 20%(3)	≤ 10%(5)	§15.00
Amount available (gallons/month)	2000	1500	3000	

*Parenthetical number refers to the number of a particular constraint associated with that specification.

the development of a strategy for making its brand of Interior latex paint. Its interior latex paint is available in two product forms known by the staff as inferior interior and superior interior. The home center purchases three qualities of latex base, A, B, and C, which it blends to make the two paint products. Blending follows a set of rules for the proportions of latex bases in each of the two paints. The specifications are given in Table 2.4, along with the monthly availability and prices of the latex bases and the prices of the two paint products as they are sold from the home center. For example, Thumbsmasher can purchase up to 2000 gallons of latex base A each month at a unit price of $2.50 per gallon. Inferior interior paint consists of at least 20% of this substance but not more than 60% of latex base C. Each gallon of Inferior interior paint produced can be sold for $9.00.

The paint staff at Thumbsmashers wishes to blend the three latex bases to produce the two paints in such a way that profit is maximum. Profit is defined here as the difference between the revenue from the sale of the paint and the cost of the base material that goes into it. The task is to formulate a linear program that will help to determine the exact blends for maximizing total profit.

Model Formulation. The objective function for this model is composed of the difference between revenue and cost. Revenue is the product of price times the gallons of superior paint and price times the gallons of interior paint. Because the cost of each gallon of product depends on the fraction of latex base of each type that is used, we must define our decision variables to be able to identify these paint mixture components:

Let: x_{AI} = the number of gallons of latex A to be used in the inferior paint;

x_{AS} = the number of gallons of latex A to be used in the superior paint;

x_{BI} = the number of gallons of latex B to be used in the inferior paint;

x_{BS} = the number of gallons of latex B to be used in the superior paint;

x_{CI} = the number of gallons of latex C to be used in the inferior paint; and

x_{CS} = the number of gallons of latex C to be used in the superior paint.

Total revenue is the product of price times the gallons of superior paint—total volume of all constituent bases—plus price times the gallons of inferior paint produced:

$$\text{total revenue} = 15.00\,(x_{AS} + x_{BS} + x_{CS}) + 9.00\,(x_{AI} + x_{BI} + x_{CI}).$$

Similarly, cost is the sum of the price of each latex base times the amount of each latex base used:

$$\text{total cost} = 2.5\,(x_{AI} + x_{AS}) + 2.00\,(x_{BI} + x_{BS}) + 1.00\,(x_{CI} + x_{CS}).$$

When revenue is subtracted from cost, the profit objective function can be written as

$$\text{Minimize } Z = 12.50x_{AS} + 6.50x_{AI} + 13.00x_{BS} + 7.00x_{BI} + 14.00x_{CS} + 8.00x_{CI}.$$

Next, a set of constraints must be provided to control the quality of the paint mixtures: at least 20% of Inferior interior paint must be made up of latex base A:

$$\frac{x_{AI}}{x_{AI} + x_{BI} + x_{CI}} \geq 0.2,$$

at least 50% of Superior interior paint must be made up of latex base A:

$$\frac{x_{AS}}{x_{AS} + x_{BS} + x_{CS}} \geq 0.5,$$

at least 20% of Superior interior paint must be made up of latex base B:

$$\frac{x_{BS}}{x_{AS} + x_{BS} + x_{CS}} \geq 0.2,$$

not more than 60% of Inferior interior paint can be made up of latex base C:

$$\frac{x_{CI}}{x_{AI} + x_{BI} + x_{CI}} \leq 0.6,$$

and not more than 10% of Superior interior paint can be made up of latex base C:

$$\frac{x_{CS}}{x_{AS} + x_{BS} + x_{CS}} \leq 0.1.$$

Limits on the availability of the latex bases must also be considered. That is,

$$x_{AI} + x_{AS} \le 2000$$
$$x_{BI} + x_{BS} \le 1500$$
$$x_{CI} + x_{CS} \le 3000.$$

The first five constraints must also be expressed in linear form using the following translation:

$$\frac{x_{AI}}{x_{AI} + x_{BI} + x_{CI}} \ge 0.2 \quad \leftrightarrow \quad x_{AI} \ge 0.2(x_{AI} + x_{BI} + x_{CI})$$

or

$$0.8x_{AI} - 0.2x_{BI} - 0.2x_{CI} \ge 0.$$

Hence the entire problem formulation can be written as

$$\text{Minimize } Z = 12.50x_{AS} + 6.50x_{AI} + 13.00x_{BS} + 7.00x_{BI} + 14.00x_{CS} + 8.00x_{CI}$$

$$\begin{aligned} \text{Subject to:} \quad 0.8x_{AI} - 0.2x_{BI} - 0.2x_{CI} &\ge 0 \\ 0.5x_{AS} - 0.5x_{BS} - 0.5x_{CS} &\ge 0 \\ -0.2x_{AS} + 0.8x_{BS} - 0.2x_{CS} &\ge 0 \\ -0.6x_{AI} - 0.6x_{BI} + 0.4x_{CI} &\le 0 \\ -0.1x_{AS} - 0.1x_{BS} + 0.9x_{CS} &\le 0 \\ x_{AI} + x_{AS} &\le 2000 \\ x_{BI} + x_{BS} &\le 1500 \\ x_{CI} + x_{CS} &\le 3000 \\ x_{AI}, x_{AS}, x_{BI}, x_{BS}, x_{CI}, x_{CS} &\ge 0. \end{aligned}$$

The optimal solution to this problem is $x_{AS} = 900$, $x_{AI} = 1100$, $x_{BS} = 720$, $x_{BI} = 780$, $x_{CS} = 180$, and $x_{CI} = 2820$, with a resulting total profit of approximately \$58,300.

2.C.3 Example 2-5: Mixing Gravels to Produce an Aggregate: A Multiobjective Optimization Problem

Problem Statement. An aggregate mixture of a defined composition is needed for a very large construction project. The composition is not exact, but consists of target ranges of coarse material, fine material, and sand. Coarse material should be present in no greater proportion than 55%, but should at least compose 48% of the mixture by weight. Fine materials should be in the range of 30–35% by weight, and sand should be in the range of 15–20% by weight.

TABLE 2.5 COMPOSITION OF SOURCE AGGREGATE MATERIALS

	% Coarse	% Fine	% Sand	Cost ($)
Source 1	35	30	35	.20
Source 2	10	50	40	.30
Source 3	60	30	10	.45
Source 4	50	35	15	.65

No single source of material that is available for the project meets these requirements, but the material from a number of sources can be blended to produce the desired composition. The sources from which the raw material for blending is purchased are listed in Table 2.5 along with the costs for each unit of raw material and the composition as taken from the source. Because it would be too costly to separate the raw materials at each source into their constituent parts and reblend them into their desired mixture, raw input materials from each source are to be purchased. For example, each pound of material from Source 3 costs $0.45 and is guaranteed to contain 60% coarse material, 30% fine material, and 10% sand. The costs include the cost to dig, refine, and transport the material from each of the sources to the project. Materials from one or more sources will be purchased and used to blend construction products whose composition falls within the specified ranges. Formulate a linear program that will suggest an optimal strategy for acquiring raw materials.

Model Formulation. The decision variables are the amounts of raw material to be purchased from each source for every unit, say, 1000 pounds, of final aggregate mixture to be blended. For the purposes of this problem, it is too costly to separate the levels of size and then reblend them in the desired mixture.

Let: x_1 = amount in pounds of material to purchase from source 1;

x_2 = amount in pounds of material to purchase from source 2;

x_3 = amount in pounds of material to purchase from source 3; and

x_4 = amount in pounds of material to purchase from source 4.

The obvious objective is to minimize the cost of aggregate. The first and obvious constraint is that 1000 pounds of the aggregate mixture are to be made. Second, the composition in terms of coarse material must be in the range of 48 to 55%. That is, in 1000 pounds of the aggregate, coarse material must constitute at least 480 of those pounds but no more than 550 pounds. In a similar fashion, the fine materials can range from 300 to 350 pounds, and the amount of sand must be between 150 and 200 pounds. The complete model formulation is

$$\text{Minimize } Z = 0.20x_1 + 0.30x_2 + 0.45x_3 + 0.65x_4$$

$$\begin{aligned}
\text{Subject to:} \quad x_1 + x_2 + x_3 + x_4 &= 1000 \\
0.35x_1 + 0.10x_2 + 0.60x_3 + 0.50x_4 &\geq 480 \\
0.35x_1 + 0.10x_2 + 0.60x_3 + 0.50x_4 &\leq 550 \\
0.30x_1 + 0.50x_2 + 0.30x_3 + 0.35x_4 &\geq 300 \\
0.30x_1 + 0.50x_2 + 0.30x_3 + 0.35x_4 &\leq 350 \\
0.35x_1 + 0.40x_2 + 0.10x_3 + 0.15x_4 &\geq 150 \\
0.35x_1 + 0.40x_2 + 0.10x_3 + 0.15x_4 &\leq 200 \\
x_1, x_2, x_3, x_4 &\geq 0.
\end{aligned}$$

The solution to this problem is $x_1^* = 400$, $x_2^* = 0$, $x_3^* = 600$, and $x_4^* = 0$ with $Z^* = \$350$.

Although the cost minimization objective function used in this model is somewhat obvious, the supervisor of the aggregate plant has other management concerns. Sources 1 and 3 are, in fact, only reachable via a disputed right-of-way across private land. The manager would like to create a least cost aggregate, but would also like to minimize the amount of material taken from sources 1 and 3 because each truckload from these sources creates a risk of confrontation. In particular, the manager would like to know the shape of the trade-off surface between cost and amount drawn from sources 1 and 3. If total cost does not increase substantially with a decrease in the volume purchased from sources 1 and 3, the manager might prefer a more costly but more politically correct solution.

While the constraint set and, therefore, the feasible region resulting from this modification to the original problem are unchanged, the problem has two distinct objective functions:

$$\text{minimize total cost } (Z_1) = 0.20x_1 + 0.30x_2 + 0.45x_3 + 0.65x_4$$

and

$$\text{minimize amount purchased from sources 1 and 3 } (Z_2) = x_1 + x_3.$$

The trade-off relationship between these objectives is shown in Figure 2.6. Solution A is the cost minimization solution ($Z_1^* = 350$ with $x_1^* = 400$ and $x_3^* = 600$) found previously. The value for Z_2 at this solution is 1000 ($x_1 = 400$, $x_3 = 600$). Solution F is the optimal solution found when objective Z_2 is substituted for Z_1 and the problem resolved ($Z_2^* = 0$ because $x_1^*, x_3^* = 0$ but the value of Z_1 at this solution is \$650). Solutions B, C, D, and E represent other possible management strategies that reflect trade-offs between these two objectives and may be selected by the engineering manager as *best compromise* rather than optimal solutions.

Problems having multiple and conflicting management objectives are common in the realm of public sector engineering management. In fact it is difficult

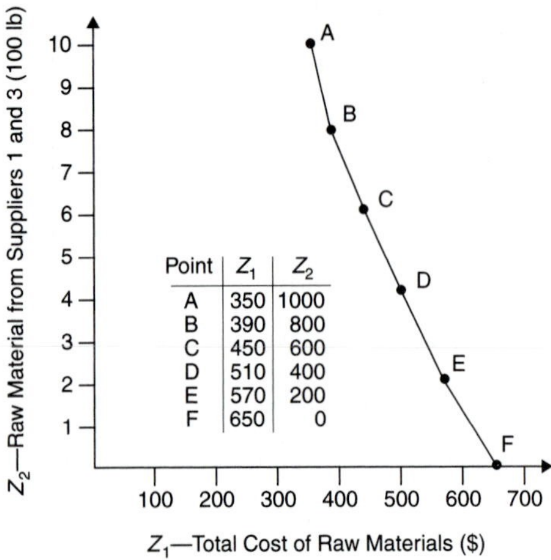

Figure 2.6 Trade-off relationship between objectives in the gravel mixing problem.

to cite a major civil or environmental investment that does not have multiple management objectives and usually multiple decision makers as well. The focus of Chapter 4 is a framework for analyzing problems having multiple objectives, and presents a systematic procedure for finding a set of best compromise solutions for which no better solutions exist.

2.C.4 Example 2-6: Selecting Projects for Bidding: An Integer Programming Problem

Problem Statement. A contractor has identified a set of eight major construction projects to be awarded on the basis of sealed bids and for which he/she is reasonably confident of being able to submit the low bid. The projects are large in scale, and each will require different levels of labor, materials, and equipment to complete. Each bid also requires time and special expertise to prepare, both of which are limited. The question facing the contractor is which projects should be selected for bidding.

The relevant data for each of the eight projects are presented in Table 2.6, including the time necessary to prepare a bid and the expected profit that would result from being awarded and completing each project. Estimates for the major resources needed for a particular project are also included. For example, project 3—the redecking of an urban bridge crossing a major river—would require a skilled labor input of about 7000 hours and the use of a very large crane during almost the entire project. The bid for this project would be the first of its kind undertaken by this contractor, and would thus take approximately 13 person weeks to prepare. The profit resulting from being awarded and successfully completing this project is estimated to be $100,000.

The resources available to the contractor for preparing bids and completing projects that might be awarded are limited. The company owns three cranes, each

TABLE 2.6 ESTIMATED RESOURCES NEEDED AND PROFIT FOR THE BIDDING PROBLEM

Project #	Time to Prepare Bid (person-weeks)	Thousands of Hours of Skilled Labor Needed	Crane Required for the Project? (1 = yes)	Profit IF Project Won ($1000)
1	8	6	1	80
2	12	5	0	110
3	13	4	1	100
4	11	7	0	90
5	9	8	0	70
6	7	3	1	80
7	8	4	1	90
8	8	5	1	60

of which is suitable for any of the projects. Only 30,000 hours of skilled labor are expected to be available to the contractor during the period of the projects. A team of four project estimators will prepare all bids so that over the next three months, a total of only 50 person-weeks is available. The company confidently assumes that it will win any bid that it prepares. Which projects should be pursued?

Model Formulation. The objective function for a model that will help analyze this problem is the maximization of expected profit, and like previous models, the decision variables that we define must allow the computation of overall profit. Unlike those previous models, these decision variables are different in that at optimality, their values must be discrete (integer):

$$\text{Let:} \quad x_j = \begin{cases} 1, \text{ if project j is selected for bidding,} \\ 0, \text{ otherwise.} \end{cases}$$

These decision variables allow formulation of the maximum profit objective and constraints on bid preparation time and equipment and labor resources:

$$\text{Maximize } Z = 80x_1 + 110x_2 + 100x_3 + 90x_4 + 70x_5 + 80x_6 + 90x_7 + 60x_8$$

$$\text{Subject to:} \quad 8x_1 + 12x_2 + 13x_3 + 11x_4 + 9x_5 + 7x_6 + 8x_7 + 8x_8 \leq 50$$

$$6x_1 + 5x_2 + 4x_3 + 7x_4 + 8x_5 + 3x_6 + 4x_7 + 5x_8 \leq 30$$

$$x_1 + x_3 + x_6 + x_7 + x_8 \leq 3$$

$$x_1, x_2, x_3, x_4, x_5, x_6, x_7, x_8 \in \{0, 1\}$$

The optimal solution is $x_2^* = x_3^* = x_5^* = x_6^* = x_7^* = 1$, $x_1^* = x_4^* = x_8^* = 0$, resulting in a total expected profit of $450,000.

The requirement that decision variables in a linear program be restricted to discrete values at optimality allows the formulation of more realistic models and provides the opportunity for the analyst to incorporate more complicated logic into the model structure. For example, adding the following constraint to the bid selection model

$$x_3 + x_5 + x_6 \leq 2$$

would enforce the condition that at most two of projects 3, 5, and 8 could be selected for bid preparation, but that none of these would be required to be selected. A complete discussion of integer linear programs, including their formulation and solution, is the topic of Chapter 5.

2.C.5 Example 2-7: The Thumbsmasher Lumber and Home Center Part II—Cutting Plywood: An Integer Programming Problem with Hidden Variables

Problem Statement. The lumberyard department of Thumbsmasher, Inc., provides precut pieces of 1/2-inch birch veneer plywood for the do-it-yourselfer furniture hobbyist. The weekly estimated demand for components and the dimensions of each are presented in Table 2.7. Management wishes to provide these sizes by cutting up the fewest possible number of 4 × 8 plywood sheets. Pieces cut to nonrequired sizes will be scrapped.

Model Formulation. This problem is starkly different from the previous problems you have seen. Those problems had very obvious decision variables: the number of tons, the amount of treatment, the number of chairs, and so on. This problem's variables are generally not evident at first glance unless you have encountered such problems before. Accordingly, we call such a situation a problem with hidden variables.

Each furniture component has a fixed size and shape, and all pieces are to be cut from individual plywood boards that also have a uniform 4-foot by 8-foot

TABLE 2.7 REQUIREMENTS FOR DIFFERENT SIZED PRECUT COMPONENTS

Furniture Component	Size (dimensions in feet)	Number Needed Each Week
Shelving	1 × 4	200
Small desk tops	2 × 4	40
Large desk tops	3 × 6	30
Table tops	4 × 4	20

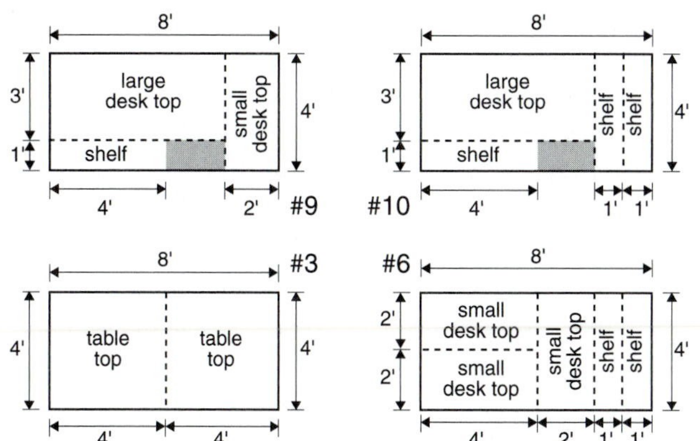

Figure 2.7 Four possible cutting patterns for plywood sheets.

TABLE 2.8 CUTTING PATTERNS AND THE NUMBER OF COMPONENTS RESULTING FROM EACH

Furniture Component	Size	Cutting Pattern and Yield									
		1	2	3	4	5	6	7	8	9	10
shelves	1 × 4	8	6	0	4	2	2	0	0	1	3
small desk tops	2 × 4	0	1	0	2	1	3	2	4	1	0
large desk tops	3 × 6	0	0	0	0	0	0	0	0	1	1
table tops	4 × 4	0	0	2	0	1	0	1	0	0	0

size and shape. Consequently each plywood board will generate a fixed number of pieces, with some possible residual waste which is too small or not of sufficient shape to be used. Four possible cutting patterns are shown in Figure 2.7. There are, in fact, a finite number of cutting patterns for the plywood stock, each producing a different mix of pieces. Table 2.8 lists most of these cutting patterns (including those shown in Figure 2.7) and the corresponding yield of each in terms of furniture components. (For a number of these patterns there may be several geometric ways to give the same result in terms of the number of pieces of each size.) By enumerating all possible unique cutting patterns and computing the yield from each, the critical decision becomes determining how many of each cutting pattern to use—how many plywood boards to cut according to each. Hence, the decision variables are defined as

$$x_j = \text{number of plywood boards to cut according to the } j\text{th pattern,}$$
$$j = 1, 2, \ldots, n,$$

where n is the total number of cutting patterns that have been identified–ten in this example. The objective function is simply to minimize the total number of boards cut with constraints to ensure that a sufficient number of each furniture component is produced. The complete formulation is:

$$\text{Minimize } Z = x_1 + x_2 + x_3 + x_4 + x_5 + x_6 + x_7 + x_8 + x_9 + x_{10}$$

$$\begin{aligned}
\text{Subject to:}\quad 8x_1 + 6x_2 + 4x_4 + 2x_5 + 2x_6 + x_9 + 3x_{10} &\geq 200 \\
x_2 + 2x_4 + x_5 + 3x_6 + 2x_7 + 4x_8 + x_9 &\geq 40 \\
x_9 + x_{10} &\geq 30 \\
2x_3 + x_5 + x_7 &\geq 20 \\
x_1, x_2, x_3, x_4, x_5, x_6, x_7, x_8, x_9, x_{10} &\in \{\text{Positive Integers}\}
\end{aligned}$$

The solution is obtained by integer linear programming (see Chapter 5) and requires a total of 64 sheets of plywood to be cut: $x_1^* = 3$, $x_2^* = 18$, $x_3^* = 10$, $x_8^* = 3$, $x_9^* = 11$, and $x_{10}^* = 19$, and all other patterns unused.

This problem resembles the previous integer programming problem in the sense that all variables must be integers, but here any feasible nonnegative integer is eligible and in the prior problem only zeros and ones were admissable solutions. In Chapter 5, on integer programming, we will see that the methodologies for solving the two problems, while based on a similar concept, are somewhat different.

CHAPTER SUMMARY

The solution of very small linear programs using the graphical method emphasizes that feasible solutions to optimization problems are determined by model constraints, while optimality is determined by the objective function. For problems having a single optimal solution, that solution will be a corner point of the feasible region. Problems with more than one optimal solution are common, and are said to have alternate optima. Problems having no feasible region are said to be infeasible, while problems with an infinite feasible region may be unbounded.

A number of important variations to the basic linear program formulation are possible, allowing special characteristics to be modeled including those having nonlinear objective functions, those having requirements for integer valued decision variables, or those displaying multiple objective functions. A general-purpose algorithm for solving linear programs is presented in the next chapter.

EXERCISES

2.1. Solve the following linear program using the graphical method. Compute the value of the objective function and decision variables at optimality, and indicate which statement best describes the solution:

Maximize $Z = 12x_1 + 18x_2$	(a) this linear program has a unique optimal solution
Subject to: $4x_1 + 5x_2 \geq 10$	(b) this linear program has alternate optima
$x_1 + 4x_2 \leq 12$	(c) this linear program is infeasible
$x_1 \leq 4$	(d) this linear program is unbounded
$x_2 \geq 1$	
$x_1, x_2, \geq 0$	

2.2. Solve the following linear program using the graphical method. Compute the value of the objective function and decision variables at optimality, and indicate which statement best describes the solution:

Minimize $Z = 2x_1 - 13x_2$	(a) this linear program has a unique optimal solution
Subject to: $-x_1 + 2x_2 \geq 8$	(b) this linear program has alternate optima
$2x_1 + 2x_2 \leq 16$	(c) this linear program is infeasible
$3x_1 \geq 9$	(d) this linear program is unbounded
$2x_2 \geq 6$	
$x_1, x_2, \geq 0$	

2.3. Solve the following linear program using the graphical method. Compute the value of the objective function and decision variables at optimality, and indicate which statement best describes the solution:

Minimize $Z = -4x_1 - 6x_2$	(a) this linear program has a unique optimal solution
Subject to: $-4x_1 - 5x_2 \leq -10$	(b) this linear program has alternate optima
$-x_1 - 4x_2 \geq -12$	(c) this linear program is infeasible
$-x_1 \geq -4$	(d) this linear program is unbounded
$x_2 \geq 1$	
$x_1, x_2, \geq 0$	

2.4. Solve the following linear program using the graphical method. Compute the value of the objective function and decision variables at optimality, and indicate which statement best describes the solution:

Minimize $Z = 2x_1 + x_2$	(a) this linear program has a unique optimal solution
Subject to: $2x_1 + 3x_2 \geq 3$	(b) this linear program has alternate optima
$x_1 + 5x_2 \leq 10$	(c) this linear program is infeasible
$2x_1 + x_2 \leq 4$	(d) this linear program is unbounded
$x_1 - x_2 \leq 1$	
$4x_2 \geq 2$	
$x_1, x_2, \geq 0$	

2.5. Solve the following linear program using the graphical method. Compute the value of the objective function and decision variables at optimality, and indicate which statement best describes the solution:

Minimize $Z = 4x_1 + 6x_2$

Subject to:

$$x_1 - x_2 \geq -4$$
$$3x_1 - 2x_2 \leq 6$$
$$x_1 + x_2 \geq 5$$
$$x_1 + x_2 \leq 10$$
$$x_2 \leq 6$$
$$x_1, x_2, \geq 0$$

(a) this linear program has a unique optimal solution
(b) this linear program has alternate optima
(c) this linear program is infeasible
(d) this linear program is unbounded

2.6. Solve the following linear program using the graphical method. Compute the value of the objective function and decision variables at optimality, and indicate which statement best describes the solution:

Maximize $Z = 24x_1 + 20x_2$

Subject to:

$$x_1 - x_2 \geq -4$$
$$3x_1 - 2x_2 \leq 6$$
$$x_1 + x_2 \geq 5$$
$$6x_1 + 5x_2 \leq 10$$
$$x_2 \leq 6$$
$$x_1, x_2, \geq 0$$

(a) this linear program has a unique optimal solution
(b) this linear program has alternate optima
(c) this linear program is infeasible
(d) this linear program is unbounded

2.7. Solve the following linear program using the graphical method. Compute the value of the objective function and decision variables at optimality, and indicate which statement best describes the solution:

Maximize $Z = 7x_1 + 9x_2$

Subject to:

$$2x_1 - 2x_2 \leq 2$$
$$4x_1 - 3x_2 \geq -6$$
$$4x_1 + 2x_2 \geq 8$$
$$4x_1 \geq 4$$
$$4x_2 \geq 6$$
$$x_1, x_2 \geq 0$$

(a) this linear program has a unique optimal solution
(b) this linear program has alternate optima
(c) this linear program is infeasible
(d) this linear program is unbounded

2.8. Solve the following linear program using the graphical method. Compute the value of the objective function and decision variables at optimality, and indicate which statement best describes the solution:

$$\begin{aligned} \text{Minimize } Z &= 8x_1 + 4x_2 \\ \text{Subject to: } 2x_1 - 2x_2 &\le 2 \\ 4x_1 - 3x_2 &\ge -6 \\ 4x_1 + 2x_2 &\ge 8 \\ 4x_1 &\ge 4 \\ 4x_2 &\ge 6 \\ x_1, x_2 &\ge 0 \end{aligned}$$

(a) this linear program has a unique optimal solution
(b) this linear program has alternate optima
(c) this linear program is infeasible
(d) this linear program is unbounded

2.9. Solve the following linear program using the graphical method. Compute the value of the objective function and decision variables at optimality, and indicate which statement best describes the solution:

$$\begin{aligned} \text{Maximize } Z &= 3x_1 + 6x_2 \\ \text{Subject to: } 4x_1 + 3x_2 &\ge 4 \\ 6x_1 - 5x_2 &= 6 \\ 3x_1 + 8x_2 &\le 18 \\ x_2 &\le 2 \\ x_1, x_2 &\ge 0 \end{aligned}$$

(a) this linear program has a unique optimal solution
(b) this linear program has alternate optima
(c) this linear program is infeasible
(d) this linear program is unbounded

2.10. Solve the following linear program using the graphical method. Compute the value of the objective function and decision variables at optimality, and indicate which statement best describes the solution:

$$\begin{aligned} \text{Maximize } Z &= 7x_1 + 9x_2 \\ \text{Subject to: } 4x_1 + 3x_2 &\le 36 \\ x_1 + 2x_2 &\le 12 \\ x_1 &\ge 4 \\ x_2 &\ge 1 \\ x_1, x_2 &\ge 0 \end{aligned}$$

(a) this linear program has a unique optimal solution
(b) this linear program has alternate optima
(c) this linear program is infeasible
(d) this linear program is unbounded

2.11. Solve the following linear program using the graphical method. Compute the value of the objective function and decision variables at optimality, and indicate which statement best describes the solution:

$$\begin{aligned} \text{Minimize } Z &= 10x_1 + 4x_2 \\ \text{Subject to: } 5x_1 - 6x_2 &\le 30 \\ 5x_1 + 2x_2 &\ge 30 \\ x_1 &\ge 5 \\ x_2 &\ge 2 \\ x_1, x_2 &\ge 0 \end{aligned}$$

(a) this linear program has a unique optimal solution
(b) this linear program has alternate optima
(c) this linear program is infeasible
(d) this linear program is unbounded

2.12. Solve the following linear program using the graphical method. Compute the value of the objective function and decision variables at optimality, and indicate which statement best describes the solution:

Maximize $Z = 10x_1 + 4x_2$

Subject to:

$$5x_1 - 6x_2 \geq 30$$
$$5x_1 + 2x_2 \leq 30$$
$$x_1 \geq 5$$
$$x_2 \geq 2$$
$$x_1, x_2 \geq 0$$

(a) this linear program has a unique optimal solution

(b) this linear program has alternate optima

(c) this linear program is infeasible

(d) this linear program is unbounded

2.13. Solve the following linear program using the graphical method. Then write an alternate objective function for this model formulation such that the corner points (6, 5) and (4, 6) would be alternate optimal solutions.

Maximize $Z = 3x_1 + 2x_2$

Subject to:

$$-3x_1 + 4x_2 \leq 12$$
$$2x_1 + 4x_2 \leq 32$$
$$x_1 \leq 4$$
$$x_1, x_2 \geq 0$$

2.14. You are the manager of a firm that provides fill gravel to construction sites. You purchase the material from three different suppliers (A, B, and C) for \$160/ton, \$140/ton and \$170/ton, respectively and have a contract to meet weekly demand for gravel at four different sites. Transport costs and the available supply, demand, and per unit shipping costs for your operation are presented in the table below:

Supplier	Site 1	Site 2	Site 3	Site 4	Supply
A	\$32	\$19	\$42	\$38	730
B	\$42	\$38	\$27	\$36	1206
C	\$24	\$18	\$26	\$29	672
Demand	460	575	720	670	

For example, Supplier B can provide you a maximum of 1206 tons of gravel per week at \$140/ton and for every ton of that material you ship to Site 3, you incur a cost of \$27.00. You are contracted to supply a total of 420 tons each week to Site 3.

Formulate a linear program that will identify the optimal shipment strategy for your operation so as to keep your overall costs as low as possible.

2.15. As county engineer, it is your responsibility to design a master plan for refuse disposal for a 4-city area. All waste is delivered to one of two transfer stations where it is compacted and placed into large transport vehicles for shipment to one of three landfills. Each transfer station can process 600 tons of refuse daily, and the capacities of the three landfills are 500, 420, and 600 tons of waste per day, respectively. All shipping costs are directly proportional to the travel distances shown in the table below, but shipments from the transfer stations are 50% lower per mile than are those by the route trucks that transport refuse directly from residence to transfer station:

From City	Daily Refuse Production	Transport Distances	
		Transfer Station 1	Transfer Station 2
Pleasantville	205	45	58
Odorsburg	369	39	79
Scentington	340	74	93
Smellberry	408	25	44
To Landfill	Daily Capacity		
Landfill No. 1	500	157	175
Landfill No. 2	420	234	136
Landfill No. 3	600	146	192

For example, a unit of refuse transported by route truck from Pleasantville to Transfer Station Number 1 travels a distance of 45 miles. If that unit of refuse is then transported to Landfill No. 3, it will be moved another 146 miles but at a 50% lower transport cost. Pleasantville generates 205 tons of waste each day, and Landfill No. 3 can process a total of 600 tons per day from both Transfer Stations.

Formulate a linear program that will help you plan a policy for managing the disposal operation at least cost.

2.16. Suppose that all four communities from Problem 2.15 were concerned about the ultimate capacities of the landfills. Modify your formulation so that the optimal transshipment strategy results in each city utilizing each landfill in proportion to their daily refuse production.

2.17. You are responsible for scheduling the activities required for construction of a small commercial establishment. These activities, the expected duration of each, and an indication of which other activities must precede each are shown in the table below:

Job	Description	Duration	Must follow Job(s)
1	Land leveling and excavation	9	—
2	Pour retaining walls and foundation	7	1
3	Install basement support structure	5	2
4	Install floor joists	3	2
5	Construct exterior walls	6	3, 4
6	Install walls and flooring	7	4
7	Install ceiling/roof superstructure	4	5, 6
8	Install electrical/mechanical/plumbing	9	6
9	Rough finish interior (wallboard, etc.)	7	6
10	Install roofing material	5	7
11	Finish interior	8	9
12	Landscaping	11	1

For example, construction of the exterior walls of the facility is estimated to take 6 crew days, and may not begin until the basement support structure has been completed—Job 3—and the floor joists have been installed—Job 4.

Formulate a linear program that will determine the starting time for each individual activity such that the facility may be in operation as soon as possible.

2.18 An alternate approach to solving the job scheduling problem presented in Problem 2.17 is one that seeks to minimize total project costs. This assumes that the costs for each job are also known in advance. Furthermore, it may be possible to accelerate some or all of the construction activities by allocating more resources to those jobs.

Consider the costs indicated for each of the construction activities presented in the table below. The description of each job is the same as in Problem 2.17 with the base costs for each job indicated. An alternate completion time and cost is also included in the table. For example, the installation of the ceiling superstructure, which normally takes 4 days at a total cost of $2,100 can be completed in 2 days if you are willing to increase that cost to $3,600.

Job	Base Cost	Base Duration	Crash Cost	Crash Duration
1	$2,500	9	$3,100	6
2	$2,700	7	$3,200	5
3	$1,800	5	$2,600	3
4	$1,400	3	$1,700	2
5	$4,200	6	$6,300	3
6	$3,200	7	$5,600	4
7	$2,100	4	$3,600	2
8	$4,200	9	$8,300	5
9	$2,300	7	$4,600	4
10	$4,200	5	$4,900	4
11	$3,800	8	$5,200	5
12	$4,000	11	$6,400	7

Your contract also has a penalty clause such that if the project is not finished in 20 days, you must pay a penalty computed using the following formula:

$$\text{Penalty} = 6.38\,(\text{number of days over } 20)^2$$

Formulate a linear program that determines the scheduled starting time for each job so as to minimize total project cost, including penalty that might be owed.

REFERENCES

BAZARAA, M. and J. JARVIS, 1990. *Linear Programming and Network Flows*, New York: Wiley, 565 pages.

DE NEUFVILLE, R., 1990. *Applied Systems Analysis*, McGraw-Hill, New York.

GASS, S., 1985. *Linear Programming: Methods and Applications*, Fifth Edition, McGraw-Hill.

HILLIER, F. S., and LIEBERMAN, G. J., 1995. *Introduction to Mathematical Programming*, Second Edition, McGraw-Hill.

MURTY, K. G., 1983. *Linear Programming*, Wiley-Interscience, 482 pages.

REKLAITIS, G. V., RAVINDRAN, A., and K. M. RAGSDELL, 1983. *Engineering Optimization: Methods and Applications*, Wiley-Interscience, 684 pages.

CHAPTER

3 The Simplex Algorithm for Solving Linear Programs

3.A PROPERTIES OF THE FEASIBLE REGION

The graphical solution to the Homewood Masonry linear programming formulation presented as Example 2-1 in the previous chapter was possible because the feasible region in decision space for problems having three or fewer variables can be shown graphically. Larger problems can be solved using special-purpose solution algorithms that usually require the use of a computer program. In this section, we describe in detail the most widely used algorithm for solving linear programs—the *simplex algorithm*.

3.A.1 Characterization of Extreme Points

Recall that the optimal solution to the Homewood Masonry problem was that solution providing the best (largest, in this case) value of the objective function as it was moved through the feasible region. It was shown graphically in Figure 2.2 that when the feasible region exists, and is bounded in the direction of improving the objective function, the optimal solution or solutions to the model will lie on the boundary of the feasible region. This results from three important properties for any feasible region of a linear program:

1. The feasible region of a linear program is *convex*. Any point on the interior of a line segment connecting two points in the feasible region is feasible.

2. The feasible region of a linear program is *compact.* The feasible region of a linear program contains its boundary. All solutions on the boundary of a feasible region are themselves feasible.
3. The feasible region of a linear program is *continuous.* There are no "holes" or "gaps" in the feasible region of a linear program.

The significance of these properties is that to find solutions which optimize the value of any objective function, we need only evaluate those on the boundary of the feasible region. For any point in the interior of the feasible region, there will always exist another point providing a better value for the objective function.

Note also from the graphical solution for Example 2-1 (Figure 2.2), that the optimal solution was one for which two of the model's original constraint equations were met with strict equality. At the optimal solution—point ($x_1 = 6, x_2 = 4$)—the constraints on (1) the availability of Wabash red clay and (2) the availability of the blending machine are satisfied as follows:

$$2x_1 + 4x_2 = 28 \qquad \text{and} \qquad 5x_1 + 5x_2 = 50.$$

These constraints are said to be *binding* because the right-hand sides of these equations at this solution are limiting the value of the objective function. If either of these right-hand side values could be increased, the feasible region in decision space would be increased and the objective function could be improved. The other constraints, however—storage vat capacities for HYDIT and FILIT—are not constraining the current solution; the preparation of (6) tons of HYDIT and (4) tons of FILIT will not require the use of all available capacity of the storage vats:

$$x_1 \le 8 \qquad \text{and} \qquad x_2 \le 6.$$

If increased storage in either vat were available to the production manager for Homewood Masonry, it should have no effect on the optimal production strategy. The feasible region shown in Figure 2.2 would be larger, but the value of the objective function could not be increased. In the language of linear programming, these constraints are *nonbinding.*

All solutions that lie at the intersection of constraint equations that are satisfied with strict equality are called *extreme points* or *corner points* of the feasible region. More precisely, if the linear program consists of n decision variables and m constraints, any point at which exactly n constraints are satisfied with strict equality represents a unique extreme point of the feasible region for that problem.

The significance of extreme points should become clear by further reviewing the graphical solutions to the Homewood Masonry problem presented in Chapter 2. Whenever the solution to a linear program is a unique optimal solution that solution lies at an extreme point of the feasible region (see Figure 2.3a). When alternate optima exist, at least two of the alternate optimal solutions will be extreme points of the feasible region (see Figure 2.3b). The fundamental rationale behind the simplex algorithm is that the set of optimal solutions to a given linear program will *always* include at least one corner point solution.

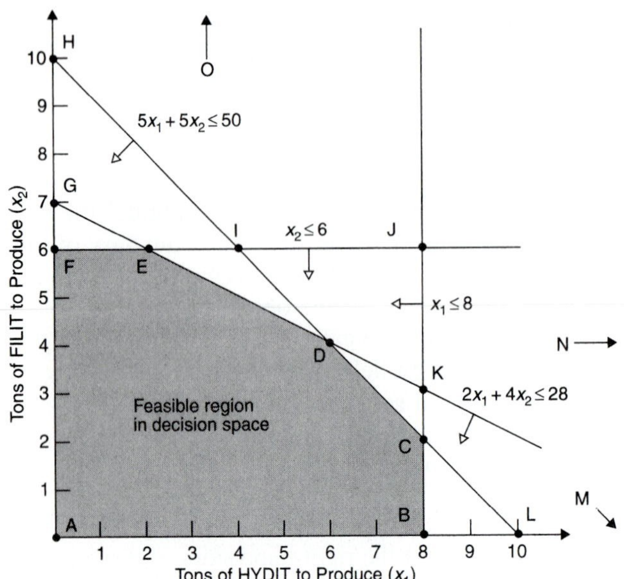

Figure 3.1 Extreme points for the Homewood Masonry problem fall at the intersections of constraint equations. There are 15 such extreme points; points M, N, and 0 are not plotted because they lie outside the figure.

More important, for any feasible region determined by a set of linear constraints, the number of extreme points is finite and relatively easy to compute. For example, for the Homewood Masonry problem, the extreme points are shown and labeled A–O in decision space (Figure 3.1); there are exactly 15 extreme points for this problem. (Points M, N, and O are not plotted on Figure 3.1: point M lies at the intersection of the red clay constraint and the x_2 axis; points N and O lie at the (theoretical) intersections of the third and fourth constraints with the x_2 and x_1 axes, respectively. The verification of the existence of these points will be made clear.)

Optimal solutions to linear programs always lie at extreme points of the feasible region, and extreme points of the feasible region always lie at the intersection of constraint equations expressed as strict equalities. Thus, to be able to identify all possible solutions that could be optimal for a given problem, we could simply identify all possible extreme points. One way this could be done is to translate the constraints of our model into a form that identically represents the feasible region in decision space, but does so with a set of equality equations. Fortunately, this transformation is easy to accomplish.

Consider our constraint on clay material from the Homewood Masonry problem, which restricted the amount of Wabash red clay that was available:

$$2x_1 + 4x_2 \leq 28$$

For any solution in decision space, we can create a new variable defined as:

$$S_1 = \text{the amount of Wabash red clay } \underline{\textit{not used}}$$

such that

$$\text{Wabash red clay not used} = 28 - (2x_1 + 4x_2)$$

or

$$S_1 = 28 - 2x_1 - 4x_2,$$

in standard form

$$2x_1 + 4x_2 + S_1 = 28.$$

Verify that, whenever this constraint is met with strict equality, it is functionally identical to our original (inequality) constraint on clay material and the value of S_1 is zero. When this new constraint is not satisfied with strict equality, S_1 takes on a value equal to the difference between how much red clay is available and how much would be used at that solution. Consequently, S_1 is referred to as a *slack* variable. For constraints of the greater than-or-equal to variety, the same thing could be accomplished by *subtracting* a *surplus* variable.

By augmenting the original model formulation using the appropriate variables—one for each constraint—we can transform any model into one whose constraint set is comprised only of equality constraints. The Homewood Masonry problem presented on page 30 in Chapter 2 becomes identically written as:

$$\text{Maximize } Z = 140x_1 + 160x_2$$

$$\begin{aligned} \text{Subject to: } 2x_1 + 4x_2 + S_1 \qquad\qquad &= 28 \\ 5x_1 + 5x_2 \quad + S_2 \qquad &= 50 \\ x_1 \qquad\qquad + S_3 \quad &= 8 \\ x_2 \qquad\qquad + S_4 &= 6 \\ x_1, x_2, S_1, S_2, S_3, S_4 &\geq 0. \end{aligned}$$

This *fully augmented* problem consists of n *structural variables* (original decision variables), m constraints, and $n + m$ total variables (structural, plus slack and surplus variables). We have also seen that whenever exactly n of these variables are set to zero, the resulting solution represents a unique extreme point of the feasible region. Finally, because we have m constraints and $(n + m) - n = m$ nonzero-valued variables at extreme points, we can solve these equations deterministically, to get an exact solution for any extreme point.

At any extreme point solution, the set of variables having nonzero values are called *basic* variables and, collectively, this set is referred to as the *basis*. Each variable whose value is zero at an extreme point solution is called a *nonbasic* variable; these variables are not "in" the basis at that solution. Each extreme point solution can be characterized by a unique collection of m basic and n nonbasic variables.

By transforming the constraint set to a system of equality equations made up of structural and slack and surplus variables, we can easily identify unique extreme point solutions; it is just a matter of solving m equations for m unknown values for each basis. One approach to the solution of the linear program might be to identify the set of all such solutions (we know that this set must contain the optimal feasible solution), solve the objective function at each solution, and simply select the one providing the best value of the objective function. While this approach might sound reasonable, it is impractical for any reasonably sized problem for two important reasons.

3.A.2 Feasibility of Extreme Points

First, as can be seen in Figure 3.1, not all extreme points are feasible, only those incident to the feasible region in decision space satisfy all problem constraints. An infeasible extreme point solution might provide a better value for our objective function, yet would not be a practical production strategy. For example, point H is an extreme point having a basis $\mathbf{B}^{\mathrm{H}} = \{x_2 = 10, S_1 = -12, S_3 = 8, \text{and } S_4 = -4\}$ but is infeasible; the constraints on the availability of Wabash red clay and FILIT curing vat availability are violated. Yet the value of of the objective function at this extreme point is \$1,600—considerably *better* than the optimal solution for this problem determined graphically (\$1,440). Points G, I, J, K, L, M, N, and O are also infeasible; at each of these points, at least one constraint is violated in the original problem. How can we identify these infeasible solutions without the benefit of a graphical solution space?

All extreme point solutions for the Homewood Masonry problem are presented in Table 3.1. Each row corresponds to a unique solution labeled on Figure 3.1, and the entry in each column shows the values for all variables (structural, and slack and surplus) at that solution. Recall that for each solution, exactly n variables (two, in this case) are nonbasic at that solution (zero values are shaded for each nonbasic variable, and there are exactly two shaded cells in each row). For example, the solution $x_1 = 0$, $x_2 = 0$ (point A) represents the solution where no production takes place such that the difference between the left-hand sides and the right-hand sides of all constraints in the original model are simply equal to their right-hand sides: 28, 50, and 4, respectively. The basis $\mathbf{B}^{\mathrm{A}}$ consists of the set $\{S_1, S_2, S_3, S_4\}$ for this solution. At point D—the optimal solution from our graphical procedure (see Figure 2.1)—the values of S_1 and S_2 are zero because the constraints corresponding to those slack variables are met with strict equality, the value of S_3 is equal to the unused storage vat capacity for HYDIT (2 tons), the value of S_4 is equal to the unused FILIT vat storage capacity (2 tons), and x_1 and x_2 are the optimal production levels (6 tons) for HYDIT and 4 tons for FILIT, respectively. More precisely, ${}^*\mathbf{B}^{\mathrm{D}} = \{x_1, x_2, S_3, S_4\}$, with $\mathbf{NB}^{\mathrm{D}} = \{S_1, S_2\}$. The asterisk denotes this basis as optimal.

At point H $\mathbf{B}^{\mathrm{H}} = \{x_2, S_1, S_3, S_4\}$, the constraints on the availability of Wabash red clay and FILIT vat storage in our original model are violated. Yet the corresponding constraints in our augmented problem formulation are satisfied if $S_1 = -12$ and $S_4 = -4$ at that solution. By identifying at all extreme point solutions

TABLE 3.1 VARIABLES AND THEIR VALUES FOR ALL EXTREME POINT SOLUTIONS OF THE HOMEWOOD MASONRY PROBLEM: EXAMPLE 2-1

Point	x_1	x_2	S_1	S_2	S_3	S_4	Z
A	0	0	28	50	8	6	0
B	8	0	12	10	0	6	1120
C	8	2	4	0	0	4	1440
D	6	4	0	0	2	2	1480
E	2	6	0	10	6	0	1240
F	0	6	4	20	8	0	960
G	0	7	0	15	8	−1	1120
H	0	10	−12	0	8	−4	1600
I	4	6	−4	0	4	0	1520
J	8	6	−12	−20	0	0	2080
K	8	3	0	−5	0	3	1600
L	10	0	8	0	−2	6	1400
M	10	0	0	−20	−6	6	1960
N	∞	0	−∞	−∞	−∞	0	∞
O	0	∞	−∞	−∞	0	−∞	∞

that are clearly infeasible in Figure 6 (points F, G, H, I, J, K, L, M, N, and O), and examining the corresponding values of variables at those solutions, it can be seen that infeasible solutions are characterized by the presence of at least one negative valued variable. Any solution to a linear program is infeasible if any variable in the augmented problem representation is negative at that solution. The optimal solution to a linear program must always be a basic feasible solution.

3.A.3 Adjacency of Extreme Points

The second difficulty with an exhaustive enumeration approach to solving linear programs is computational: for all but very small problems, there are simply too many extreme point solutions that would have to be evaluated. Recall from your basic statistics course that the number of ways of selecting a unique subset m, from a superset of elements $n + m$, is given as

$$\binom{n+m}{n} = \frac{(n+m)!}{n\,((n+m)-n)!}.$$

For the Homewood Masonry problem, the number of extreme point solutions may be computed:

$$\binom{2+4}{2} = \frac{(2+4)!}{2!\,((2+4)-2)!} = \frac{6!}{2!4!} = \frac{720}{48} = 15.$$

How many extreme points would there be for a relatively small problem consisting of, say, 50 structural variables and 200 constraints:

$$\binom{250}{50} = \frac{250!}{200!50!} = 1.35 \times 10^{53} \quad \text{possible solutions}.$$

How many extreme points would there be for a moderately sized problem consisting of, say, 1000 variables and 5000 constraints?

As will be shown in the following section, the simplex algorithm operates by first identifying a feasible extreme point, then systematically moving from that basic feasible solution to another basic feasible solution until no further improvement can be made to the objective function. Adjacency of extreme points in this context describes the relationship between solutions in terms of their respective bases, rather than a physical interpretation.

Consider the list of basic feasible solutions for the Homewood Masonry problem as presented in Table 3.1. As discussed previously, each extreme point solution can be determined by setting exactly two variables to zero and solving for the remaining (basic) variables; any basic solution will have exactly n basic variables and m nonbasic variables. Whenever any two extreme point solutions share all except one nonbasic variable, those solutions are said to be *adjacent*. By this definition, solution A is adjacent to solution B, F, G, H, L, M, N, and O. Solution D is adjacent to eight extreme points too, four of which are not adjacent to solution A and four of which are. How many of the extreme points adjacent to point E are feasible? Adjacency for these, and all other extreme points, is easily determined from the information in Table 3.1.

Adjacency is important in the simplex algorithm because it allows the identification of a subset of all feasible extreme points to which a given solution can "move" in such a way as to improve the value of the objective function. In addition, an evaluation of adjacent solutions allows us to identify when such a move would cause a solution to become infeasible. Suppose that we begin at the origin in Figure 3.2 (point A_0 in Table 3.2), which is the basic feasible solution for the Homewood Masonry problem in which the set of slack variables comprises the basis: $\mathbf{B}^{\mathrm{A}} = \{S_1, S_2, S_3, S_4\}$. As we move away from point A toward point F, x_2 in-

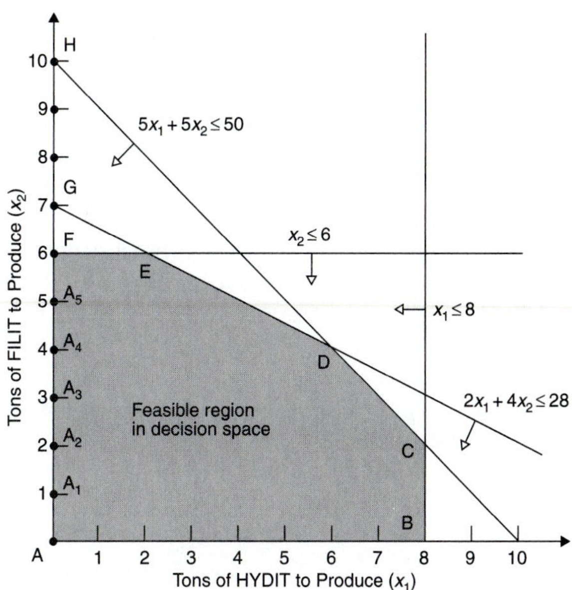

Figure 3.2 Incrementally increasing the value of a nonbasic variable moves a solution away from a corner point, and towards an adjacent basic solution. If all variables are nonnegative for any solution, that solution is feasible.

TABLE 3.2 CHANGES IN THE VALUES OF VARIABLES WHILE MOVING INCREMENTALLY FROM ONE EXTREME POINT TO ANOTHER

Point	x_1	x_2	S_1	S_2	S_3	S_4	Z
A	0	0	28	50	8	6	0
A_1	0	1	24	45	8	5	160
A_2	0	2	20	40	8	4	230
A_3	0	3	16	35	8	3	480
A_4	0	4	12	30	8	2	540
A_5	0	5	8	25	8	1	640
F	0	6	4	20	8	0	960
G	0	7	0	15	8	−1	1120

creases in value. At point A_1 $x_2 = 1$, and x_1 becomes the only zero-valued variable. This is consistent with the argument that extreme point solutions have exactly two zero-valued variables; A_1 is clearly not an extreme point solution. Note also that at point A_1, the values for the other basic variables have changed; verify that the value of the objective function has improved from \$0 to \$180. As production activity increases, the amount of unused resources (as measured by the slack variables) decreases. As we continue to increase the value of x_2 (moving to points A_2 ... A_3 ... etc.), the values for the basic variables continues to decrease until we reach point A_6, which corresponds to the extreme point F in Figure 3.2 and Table 3.2. At this point, x_2 has been increased to a value of 6, while S_4 has been reduced to a value of zero. This is reinforced by the fact that solution F lies on the line in which the constraint on FILIT curing vat capacity is met with strict equality. We say that at solution F, variable x_2 *enters* the basis while S_4 *leaves* the basis, and x_1 is a nonbasic variable in both solutions. Therefore, solutions A and F share all except one nonbasic variable; they are adjacent. All solutions on the line segment connecting points A and F are feasible because all variables are always nonnegative, but these intermediate feasible solutions are not basic, or extreme point, solutions.

By increasing further the value of x_2, the solution begins to move outside the feasible region. The variable S_4 must take on a negative value to satisfy the condition:

$$x_2 + S_4 = 6.$$

At extreme point G, the value of S_4 will be -1, and as we continue to increase the value of x_2 toward extreme point H, S_4 becomes increasingly negative. Had we instead elected to move from point F to point E, variable x_1 would have entered the basis, and variable S_1 would have been driven to zero (leaving the basis). Verify that it would not have been possible to move from point A to point E with the exchange of a single nonbasic variable in A, for a single basic variable in E; these solutions are not adjacent.

The important concept is that for any given basic feasible solution, we need only evaluate a relatively small subset of basic feasible solutions to determine how we might improve the value of the objective function. Furthermore, we can ensure that the problem will remain feasible with any such exchange we might make. Lastly, when we can make no additional moves without either (1) causing the problem to become infeasible or (2) worsening the value of the objective function, the current solution is optimal. This is precisely what is achieved for any linear programming model by using the simplex algorithm.

3.B THE SIMPLEX ALGORITHM

The simplex algorithm is rather easy to describe:

> Begin with any basic feasible solution. If one cannot be found, the problem is infeasible. If one can be found, it can be evaluated using the objective function. Given this current basic feasible solution, move to any adjacent basic feasible solution that improves the value for the objective function. Continue moving to adjacent basic feasible solutions until no additional improvement in the value of the objective function can be made. If, at this solution, it is possible to move to another basic feasible so-

lution without worsening the value of the objective function, alternate optimal solutions are present. If the move to an adjacent feasible solution can be made such that the value of the objective function can be improved without limit, the problem is unbounded.

In this section, we present the simplex algorithm in the context of the Homewood Masonry problem as a set of answers to several basic questions that are asked as we "search" for an optimal solution:

1. Where (from which basic solution) do we begin the procedure?

Assuming that the current basic solution is feasible:

2. Is the current solution optimal?
3. If not, which variable should enter the basis so as to improve the current solution?
4. How big should the value of the incoming variable be upon entering the basis so that the solution remains feasible?
5. Which variable should leave the basis (be driven to zero)?
6. What are the new values for the other basic variables?
7. How do we recognize an infeasible solution?
8. How do we recognize an unbounded solution?
9. How do we detect alternate optima?

A thorough understanding of how these questions are answered within the context of the linear programming formulation and characterization of basic solutions will provide a sufficient understanding of the simplex algorithm. A description of the computational implementation of the algorithm is beyond the scope of this text.

3.B.1 Where Do We Begin the Simplex Algorithm?

The simplex algorithm is usually presented as consisting of two phases. Phase I starts at a basic solution, and tries to find a basic feasible solution. Phase II begins at that basic feasible solution, and tries to improve it. For our purposes, we will begin with a discussion of the phase II component of the algorithm, and assume that our starting solution is feasible. A second example problem will later be used to demonstrate the phase I component.

A convenient starting solution for the Homewood Masonry problem is the basic feasible solution represented by an *all-slack basis*. That is, we select our n structural variables to be set equal to zero—$\mathbf{NB}^{\mathrm{A}} = \{x_1, x_2\}$—and solve for the basis consisting of the variables that we used to augment the constraint set, $\mathbf{B}^{\mathrm{A}} = \{S_1, S_2, S_3, S_4\}$. Actually, it really does not matter which basic feasible solution we use to begin the simplex algorithm. The algorithm is guaranteed to find an optimal solution (assuming that there **are** feasible solutions) regardless of the starting solution. Generally, however, the all-slack/surplus basis offers a convenient place to begin.

We discuss the algorithm using a bookkeeping method that allows us to conveniently portray any basic solution (basis representation) by writing the basic variables and the objective function as a function of the nonbasic variables at any particular extreme point. For the Homewood Masonry problem, the basic feasible solution corresponding to the all-slack basis $\mathbf{B}^A = \{S_1, S_2, S_3, S_4\}$ is written:

Basis Representation for $\mathbf{B}^A = \{S_1, S_2, S_3, S_4\}$.

$$S_1 = 28 - 2x_1 - 4x_2$$
$$S_2 = 50 - 5x_1 - 5x_2$$
$$S_3 = 8 - x_1$$
$$S_4 = 6 - x_2$$
$$Z = 0 + 140x_1 + 160x_2.$$

The format for this basis representation is that each basic variable is written as a function of the nonbasic variables, whose values are, of course, zero. The value of the basic variables are simply the constant terms in the right-hand side of the respective equations written above a separator line. Below that line is an expression for the objective function, again, in terms of the nonbasic variables.[1] For example, the all-slack basic feasible solution (corresponding to point A in Figure 3.1 and the Table 3.1, contains the basis $S_1 = 28$, $S_2 = 50$, $S_3 = 8$, and $S_4 = 6$, and nonbasic variables $x_1 = 0$ and $x_2 = 0$, with an objective function value of zero ($140x_1 + 160x_2$). Once again, the basic variables, their values at this solution, and the coefficients that multiply the nonbasic variables in the basis representation, are unique for this solution.

3.B.2 Is the Current Solution Optimal?

For each basis representation, we first must determine if the current solution is optimal. This is done by examining the coefficients that multiply the nonbasic variables in the objective function row. These coefficients indicate by how much the objective function would be changed if the corresponding nonbasic variable is brought into the basis at a value of one. For example, in the current solution, the value of the objective function is zero, and the coefficients that multiply x_1 and x_2 in the objective function expression are both positive: 140 and 160, respectively. This means that if x_1 (currently zero) could be brought into the basis at a value of 1, the objective would increase by 140. Similarly, if x_2 (currently zero) could be

[1] In no way does this line mean that the constraints are to be summed or operated on by some other arithmetic operation across constraints. It's just a separator.

brought into the basis at a value of 1, the objective would increase by 160. Because the objective function is being maximized, either of these "moves" would result in an improvement over the current solution. Thus, the current solution is not optimal. Verify that for a minimization problem, improvement in the current value of the objective function would be indicated by negative coefficients on the nonbasic variables in the objective function.

3.B.3 If Not, Which Variable Should Enter the Basis?

The coefficient that multiplies a variable in the objective function row of the basis representation is referred to as the *reduced cost* of the associated nonbasic variable, and is defined as the amount by which the value of the objective function will change if the corresponding nonbasic variable is brought into the basis at a value of one. If the problem is a maximization, any nonbasic variable having a positive reduced cost is a candidate for entering the basis; doing so will improve the current solution. If the formulation is a minimization, selecting a variable with a positive reduced cost to enter the basis will result in a worse solution: a larger objective function value. What do you suppose is indicated by a basis representation indicating one or more nonbasic variables having zero reduced cost?

If more than one nonbasic variable in the objective function expression has a reduced cost indicating that improvement would result from bringing this variable into the basis, then choosing any such variable to enter the basis will be sufficient. In general, there may be a very large number of such coefficients, and not all need to be evaluated. In this case, bringing either x_1 or x_2 into the basis would improve the value of the objective function, but at different rates. As will be shown in the current example, the incoming variable that will improve the value of the objective function the fastest will not necessarily be the one that will improve it the most.

The same procedure is used if the problem is a minimization. If this were the case for the current basis representation, then either of these "moves" would result in a solution that is worse than the current solution (examine Figure 3.2 and Table 3.2 to verify that in this case, the current solution would be optimal). If the coefficient were zero, bringing the corresponding nonbasic variable into the basis would have no impact on the value of the objective function.

3.B.4 How Big Should the Value of the Incoming Variable be upon Entering the Basis so That the Solution Remains Feasible?

Bringing a nonbasic variable into the basis will improve the objective function at a rate consistent with its reduced cost. At the current solution, the reduced cost of x_1 is 140 and the reduced cost of x_2 is 160. Because this problem is being modeled as a maximization, it is desirable to bring in the entering variable at a value that is as large as possible, without causing the solution to become infeasible. In other words, bring the entering variable into the basis at the largest possible value, while ensuring that the variable that will leave the basis does not become negative.

For example, suppose that after examining the basis representation $\mathbf{B}^A$ and selecting x_1 as the incoming variable, we arbitrarily select basic variable S_1 to leave the basis. That is, S_1 would be driven from the basis (reduced in value to zero) if the value of x_1 were increased sufficiently. If x_1 were increased beyond that point, then S_1 would have to take on a negative value, and the new solution would be infeasible:

$$\text{For } S_1 = 28 - 2x_1 - 4x_2, \qquad S_1 \geq 0 \text{ requires that } x_1 \leq 14$$

This means that if x_1 replaces S_1 in the basis from the current solution, its value could not exceed 14 or else S_1 would have to assume a negative value, and the new solution would become infeasible. The same condition must be satisfied for all possible leaving variables:

$$\text{For } S_2 = 50 - 5x_1 - 5x_2, \qquad S_2 \geq 0 \text{ requires that } x_1 \leq 10.$$

$$\text{For } S_3 = 8 - x_1, \qquad S_3 \geq 0 \text{ requires that } x_1 \leq 8.$$

$$\text{For } S_4 = 6 - x_2, \qquad S_4 \geq 0 \text{ requires that } x_1 \leq \infty.$$

Clearly, if x_1 is chosen to enter the basis, the largest its value can be is the smallest of these limits established by the requirement to remain feasible (8). The same analyses would have been required if x_2 is selected as the incoming basic variable:

$$\text{For } S_1 = 28 - 2x_1 - 4x_2, \qquad S_1 \geq 0 \text{ requires that } x_2 \leq 7.$$

$$\text{For } S_2 = 50 - 5x_1 - 5x_2, \qquad S_2 \geq 0 \text{ requires that } x_2 \leq 10.$$

$$\text{For } S_3 = 8 - x_1, \qquad S_3 \geq 0 \text{ requires that } x_2 \leq \infty.$$

$$\text{For } S_4 = 6 - x_2, \qquad S_4 \geq 0 \text{ requires that } x_2 \leq 6.$$

In this case, x_1 would replace S_4 as the new basic variable brought in at a value of 6.

Reexamine Figure 3.2 and Table 3.2 where we moved from the origin of the feasible region (point A) in the direction $x_2 > 0$ until, at point F, the basic variable S_4 was driven to zero. Any additional increase in the value of x_2 would have resulted in an infeasible solution.

3.B.5 Which Variable Should Leave the Basis (Be Driven to Zero)?

The (nonbasic) variable that will leave the basis is the one that first goes to zero as the value of the incoming variable is increased from zero. The result will be a simplex *move* or *pivot* from one basic feasible solution to an adjacent basic feasible solution.

3.B.6 What Are the New Values for the Other Basic Variables?

The new values for the basic variables at the new solution are computed by first determining the value for the new incoming basic variable as a function of the new nonbasic variables. Bringing x_1 into the basis in place of S_3 gives:

$$X_1 = 8 - S_3.$$

This new expression for x_1 can be substituted into the previous functions defining the basic variables in terms of the new nonbasic variables. For example, the expression for the basic variable S_1 becomes:

$$S_1 = 28 - 2(8 - S_3) - 4x_2 \quad \text{or} \quad S_1 = 12 - 4x_2 + 2S_3.$$

Similarly, the new expressions for basic variables S_2 and S_4 are now:

$$S_2 = 50 - 5(8 - S_3) - 5x_2 \quad \text{or} \quad S_2 = 10 - 5x_2 + 5S_3.$$

and

$$S_4 = 6 - x_2.$$

Last, we compute the new value for the objective function at our new solution:

$$Z = 0 + 140(8 - S_3) + 160x_2 \quad \text{or} \quad Z = 1120 - 140S_3 + 160x_2.$$

We have successfully completed our first *simplex pivot*, having moved from point A to point B in Figure 3.1. That move involved the exchange of exactly one nonbasic variable (x_1) in solution A for exactly one basic variable (S_3) resulting in a new feasible extreme point solution as reflected by the following basis representation:

Basis Representation for $B^B = \{S_1, S_2, x_1, S_4\}$.

$$\begin{array}{ll} S_1 = 12 - 4x_2 + 2S_3 & S_1 \geq 0 \text{ requires } x_2 \leq 3 \\ S_2 = 10 - 5x_2 + 5S_3 & S_2 \geq 0 \text{ requires } x_2 \leq 2 \\ x_1 = 8 - S_3 & x_1 \geq 0 \text{ requires } x_2 \leq \infty \\ S_4 = 6 - x_2 & S_4 \geq 0 \text{ requires } x_2 \leq 6 \\ \hline Z = 1120 + 160x_2 - 140S_3. & \end{array}$$

As anticipated, bringing x_1 into the basis at a value of 8 resulted in an improvement in the value of the objective function: 140 (8) = 1,120. Compare this improvement with that resulting from bringing x_2 into the basis. Even though the rate of change in the objective function from a unit increase in x_2 was greater (160 per unit increase) than that for a unit increase of x_1 (140 per unit increase) brought into the basis, feasibility requirements allowed x_1 to be brought into the basis at a higher value.

Returning to question 2, we can check the current solution for optimality. In this case, the coefficients that multiply the nonbasic variables in the objective function expression are +160 for x_2 and −140 for S_3. This means that a move which would bring x_2 into the basis at a value of 1 would improve the value of the objective function by 160. Because we are maximizing, this would be an improvement in the current solution. However, bringing S_3 into the basis at a value of 1 would reduce the value of the objective function by 140 units, and would thus be undesirable. We therefore choose to bring x_2 into the basis.

Again, we would like the incoming variable to be as large as possible. Checking our feasibility conditions for the current basic variables in the current basis representation, we find that the largest increase possible for the value of x_2 as it enters the basis is 2, constrained by the feasibility requirement for basic variable S_2. Thus, x_2 will enter the basis at a value of 2, the objective function will increase in value by 160 (2) = 320, and the new basic feasible solution will be $\mathbf{B}^C = \{S_1, x_2, x_1, S_4\}$, with Z = 1,440. Once again, we can solve for the incoming basic variable (x_2) and write expressions for all basic variables as functions of the new nonbasic variables (S_2 and S_3):

$$x_2 = \frac{10 - S_2 + 5S_3}{5} = 2 - \frac{1}{5}S_2 + S_3$$

$$S_1 = 12 - 4\left(2 - \frac{1}{5}S_2 + S_3\right) + 2S_3 = 4 + \frac{4}{5}S_2 - 2S_3$$

$$x_1 = 8 - S_3$$

$$S_4 = 6 - \left(2 - \frac{1}{5}S_2 + S_3\right) = 4 + \frac{1}{5}S_2 - S_3.$$

The new value for the objective function is computed in the same manner:

$$Z = 1120 + 160\left(2 - \frac{1}{5}S_2 + S_3\right) - 140S_3 = 1440 - 32S_2 + 20S_3$$

resulting in the new basis representation.

Basis Representation for $\mathbf{B}^C = \{S_1, x_2, x_1, S_4\}$.

$S_1 = 4 + \frac{4}{5}S_2 - 2S_3$	$S_1 \geq 0$ requires $S_3 \leq 2$
$x_2 = 2 - \frac{1}{5}S_2 + S_3$	$x_2 \geq 0$ requires $S_3 \leq \infty$
$x_1 = 8 - S_3$	$x_1 \geq 0$ requires $S_3 \leq 8$
$S_4 = 4 + \frac{1}{5}S_2 - S_3$	$S_4 \geq 0$ requires $S_3 \leq 4$
$Z = 1440 - 32S_2 + 20S_3.$	

By examining the objective function row, we find that the new basic feasible solution is still not optimal. The positive reduced cost of the nonbasic variable S_3 in the objective function indicates that for every unit of this variable brought into the basis, the objective function would increase by 20 units. Furthermore, we see that the largest value possible for S_3 as it enters the basis is 2, when S_1 would leave the basis. Make this simplex pivot and compute our new basis representation.

Basis Representation for $B^D = \{S_3, x_2, x_1, S_4\}$.

$$
\begin{aligned}
S_3 &= 2 - \tfrac{1}{2}\,S_1 + \tfrac{2}{5}\,S_2 \\
x_2 &= 4 - \tfrac{1}{2}\,S_1 + \tfrac{1}{5}\,S_2 \\
x_1 &= 6 + \tfrac{1}{2}\,S_1 - \tfrac{2}{5}\,S_2 \\
S_4 &= 2 + \tfrac{1}{2}\,S_1 - \tfrac{1}{5}\,S_2 \\
\hline
Z &= 1480 - 10S_1 - 24S_2.
\end{aligned}
$$

By examining our optimality conditions present in this basis representation, we see that the coefficients that multiply the nonbasic variables S_1 and S_3 are both negative. This means that if we bring either of these variables into the basis at a value of 1, the value of the objective function would decrease. Because we are maximizing this problem, such a solution would be worse than the current solution. Because the current basic feasible solution cannot be improved by moving to an adjacent basic feasible solution, the current solution is optimal.

Even though the feasible region of the Homewood Masonry problem included six feasible extreme points, we needed to perform only three simplex pivots to find an optimal solution. Compare each of the basis representations for the extreme points that we visited during our progression toward the optimal solution to the points plotted on Figure 3.1 and listed in Table 3.1. In particular, review how the level of slack for each constraint shifted as we moved from solution to solution. Does it make sense to you that S_3 left the basis during one simplex pivot, and reentered during another? Suppose we had chosen x_2 to enter the basis for the first pivot instead of x_1. How many pivots would have been required to reach the optimal solution? The simplex algorithm is dramatically efficient in finding optimal solutions in this manner; usually, only a small fraction of extreme point solutions need to be considered. As you formulate and solve larger and larger problems, notice the number of iterations that are required (most linear program solvers report iterations or pivots as a normal output from the solution process, though some require the user to request such information explicitly).

3.B.7 How Do We Recognize an Unbounded Solution?

The Homewood Masonry problem is clearly bounded. We saw this from the graphical solution, and the optimal basis representation indicates that it is not possible to improve the current value of the objective function. Indeed, as shown in

Figure 2.3d, an unbounded solution is one in which the value of the objective function can be improved without limit. So where within the simplex algorithm described, would we have been able to identify an unbounded solution?

During our last simplex pivot in the problem we just solved, we had selected S_3 as the variable that would be brought into the basis so that we could improve the value of the objective function. When we checked feasibility conditions in order to determine the appropriate variable to leave the basis, we found that it would not be possible to drive x_2 from the basis; regardless of how large we might make S_3, x_2 would never be forced to a value of zero. Had all possible leaving variables been unable to be driven from the basis, the problem would have been unbounded.

3.B.8 How Do We Detect Alternate Optima?

Alternate optimal solutions exist when more than one basic feasible solution gives the same value of the objective function at optimality. Recall that from Figure 2.3b, when the objective function for the Homewood Masonry problem was changed to

$$\text{Maximize } 140x_1 + 140x_2$$

solutions C and D were alternate optimal solutions. Verify that the basis representation shown next is that for solution C of such a modified problem:

Basis Representation for (Alternate) $B^C = \{S_1, x_2, x_1, S_4\}$.

$$\begin{aligned} S_1 &= 4 + \tfrac{4}{5}S_2 - 2S_3 \\ x_2 &= 2 - \tfrac{1}{5}S_2 + S_3 \\ x_1 &= 8 - S_3 \\ S_4 &= 4 + \tfrac{1}{5}S_2 - S_3 \\ \hline Z &= 1400 - 32xS_2 + 0S_3. \end{aligned}$$

At this solution, optimality is indicated because there does not exist a positive reduced cost for any nonbasic variable in the objective function. This means that no adjacent feasible extreme point can provide a better value for the objective function. However, note that the coefficient on nonbasic variable S_2 in the objective function is zero, indicating that if S_2 is brought into the basis at a value of one, the value for the objective function would not change; that new basic feasible solution would be no worse than the current solution. These two solutions would thus be alternate optima. If, at optimality, it is possible to move to an adjacent basic feasible solution without worsening the value of the objective function, alternate optimal solutions are present.

3.B.9 How Do We Recognize an Infeasible Solution?

As discussed in Section 3.A.2, any basic solution having one or more zero-valued basic variables is infeasible. In some cases, the all-slack/surplus basis—the convenient starting point that we used to begin our solution of Example 2-1—is infeasible. For example, suppose that we added a production constraint to our model that required at least 3 tons of material be produced each week:

$$x_1 + x_2 \geq 3$$

or in augmented form

$$x_1 + x_2 - S_5 = 3.$$

The all-slack/surplus basis representation for this problem would become the following:

Basis Representation for $B^A = \{S_1, S_2, S_3, S_4, S_5\}$.

$$\begin{aligned} S_1 &= 28 - 2x_1 - 4x_2 \\ S_2 &= 50 - 5x_1 - 5x_2 \\ S_3 &= 8 - x_1 \\ S_4 &= 6 - x_2 \\ S_5 &= -3 + x_1 + x_2 \\ \hline Z &= 0 + 140x_1 + 160x_2. \end{aligned}$$

The feasible region for this modified formulation is shown as Figure 3.3.

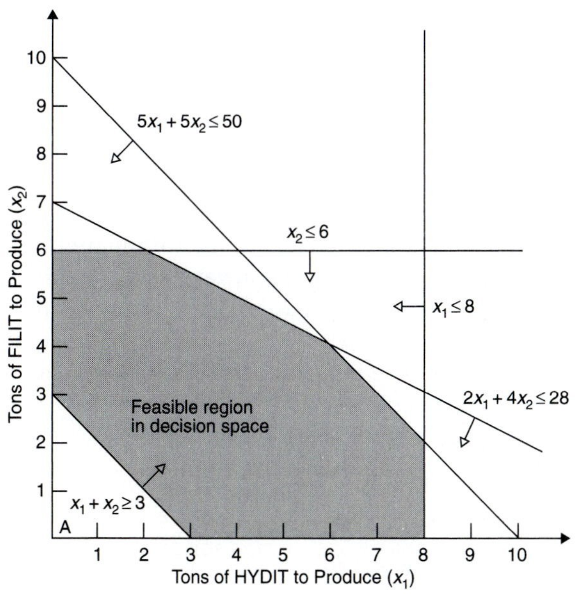

Figure 3.3 Adding the requirement that at least three units of material be produced each week results in a modification to the feasible region such that the origin, which corresponds to the all-slack/surplus basis, is no longer a feasible extreme point solution.

Note that at this solution, the initial basic solution is infeasible; the value of the surplus variable associated with the new constraint is negative at this solution. Before we can initiate the simplex algorithm, we must find a basic feasible solution from which to begin. To do this, we must apply the phase I simplex procedure.

3.C FINDING AN INITIAL FEASIBLE EXTREME POINT SOLUTION

To demonstrate a method for finding an initial feasible solution when the all-slack/surplus basis is infeasible, we will solve the problem of allocating work effort between two mines that was presented as Example 2-2 in Section 2.B.5 of the previous chapter. That problem formulation is repeated here:

$$\text{Minimize } Z = 3x_1 + 9x_2$$

$$\begin{aligned} \text{Subject to:} \quad & 40x_1 + 15x_2 \geq 100 \\ & 14x_1 + 35x_2 \geq 140 \\ & x_1 \leq 5 \\ & x_2 \leq 5 \\ & x_1, x_2 \geq 0. \end{aligned}$$

Recall the feasible region for this problem, shown in Figure 2.4, and notice that the extreme point that corresponds to the all-slack/surplus basis is not feasible; the first two constraints are violated at this origin. Because we can graph the feasible region for this problem, it is clear that feasible solutions do exist. For larger problems, when the origin solution is infeasible, (1) the problem may have no feasible solutions and (2) if the problem does have a feasible region, a feasible basis representation may not be obvious. In order for the simplex algorithm to be used as described earlier, a basic feasible solution must be available.

One method for finding a basic feasible solution (if one exists) for a problem having an infeasible all-slack/surplus basis is to transform the formulation into one that is guaranteed to have a feasible region. This can always be accomplished by adding "artificial" variables to the problem. If a feasible solution to this modified problem can be found such that the value of the artificial variables is zero, this solution is also a basic feasible solution to the original problem. If one cannot be found, then the original problem is infeasible.

Let A be an artificial variable that can be set arbitrarily large such that there will certainly be a feasible solution for the modified constraint set shown here.

$$40x_1 + 15x_2 + A \geq 100$$

$$14x_1 + 35x_2 + A \geq 140$$

$$x_1 \qquad - A \le 5$$

$$x_2 - A \le 5.$$

If a feasible solution can be found such that $A = 0$ for this modified constraint set, that solution is feasible for the original constraint set. Thus the objective function for this auxiliary problem is to minimize A; the augmented auxiliary formulation is shown here:

$$\text{Minimize } A$$

$$\begin{aligned} \text{Subject to:}\quad 40x_1 + 15x_2 + A - S_1 &= 100 \\ 14x_1 + 35x_2 + A - S_2 &= 140 \\ x_1 \qquad - A + S_3 &= 5 \\ x_2 - A + S_4 &= 5 \\ x_1, x_2, S_1, S_2, S_3, S_4, A &\ge 0. \end{aligned}$$

The notation used does not make a distinction between slack and surplus variables because they are treated identically throughout the simplex algorithm. The basis representation for the all slack/surplus basis for this auxiliary problem is the following:

Basis Representation for $B^{\text{Origin}} = \{S_1, S_2, S_3, S_4\}$.

$$\begin{aligned} S_1 &= -100 + 40x_1 + 15x_2 + A \\ S_2 &= -140 + 14x_1 + 35x_2 + A \\ S_3 &= 5 - x_1 + A \\ S_4 &= 5 - x_2 + A \\ \hline Z &= 0 + A. \end{aligned}$$

which is clearly infeasible; the basic variables S_1 and S_2 are negative at this solution. However, if the artificial variable A were to enter the basis in place of the most negative of these basic variables, the resulting basis representation would be feasible for the auxiliary problem:

Basis Representation for $B = \{S_1, A, S_3, S_4\}$.

$$\begin{aligned} S_1 &= 40 + 26x_1 - 20x_2 + S_2 \\ A &= 140 - 14x_1 - 35x_2 + S_2 \\ S_3 &= 145 - 15x_1 - 35x_2 + S_2 \\ S_4 &= 145 - 14x_1 - 36x_2 + S_2 \\ \hline Z &= 140 - 14x_1 - 35x_2 + S_2. \end{aligned}$$

Because the auxiliary problem has an objective function that is being minimized, the negative reduced costs for x_1 and x_2 in this basis representation indicate that the current solution can be improved by bringing either of these variables into the basis. If x_2 is selected to enter the basis, it would replace S_1, and the resulting basis representation would be the following:

Basis Representation for B = {x_2, A, S_3, S_4}.

$$
\begin{aligned}
x_2 &= 2 + \tfrac{13}{10}\,x_1 - \tfrac{1}{20}\,S_1 + \tfrac{1}{20}\,S_2 \\
A &= 70 - \tfrac{119}{2}\,x_1 + \tfrac{7}{4}\,S_1 - \tfrac{3}{4}\,S_2 \\
S_3 &= 75 - \tfrac{121}{2}\,x_1 + \tfrac{7}{4}\,S_1 - \tfrac{3}{4}\,S_2 \\
S_4 &= 73 - \tfrac{304}{5}\,x_1 + \tfrac{9}{5}\,S_1 - \tfrac{4}{5}\,S_2 \\
\hline
Z &= 70 - \tfrac{119}{2}\,x_1 + \tfrac{7}{4}\,S_1 - \tfrac{3}{4}\,S_2.
\end{aligned}
$$

This solution is still not optimal for the auxiliary problem because it could be improved by bringing either x_1 or S_2 into the basis. Verify that if we select x_1 to enter the basis, it would replace the artificial variable A. That is, A would become nonbasic, and therefore take on a value of zero. If we select S_2 to enter the basis, it would replace S_4, and the artificial variable A would remain basic. Driving the artificial variable from the basis while maintaining feasibility is exactly what we are trying to accomplish; we select x_1 to enter the basis and recompute the basis representation:

Basic Representation for B = {x_2, x_1, S_3, S_4}.

$$
\begin{aligned}
x_2 &= \tfrac{60}{17} - \tfrac{1}{85}\,S_1 + \tfrac{4}{119}\,S_2 - \tfrac{13}{595}\,A \\
x_1 &= \tfrac{20}{17} + \tfrac{1}{34}\,S_1 - \tfrac{3}{238}\,S_2 - \tfrac{2}{119}\,A \\
S_3 &= \tfrac{65}{17} - \tfrac{1}{34}\,S_1 + \tfrac{3}{238}\,S_2 - \tfrac{121}{119}\,A \\
S_4 &= \tfrac{25}{17} + \tfrac{1}{85}\,S_1 - \tfrac{4}{119}\,S_2 - \tfrac{608}{595}\,A \\
\hline
Z &= 0 + A.
\end{aligned}
$$

This solution is optimal for the auxiliary problem, it is feasible, and the value of the objective function is zero. We have thus found a basic feasible solution for the original problem. Had we found an optimal solution for the auxiliary problem such that the artificial variable remained in the basis with a value strictly greater than zero, we would conclude that the original problem is infeasible. In this case, however the optimal solution to the auxiliary problem can be used as a starting solution for the phase II simplex algorithm.

The basis representation for the initial phase II basic feasible solution of the original problem may be found by (1) deleting all terms in the optimal auxiliary problem containing the artificial variable A (because it is nonbasic, the artificial variable has value of zero, it is of no further use) and (2) substituting the original objective function written in terms of the optimal basic variables from the auxiliary problem:

$$Z = 3x_1 + 9x_2$$

or

$$Z = 3\left(\frac{20}{17} + \frac{1}{34}S_1 - \frac{3}{238}S_2\right) + 9\left(\frac{60}{17} - \frac{1}{85}S_1 + \frac{4}{119}S_2\right)$$

or

$$Z = \frac{600}{17} - \frac{3}{170}S_1 + \frac{9}{34}S_2.$$

The initial basic feasible solution for the phase II simplex algorithm becomes the following:

Basis Representation for $B^A = \{x_2, x_1, S_3, S_4\}$.

$$\begin{aligned}
x_2 &= \tfrac{60}{17} - \tfrac{1}{85}S_1 + \tfrac{4}{119}S_2 \\
x_1 &= \tfrac{20}{17} + \tfrac{1}{34}S_1 - \tfrac{3}{238}S_2 \\
S_3 &= \tfrac{65}{17} - \tfrac{1}{34}S_1 + \tfrac{3}{238}S_2 \\
S_4 &= \tfrac{25}{17} + \tfrac{1}{85}S_1 - \tfrac{4}{119}S_2 \\
\hline
Z &= \tfrac{600}{17} - \tfrac{3}{170}S_1 + \tfrac{9}{34}S_2.
\end{aligned}$$

This solution is not optimal because the reduced cost of S_1 at this solution indicates that the objective function could be improved; because we are minimizing the objective function, a negative reduced cost indicates the potential for improvement. Performing this pivot results in the optimal solution, which can be verified by examining Figure 2.4 and Table 2.3.

Basis Representation for $B^D = \{x_2, x_1, S_1, S_4\}$.

$$\begin{aligned}
x_2 &= 2 + \tfrac{1}{35}S_2 - \tfrac{2}{5}S_3 \\
x_1 &= 5 - S_3 \\
S_1 &= 130 + \tfrac{3}{7}S_2 - 34S_3 \\
S_4 &= 3 - \tfrac{1}{35}S_2 - \tfrac{2}{5}S_3 \\
\hline
Z &= 33 + \tfrac{9}{35}S_2 + \tfrac{3}{5}S_3.
\end{aligned}$$

Optimality is indicated by this basis representation because the reduced costs of the nonbasic variables are positive for this minimization problem. The interpretation of reduced cost is the same as when we solved the maximization problem (Example 2-2); it is the unit change in the value of the objective function if the corresponding nonbasic variable is brought into the basis. As will be shown in the next section, this economic interpretation of the relative value of model variables at different solutions is one of several important characteristics of linear program solutions.

3.D SENSITIVITY ANALYSIS

The benefits from being able to formulate an engineering management model as a linear program result from the ability of solution procedures such as the simplex algorithm to evaluate a very large number of solutions, and to be able to determine the optimal solution(s). These techniques can guarantee that, if the feasible region exists and is bounded in the direction of optimization, an optimal solution can be found within a time that is practical for most problems. The weakness of this modeling procedure is that, frequently, the problem domain must be abstracted in order to conform to restrictions on linearity; model parameters are assumed to be such that linear constraints and objective functions can be written. Fortunately, in addition to finding the optimal solution to a linear program formulation, the simplex algorithm also provides explicit information about the quality of any given optimal solution. If a model is known to be extremely sensitive to changes in model coefficients in the vicinity of the optimal solution, then greater care should be taken in estimating model parameters. If the model can be shown to be less sensitive to such changes, then the analyst may be more confident about the assumptions that were made in developing the model.

This information results from something called *sensitivity analysis*. In this section, we discuss the basic approach to sensitivity analysis using the graphical presentation of the Homewood Masonry problem presented in Chapter 2.

3.D.1 An Overview of Sensitivity Analysis

Two different types of sensitivity analysis are particularly important: *right-hand side* (RHS) sensitivity analysis and *objective function* sensitivity analysis. In both cases we are interested in two types of information: (1) What is the result of a modification to our original model formulation? and (2) Over what range of such a change is this result valid?

Right-hand side sensitivity analysis is used to measure the sensitivity of an optimal solution to changes in the original right-hand side vector of model constants. In the Homewood Masonry problem, for example, the linear program formulation was based on the assumption that the amount of Wabash red clay available to the production process was 28 cubic meters per week. When the simplex algorithm was used to solve the formulation, the optimal solution required the use of all this resource. This constraint is thus binding in the sense that the availability of this resource limits the value of the objective function.

But suppose that this assumption were found to be incorrect, and we actually had access to 31 cubic meters of this material. Would this mean that the solution we computed is no longer optimal? Would a different production policy result in a greater return and a different strategy for using resources? It would be reasonable to assume that having more of a limiting resource would allow us to find a better solution. But beyond some additional availability having more may not be desirable.

Similarly, the limit on blending machine time was assumed to be 50 hours per week, yet our optimal solution required the use of only 40 hours. Because this constraint was not satisfied with strict equality, it was nonbinding at optimality. What if instead of our assumed 50 hours of blending time available each week, we could only manage 45 hours per week. Would this change our management strategy? Not only would it be desirable to be able to judge the impact of such changes to our model, it would be desirable to have a mechanism for valuing individual resources in the production process.

Objective function sensitivity analysis is used to measure the sensitivity of an optimal solution to changes in the values of the coefficients that multiply the decision variables in the objective function of a linear program. For example, in the Homewood Masonry problem, return from the manufacture and sale of one unit of HYDIT was \$140 per ton, and for FILIT, \$160 per ton. But suppose that the return on HYDIT increased by \$1 relative to the return on FILIT (return from the sale of HYDIT = \$140.50, return from the sale of FILIT = \$159.50, for example). Clearly, if our production strategy remained the same ($x_1 = 6, x_2 = 4$), the optimal value of the objective function would change ($\$140.50x_1 + \$159.50x_2 = \$1481$). But would the optimal solution strategy change? If so, how would it change? If not, at what point would it change? (If the return on FILIT dropped to zero, would not the optimal solution certainly change?).

Objective and RHS sensitivity analysis provide explicit information about the effects of such changes. To understand the rationale for these analyses, we return to the graphical representation of the Homewood Masonry problem.

3.D.2 Graphical Interpretation of Sensitivity Analysis

A graphical depiction of the results of varying parameters and data found in a linear program provides a good framework for understanding sensitivity analysis and its vocabulary. We will use the Homewood Masonry production strategy example to show how sensitivity analysis information is derived and used in linear programming postoptimality analysis.

Right-Hand Side Sensitivity and Dual Prices. The rationale behind RHS sensitivity analysis is that the RHSs of a model's constraints are often subject to change, or are based on uncertain data. Consequently, the analyst should be prepared to answer questions about the validity of his or her model based on changing or uncertain conditions and parameters that the model assumes to be constant and certain.

Consider the constraint from the original model that restricted the amount of Wabash red clay that was available to 28 cubic meters each week:

$$2x_1 + 4x_2 \leq 28.$$

A natural question that might arise in the development of the production strategy is: What would happen if we could get access to more clay or if the amount we currently receive were reduced? Recall from the way in which the feasible region for our solution to the Homewood Masonry problem was constructed in Figure 2.1, that the right-hand side coefficient determined the location of its corresponding constraint. The coefficients that multiply the decision variables in a given constraint determined the angle or orientation of that constraint. And the sense of the inequality determined which half-space created by that constraint was feasible. Consequently, a change in the right-hand side of a constraint will result in a corresponding shift in the position of the constraint, but not orientation.

Increasing or decreasing the amount of red clay available simply moves the constraint up or down, respectively, in the two-dimensional decision space with its slope remaining constant. When a constraint is moved in decision space, we must be concerned with how its movement affects the extreme points associated with it; as we have seen, it is the set of feasible extreme points that define the possible locations for the optimal solution. So when extreme points are changed due to modifications in the RHSs of constraints, we must determine what effect these changes have on the optimal solution to the problem. The movement of a constraint could have no effect on the optimal solution (basis), could change the value of the optimal solution but not the extreme point (basis) associated with it, or could cause the optimal solution to move to a new extreme point.

At the optimal solution to the Homewood Masonry problem, the constraint on the use of Wabash red clay was binding because the optimal solution to that problem was such that the slack variable associated with that constraint—S_1—was zero. That solution is labeled as solution D in Figure 3.4. A change in the position of this constraint will change the value of the optimal solution. For example, if the amount of clay were increased by one unit, to 29 cubic meters, the constraint would shift upwards and to the right (indicated by the dashed line in Figure 3.4), the size of the feasible region would increase by the shaded area, and the new optimal solution would shift to point D^1 ($x_1 = 5.5$, $x_2 = 4.5$, and $Z^* = 1490$). But the optimal basic feasible solution—the set of basic variables at optimality—is unchanged: $\mathbf{B}^D = \{x_1, x_2, S_3, S_4\}$. This means that for one extra cubic meter of red clay, Homewood Masonry revenues would increase by $10. We call this parameter the *shadow price* (or *dual price*) of this resource. Stated another way, Homewood Masonry would be *indifferent* between its present production strategy, and this new strategy if the price for an additional cubic meter of Wabash red clay is $10. In the terminology of sensitivity analysis it is stated as follows: given the cur-

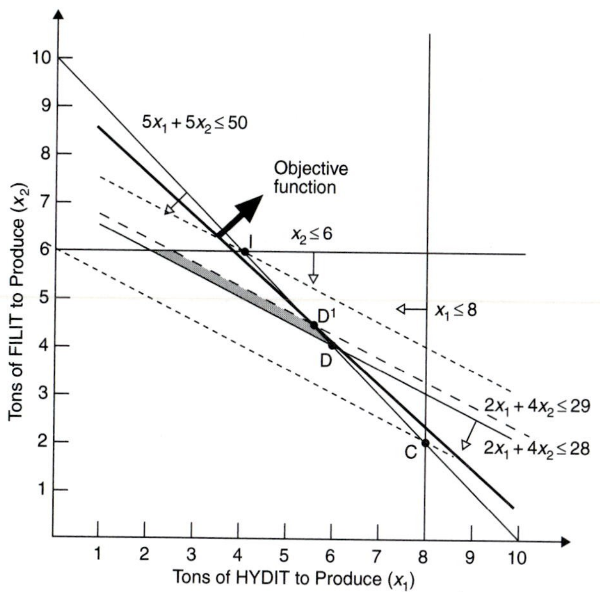

Figure 3.4 A change in the RHS of a constraint will change the location, but not the orientation of that constraint. This may change the size of the feasible region, and if this constraint is binding, the value of the objective function may change as well.

rent optimal production strategy, the *marginal value* of an additional unit of the red clay resource is \$10. If the price of additional clay were less than \$10, Homewood Masonry would be better off to purchase additional units. If the price were greater than \$10 for additional units, Homewood Masonry would not be willing to do so.

The next logical question is: "For how many extra units of red clay would Homewood Masonry be willing to pay up to \$10 per unit?" This question can be answered by again referring to Figure 3.4 and noting that we can continue to increase the right-hand side of the clay supply constraint until the constraint intersects extreme point I (shown as a dotted line). If this right-hand side is increased further, the constraint moves *through* point I, and a new basic feasible solution becomes optimal: ($\mathbf{B}^{\mathrm{I}} = \{x_1, x_2, S_1, S_3\}$). This new optimal solution is bounded by two different constraints. Specifically, S_4 leaves the basis and S_1 enters. Increasing the right-hand side of this constraint beyond the point at which all three constraints intersect at point I would not improve the value of the objective function; this resource would no longer be binding. This upper bound on the right-hand side of the red clay resource constraint is 32 units.

The logic that applied to a possible increase in the right-hand side value, also applies to a decrease in the right-hand side of the supply constraint. Referring to Figure 3.4, a reduction in red clay would shift the constraint downward and to the left. The same solution would remain optimal until a lower limit of 24 units was reached at point C (lowermost dotted line). Any further decrease in the right-hand side of this constraint would result in a new optimal solution (basis). Note that in this case however, the constraint on red clay would continue to be binding, though its shadow price would change. Each unit decrease results in the same rate of change in objective function value that was calculated for unit increases, $10 per cubic meter. For example, if the supplier of red clay to Homewood Masonry failed to deliver the full contracted amount of 28 cubic meters of material, the shadow price would indicate the minimum compensation that Homewood Masonry should be willing to accept for each unit of shortfall.

Right-hand side sensitivity analysis of the red clay resource can be summarized as follows. If the availability of red clay for the Homewood Masonry production process is between 24 and 32 units, the current basic feasible solution will remain optimal and the value of the objective function will increase or decrease by $10 per unit for every unit of increase or decrease from the base level of 28 cubic meters, respectively. The fact that the optimal basis does not change over this range means that the same constraints will be binding over this range of change; the present capacities of the curing vats will not impact the overall production policy, but the availability of blending time will remain a limiting factor. The same analyses can be performed for the other model constraints, and are summarized in Table 3.3. Because, at the current optimal solution, the constraints on curing vat capacities are not binding, their shadow prices are $0. The manager of Homewood Masonry should not be willing to pay anything to increase these capacities, because doing so would not increase the profitability of the production operation. In fact both capacities could be reduced by 2 cubic meters without changing the optimal solution or production policy. Study the graphical solution for Example 2-1 until you are comfortable with this interpretation of right-hand side sensitivity analysis. Later we will discuss how this information can be obtained from the optimal basis representation for problems of any size.

For each of the three constraints above the dual price had a definite economic interpretation. This is because the units of the objective function are in dollars. In cases where the objective function units are not in dollars the dual may have no direct economic interpretation, but may be extremely valuable nonetheless. An example of such a case would be maximizing hydropower production potential from a reservoir subject to capacity limits where the dual price might be kilowatt-

TABLE 3.3 SUMMARY OF RIGHT-HAND SIDE SENSITIVITY INFORMATION FOR EXAMPLE 2.1

Constraint	Current RHS	Optimal Usage	Allowable Decrease	Allowable Increase	Shadow Price
Red clay resource	28 m^3	28 m^3	4 m^3	4 m^3	$10
Blending machine time	50 hr	50 m^3	10 hr	5 hr	$24
HYDIT curing vat capacity	8 m^3	6 m^3	2 m^3	unlimited	$0
FILIT curing vat capacity	6 m^3	4 m^3	2 m^3	unlimited	$0

hours per cubic meter of water storage. In any case, the dual price reflects the relative value of the corresponding constraint in terms of both the validity of the current optimal solution (basic feasible solution), as well as on the value of the objective function at optimality.

Objective Function Sensitivity. Like right-hand side sensitivity analysis, objective function sensitivity analysis can be obtained from the optimal solution of a linear program. The rationale behind objective sensitivity analysis is that the coefficients on the decision variables in the objective function could be based on uncertain data or a decision maker's subjective judgment of the situation being modeled. When we changed the RHS value of a constraint, the constraint moved in decision space with a constant slope. When we change a coefficient of a decision variable in the objective function, the objective function slope will vary with the changing coefficient, but it will graphically stay anchored at the optimal extreme point and rotate around it.

Consider the objective function of the original Homewood Masonry problem:

$$\text{Maximize } 140x_1 + 160x_2$$

The coefficients that multiplied x_1 and x_2 are the estimated dollar return from a unit production of products: \$140 per unit for HYDIT and \$160 per unit of FILIT, respectively. The optimal solution was found graphically by moving the objective function through the feasible region until it last intersected that feasible region at one or more extreme point solutions that were optimal. These coefficients determined the gradient or slope of that objective function, and thus determined which extreme point solution(s) was optimal.

Suppose, for example, that instead of \$140 per ton of HYDIT produced, Homewood Masonry received only \$139. Would the policy of producing 6 units of HYDIT and 4 units of FILIT still be optimal? The graphical solution for this problem is repeated in Figure 3.5 with the original objective function labeled as line OF1, and this new objective function as line OF2. This change in profit level, while changing slightly the orientation of the objective function, does not result in a different optimal extreme point solution, though the optimal value of the objective function would be reduced by \$6. But what if the profitability of HYDIT continued to fall? The slope of the objective function would continue to fall. How much of a reduction could be tolerated before a new production strategy would be optimal? By continuing to rotate the objective function in Figure 3.5 in this counterclockwise direction by reducing the coefficient on x_1, the objective function would continue to pivot around point ($x_1 = 6$, $x_2 = 4$) until when the return on x_1 is reduced to \$80, alternate optimal solutions result as indicated by OF3; both solutions D and E would result in a total return of \$1120. Any further decrease of the return on HYDIT would result in solution D becoming a new unique optimal solution as shown by OF4.

Similarly, if the marginal return on x_1 increased relative to that of x_2, the gradient of the objective function would rotate in the clockwise direction, with the

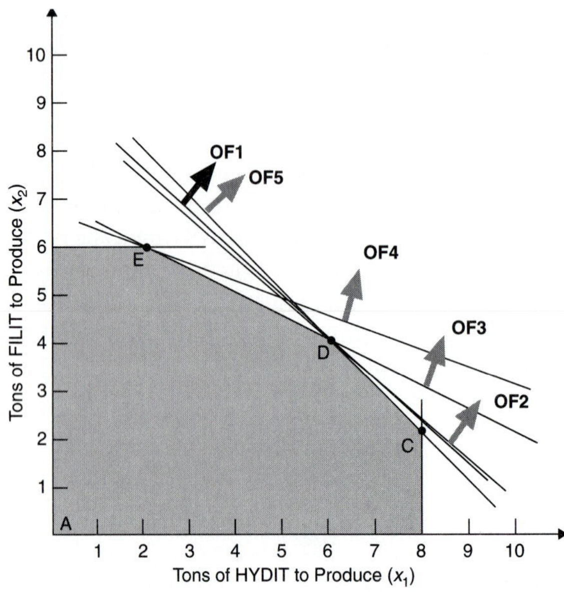

Figure 3.5 A change in the coefficients that multiply the decision variables in the objective function will change the gradient of the objective function, which may in turn result in a different extreme point solution being optimal.

TABLE 3.4 SUMMARY OF OBJECTIVE FUNCTION SENSITIVITY INFORMATION FOR EXAMPLE 2-1

Structural Variable	Optimal Value	Current Coefficient	Allowable Coefficient Decrease	Allowable Coefficient Increase	Unit Change in Profit
x_1	6 units	\$140/unit	\$60	\$20	\$6
x_2	4 units	\$160/unit	\$20	\$120	\$4

same solution remaining optimal until the gradient of the objective function became the same as the constraint on blending machine time, or

$$\text{Maximize } 160x_1 + 160x_2.$$

At this point, solutions D and C would be alternate optima, and the total return to Homewood Masonry would be \$1600. Objective sensitivities for both objective function coefficients are summarized in Table 3.4.

The graphical explanation that we have just presented may lead the reader to believe that sensitivity analysis is a simplistic and trivial concept. We have merely laid the foundation for the concepts that are vital for an understanding of the sensitivity analysis information that linear programming software will present to the modeler. When the problem size reaches hundreds or thousands of variables and there are no graphs to help in the explanation or understanding of the model results, then the systems analyst must often rely on the seemingly cryptic computer output to unravel the complex, multidimensional relationships between problem data, variable values, and the optimal solution.

3.D.3 Analytical Interpretation of Sensitivity Analysis

The optimal basis representation contains a wealth of sensitivity information that provides valuable insights for the analyst about the robustness of the optimal solution as well as the reliability of the assumptions that were used in model formulation. Many important management questions can be answered using this kind of information. In this section we'll look at how to extract such information, by returning to the familiar optimal basis representation for the Homewood Masonry problem:

Basis Representation for $B^D = \{S_3, x_2, x_1, S_4\}$.

$$S_3 = 2 - \tfrac{1}{2} S_1 + \tfrac{2}{5} S_2$$

$$x_2 = 4 - \tfrac{1}{2} S_1 + \tfrac{1}{5} S_2$$

$$x_1 = 6 + \tfrac{1}{2} S_1 - \tfrac{2}{5} S_2$$

$$S_4 = 2 + \tfrac{1}{2} S_1 - \tfrac{1}{5} S_2$$

$$Z = 1480 - 10S_1 - 24S_2.$$

Reduced Cost of Nonbasic Variables. Recall that when we solved the Homewood Masonry problem using the simplex method, optimality conditions were evaluated for each new basis representation. If a nonbasic variable had a positive coefficient for a maximization problem (negative coefficient for a minimization problem), then the current solution was not optimal; bringing that nonbasic variable into the basis would improve the value of the objective function. The magnitude of the coefficient indicated the *marginal change* in the value of the objective function that could be expected; the change in the value of the objective function for a unit increase in the value of that nonbasic variable as it enters the basis. This is more commonly referred to as the *reduced cost* of that variable.

The optimal solution for Example 2-1 shows the reduced costs for the nonbasic variables at that solution: -10 and -24 for S_1 and S_2, respectively. This means that if we bring S_1 into the basis, we will pivot to a feasible extreme point that would decrease the value of the objective function by \$10 multiplied by the resulting value of S_1. Because in this case S_1 is a slack variable, the economic interpretation of such a requirement pertains to the right-hand side limit of the associated constraint equation. The reduced cost of a slack or surplus variable that is nonbasic at optimality is the marginal value of a unit change in the right-hand side of the corresponding constraint—the shadow price for that resource, as discussed earlier. Because the reduced cost of all basic variables is zero, the marginal value of a change to the right-hand side of a nonbinding constraint is also shown to be zero. This is also consistent with our earlier observation that if a slack or surplus variable is basic at optimality, the associated constraint is nonbinding.

You may have recognized that it is possible that a basic variable can have a value of zero. If there is a tie for which variable will leave the basis during a simplex pivot, the basic variable that remains in the basis will be zero following that pivot. A graphical interpretation of such a condition in two dimensions might be the case where three constraint equations intersect at a single extreme point. For the basis representation of such a point, all three of the slack/surplus variables associated with these three constraints have a value of zero, but only two would be nonbasic—and thus have nonzero reduced costs. The other would be basic with value zero, but would have a reduced cost of zero. Do you see that such a situation would actually be one where the basis representation would indicate alternate optima? This causes only minor practical problems in the simplex algorithm as will be discussed briefly later in this chapter.

The same reasoning applies to structural variables that are nonbasic at optimality, but the economic interpretation is usually viewed from a different perspective. If the optimal solution to the Homewood Masonry problem, for example, had been extreme point B, then x_2 would have been nonbasic at optimality (no FILIT material produced), and its reduced cost would have been nonpositive. From a management perspective, this reduced cost would be interpreted as the amount by which the value of the objective function would decrease if Homewood Masonry were forced to produce a unit of the FILIT product.

Right-Hand Side Sensitivity Range. Our graphical interpretation of the right-hand side sensitivity information presented in Figure 3.4 demonstrated that for Example 2.1, the right-hand side of the blending machine constraint, which was originally 50 cubic meters per week, could have fallen within a range of values for which the current solution $\mathbf{B}^D = \{x_1, x_2, S_3, S_4\}$ remains optimal. This range for the value of that right-hand side was shown to be from 40 to 55 hours, with a constant rate of change over that range equal to the shadow price of that constraint: \$24 per hour. How would we extract the endpoints of this range from the basis representation above?

For slack and surplus variables that are nonbasic at optimality—such as S_1 and S_2 in the current example—the adjacent basic feasible solutions to which we would pivot by increasing or decreasing the right-hand side of the associated constraint are the points at which the optimal basis would change. For example, if the nonbasic variable S_2 were forced into the basis from the current optimal solution, its value would be limited by feasibility conditions, S_4 would leave the basis, and the value of S_2 when it entered the basis would be 10. This is the same as saying that the right-hand side associated with the blending time constraint could decrease by 10 units before the optimal basis would change. Verify this by reexamining Figure 3.4 and Table 2.3. Similarly, if S_2 entered the basis and its value were *decreased* from its present value of zero until a basic variable is driven to zero, this would be paramount to asking: "By how much could the right-hand side of the original constraint increase without changing the optimal solution. In this example, S_3 would leave the basis if S_2 entered the basis at a value of -5. From another perspective, if the value of S_2 in our augmented problem were -5, then the right-hand side of that constraint could be increased by 5 units without changing the extreme point solution that would be optimal. So over the range 40 to 55 hours of blending machine time, the current solution would remain optimal.

For slack and surplus variables that are basic at optimality (S_3 and S_4 in this example), the reasoning is similar, but a bit more straightforward. For slack variables that are basic at optimality, the amount by which the right-hand side of the associated constraint could be decreased without changing the current optimal solution is the value of the slack variable, while the amount by which it could be increased without a new solution being optimal would be infinity. For surplus variables that are basic, their value at optimality is an upper limit on the associated right-hand side, while the lower limit is infinity.

Objective Function Sensitivity Range. From the discussion surrounding Figure 3.5, it should be clear that at the limits over which a coefficient multiplying a decision variable can vary without changing the optimal solution, alternate optimal solutions exist. To identify these limiting values for a particular objective function coefficient, all other parameters remaining unchanged, we need only solve for the value of this coefficient such that these two adjacent basic feasible solutions give the same value for the objective function.

In the optimal basis representation for Example 2-1 presented earlier in this section, an adjacent extreme point solution could be identified by bringing S_1 into the basis; doing so would force S_3 from the optimal basis, and S_1 would enter the basis at a value of 4. The reader should verify that the new extreme point solution would have a basis that would include $x_1 = 8$ and $x_2 = 2$. This new extreme point solution would be an alternate optimal solution for Example 2-1 if it would give the same value for the objective function as the original optimal basis, which included $x_1 = 6$ and $x_2 = 4$:

$$c_1(6) + c_2(4) = c_1(8) + c_2(2).$$

With c_2 unchanged,

$$c_1(6) + 160(4) = c_1(8) + 160(2),$$

or

$$c_1 = 160.$$

That is, c_1 could increase from a value of \$140 per unit production to \$160 per unit without changing the optimal basis. Similarly, if c_1 were unchanged from at \$140, then the value of c_2 could likewise change between \$160 and \$280 without changing the optimal basis:

$$c_1(6) + c_2(4) = c_1(8) + c_2(2)$$

With c_2 unchanged,

$$140(6) + c_2(4) = 140(8) + c_2(2),$$

or

$$c_2 = 280.$$

Of course the value of the objective function would change if these cost coefficients are changed over this range. Verify the lower limits for objective function sensitivity as presented in Figure 3.5 and Table 3.4 if S_2 is made basic. When this analysis is performed for problems having more than three structural variables, an additional step is required to determine the incoming nonbasic variable that will result in the smallest change to the value of the objective function and using that to determine the values of structural variables for the adjacent solution.

3.E THE TABLEAU METHOD FOR SIMPLEX PIVOTING

An efficient way of displaying the status of the simplex algorithm through the pivoting process is called the *tableau method.* Rather than write each variable for every constraint, only the coefficients need be displayed at each iteration. Simplex pivoting can be achieved by manipulating this tabular display of each basic feasible solution.

The augmented version of the Homewood Masonry problem will be written in tabular form with columns for each problem variable—structural and slack/surplus—and rows for each basic variable. For convenience of expression, the objective function is treated in the same manner as a basic variable. At the all-slack/surplus basis,

$$\begin{aligned}
Z - 140x_1 - 160x_2 \qquad\qquad\qquad &= 0\\
2x_1 + 4x_2 + S_1 \qquad\qquad &= 28\\
5x_1 + 5x_2 \qquad + S_2 \qquad &= 50\\
x_1 \qquad\qquad\qquad + S_3 \quad &= 5\\
x_2 \qquad\qquad\qquad + S_4 &= 5.
\end{aligned}$$

Maintaining this gridlike structure with the rightmost column displaying the values of the basic variables (for the all-slack/surplus basis, these are just the values of the corresponding right-hand sides), we can produce a table—or tableau—that acts as a bookkeeping procedure for simplex iterations. The initial simplex tableau for the Homewood Masonry problem is presented as Table 3.5. Note that the columns associated with the basic variables have an entry of one in the row for which it is a basic variable, and entries of zero in the other rows of that column. This format is essentially the same as the basis representation used in our previous

TABLE 3.5 INITIAL SIMPLEX TABLEAU FOR THE HOMEWOOD MASONRY PROBLEM

Basis	Z	x_1	x_2	S_1	S_2	S_3	S_4	Value	Ratio
Z	1	−140	−160	0	0	0	0	0	
S_1	0	2	4	1	0	0	0	28	28/2 = 14
S_2	0	5	5	0	1	0	0	50	50/5 = 10
S_3	0	1	0	0	0	1	0	8	8/1 = 8
S_4	0	0	1	0	0	0	1	6	∞

description of the simplex algorithm, except that the variables are not included explicitly in each equation.

The entries in the Z row of the tableau are the negative of the reduced costs, and will only be nonzero for nonbasic variables. For maximization problems, optimality is indicated when all coefficients in the Z row are nonnegative. For minimization problems, the stopping criterion is the opposite; all coefficients in the Z row will be nonpositive. In this example, bringing either x_1 or x_2 into the basis from this initial solution will improve the value of the objective function at the corresponding rate of improvement. So that we can compare this process with the previous simplex iterations, let's again select x_1 to enter the basis during this iteration. The column of the entering variable—the *pivot column*—is shaded for illustrative purposes.

Next, we determine the variable that will leave the basis as x_1 enters. Recall from our previous discussion, that in order to maintain feasibility, the leaving variable will be the one that first goes to zero as the value of the new basic variable is increased. Each basic variable having a positive element in the pivot column must be tested; those having nonpositive values will not be decreased in value as the new variable enters the basis, hence they will never be driven out of the basis. Divide the current value of each basic variable by its corresponding element in the pivot column. This ratio is computed in the rightmost column of Table 3.5. The ratio that is the smallest indicates the leaving variable, shaded as the *pivot row*. In this case the leaving variable will be S_3. The intersection of the S_3 row—the row corresponding to the leaving variable—and the x_1 column—the column corresponding to the entering variable—identifies the *pivot element.*

We can now begin to construct the simplex tableau for the adjacent feasible extreme point to which we will pivot. This, and the remaining simplex tableaus for solving the Homewood Masonry problem are presented in Table 3.6; each tableau corresponds to a basic feasible solution labeled on Figure 3.1 and in Table 3.1. The iterative simplex pivoting procedure can be specified as a simple seven-step accounting procedure:

TABLE 3.6 THE SIMPLEX TABLEAUS FOR THE HOMEWOOD MASONRY PROBLEM

Tableau	Basis	x_1	x_2	S_1	S_2	S_3	S_4	Value	Ratio
B^A	Z	−140	−160	0	0	0	0	0	
	S_1	2	4	1	0	0	0	28	28/2 = 14
	S_2	5	5	0	1	0	0	50	50/5 = 10
	S_3	1	0	0	0	1	0	8	8/1 = 8
	S_4	0	1	0	0	0	1	6	∞
B^B	Z	0	−160	0	0	140	0	1120	
	S_1	0	4	1	0	−2	0	12	12/3 = 3
	S_2	0	5	0	1	−5	0	10	10/5 = 2
	x_1	**1**	**0**	**0**	**0**	**1**	**0**	**8**	∞
	S_4	0	1	0	0	0	1	6	6/1 = 6
B^C	Z	0	0	0	32	−20	0	1440	
	S_1	0	0	1	−4/5	2	0	4	4/2 = 2
	x_2	0	1	0	1/5	−1	0	2	∞
	x_1	1	0	0	0	1	0	8	8/1 = 8
	S_4	0	0	0	−1/5	1	1	4	4/1 = 4
B^D	Z	0	0	10	24	0	0	1480	Optimal
	S_3	0	0	1/2	−2/5	1	0	2	
	x_2	0	1	1/2	−1/5	0	0	4	
	x_1	1	0	−1/2	2/5	0	0	6	
	S_4	0	0	−1/2	1/5	0	1	2	

Step 1. Form the row for the new basic variable by dividing the pivot row in the current tableau by the pivot element. Write this row in the new tableau.

Step 2. Complete the columns corresponding to the basic variables in the new tableau by entering a 0 in all rows except for the identity row; place a 1 at that location.

Step 3. Compute the remaining elements in the new tableau using the following formula:

$$\text{new row element} = \text{old row element} - \left(\begin{matrix} \text{element in the pivot} \\ \text{column of previous tableau} \end{matrix} \times \begin{matrix} \text{element in the} \\ \text{new pivot row} \end{matrix} \right)$$

Step 4. Determine if the current tableau (basic feasible solution) is optimal by examining the elements in the Z row. Optimality is indicated if all such elements are nonnegative for maximization problems or nonpositive for minimization problems. If the current tableau is optimal, go to step 7.

Step 5. Select the variable that will enter the basis from among those having negative entries in the Z row for maximization problems, or those having positive entries for minimization problems. Label the corresponding column of the current tableau as the pivot column.

Step 6. Determine the variable that will leave the basis by dividing the current value of each basic variable by the corresponding value in the pivot column, but only for positive pivot column entries. Label the row having the lowest such value as the pivot row. Label the element at the intersection of the pivot column and the pivot row as the pivot element. Go to step 1. If there are no positive pivot column entries, the problem is unbounded; the formulation is in error.

Step 7. If one or more nonbasic columns have an entry of 0 in the Z row of the current tableau, alternate optima are present.

For the first iteration (pivot) using this procedure, we begin by computing the new tableau row for the incoming variable, x_1, by dividing the old pivot row by the pivot element (step 1). This results in the x_1 row of tableau 2 having entries 1, 0, 0, 0, 1, 0, and 8 and is designated the new pivot row. For clarity, these entries are bolded in Table 3.6. Next (step 2), we complete the columns associated with the new basis or the new tableau by placing a 0 in each row of those columns except for the identity column; a 1 is placed at that location. For example, the column in the new tableau for the new basic variable x_1 has entries 0, 0, 0, 1, 0 (including the 0 entry in the Z row). (This step is actually not necessary as the same values would result from applying the formula in step 3 to these column entries; it is merely a convenience.) Complete the rest of the new tableau by applying the formula stated above (step 3). For example, the entry in the Z row of the S_3 column in the new tableau is computed as

$$\text{new } (Z, S_3) \text{ element} = 0 - (-140 \times 1) = 140$$

The new values for the basic variables are computed in the same way, for example,

$$\text{new } (S_1, \text{value}) \text{ element} = 28 - (2 \times 8) = 12$$

and so on for all remaining entries in the new tableau. You may have correctly recognized this process as Gaussian elimination from your previous experience with matrix algebra.

Next, we determine if the new tableau represents an optimal solution (step 4) by examining the entries in the Z row for the nonbasic variables. The current solution is not optimal because it can be improved if x_2 is brought into the basis; this is our only option at this point so the x_2 column becomes the pivot column of the current tableau (step 5). As x_2 enters the basis and is increased in value, S_2 first becomes 0 and will leave the basis when $x_2 = 2$ (step 6). S_2 is therefore labeled as the pivot row; the pivot element is element (S_2, x_2). We repeat the process of constructing a new simplex tableau by returning to step 1. This set of steps will be repeated until the optimality conditions are identified in step 4.

Alternate optima are readily identified when solving linear programs using the tableau accounting procedure. If optimality is indicated for a given tableau, but at least one nonbasic variable has a coefficient of 0 in the Z row, then bringing that variable into the basis would not change the value of the objective function in the next tableau. Such a new tableau would be an alternate optimal solution. Suppose that the reduced cost for a nonbasic variable is 0 but the current solution is not optimal; there are other entries in the Z row of both positive and negative sign. This might happen, for example, if two potential incoming variables have the same value in the Z row of a tableau. Verify that bringing one such variable into the basis will result in the Z row entry of the other in the next tableau being zero, even though that variable will remain nonbasic. In a two-variable problem, this condition would result when three constraint equations pass through the same extreme point; one of those constraints is not binding even though it is satisfied with strict equality. These solutions may or may not be optimal.

Unbounded solutions and degenerate solutions are also easy to identify during the processing of simplex tableaus. If, during the pivoting process for a maximization problem, a nonbasic variable has a negative coefficient in the Z row (positive coefficient for minimization problems), but all other coefficients in that column are nonpositive, then the problem is unbounded; the ratio test will fail because it will not be possible to drive any basic variable from the basis if that nonbasic variable enters. If the ratio test for determining the leaving variable results in a tie between two or more variables, then the variable that does not leave the basis will have a value of 0 in the subsequent simplex tableau. A basic variable with a zero value is called *degenerate.* It is theoretically possible that such a condition will result in a cycling of the simplex process that will never converge on an optimal solution. Modern computer codes for performing the simplex algorithm all contain safeguards that prevent this anomalous condition.

CHAPTER SUMMARY

The simplex algorithm can find the optimal solution to a linear program by exploiting three special properties of the feasible region: convexity, compactness, and contiguity. Because these properties are characteristic of all linear programs, optimal model solutions will be found among the set of feasible extreme point solutions, which is a finite and countable set.

A transformation of the original model constraints into a mathematically identical set of equality equations allows the identification of all feasible extreme point solutions. The simplex algorithm begins by finding one such solution that is feasible, then moves between feasible extreme point solutions until no further improvement can be made in the value of the objective function. At this point, the last extreme point solution obtained is optimal, or at least a member of the set of optimal solutions in the case where the problem has alternate optimal solutions. If no feasible extreme point solution can be found, the problem is said to be infeasible, and if the value of the objective function can be improved without limit, the problem is said to be unbounded.

Two accounting procedures were presented for implementing the simplex algorithm for small problems by hand. Both procedures begin with the augmentation of the original problem equations with variables that allow these equations to be written as equality equations. These *slack* or *surplus* variables account for the difference between the left-hand side and the right-hand side of the original equations. Extreme point solutions may then be characterized as solutions in which this system of m equality equations can be solved to find a unique solution for exactly m variables, called the *basic* variables for that solution. The basis representation method allows expressions for these basic variables to be written in terms of the remaining (nonbasic) variables. At any such extreme point solution, coefficients for the nonbasic variables indicate the amount by which the objective function would change if that nonbasic variable were exchanged with a basic variable. This exchange would identify an *adjacent* basic feasible solution. If the objective function could be improved by moving to such an adjacent solution, the algorithm continues. If not, the current basic feasible solution is optimal.

The tableau method is a streamlined version of the basis representation method that uses a tabular representation for each basic feasible solution. The basis representation method facilitates the explanation of why the simplex algorithm works, while the tableau method is somewhat less tedious to use.

Problems of any size could be solved using these methods, but the level of computational effort required exceeds the patience of most humans. Computer programs are widely available for virtually any hardware platform for solving such problems.

EXERCISES

3.1. Solve the following linear program using the simplex method. Explain each simplex pivot by showing a complete basis representation or simplex tableau for each extreme point visited. Compute the value of the objective function and decision variables at optimality, and indicate which statement best describes the solution and why:

$$\begin{aligned} \text{Maximize } Z &= 12x_1 + 18x_2 \\ \text{Subject to: } 6x_1 + 5x_2 &\le 60 \\ x_1 + 3x_2 &\le 15 \\ x_1 \qquad &\le 9 \\ x_2 &\le 4 \\ x_1, x_2 &\ge 0 \end{aligned}$$

(a) this linear program has a unique optimal solution
(b) this linear program has alternate optima
(c) this linear program is infeasible
(d) this linear program is unbounded

3.2. Solve the following linear program using the simplex method. Explain each simplex pivot by showing a complete basis representation or simplex tableau for each extreme point visited. Compute the value of the objective function and decision variables at optimality, and indicate which statement best describes the solution and why:

$$\begin{aligned} \text{Minimize } Z &= 4x_1 - 5x_2 \\ \text{Subject to: } -6x_1 + 3x_2 &\le 12 \\ 4x_1 - 2x_2 &\le 24 \\ 3x_1 + 2x_2 &\le 30 \\ x_2 &\le 6 \\ x_1, x_2 &\ge 0 \end{aligned}$$

(a) this linear program has a unique optimal solution
(b) this linear program has alternate optima
(c) this linear program is infeasible
(d) this linear program is unbounded

3.3. Solve the following linear program using the simplex method. Explain each simplex pivot by showing a complete basis representation or simplex tableau for each extreme point visited. Compute the value of the objective function and decision variables at optimality, and indicate which statement best describes the solution and why:

$$\begin{aligned} \text{Maximize } Z &= 3x_1 + 2x_2 \\ \text{Subject to: } -x_1 + 2x_2 &\le 6 \\ 2x_1 - 5x_2 &\le 10 \\ -2x_1 + 2x_2 &\le 2 \\ x_1, x_2 &\ge 0 \end{aligned}$$

(a) this linear program has a unique optimal solution
(b) this linear program has alternate optima
(c) this linear program is infeasible
(d) this linear program is unbounded

3.4. Solve the following linear program using the simplex method. Explain each simplex pivot by showing a complete basis representation or simplex tableau for each extreme point visited. Compute the value of the objective function and decision variables at optimality, and indicate which statement best describes the solution and why:

$$\begin{aligned} \text{Maximize } Z &= 4x_1 + 6x_2 \\ \text{Subject to: } x_1 + x_2 &\ge -4 \\ 3x_1 - 2x_2 &\le 6 \\ x_1 + x_2 &\ge 5 \\ x_1 + x_2 &\le 10 \\ x_1, x_2 &\ge 0 \end{aligned}$$

(a) this linear program has a unique optimal solution
(b) this linear program has alternate optima
(c) this linear program is infeasible
(d) this linear program is unbounded

3.5. Solve the following linear program using the simplex method. Explain each simplex pivot by showing a complete basis representation or simplex tableau for each extreme point visited. Compute the value of the objective function and decision variables at optimality, and indicate which statement best describes the solution and why:

Maximize $Z = x_1 + x_2$	(a) this linear program has a unique optimal solution
Subject to: $x_1 + x_2 \geq -4$	(b) this linear program has alternate optima
$3x_1 - 2x_2 \leq 6$	(c) this linear program is infeasible
$x_1 - x_2 \leq 10$	(d) this linear program is unbounded
$x_1, x_2 \geq 0$	

3.6. Solve the following linear program using the simplex method. Explain each simplex pivot by showing a complete basis representation or simplex tableau for each extreme point visited. Compute the value of the objective function and decision variables at optimality, and indicate which statement best describes the solution and why:

Maximize $Z = x_1 + x_2$	(a) this linear program has a unique optimal solution
Subject to: $x_1 + 2x_2 \leq 14$	(b) this linear program has alternate optima
$2x_1 + x_2 \leq 16$	(c) this linear program is infeasible
$x_1 \leq 7$	(d) this linear program is unbounded
$x_2 \leq 5$	
$x_1, x_2 \geq 0$	

3.7. Solve the following linear program using the simplex method. Explain each simplex pivot by showing a complete basis representation or simplex tableau for each extreme point visited. Compute the value of the objective function and decision variables at optimality, and indicate which statement best describes the solution and why:

Maximize $Z = 9x_1 + 6x_2$	(a) this linear program has a unique optimal solution
Subject to: $3x_1 + 2x_2 \leq 30$	(b) this linear program has alternate optima
$-6x_1 + 3x_2 \leq 12$	(c) this linear program is infeasible
$4x_1 - 2x_2 \leq 24$	(d) this linear program is unbounded
$x_2 \leq 6$	
$x_1, x_2 \geq 0$	

3.8. Solve the following linear program using the simplex method. Explain each simplex pivot by showing a complete basis representation or simplex tableau for each extreme point visited. Compute the value of the objective function and decision variables at optimality, and indicate which statement best describes the solution and why:

Minimize $Z = 2x_1 + x_2$	(a) this linear program has a unique optimal solution
Subject to: $4x_1 - 12x_2 \leq -6$	(b) this linear program has alternate optima

$$-4x_1 + 6x_2 \leq 12$$ (c) this linear program is infeasible

$$4x_1 + 2x_2 \geq 8$$ (d) this linear program is unbounded

$$x_1 + x_2 \leq 9$$

$$4x_2 \leq 16$$

$$x_1, x_2 \geq 0$$

3.9. Solve the following linear program using the simplex method. Explain each simplex pivot by showing a complete basis representation or simplex tableau for each extreme point visited. Compute the value of the objective function and decision variables at optimality, and indicate which statement best describes the solution and why:

Minimize $Z = 7x_1 - 5x_2$ (a) this linear program has a unique optimal solution

Subject to: $2x_1 - x_2 \geq 4$ (b) this linear program has alternate optima

$10x_1 + 3x_2 \leq 30$ (c) this linear program is infeasible

$x_1 + 2x_2 \geq 10$ (d) this linear program is unbounded

$x_1, x_2 \geq 0$

3.10. Solve the following linear program using the simplex method. Explain each simplex pivot by showing a complete basis representation or simplex tableau for each extreme point visited. Compute the value of the objective function and decision variables at optimality, and indicate which statement best describes the solution and why:

Minimize $Z = 2x_1 + 3x_2 + x_3$ (a) this linear program has a unique optimal solution

Subject to: $2x_1 + x_2 - x_3 \geq 3$ (b) this linear program has alternate optima

$x_1 + x_2 + x_3 \geq 2$ (c) this linear program is infeasible

$x_1, x_2, x_3 \geq 0$ (d) this linear program is unbounded

3.11. Solve problem 3.10 by enumerating all possible extreme point solutions and comparing them to determine the optimal solution.

3.12. Solve the following linear program using the simplex method. Explain each simplex pivot by showing a complete basis representation or simplex tableau for each extreme point visited. Compute the value of the objective function and decision variables at optimality, and indicate which statement best describes the solution and why:

Maximize $Z = 2x_1 + 3x_2 + 2x_3$ (a) this linear program has a unique optimal solution

Subject to: $2x_1 + x_2 + x_3 \leq 4$ (b) this linear program has alternate optima

$x_1 + 2x_2 + x_3 \leq 7$ (c) this linear program is infeasible

$x_1 + 2x_2 + x_3 \leq 12$ (d) this linear program is unbounded

$x_1, x_2, x_3 \geq 0$

3.13. Solve the following linear program using the simplex method. Explain each simplex pivot by showing a complete basis representation or simplex tableau for each extreme point visited. Compute the value of the objective function and decision variables at optimality, and indicate which statement best describes the solution and why:

Maximize $Z = 2x_1 + 4x_2 + x_3 + x_4$	(a) this linear program has a unique optimal solution
Subject to: $x_1 + 3x_2 + x_4 \leq 2$	(b) this linear program has alternate optima
$x_2 + 4x_3 + x_4 \leq 5$	(c) this linear program is infeasible
$x_1, x_2, x_3, x_4 \geq 0$	(d) this linear program is unbounded

3.14. Solve problem 3.13 by enumerating all possible extreme point solutions and comparing them to determine the optimal solution.

3.15. Solve the following linear program using the simplex method. Explain each simplex pivot by showing a complete basis representation or simplex tableau for each extreme point visited. Compute the value of the objective function and decision variables at optimality, and indicate which statement best describes the solution and why:

Maximize $Z = 3x_1 + 2x_2 - x_3 + 2x_4$	(a) this linear program has a unique optimal solution
Subject to: $2x_1 - 4x_2 - x_3 + x_4 \leq 10$	(b) this linear program has alternate optima
$x_1 + x_2 - 2x_3 - 3x_4 \leq 12$	(c) this linear program is infeasible
$x_1 - x_2 - 4x_3 + x_4 \leq 3$	(d) this linear program is unbounded
$x_1, x_2, x_3, x_4 \geq 0$	

3.16. A company manufactures three different types of pipe fitting: tees, elbows, and splicers. Daily production of these parts are limited by the availability of lathe time, grinder time, and labor availability as indicated in the table below:

Resources	Products			Availability of Resource
	100 Tees	100 Elbows	100 Splicers	
Person hours	6	4	5	24
Lathe hours	1	2	1	8
Grinder hours	2	1	0	12
Profit per 100 units	$700	$550	$480	

For example, each 100 units of tees requires 6 person hours to produce, including 1 hour of lathe time and 2 hours of grinder time. All the tees that are made can be sold for $700 per 100 units. A total of 24 person hours, 8 lathe hours, and 12 grinder hours are available on a given day.

(a) Formulate a linear program that will suggest a production policy for maximizing daily profit.

(b) Augment your constraint set by adding the appropriate slack and surplus variables. List in tabular form, all extreme point solutions of the solution space for this problem.

(c) Graph the feasible region in decision space for this problem and label the feasible extreme points to correspond to those presented in part (b).

(d) Solve this linear program using the simplex algorithm summarizing the results of each simplex pivot as a basis representation or simplex tableau. Explain fully your justification for selecting variables to enter and leave the basis for each iteration.

3.17. From the optimal basis representation for Problem 3.16, answer the following questions:

(a) What is constraining your present production level?

(b) Suppose that one of your clients approached you about a modification to your contract agreement that required you to deliver at least 100 splicers each day. How would this change your optimal production policy, and what concessions might you require from this customer?

(c) Suppose a local equipment rental company is willing to rent you additional machine time on an hourly basis. How much would you be willing to pay per hour for additional lathe and grinding machine work?

(d) Suppose that the profit per 100 units of Elbows fell to $500. How would this change your production policy, and what would be your new optimal daily operating profit?

(e) Suppose the local labor union demanded an extension to mandatory rest periods each shift, which would reduce your available work time to 21 person hours per day. How would this affect your production policy and daily profit margin?

3.18. A poultry farmer owns 100 laying hens. Each week a hen can either lay 12 eggs or hatch 4 eggs, but not both. At the end of the four week period all eggs and chicks will be sold; an egg will bring 10 cents and a chick will bring 60 cents. Assume that an egg can be hatched only in the week after it is laid. Otherwise it must be saved and sold for 10 cents. Also, assume that there are no eggs available for hatching in the first week.

Formulate a linear program that will suggest a hen-management strategy that will maximize profit for this operation.

3.19. You are the production manager for a very large construction company that is about to start a major concrete poor that is estimated to last for 10 days. Your primary job during this time is to provide the necessary forms each day:

Day	1	2	3	4	5	6	7	8	9	19
Forms Needed	50	60	80	70	50	60	90	80	50	100

To meet these requirements you have three options: 1) you can purchase new forms for $200 each; 2) you can recondition used forms using a standard process that takes 4 days and costs $25 per form; or 3) you can recondition forms using a special "fast" process that takes 2 days and costs $75 each. Assume that the company supplying you with forms has sufficient inventory to supply all new forms if necessary. Also, you have no forms prior to the start of this operation.

Formulate a linear program that will suggest a strategy for meeting the demand for forms at least total cost.

3.20. Homewood Masonry is considering expanding its operation and entering into the local structural concrete market. Management has decided to produce and distribute one batch of each of two different grades of concrete. Each grade is a different mixture of cement, sand, and gravel as indicated by the amounts of each material needed for a given week's production as specified in the following table.

Grade	Cement (tons)	Sand (tons)	Gravel (tons)
I	1	2	4
II	1	3	6

For example, this week's batch of Grade I concrete produced requires 1 ton of cement, 2 tons of sand, and 4 tons of gravel. If this new business is to be profitable for Homewood Masonry, these materials must be procured as cost effectively as possible. Your job is to develop a strategy for purchasing these materials. You have received bids from two regional suppliers that include the cost/ton (including transportation) shipped to your site:

	Supplier A		Supplier B	
Material	Cost/ton ($)	Available	Cost/ton ($)	Available
Cement	$150.00	3	$175.00	6
Sand	$10.00	4	$7.00	5
Gravel	$17.00	4	$15.00	6

For example, a ton of cement from Supplier A costs $150.00, and Supplier A has a total of 3 tons available for your operation.

Formulate a linear program that will suggest an optimal strategy for purchasing raw materials for the production of these concrete mixtures.

3.21. You have just received word that a wealthy relative who died very recently left a provision in her will for an endowment to support your education, and that of any children you have or may have in the future. Starting on August 1st of each year, and every two months thereafter, the estate issues a check for $25,000 payable to you. You have three options for investing these funds.

1. A special 2-month treasury note that returns 1.8% on the invested amount,
2. A 4-month money market fund that returns 4.0% on the invested amount, or
3. A seasonal 6-month futures option returning 8.1% on the invested amount available only on January 1 of each year.

All funds deposited in any investment instrument must remain in the account through the entire investment period, but then may be withdrawn and held, or reinvested during the next period. And to insure diversification of investment, not more than 60% of total assets may reside in any one vehicle at any time. The will further states that on July 31 of each year, all money resulting from these contributions must be withdrawn and spent on your education.

Formulate a linear program that will suggest an investment strategy that will maximize the amount you have available to spend on education.

3.22. Consider the simply supported beam shown below:

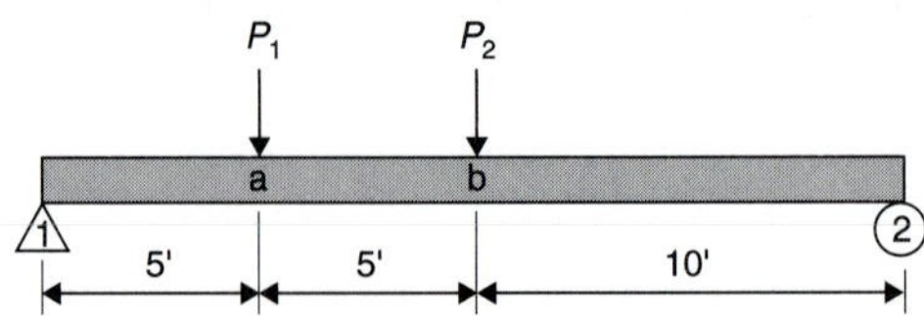

The load carrying capacities of supports 1 and 2 are 12 and 19 kips, respectively. The beam itself can withstand a bending moment of 65 kip-feet, and the maximum stress created by any loading is assumed to occur at one of the load points (a or b).

Formulate a linear program that will determine the maximum possible loads (P_1 and P_2) that this system can carry. Solve your model using the simplex algorithm, and summarize each simplex pivot by showing the appropriate basis representation or simplex tableau.

3.23. Referring to your optimal basis representation or tableau for Exercise 3.22, and assuming that each additional pound of loading is worth ten dollars, how much would you be willing to spend to strengthen support 1, and how much strength would you be willing to add for that price or less?

3.24. Modify your formulation for Exercise 3.22 to find the maximum permissable loads (P_1 and P_2) such that the force exerted at supports 1 and 2 are equal.

REFERENCES

COOKE, W. P., 1985. *Quantitative Methods for Management Decisions*, McGraw-Hill, 686 pages.

HADLEY, G., 1962. *Linear Programming*, Reading Ma: Addison-Wesley, 520 pages.

IGNIZIO, J. P. and T. M. Cavalier, 1994. *Linear Programming*, Prentice Hall, 666 pages.

MURTY, K. G., 1994. *Operations Research: Deterministic Optimization Models*, Prentice Hall, 608 pages.

WILLIAMS, H. P., 1993. *Models Solving in Mathematical Programming*, Wiley, 359 pages.

WINSTON, W. L., 1994. *Operations Research: Applications and Algorithms, Third edition*, Belmont CA: Duxbury, 1318 pages.

CHAPTER 4

Linear Programs with Multiple Objectives

4.A RATIONALE FOR MULTIOBJECTIVE DECISION MODELS

For many engineering management problems, particularly those in the public sector, more than one objective is generally important. Consider the problem of developing an operating strategy for a large, multipurpose water reservoir. It is not uncommon for such a facility to be used to meet a variety of societal needs including municipal water supply, agricultural water supply, flood control and streamflow management, hydroelectric power production, outdoor recreation, and protection of fragile environmental habitat. From a management perspective, these uses of such a facility conflict with one another. For example an optimal management strategy with respect to insuring a reliable supply of water from a reservoir for municipal use might have as its objective function: *maximize storage in the reservoir at all times*. Yet for purposes of containing an extreme flood event, the objective function could be just the opposite: *minimize storage in the reservoir at all times*. Several optimization techniques have been developed to capture explicitly, the trade-offs that may exist between conflicting, and possibly noncommensurate, objective. In this chapter we lay the foundation for multiobjective analysis that can be applied across the spectrum of public sector engineering management problems and demonstrate the application of two of the most useful multiobjective optimization methodologies.

Multiobjective programming deals with optimization problems with two or more objective functions. The multiobjective programming formulation differs from the classical (single-objective) optimization problem only in the expression of their respective objective functions; the multiobjective formulation accommodates explicitly more than one. Yet the evaluation of management solutions is significantly different: instead of seeking an optimal or best overall solution, the goal of multiobjective analysis is to quantify the degree of conflict, or *trade-off* among objectives. From another perspective, we seek to find the set of solutions for which we can demonstrate that no better solutions exist. This best available set of solutions is referred to as the set of *noninferior* solutions. It is from this set that the person or persons responsible for decision making should choose; the role of the systems analyst is to describe as accurately and completely as possible the range for that choice and the trade-offs among objectives between members of that set of management solutions. Noninferiority is the metric by which we include or exclude solutions in this set.

4.A.1 A Definition of Noninferiority

In single-objective problems our goal is to find the single feasible solution that provides the optimal value of the objective function. Even in cases where alternate optima exist, the optimal value of the objective function is the same for each alternate optima (extreme point) as can be seen in Figure 2.3. For problems (models) having multiple objectives, the solution that optimizes any one objective will not, in general, optimize any other. In fact for decision-making problems that are most challenging from an engineering management perspective, there is usually a very large degree of conflict between objectives such as in the example of reservoir management. Another example might be in the area of structural design, where objectives might include the maximization of strength concurrent with the desire to minimize weight or cost. In managing environmental resources, we might seek to trade off environmental quality and economic efficiency concerns, or even conflicting environmental quality goals; minimizing the volume of landfill disposal against discharges to the atmosphere by incineration of municipal refuse. Note that in these latter examples, the units of measure for different objective functions may be quite different as well. We call such objective functions *noncommensurate*.

When dealing with objectives that are in conflict, the concept of optimality may be inappropriate; a strategy that is optimal with respect to one objective may likely be clearly inferior for another. Consequently, a new concept is introduced by which we can measure solutions against multiple, conflicting, and even noncommensurate objectives; the concept of noninferiority:

> A solution to a problem having multiple and conflicting objectives is *noninferior* if there exists no other feasible solution with better performance with respect to any one objective, without having worse performance in at least one other objective.

Noninferiority is similar to the economic concept of *dominance* and is even called *nondominance* by some mathematical programmers, *efficiency* by statisticians and economists, and *Pareto optimality* by welfare economists. A simple extension to the Homewood Masonry problem presented as Example 2-1 will help the reader understand this often nebulous concept.

4.A.2 Example 4-1: Environmental Concerns for Homewood Masonry

Problem Statement. The management of Homewood Masonry has long been concerned about the local environmental impacts of their production operation both as a responsible member of the community within which their plant resides, and in anticipation of increasingly tighter standards and governmental controls. You have been asked to study the operation of the plant and to identify from a technical perspective the level of conflict that exists between these two management objectives.

After analyzing the results of a comprehensive air monitoring program, you discover that the major environmental impact of the operation results from a release of contaminated dust during the blending process; the binder used in manufacturing both HYDIT and FILIT attaches to these dust particles, and is thereby released to the environment during production. Laboratory tests suggest the release of this pollutant from the plant amounts to 500 milligrams for each ton of HYDIT produced and 200 milligrams for each ton of FILIT produced. A second objective function, one that seeks to minimize total plant emissions, can now be specified as

$$\text{Minimize } Z_2 = 500x_1 + 200x_2$$

The feasibility of solutions (the feasible region in decision space) is not affected by the consideration of this objective function. Note that the sense of this objective function is opposite that of our original production objective function (*maximize total weekly revenue*), and the units (*milligrams discharged*) are different as well (*dollars*). Yet both objective functions are related through the same set of decision variables.

By the same argument presented in the previous chapter, the solution that optimizes this second objective function must be a basic feasible (extreme point) solution in the original problem. The values for the decision variables at each of these solutions are repeated in Table 4.1, and the values of both objective functions at each of those solutions are included. Not surprisingly, the solution that optimizes the environmental objective is the "do nothing" solution: $x_1 = 0$ $x_2 = 0$.

Because both objective functions have been previously specified, and are thus assumed to reflect the overall goals of production and environmental concern (implicitly, we assume that there are no other management objectives), we can apply the concept of noninferiority as defined above to each of these solutions (production alternatives). Notice that the solutions that optimize the individual objective

TABLE 4.1 DECISION VARIABLES AND THEIR VALUES FOR ALL FEASIBLE EXTREME POINT SOLUTIONS FOR THE TWO-OBJECTIVE HOMEWOOD MASONRY PROBLEM: EXAMPLE 4-1

Alternative	x_1	x_2	Max Z_1	Min Z_2	Noninferiority
A	0	0	0	0	Noninferior
B	8	0	1120	4000	Dominated by D, E
C	8	2	1440	4400	Dominated by D
D	6	4	1480	3800	Noninferior
E	2	6	1240	2200	Noninferior
F	0	6	960	1200	Noninferior

functions Z_1 and Z_2—Alternative A and Alternative D, respectively—are indicated as being noninferior. In fact, for any multiobjective optimization model, the solution that optimizes any single objective function is always noninferior, unless there are alternate optima at that solution with respect to that objective function (this qualification will be clarified later). By the definition of noninferiority, if a solution is optimal for a given objective function, it is not possible to find a clearly better feasible solution regardless of how that solution might perform with respect to any (or all) other objective functions.

Consider alternative B. It is not the worst solution with respect to profit, it is clearly better than alternative F by this measure. Nor is it the worst solution environmentally; it is better than alternative C. But given the stated objective functions and awareness of this set of alternatives, would you ever select alternative B? Would anybody ever select alternative B? Stated another way, is there any alternative that would always be preferred to alternative B by anybody having preferences represented by this specific set of objective functions? The answer, of course, is that both alternatives D and E perform better with respect to both objectives than does alternative B, so that no decision maker would ever implement alternative B if he or she were aware of the availability of alternatives D or E. Similarly, alternative C is clearly *dominated* by alternative D.

Now let's compare alternative D—an alternative that has already been shown to be noninferior—with alternative E. While alternative D represents a production strategy that maximizes profit for Homewood Masonry, it would also have a more adverse impact on the local environment than would E. Therefore, the choice between these alternatives is not obvious, and probably depends on the specific preferences of the decision maker. It is easy to envision a scenario in which the board of directors of Homewood Masonry might themselves be divided over which strategy to implement, particularly if they reside in the vicinity of the plant, for instance. Can you see that the same logic applies to the determination that alternative F is also noninferior?

The goal of such an analysis is thus to identify all solutions that are noninferior; the set of solutions for which there does not exist another solution that would always be preferable to any of those solutions. This set of alternatives is referred to as the *noninferior set*, or sometimes the *Pareto frontier*. It is then the responsibility of the decision maker to select from among these solutions, that which represents their *best compromise solution* among the stated objectives.

The definition of noninferiority seems more difficult to state than to comprehend. Make sure you understand the logic used to determine dominance and nondominance with respect to the solutions presented in Table 4.1, then review carefully the definition of noninferiority given above. When you feel that you' ve mastered the concept, examine the data for a three-objective, eight-solution multiobjective program presented in Table 4-2 and try to verify that the noninferior set for this problem consists of points A, B, C, D, E, H. Note that objective Z_2 has alternate optima—both solution E and solution F provide an objective function value of 9—but for the three objective problem, solution E dominates solution F.

You might also try writing your own objective function that depends on those values of the decision variables x_1 and x_2, and see how the inclusion of this forth objective function changes the noninferior set. You should start to realize that the determination of noninferiority gets increasingly complicated as the problem grows in size; both number of basic feasible extreme point solutions and number of objective functions. Most real engineering problems in the public sector have hundreds of thousands of feasible extreme points, and may have tens of objective functions. Before we discuss a general-purpose algorithm for identifying the

TABLE 4.2 DECISION VARIABLES AND THEIR VALUES FOR ALL FEASIBLE EXTREME POINT SOLUTIONS FOR A MORE COMPLICATED THREE-OBJECTIVE PROBLEM: EXAMPLE 4-2

Alternative	x_1	x_2	Max Z_1	Max Z_2	Min Z_3	Noninferiority
A	2	0	6	−2	2	Noninferior
B	4	1	10	−2	5	Noninferior
C	6	5	8	4	11	Noninferior
D	6	7	4	8	13	Noninferior
E	3	6	−3	9	5	Noninferior
F	1	5	−7	9	6	Dominated by E
G	0	3	−6	6	12	Dominated by D, E
H	0	1	−2	2	1	Noninferior

noninferior set, it is useful to develop a graphical framework within which to study further the concept of noninferiority.

4.A.3 A Graphical Interpretation of Noninferiority

Consider the two-objective mathematical program presented below:

$$\text{Maximize } [Z_1(x_1, x_2), Z_2(x_1, x_2)]$$

$$\text{where:} \quad Z_1 = 3x_1 - 2x_2$$

$$Z_2 = -x_1 + 2x_2$$

$$\text{Subject to:} \quad 4x_1 + 8x_2 \geq 8$$

$$3x_1 - 6x_2 \leq 6$$

$$4x_1 - 2x_2 \leq 14$$

$$x_1 \leq 6$$

$$-x_1 + 3x_2 \leq 15$$

$$-2x_1 + 4x_2 \leq 18$$

$$-6x_1 + 3x_2 \leq 9$$

$$x_1, x_2 \geq 0.$$

The feasible region in decision space and the objective functions are plotted in Figure 4.1 with each basic feasible solution labeled A–H.

The most astute of readers will have noticed that this two-objective problem uses the same feasible region specified in Table 4.2 as well as the first two objective functions listed in that table (we will ignore the third minimization objective for the time being). The shaded cells in that table indicate the optimal solutions.

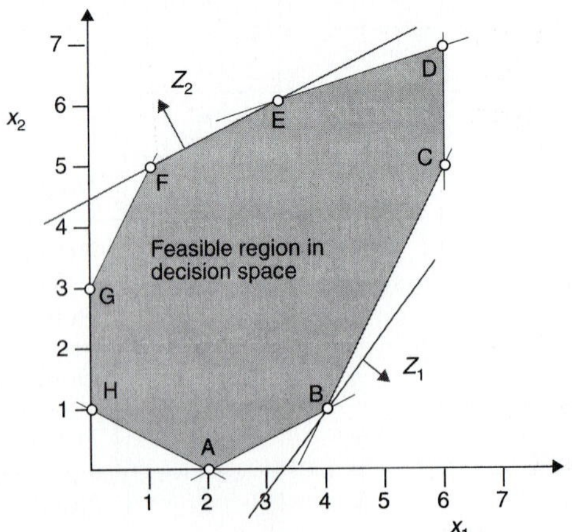

Figure 4.1 The feasible region in decision space for the problem presented in Table 4.2 with solutions that optimize Z_1 and Z_2 shown passing through their respective optima—points B and F, respectively.

The presence of alternate optima for Z_2 is not surprising if we note that the coefficients that multiply the decision variables in that objective $(-x_1 + 2x_2)$ result in an objective function having a slope that is identical to one of the binding constraints $(2x_1 + 4x_2 \leq 18)$.

Because we have limited our example problem to not more than three objectives, we can map the feasible region in decision space to a corresponding *feasible region in objective space*; we simply plot the ordered (Z_1, Z_2) pairs as presented in Figure 4.1. Using the common reference provided by Table 4.2, each basic feasible solution labeled in Figure 4.2 has a corresponding solution in objective space using the same letter designator. For example, point B in Figure 4.1 corresponds to Point B in Figure 4.2, with the corresponding coordinates taken from Table 4.2. Significantly, adjacent feasible extreme points in decision space map to adjacent solutions in objective space. Whereas the shape of the feasible region in decision space depends on the constraint set for a particular problem, the shape of the feasible region in objective space depends on the objective functions, which serve as "mapping functions" for a particular set of objectives.

This graphical representation provides a much easier means for identifying noninferior solutions. First, it should be obvious that all interior points must be inferior, because given any such point, one would always be able to find another feasible solution that would improve both objectives simultaneously. For example, consider interior point P in Figure 4.2, which is inferior. Alternative D gives more Z_1 than does P without decreasing the amount of Z_2. Similarly, D gives more Z_2 without decreasing Z_1. In fact, any alternative in the shaded wedge shape to the "northeast" of point P *dominates* alternative P. We can generalize this notion in the form of a rule having this directional analog.

A feasible solution to a two objective optimization problem in which both objective functions are to be maximized is noninferior if there does not exist a feasible solution in the northeast corner of a quadrant centered at that point.

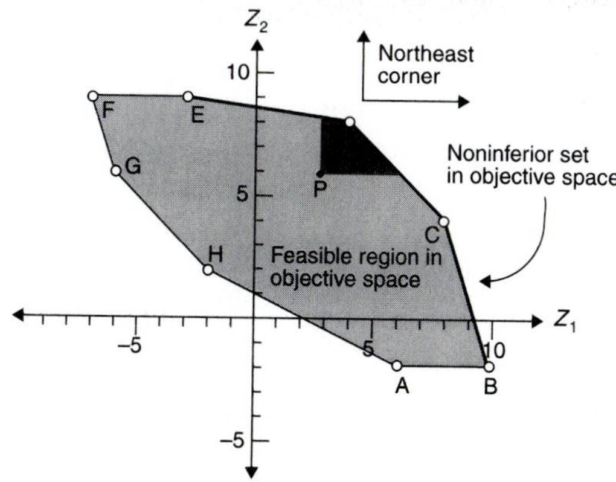

Figure 4.2 The feasible region in objective space is defined by plotting all basic feasible solutions from decision space mapped through the objective functons Z_1 and Z_2. Noninferiority is then easily determined using the *northeast corner rule*.

Applying the northeast corner rule to the rest of the entire feasible region in Figure 4.2 leads to the conclusion that any point on the boundary that is not on the northeastern side of the feasible region is inferior. The noninferior solutions for the feasible region in Figure 4.2 are found in the thickened portion of the boundary between points B and E. Use the northeast corner rule to convince yourself that solution F is indeed dominated (by solution E) and is thus not a member of the noninferior set even though it was shown to be an alternate optima when we solved Z_2 as a single objective optimization.

We can, of course, generalize this result to evaluating solutions for problems with any combination of objective function sense. For example, in the current problem, if, instead of both objectives being maximized, they were minimized. Can you see that the noninferior set would then consist of those solutions on the southwest border of the feasible region in objective space between points A and F? What if one objective is a maximization and one a minimization? What should you conclude if the trade-off surface (noninferior set in objective space) reduces to a single point?

The noninferior set for the two-objective problem that we just solved consisted of the points labeled B, C, D, and E. Yet when considering a third objective Z_3 in Table 4.2 points A and H are also included in the noninferior set. An important assumption underlying multiobjective analyses is that the decision maker(s) must be able to articulate all relevant objectives for a particular problem. Otherwise, solutions that are noninferior may be excluded from consideration in the same way that for our hypothetical example, we would not consider alternative H for implementation without a consideration of objective Z_3.

Now that we are comfortable with the concept of noninferiority, let's examine two methodologies that will allow us to identify efficient solutions when it is not possible to graph our solution space. We will demonstrate these techniques with the sample problem we just studied, but the reader should appreciate that the methodologies are applicable to any multiobjective model.

4.B METHODS FOR GENERATING THE NONINFERIOR SET

A number of methodologies have been devised to portray the noninferior set among conflicting objectives. We will confine our treatment of this topic to a class of techniques that enjoys widespread use among engineers. *Generating techniques*, as they are commonly called, do not require (or allow) decision makers' preferences to be incorporated into the solution process. The relative importance of one objective in comparison to another is not considered when identifying the noninferior set, but used later on to compare noninferior solutions and to quantify the trade-offs between them. Typically, analyst(s) will work iteratively with the decision maker(s) to identify a complete set of objective functions for a particular problem domain and to specify the appropriate set of decision variables to relate these objectives to one another and to problem constraint conditions. The noninferior set is then generated by the appropriate technique, such as those presented below, and presented to the decision maker for further consideration.

The selection of a solution to be implemented from among those solutions in the noninferior set is the responsibility of the decision maker(s). The strength of the use of generating methods for multiobjective optimization is that the roles of the analyst(s) versus the decision maker(s) are as they should be: the analyst provides comprehensive information about the best available choices in a given problem domain, and the decision maker assumes the responsibility for selecting among those choices. The analyst is not involved with making value judgments about the relative importance of one objective over another, and the decision maker need not worry about the technical aspects of the physical system nor fear that better solutions are being overlooked.

We will present two methods for generating the noninferior, set; the weighting method and the constraint method. There are strengths and weaknesses of each method for a given application, but they both rely on the repeated solution of linear programs. The general single-objective optimization with n decision variables and m constraints was presented in Chapter 1 (Section 1.E.14). The general multiobjective optimization problem with n decision variables, m constraints, and p objectives is:

$$\text{Optimize } Z = Z_1(x_1, x_2, \ldots, x_n)$$
$$Z_2(x_1, x_2, \ldots, x_n)$$
$$\ldots\ Z_p(x_1, x_2, \ldots, x_n)$$

$$\begin{aligned} \text{Subject to:} \qquad & g_1(x_1, x_2, \ldots, x_n) \le b_1 \\ & g_2(x_1, x_2, \ldots, x_n) \le b_2 \\ & \ldots \\ & \ldots \\ & \ldots \\ & g_m(x_1, x_2, \ldots, x_n) \le b_m \\ & x_j \ge 0\ \forall j \quad \text{(for all } j). \end{aligned}$$

where $\mathbf{Z}(x_1, x_2, \ldots, x_n)$ is the multiobjective objective function and $Z_1()$, $Z_2()$, ..., $Z_p()$ are the p individual objective functions. Note that the individual objective functions are merely listed; they are not added, multiplied, or combined in any way. For convenience of illustration, we will assume that all objectives in the model are being maximized.

4.B.1 The Weighting Method of Multiobjective Optimization

The weighting method is acknowledged as being the oldest, and probably most frequently used multiobjective solution technique. Once the objectives, decision variables, and constraint equations have been fully specified, the weighting method can be accomplished as follows:

1. Solve p linear programs, each having a different objective function. Each of these solutions is a noninferior solution for the p objective function problem provided that alternate optima do not exist at that solution. If alternate optima are indicated, at least one of the optimal basic feasible extreme points will be noninferior (it is possible that more than one will be noninferior, but likely that some will be dominated).
2. Combine all objective functions into a single-objective function by multiplying each objective function by a weight and adding them together such that

$$\text{Maximize } Z = [Z_1, Z_2, \ldots, Z_p]$$

becomes

$$\text{Maximize } Z(w_1, w_2, \ldots, w_p) = w_1 Z_1 + w_2 Z_2 + \cdots + w_p Z_p.$$

This objective function is often referred to as the *grand* objective.

3. Solve a series of linear programs using the grand objective while systematically varying the weights on the individual objectives. Each of these solutions will be a noninferior solution for the multiobjective problem. The number of different sets of weights and the number of linear programs solved depend on the complexity of the trade-off surface and the time available to the analyst.

The weighting method will be used to solve the two-objective problem that was solved graphically (and exhaustively) in Section 4.A.3.

Solve p Individual Linear Programs. Solving our model Z_1 as the only objective function may be viewed as moving a vertical line through objective space similar to how we solved graphical problems for which we could plot decision space. The feasible region for objective space for the current problem is reproduced in Figure 4.3, with the objective function gradient for Z_1 shown as the vertical line passing through point B—the optimal solution for that single objective problem. We refer to this gradient of the objective function as gradient 1.

The same procedure is then used to solve the single-objective problem using Z_2,which is also plotted on Figure 4.3—the horizontal line labeled gradient 2. Recall from our previous experience that alternate optima exist for this solution—points E and F. Analytical procedures for determining which of these optimal solutions is noninferior will be presented later.

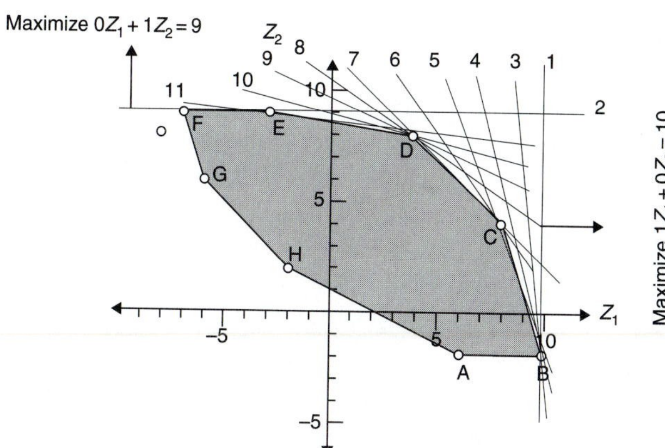

Figure 4.3 Optimizing individual objective problems may be viewed as moving that objective function through objective space with a weight of one on that objective, and a weight of zero on all other objective functions. Changing the weights on objectives changes the gradient of the grand objective function.

Set Up Grand Objective Function. The grand objective function is formed by multiplying each objective function by a weighting factor and adding these weighted objective function terms together. For our problem, the grand objective function is written

$$\text{Maximize } Z^G = w_1 Z_1 + x_2 Z_2$$
$$= w_1(3x_1 - 2x_2) + w_2(-x_1 + 2x_2).$$

For minimization objectives, one can multiply that objective function by -1 to change its sense to a maximization. The weight is a variable whose value will changed systematically during the solution process. It will be clear as we begin to generate noninferior solutions that it is the *relative* weights on these objective functions that is important, not their specific values.

Generating the Noninferior Set. For each set of positive weights used in the grand objective function, the resulting solution will be a noninferior solution. Grand objective functions for sets of weights are shown in Table 4.3. For example, suppose we set weights of 0.9 for w_1 and 0.1 for w_2 in the grand objective that we just constructed. The resulting single-objective function in solving a normal linear program would be:

$$\text{Maximize } Z^G = 0.9(3x_1 - 2x_2) + 0.1(-x_1 + 2x_2)$$
$$= 2.6x_1 - 1.6x_2$$

This objective function is labeled as gradient 3 in Table 4.3 and as plotted on Figure 4.3. Gradients 1 and 2 were those that resulted from optimizing each objective function individually and are also included in that table and plotted in objective space. We complete the table by ranging the weights w_1 and w_2 from 1 to 0 and 0 to 1, respectively, and plotting these gradients in objective space as well.[1]

[1]It is the relative values of these weights that is important; the convention that the weights sum to one is a convenience.

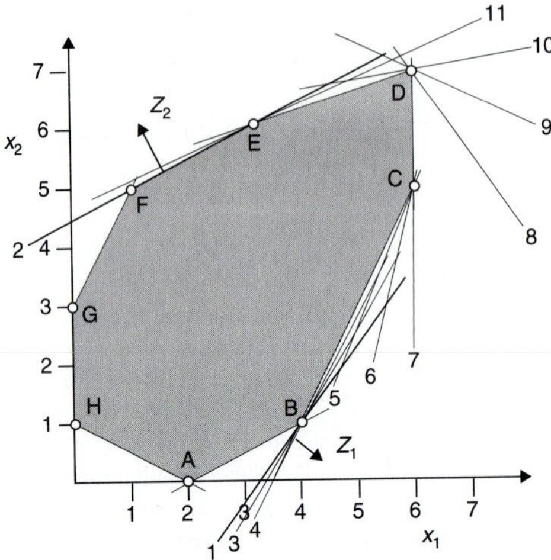

Figure 4.4 The feasible region in decision space for the problem presented in Table 4.2 with solutions that optimize Z_1 and Z_2 shown passing through their respective optima—points B and E–F, respectively, and gradients in decision space for the grand objective functions shown in Table 4.3.

Note that each time we solve a linear program using the resulting grand objective function, the solution is a noninferior feasible extreme point. Because it is important to recognize the relationship between objective space and decision space as we generate the noninferior set, we also reproduce the graph of the feasible region in decision for this problem as Figure 4.4 and plot gradients 1 through 11.

Care when selecting a strategy for varying these weights is essential. In solving this problem using the weighting method we had the advantage of knowing what the noninferior set was because we had a graphical representation of the solution space. The convention that the weights sum to one was not necessary, but is common practice. The important thing is to develop a procedure having fine enough resolution so that all solutions can be found. In this case we could have incremented/decremented the values of the weights by 0.2 instead of 0.1 and still have found all noninferior solutions with half as much computational effort. For relatively flat portions of a trade-off curve, a smaller increment for relative weights may be necessary. And as the number of objective functions increases, the number of combinations of weights on those objectives in the grand objective function increases as well. For larger problems, the analyst must always be concerned about balancing the computational effort required to find all noninferior solutions, which may be prohibitive, against the necessity of finding *all* noninferior solutions. In general, however, computation is very cheap when measured against the value of being able to hand a decision maker a true and complete noninferior set of management alternatives.

4.B.2 Dealing with Alternate Optima When Using the Weighting Method

While performing the weighting method of generating the noninferior set for a multiobjective optimization having p individual objectives, alternate optima may be encountered (1) when solving one or more problems using the individual p objectives or (2) when solving a linear program using the grand objective function constructed by weighting and combining all p objective functions. In the first case, the analyst must realize that only one of the alternate optima is a noninferior solution as was shown by using the northeast corner rule on Figure 4.2. In the second case, all alternate optima are also noninferior; notice, for example, that gradient 7 in Table 4.3 and Figure 4.3 found noninferior points D and C, which would have been indicated as alternate optima when solving the grand objective function indicated.

How can we determine which alternate optima is noninferior if such a condition is detected when solving the individual p optimizations? Recall that when we solved the linear program using only objective function Z_2 in the previous example alternate optima were indicated—points E and F in Figure 4.3 and Table 4.3, for gradient 2. This model is the same as one having a weight of 1 on objective Z_2 and a weight of 0 on objective Z_1. What if instead, we had chosen to solve the problem with a weight of 1 on objective Z_2 and a weight of ε on Z_1:

TABLE 4.3 THE GRAND OBJECTIVE FUNCTION AND THE CORRESPONDING WEIGHTS ON THE INDIVIDUAL OBJECTIVES ARE SHOWN TOGETHER WITH THE NONINFERIOR SOLUTION THAT WOULD RESULT FROM OPTIMIZING THIS OBJECTIVE FUNCTION. THE GRADIENT NUMBER REFERENCES THE CORRESPONDING OBJECTIVE FUNCTION IN FIGURES 4.3 AND 4.4

Gradient	w_1	w_2	Objective Function Z^G	Solution
1	1.0	0.0	$3x_1 - 2x_2$	B
3	0.9	0.1	$2.6x_1 - 1.6x_2$	B
4	0.8	0.2	$2.2x_1 - 1.2x_2$	B
5	0.7	0.3	$1.8x_1 - 0.8x_2$	C
6	0.6	0.4	$1.4x_1 - 0.4x_2$	C
7	0.5	0.5	x_1	C, D
8	0.4	0.6	$0.6x_1 + 0.4x_2$	D
9	0.3	0.7	$0.2x_1 + 0.8x_2$	D
10	0.2	0.8	$-0.2x_1 + 1.2x_2$	D
11	0.1	0.9	$-0.6x_1 + 1.6x_2$	E
2	0.0	1.0	$-x_1 + 0.2x_2$	F, E

$$\text{Maximize } Z_2 = \varepsilon Z_1 + 1.0Z_2$$

where ε is an infinitesimal positive weight on Z_1. This would have the effect of tilting gradient 2 in Figure 4.2 slightly clockwise such that the optimal solution to this modified problem would be the single feasible extreme point E; the extreme point that is noninferior for the multiple objective model when optimizing Z_2 by itself.

This method can be generalized for problems of any size having any number of objective functions as follows:

> For any multiobjective problem having p objective functions, if the solution obtained when solving the ith objective function displays alternate optimal solutions, the noninferior alternate optima will be that solution resulting from solving a linear program having a grand objective function with a weight of 1 on the ith objective, and a weight of ε on the remaining $p - 1$ objectives.

Of course one should not be surprised if the solution obtained after solving this second model is identical to the first solution. It is just that this time the solution will be a unique optima. Another method for finding the noninferior alternate optima will be discussed later and should serve to strengthen your understanding of this concept.

4.B.3 The Constraint Method of Multiobjective Optimization

An alternate method for generating the noninferior set in objective space having p objective is called the *constraint* method. After solving p individual models to identify the solution that optimizes each, one objective function is selected (arbitrarily) to be optimized, with the other objective functions included in the constraint set with right-hand sides set so as to restrain the value of the objective function that was selected for optimization. By iteratively solving this modified formulation, and because, as with the weighting method, each solution to the modified problem is a noninferior solution to the original problem, an approximation of the noninferior set in objective space can be generated. The same hypothetical two objective problem presented in Section 4.A.3 will be solved below using the constraint method.

Construct the Payoff Table. Solve p individual optimization problems and construct a payoff table, shown as Table 4.4. The payoff table is a $p \times p$ matrix with a column for each objective function, and a row for each optimal solution. For example, solution vector x^1 is the solution that optimizes Z_1 with a value of 10; at this solution, the value for objective function Z_2 is -2. Each solution listed in the payoff table must be noninferior. If alternate optima are detected when solving any of the p individual formulations—as in this case when we solved Z_2 (see

TABLE 4.4 PAYOFF TABLE FOR A HYPOTHETICAL TWO OBJECTIVE OPTIMIZATION PROBLEM

Solution	x_1	x_2	$Z_1 = 3x_1 - 2x_2$	$Z_2 = -x_1 + 2x_2$	Extreme Point (Figures 4.2 & 4.3)
x^1	4	1	10	−2	B
x^2	3	6	−3	9	E

Figure 4.2)—the noninferior alternate optimal solution must be determined. Consider the formulation that optimizes Z_2 for this sample problem:

$$\text{Maximize } Z_2 = -x_1 + 2x_2$$

$$\begin{aligned} \text{Subject to:} \quad 4x_1 + 8x_2 &\geq 8 \\ 3x_1 - 6x_2 &\leq 6 \\ 4x_1 - 2x_2 &\leq 14 \\ x_1 &\leq 6 \\ -x_1 + 3x_2 &\leq 15 \\ -2x_1 + 4x_2 &\leq 18 \\ -6x_1 + 3x_2 &\leq 9 \\ x_1, x_2, &\geq 0 \end{aligned}$$

which gave a solution $x_1 = 1$, $x_2 = 5$, and $Z^* = 9$, with alternate optima indicated. An alternate method for determining which of these alternate optima is noninferior is to modify this formulation so that Z_2 is constrained to a value of 9, while Z_1 is optimized:

$$\text{Maximize } Z_1 = 3x_1 - 2x_2$$

$$\begin{aligned} \text{Subject to:} \quad 4x_1 + 8x_2 &\geq 8 \\ 3x_1 - 6x_2 &\leq 6 \\ 4x_1 - 2x_2 &\leq 14 \\ x_1 &\leq 6 \\ -x_1 + 3x_2 &\leq 15 \\ -2x_1 + 4x_2 &\leq 18 \\ -6x_1 + 3x_2 &\leq 9 \\ -x_1 + 2x_2 &= 9 \\ x_1, x_2, &\leq 0 \end{aligned}$$

The optimal solution to this subproblem will always be the noninferior optimal solution to the single-objective problem. When there are more than two objective functions being modeled, this subproblem can have as its objective function, the total of all $p - 1$ other objective functions.

The significance of the payoff table is that it identifies, for each objective function the range of values each can have on the noninferior set. This range is bounded above by the largest value in the payoff table corresponding to that objective function, and below by the smallest value. In our example, there will not exist a noninferior solution in Z_1, Z_2 space that has a larger value for Z_1 than 10, nor a smaller value for Z_1 than -3:

$$L^1_{\max} = 10 \qquad L^1_{\min} = -3$$

$$L^2_{\max} = 9 \qquad L^2_{\min} = -2$$

For problems having more objectives, the size of the payoff table is larger, but the range for each objective is determined by these limits.

Set Up the Constrained Problem. The constrained problem is specified by selecting one objective function (arbitrarily) for optimization, and moving all other $p - 1$ objectives into the constraint set with the addition of a right-hand side coefficient for each. This coefficient will be between $L_{\max}$ and $L_{\min}$ for all objective functions. Selecting Z_1 to optimize and moving Z_2 into the constraint set gives the constrained problem:

$$\text{Maximize } Z_1 = 3x_1 - 2x_2$$

$$\begin{aligned}
\text{Subject to:} \quad & 4x_1 + 8x_2 \geq 8 \\
& 3x_1 - 6x_2 \leq 6 \\
& 4x_1 - 2x_2 \leq 14 \\
& x_1 \leq 6 \\
& -x_1 + 3x_2 \leq 15 \\
& -2x_1 + 4x_2 \leq 18 \\
& -6x_1 + 3x_2 \leq 9 \\
& -x_1 + 2x_2 \geq L^2_k \qquad \text{with } L^2_{\min} \leq L^2_k \leq L^2_{\max} \\
& x_1, x_2 \geq 0
\end{aligned}$$

Generate an Approximation of the Noninferior Set. By repeatedly solving the constrained problem, an approximation of the full and precise noninferior set in objective space can be generated. As with the weighting method, the optimal solution to each constrained problem is a noninferior solution to the original problem.

The precision with which one approximates the true noninferior set using the constant method depends on the number of times one is willing or able to solve the constrained problem developed in step 2. Let r be the number of noninferior solutions to be generated in such an approximation; for this example, $r = 5$. Then the (5) values for the right-hand side of objective Z_2 in the constraint set are determined using the following formula:

$$L_k = L_{\min} + \left[\frac{t}{(r-1)}\right](L_{\max} - L_{\min}), \qquad \textit{for } t = 0, 1, 2, \ldots, (r-1).$$

Solving this equation five times results in five different values for L_k:

$$L_1 = -2 + \left[\frac{0}{(4)}\right][9 - (-2)] = -2$$

$$L_2 = -2 + \left[\frac{1}{(4)}\right][9 - (-2)] = 0.75$$

$$L_3 = -2 + \left[\frac{2}{(4)}\right][9 - (-2)] = 3.5$$

$$L_4 = -2 + \left[\frac{3}{(4)}\right][9 - (-2)] = 6.25$$

$$L_5 = -2 + \left[\frac{4}{(4)}\right][9 - (-2)] = 9$$

Solving the constrained problem five times using a different value of L_k each time will result in a noninferior set of five solutions evenly distributed across the Z_2 axis between $Z_2 = L_{\min}$ and $Z_2 = L_{\max}$, inclusive. Figure 4.5 shows the feasible re-

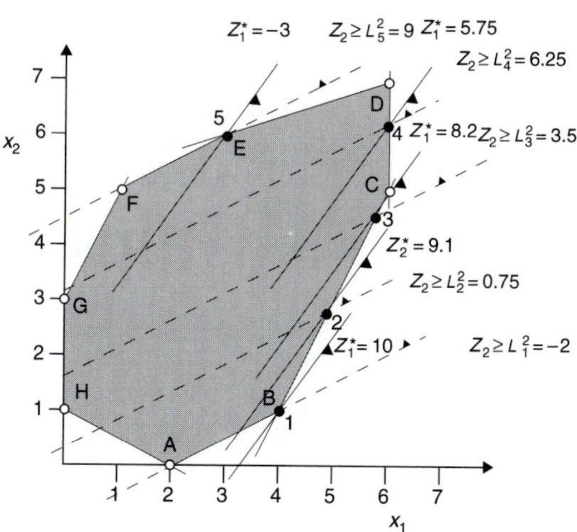

Figure 4.5 The feasible region in decision space for the example problem shown with Z_2 constrained to five different values of L_k, and the corresponding location of the objective function Z_1. The corresponding values of Z_1 and Z_2 indicate noninferior solutions.

TABLE 4.5 SUMMARY OF NONINFERIOR SOLUTIONS GENERATED BY THE CONSTRAINT METHOD

Noninferior Solution	L_k	x_1	x_2	$Z_1 = 3x_1 - 2x_2$	$Z_2 = -x_1 + 2x_2$
1	−2	4	1	10	−2
2	0.75	4.9	2.83	9.1	0.75
3	3.5	5.82	4.66	8.2	3.5
4	6.25	6	6.125	5.75	6.25
5	9	3	6	−3	9

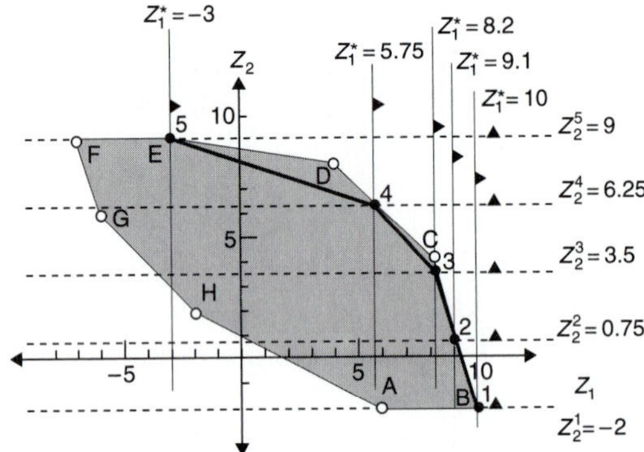

Figure 4.6 The feasible region in objective space for the example problem shown with Z_2 constrained to five different values of L_k, and the corresponding location of the objective function Z_1. The heavy black line connecting points 1 through 5 is the approximation of the noninferior set.

gion in decision space with each of these new constraints plotted as a dashed line, and the optimal gradient Z_1 when that particular problem is solved. The corresponding solid points on the graph represent the resulting noninferior solutions in decision space, numbered 1 through 5. These solutions are summarized in Table 4.5. Note that in all cases the value for Z_2 at that noninferior solution is equal to the value for L_k, suggesting that that constraint—objective Z_2 in the constraint set—is binding. The corresponding solution space in objective space is presented as Figure 4.6. The noninferior solutions generated by the constraint method are included as solid dots. Again, the objective functions Z_1 and Z_2 are shown passing through each noninferior solution, labeled 1 through 5. The heavy solid line connecting these points is an approximation of the noninferior set.

Notice that the approximation to the noninferior set generated by the constraint method does not "find" noninferior solution C and D which were found using the weighting method. Furthermore, the approximation is better in the region between points B and C than it is between points D and E. By increasing the number of points used to approximate the noninferior solution, however, we can get as close an approximation as we wish. Typically, after an initial screening of solutions that are evenly distributed within the range L_{min} and L_{max} for each objective function shifted to the constraint set, additional values can be used to search for noninferior solutions in regions of the solution space that might be more important to the decision maker.

4.B.4 Selecting a Generating Method

Two of the more important and popular methods for generating the noninferior set in objective space between two or more conflicting objectives are the weighting and the constraint methods presented in this chapter. Other methods have been proposed, and the selection of the best one for a particular analysis depends, for the most part on the experience of the analyst.

As the number of objectives and size of the solution space increases, the computational effort required to find all noninferior solutions rises dramatically. Both methods rely on the judgment of the analyst as to the configuration and shape of the noninferior set.

When the degree of conflict between objectives is expected to be significant, the weighting method is quite effective at finding noninferior solutions. But if too coarse a resolution of weights for the weighting method is specified, noninferior solutions may be missed. In fact it is possible that noninferior solutions exist that would not be found with any combination of weights.[2] If too fine a resolution is specified, any combinations of weights may find the same noninferior solution and computation costs may be excessive.

If the decision maker is able to articulate a preference for one or more objectives in terms of a range of acceptable values, then the constraint method may be more effective because these objectives can be moved to the constraint set with right-hand side values that are specifically, rather than generally specified to best "cover" a region of greatest interest for the decision maker. For example, if the decision maker has articulated a concern that costs be minimized, he/she may be further encouraged to specify an acceptable region for these costs: "Cost should be minimized, but in no case be allowed to exceed X dollars."

[2] Suppose a solution $Z_1 = 400$, $Z_2 = 1000$ was added to the set of solutions presented in Table 4.1. This solution would be noninferior even though no combination of weights for the two objective functions would find it in objective space. Plot these solutions and convince yourself of this fact. These solutions are called *gap point solutions* and are quite common when the number of objectives being considered is large, and the solution space for one or more objectives is discrete.

For analysts working in the public sector, virtually every decision, and therefore every model constructed to support those decisions, has a multiobjective context. It is incumbent upon the analyst to explore with the decision maker the entire framework within which management decisions are made. Frequently, decision makers do not appreciate the efficiency of the analytical tools available to address explicitly the trade-offs between objectives and may not even be aware of all the objectives that should or could be considered related to a specific decision. The notion of noninferiority as it pertains to public sector decisions and the ability to generate the noninferior set among conflicting objectives are among the most powerful tools of the analyst who works on public sector problems.

CHAPTER SUMMARY

More frequently than not, management decisions in the public sector must consider multiple, and often conflicting objectives. When this is the case, there may not be optimal solutions. Rather, the solution that optimizes one objective will not optimize a conflicting objective function. The least cost solution may not be one providing the best or most reliable level of service, for example. Instead of trying to find an optimal solution, we are now interested in finding the full set of *noninferior* solutions—a solution is noninferior if there does not exist another solution that performs better in terms of one objective function without performing worse with respect to at least one other. The complete set of noninferior solutions is called the *noninferior set* and defines the trade-offs between objectives.

A modification to the basic linear programming formulation, with a corresponding modification to the way in which the model is solved, can define the precise nature of the trade-offs that exists between such conflicting objectives, thereby generating the noninferior set in objective space. Two reliable methods for solving such problems are the *constraint method* and the *weighting method* of multiobjective optimization. Both methods begin by computing the solutions that optimize each individual objective being modeled. If these solutions are unique optima, they are noninferior. If not, exactly one of each alternate optima is noninferior for that objective.

The constraint method proceeds by transforming the original multiobjective problem formulation into a single-objective optimization problem by placing all except one objective function into the constraint set, with the right-hand side of each set within a range of values limited above by its value when the corresponding single-objective problem was solved, and below by the worst of its values from the other single-objective optimizations. By ranging the right-hand sides for all such constraints in the transformed problem and resolving each new problem, the noninferior set is generated; each new solution is a noninferior solution. The weighting method is similar, but the objectives are added to make up a grand objective function, with weights assigned to each.

EXERCISES

4.1. Five different late-night telemarketing programs guarantee different quantities of love, money, health, fame, and friendship—for different monthly payments, of course—as indicated by the table below:

The promise	Program A	Program B	Program C	Program D	Program E
Love	2	2	5	1	2
Money	3	4	4	2	2
Health	3	2	4	2	2
Fame	1	2	0	1	2
Friendship	2	5	1	5	2
Low Monthly Cost	29.95	49.95	19.95	19.95	39.95

Which of these programs represents a noninferior alternative? For those that do not, indicate which programs are clearly superior.

4.2. Consider the following multiple objective linear program:

$$\text{Maximize } Z_1 = 4x_1 + 6x_2$$

$$\text{Maximize } Z_2 = -4x_1 + 2x_2$$

$$\text{Subject to:} \quad x_1 + x_2 \geq -4$$

$$3x_1 - 2x_2 \leq 6$$

$$x_1 + x_2 \geq 5$$

$$x_1 + x_2 \leq 10$$

$$x_1, x_2 \geq 0$$

(a) Plot the feasible region in decision space for this problem.
(b) Plot the corresponding feasible region in objective space for this problem. For each extreme point indicate if it is a noninferior, or dominated solution.
(c) Use the constraint method (graphically) to generate an approximation of the noninferior set having 6 noninferior solutions evenly spaced along the Z_1 axis.
(d) Use the constraint method (graphically) to generate an approximation of the noninferior set having 6 noninferior solutions evenly spaced along the Z_2 axis.

4.3. Consider the following multiple objective linear program:

$$\text{Minimize } Z_1 = 2x_1 + x_2$$

$$\text{Maximize } Z_2 = 3x_1 + 7x_2$$

$$\text{Subject to:} \quad 4x_1 - 12x_2 \leq -6$$
$$-4x_1 + 6x_2 \leq 12$$
$$4x_1 + 2x_2 \geq 8$$
$$x_1 + x_2 \leq 9$$
$$4x_2 \leq 16$$
$$x_1, x_2 \geq 0$$

(a) Plot the feasible region in decision space for this problem.
(b) Plot the corresponding feasible region in objective space for this problem. For each extreme point indicate if it is a noninferior, or dominated solution.
(c) Use the constraint method (graphically) to generate an approximation of the noninferior set having 6 noninferior solutions evenly spaced along the Z_1 axis.
(d) Use the weighting method (graphically) to generate an approximation of the noninferior set having 6 noninferior solutions evenly spaced along the Z_2 axis.

4.4. Consider the following multiple objective linear program:

$$\text{Maximize } Z_1 = 12x_1 + 18x_2$$
$$\text{Minimize } Z_2 = 3x_1 + 5x_2$$
$$\text{Subject to:} \quad 6x_1 + 5x_2 \leq 60$$
$$x_1 + 3x_2 \leq 15$$
$$x_1 \leq 9$$
$$x_2 \leq 4$$
$$x_1, x_2 \geq 0$$

(a) Plot the feasible region in decision space for this problem.
(b) Plot the corresponding feasible region in objective space for this problem. For each extreme point indicate if it is a noninferior, or dominated solution.
(c) Use the constraint method (graphically) to generate an approximation of the noninferior set having 6 noninferior solutions evenly spaced along the Z_1 axis.
(d) Use the constraint method (graphically) to generate an approximation of the noninferior set having 6 noninferior solutions evenly spaced along the Z_2 axis.

4.5. Consider the following multiple objective linear program:

$$\text{Maximize } Z_1 = 2x_1 + x_2$$
$$\text{Minimize } Z_2 = -3x_1 + 2x_2$$
$$\text{Subject to:} \quad 2x_1 + 5x_2 \leq 60$$
$$x_1 + x_2 \leq 18$$
$$3x_1 + x_2 \leq 4$$
$$x_2 \leq 10$$
$$x_1, x_2 \geq 0$$

(a) Plot the feasible region in decision space for this problem.

(b) Plot the corresponding feasible region in objective space for this problem. For each extreme point indicate if it is a noninferior, or dominated solution.

(c) Use the constraint method (graphically) to generate an approximation of the noninferior set having 6 noninferior solutions evenly spaced along the Z_1 axis.

(d) Use the weighting method (graphically) to generate an approximation of the noninferior set having 6 noninferior solutions evenly spaced along the Z_2 axis.

4.6. Consider the following multiple objective linear program:

$$\text{Maximize } Z_1 = x_1 + x_2$$

$$\text{Minimize } Z_2 = 0.5x_1 + x_2$$

$$\text{Subject to:} \quad x_1 + x_2 \leq 10$$

$$x_1 - x_2 \leq 6$$

$$x_2 \leq 8$$

$$x_1, x_2 \geq 0$$

(a) Plot the feasible region in decision space for this problem.

(b) Plot the corresponding feasible region in objective space for this problem. For each extreme point indicate if it is a noninferior, or dominated solution.

(c) Use the constraint method (graphically) to generate an approximation of the noninferior set having 6 noninferior solutions evenly spaced along the Z_1 axis.

(d) Use the weighting method (graphically) to generate an approximation of the noninferior set having 6 noninferior solutions evenly spaced along the Z_2 axis.

4.7. Consider the following multiple objective linear program:

$$\text{Maximize } Z_1 = 4x_1 + 6x_2$$

$$\text{Minimize } Z_2 = 5x_1 + 2x_2$$

$$\text{Subject to:} \quad x_1 - x_2 \geq -4$$

$$3x_1 - 2x_2 \leq 6$$

$$x_1 + x_2 \geq 5$$

$$x_1 + x_2 \leq 10$$

$$x_2 \leq 6$$

$$x_1, x_2 \geq 0$$

(a) Plot the feasible region in decision space for this problem.

(b) Plot the corresponding feasible region in objective space for this problem. For each extreme point indicate if it is a noninferior, or dominated solution.

(c) Use the constraint method (graphically) to generate an approximation of the noninferior set having 6 noninferior solutions evenly spaced along the Z_1 axis.

(d) Use the weighting method (graphically) to generate an approximation of the noninferior set having 6 noninferior solutions evenly spaced along the Z_2 axis.

4.8. Consider the following multiple objective linear program:

$$\text{Maximize } Z_1 = 4x_1 + 6x_2$$

$$\text{Minimize } Z_2 = 5x_1 + 2x_2$$

$$\begin{aligned} \text{Subject to:} \quad x_1 - x_2 &\geq -4 \\ 3x_1 - 2x_2 &\leq 6 \\ x_1 + x_2 &\geq 5 \\ x_1 + x_2 &\leq 10 \\ x_2 &\leq 6 \\ x_1, x_2 &\geq 0 \end{aligned}$$

(a) Plot the feasible region in decision space for this problem.
(b) Plot the corresponding feasible region in objective space for this problem. For each extreme point indicate if it is a noninferior, or dominated solution.
(c) Use the constraint method (graphically) to generate an approximation of the noninferior set having 6 noninferior solutions evenly spaced along the Z_1 axis.
(d) Use the weighting method (graphically) to generate an approximation of the noninferior set having 6 noninferior solutions evenly spaced along the Z_2 axis.

4.9. Consider the following multiple objective linear program:

$$\text{Maximize } Z_1 = 9x_1 + 6x_2$$

$$\text{Minimize } Z_2 = 3x_1 - 2x_2$$

$$\begin{aligned} \text{Subject to:} \quad 3x_1 + 2x_2 &\leq 30 \\ -6x_1 + 3x_2 &\leq 12 \\ 4x_1 - 2x_2 &\leq 24 \\ x_2 &\leq 6 \\ x_1, x_2 &\geq 0 \end{aligned}$$

(a) Plot the feasible region in decision space for this problem.
(b) Plot the corresponding feasible region in objective space for this problem. For each extreme point indicate if it is a noninferior, or dominated solution.
(c) Use the constraint method (graphically) to generate an approximation of the noninferior set having 6 noninferior solutions evenly spaced along the Z_1 axis.
(d) Use the weighting method (graphically) to generate an approximation of the noninferior set having 6 noninferior solutions evenly spaced along the Z_2 axis.

4.10. Consider the following multiple objective linear program:

$$\text{Maximize } Z_1 = 4x_1 - 5x_2$$

$$\text{Minimize } Z_2 = 4x_1 - 2x_2$$

$$\begin{aligned} \text{Subject to:} \quad -6x_1 + 3x_2 &\leq 12 \\ 4x_1 - 2x_2 &\leq 24 \\ 3x_1 + 2x_2 &\leq 30 \end{aligned}$$

$$x_2 \leq 6$$

$$x_1, x_2 \geq 0$$

(a) Plot the feasible region in decision space for this problem.

(b) Plot the corresponding feasible region in objective space for this problem. For each extreme point indicate if it is a noninferior, or dominated solution.

(c) Use the constraint method (graphically) to generate an approximation of the noninferior set having 6 noninferior solutions evenly spaced along the Z_1 axis.

(d) Use the weighting method (graphically) to generate an approximation of the noninferior set having 6 noninferior solutions evenly spaced along the Z_2 axis.

4.11. Recall the problem of formulating a linear program to find the optimal schedule for a set of construction activities that would result in the shortest possible construction time for a small commercial establishment (Exercise 2.17). A subsequent exercise (Exercise 2.18) presented this problem from the perspective of a manager concerned about minimizing total cost of production. Explain how you would generate the trade-off surface between these two objectives.

4.12. For a particular multiple objective problem, the vertices of the feasible region in objective space are presented in the table below. Complete the table by indicating which points are noninferior, and for those that are not, by which points they are dominated.

Vertex	Max Z_1	Min Z_2	Min Z_3	Min Z_4	Noninferior or Dominated?
A	10	4	5	12	
B	12	3	6	10	
C	6	2	6	13	
D	5	4	4	4	
E	8	1	3	7	

4.13. The rolling mill in a large steel plant generates two types of water wastes: pickling waste and process water. If these wastes are discharged directly into the local city sewer without treatment, they are subject to an effluent tax. Alternatively, the plant operates a treatment facility capable of removing 90% of the pollutants from the waste streams. In this case the pickling wastes require pre-treatment (neutralization) before being processed through the plant's treatment facility.

The local municipality controls the effluent tax on untreated wastes entering city sewers; the higher the effluent tax, the more wastes the steel mill will treat on site. However, this could cause a decrease in the plant's productivity that could directly affect the economy of the community. The lower the effluent tax, the less waste the plant will treat and production will be higher. But this in turn means that the community will be subsidizing the mill's waste treatment at the city sewage treatment plant.

Let: x_1 = tons of steel to be manufactured per day

x_2 = volume of pickle waste (gallons) treated per day

x_3 = volume of process water (gallons) treated per day

Income from the production of steel = \$25/ton

Pickle waste generated = 100 gallons/ton of steel

Process water generated = 1000 gallons/ton of steel

Pre-treatment cost = \$.02/thousand gallons treated

Treatment cost = \$.10/thousand gallons treated

Pre-treatment capacity = 2,000,000 gallons/day

Treatment capacity = 50,000,000 gallons/day

Treatment efficiency = 90% (both wastes)

Tax on untreated pickle waste = \$0.15/thousand gallons

Tax on untreated process water = \$0.05/thousand gallons

Limit on pickle waste discharge to city sewer = 500,000 gallons/day

Limit on process water discharge to city sewer = 10,000,000 gallons/day

(a) Formulate an optimization model from the standpoint of the plant manager who is only interested in maximizing net profits (income less treatment related costs) for the company.

(b) Formulate the problem as a multiobjective program from the standpoint of the city council, which must set effluent discharge taxes in such a way as to prevent over-subsidizing the plant's operation, but which also wants the plant to remain fiscally healthy.

REFERENCES

CHRISTMAN, J., T. FRY, G. REEVES, H. LEWIS, and R. WEINSTEIN, 1989. "A Multiobjective Linear Programming Methodology for Public Sector Tax Planning," *Interfaces* 19(5), 13–22.

COHON, J. L., 1978. *Multiobjective Programming and Planning*, Academic Press, 333 pages.

DE NEUFVELLE, R. 1990. *Applied Systems Analysis: Engineering Planning and Technology Management*, McGraw-Hill, 496 pages.

GOICOECHEA, A., D. R. HANSEN and I. DUCKSTEIN, 1982. *Multiobjective Decision Analysis with Engineering and Business Applications*, Wiley, 518 pages.

IGNIZIO, J., 1976. *Goal Programming and Extensions*, Lexington MA: Lexington Books.

IGNIZIO, J. P. and T. M. CAVALIER, 1994. *Linear Programming*, Prentice Hall, 666 pages.

STEUER, R., 1985. *Multiple Criteria Optimization*, New York: Wiley.

CHAPTER

5 Integer Programming and Network Problems—Part I

5.A INTRODUCTION

The field of civil and environmental engineering systems is largely an applied field as opposed to a theoretical field—despite your impressions from the first few chapters of this book. Theory is taught in this book only because it is needed to understand how we can solve some problems and why we can solve them and not others. Conveying theory is not an end in itself for this book, but a means to an end. The end is the art of building engineering models, not the science of solution.

Having told the reader of the applied orientation of this text, we now begin two chapters whose titles seem to shout "theory" to those in the fields of operations research and mathematical programming. You need not worry. Our clear intent is to give emphasis to the practical aspects of integer programming, namely, formulation or problem statement. This is not to say that all theory will be absent from the chapters, only that it will not be the primary focus.

We begin this chapter with a discussion of the characteristics of integer programming (IP) problems, the features that set them apart from other optimization problems. In the process of laying out the characteristics of IP problems, we quite naturally describe some of the important situations to which integer programming applies. After discussing the nature of IP problems, we next present a set of problems, all of which apply to networks. This set of problems, when structured correctly, will always or very frequently provide integer-valued decisions when solved by linear programming alone. These problems optimize flows in networks or optimize structures on networks, where by "networks" we mean connected arcs such as occur in highway systems. The relationship between network problems and integer programming is a strong one both from a theoretical and a practical point of view. Our emphasis is on the practical connection: a number of network flow problems are natural integer programs. That is, their structure is such that the decision variables from the linear programming solution of a network problem are automatically integer valued. Constructing problems that have this special form is therefore an important goal since no further effort needs to be expended to achieve integers. Thus, the chapter focuses on a set of significant problems—the network problems—which provide us guidance on the efficient formulation of integer programs.

In the next chapter, we discuss integer programming solution methodologies. These techniques are designed for those problems which cannot be formulated in ways that favor integer solutions. We conclude that chapter with discussion of additional integer network models.

In the material that follows, we mean by integers the numbers 0, 1, 2, 3, ...; that is, we mean zero and the positive integers.

5.B ONLY THE INTEGERS ARE ADMISSIBLE ANSWERS

5.B.1 Discrete Items of Manufacture: The General Integer Programming Problem

The single most significant feature of IP problems is the need for decisions which are integer valued. That need derives from the practical nature of IP problems. If we were theorists, any value of a decision variable would suffice for problem solution. As practitioners, however, we insist that solutions have physical meaning, that they "make sense."

What are some discrete activities to which LP models apply where integer-valued decisions are the only decisions that make sense? The first example problem in the book, the profit-maximizing furniture factory, is a good case to examine. The number of chairs, desks, and tables, and so on that should be made are the decision variables. Manufacturing 35.4 chairs, 79.8 desks, and 203.5 tables makes no sense. Four-tenths of a chair is even an absurd concept. In a problem such as this one, the natural tendency is "to round." Because the problem is resource constrained, "to round down" is probably the appropriate reaction else resource limits could be violated. Further, it should be clear that the loss in profit from not making 0.4 chair, 0.8 desk, and 0.5 table out of about 300 items of furniture will be "in the noise." Conceivably, though, if the chair and table numbers were rounded down, the desk number might be rounded up. The determination of whether such rounding steps are useful and do not violate resource constraints is, of course, specific to the problem at hand. It is an easy matter to check if a rounding has produced an infeasible solution. Run the problem once again with all its constraints and with the decision variables set equal to the rounded values. If the LP terminates feasible, the rounding has not violated resource constraints.

Most problems in which the decision variables are discrete items of manufacture are naturally integer programming problems. Under certain circumstances, these problems may be amenable to rounding to integers. Under other circumstances, specialized procedures to achieve integers may be necessary. On the other hand, blending problems, like the paint mixing problem or the gravel mixing problem, may generally escape the need for integer valued activities. For these problems, ordinary linear programming will do very nicely.

The general integer programming problem can be illustrated by an overly simple furniture factory in which the workers can make only one kind of chair, only one kind of table, and no other items. Three resources constrain profits: the amount of pine available, the amount of maple available, and worker time. A profit value is given for each chair and table.

Let: x_1 = the number of chairs to be made, and

x_2 = the number of tables to be made.

The three resource constraints are labeled in Figure 5.1. Recall that the optimal solution to a linear programming problem occurs at an extreme point of the convex set carved out by the linear constraints. None of the four extreme points with positive profit, A, B, C, and D, however, are all integer. Hence, the solution lies *inside* the convex region.

Only the origin, with zero profit and zero items of manufacture, provides an all-integer solution, and the origin is clearly not the optimal solution. In this simple, two-dimensional problem, we can enumerate all of the possible solutions.

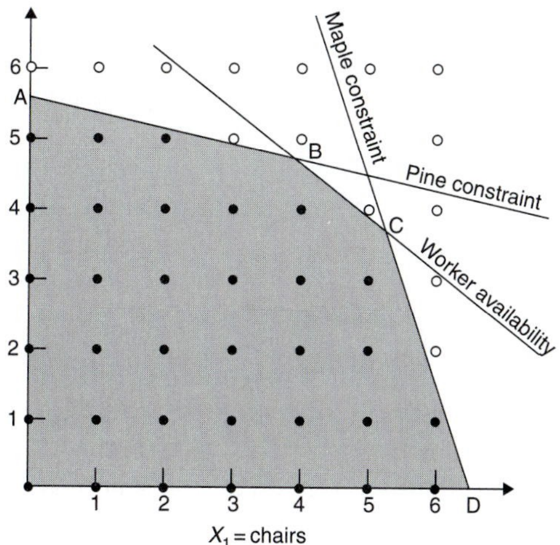

Figure 5.1 A feasible region of integer decisions. Although a convex feasible region is carved out by the three resource constraints, the real feasible region consists only of the integer lattice points in the shaded region.

That is, we can calculate profit at each and every integer lattice point and hence locate the optimal all-integer solution. In the general problem with many different activities and many resources, however, enumeration is not practical at all. In the next chapter, we describe a solution procedure for integer programming problems which is applicable to problems with many activities and which has the potential to short cut the enumeration process while determining the optimal all-integer answer. It is called branch and bound.

5.B.2 Yes or No Decisions: The Zero-One Programming Problem

Arguably, the most important application of integer programming is to decision making processes in which the choices are of a yes-or-no nature. Such settings are numerous in real-world decision making. In a general sense, the decisions are whether or not to build; whether or not to purchase; whether or not to deploy; whether or not to mine; whether or not to cut; whether or not to remove; whether or not to replace, renovate, or install; whether or not to allocate; whether or not to ship, whether or not to monitor; and so on.

Several verbal examples of these decision problems are sketched briefly, and a mathematical formulation of one of the problems is offered here.

1. The road system in a developing country is to be upgraded to allow flow of trade between its two major cities which are at roughly opposite ends of the country. A highway between these two cities will pass through a number of smaller cities (which cities are not yet specified) in order to provide logistical support en route. The path may also cross rivers and chasms and potentially tunnel through mountains. The nation's transportation agency wishes to recommend which road links should be built and which smaller cities along the way should be included on the highway to connect the two cities at the least total cost. The decisions are which links to build; that is, a yes-no decision is needed on each potential link in the system (Figure 5.2).

The problem can be formulated as a linear zero-one programming problem using decision variables which indicate whether or not a particular arc is built. We let $x_{ij} = 1, 0$; it is one if the highway is constructed between nodes i and j and, zero otherwise. Experienced highway engineers have estimated the costs to upgrade each potential link in the system to highway status. Where a bridge or tunnel is required, the cost of the link includes the cost of the bridge or tunnel. We use c_{ij} to indicate the cost of building the link between nodes i and j.

The problem objective is

$$\begin{aligned}\text{Minimize } Z = {} & c_{A1}x_{A1} + c_{A2}x_{A2} + c_{A3}x_{A3} + c_{17}x_{17} + c_{12}x_{12} \\ & + c_{21}x_{21} + c_{23}x_{23} + c_{25}x_{25} + c_{28}x_{28} \\ & + c_{32}x_{32} + c_{35}x_{35} + c_{36}x_{36} + c_{34}x_{34}\end{aligned}$$

$$+ c_{46}x_{46} + c_{52}x_{52} + c_{53}x_{53} + c_{56}x_{56}$$
$$+ c_{63}x_{63} + c_{64}x_{64} + c_{65}x_{65} + c_{6,10}x_{6,10}$$
$$+ c_{78}x_{78} + c_{79}x_{79} + c_{87}x_{87} + c_{82}x_{82} + c_{8,10}x_{8,10}$$
$$+ c_{9,10}x_{9,10} + c_{9,B}x_{9,B}$$
$$+ c_{10,8}x_{10,8} + c_{10,9}x_{10,9} + c_{10,6}x_{10,6} + c_{10,B}x_{10,B}$$

Note that x_{64} has a different meaning than x_{46}. (A similar argument can be used for x_{78} and x_{87} and other variable pairs.) If x_{64} is one, the interpretation is that the flow along the highway proceeds from node 6 to node 4 but that it enters node 6 from some node k other than node 4, say, node 5. This node k was connected back to the origin node A through one or more links. If x_{46} is one, on the other hand, the interpretation is that the flow enters node 6 from node 4, which is connected through several links back to the origin. The flow then proceeds out of node 6 to some later node in the sequence, with the sequence proceeding eventually to destination node B.

The problem formulation assumes that a single unit of flow enters node A from outside the system and that thereafter, the unit chooses a set of links in sequence in order to work its way through the network to node B where it again leaves the system. The cost of the routing chosen must be a minimum. Certain sequences will never be chosen as they could not possibly lead to least cost sequences. For instance, the sequence $\{x_{A2} = 1, x_{28} = 1, x_{87} = 1, x_{71} = 1, x_{12} = 1\}$ would never be chosen because a loop has been created and the unit has still only reached node 2. The loop could be pruned off with a cost savings and the route could proceed from node 2 without the added cost. For this reason x_{71} is not defined since it would in almost every case lead to the formation of a loop. Several other variables are left undefined for this same reason.

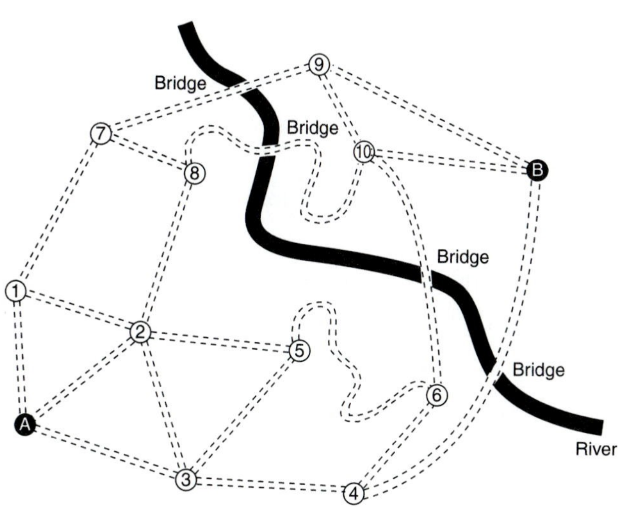

Figure 5.2 Choosing a route for a highway between cities A and B (the shortest-path problem). The double arcs between numbered nodes and between the two lettered cities indicates the position of the low-quality roads in the present transportation system. Because of mountainous terrain, the paths between nodes 8 and 10 and between nodes 5 and 6 are winding and especially expensive to construct. No matter how the highway goes from A to B, it will have to cross the river that cuts the country in two. Bridges are required, but the bridge on the arc (8, 10) is especially expensive because of the width of the river and depth of the canyon at that point in the river's course.

The constraints for this problem are input-output equations. For each node, flows enter, and for each node flows exit. The entry-exit equations are written for all 12 nodes in Figure 5.2 with node A having one unit enter and node B having one unit exit. For ease of understanding, the constraints are not written in standard form, but the reader should know how to convert these equations to standard form.

node A: $x_{A1} + x_{A2} + x_{A3} = 1$

node 1: $x_{A1} + x_{21} = x_{17}$

node 2: $x_{A2} + x_{12} + x_{32} = x_{21} + x_{23} + x_{25} + x_{28}$

node 3: $x_{A3} + x_{23} + x_{53} + x_{63} = x_{32} + x_{35} + x_{36} + x_{34}$

node 4: $x_{34} + x_{64} = x_{46} + x_{4B}$

node 5: $x_{25} + x_{35} + x_{65} = x_{52} + x_{53} + x_{56}$

node 6: $x_{36} + x_{46} + x_{56} = x_{63} + x_{64} + x_{65}$

node 7: $x_{17} + x_{87} = x_{78} + x_{79}$

node 8: $x_{28} + x_{78} = x_{87} + x_{8,10}$

node 9: $x_{79} + x_{10,9} = x_{9,10} + x_{9B}$

node 10: $x_{6,10} + x_{8,10} + x_{9,10} = x_{10,6} + x_{10,8} + x_{10,9} + x_{10B}$

node B: $x_{9B} + x_{10B} + x_{4B} = 1$

If we were willing to include possible flow directions which would lead to loops, the problem could be condensed to a form that would not be wrong but that would include possible directions that would never occur. We next offer this condensed version because it is the form most often seen in formulations of the problem in the literature.

We will need the following additional notation to complete the condensed version of the problem.

Let

N_A = {the set of nodes j with direct connections to node A}; for this case, $N_A = \{1, 2, 3\}$;

N_B = {the set of nodes j with direct connections to node B}; for this case, $N_B = \{4, 9, 10\}$; and

N_i = {the set of nodes j with direct connections to node i}.

In addition, the symbol $\in$ means "contained in." Hence the notation,

$$\sum_{j \in N_i} x_{ji}$$

can be read as the sum of flows into i from all the nodes j that are contained in the set N_i, that is, from all the nodes j that are directly connected to node i. Now, we can write the condensed version of the problem as

$$\text{Minimize } \sum_{j \in N_A} c_{Aj} x_{Aj} + \sum_{i=1}^{10} \sum_{j \in N_i} c_{ij} x_{ij} + \sum_{j \in N_B} c_{jB} x_{jB}$$

$$\text{Subject to: } \sum_{j \in N_A} x_{Aj} = 1$$

$$\sum_{j \in N_i} x_{ji} = \sum_{j \in N_i} x_{ij}, \qquad i = 1, 2, \ldots, 10,$$

$$\sum_{j \in N_B} x_{jB} = 1,$$

$$x_{Aj} \geq 0, \qquad j \in N_A,$$

$$x_{jB} \geq 0, \qquad j \in N_B,$$

$$x_{ij} \geq 0, \qquad \begin{cases} i = 1, 2, \ldots, 10, \\ j \in N_i. \end{cases}$$

The first constraint says that one unit of flow is being sent from A to one of the nodes to which it connects. The last constraint says that a unit of flow must be received at node B from one of the nodes to which it is connected. The intermediate constraints say that the flow into each node will equal the flow out of that node.

This problem, when put in standard form, can be solved using a linear programming algorithm. A typical answer, depending on costs, might be

$$x_{A3} = x_{34} = x_{46} = x_{6,10} = x_{10,B} = 1.$$

That is, the unit of flow will be moving in sequence from A to 3 to 4 to 6 to 10 to B. You are probably tempted to ask, "Why not send a half unit from A to 1 to 7 to 9 to B and another half unit the first way?" First the answer would be: "Unless the costs of the two routes are precisely equal, the algorithm would always choose to send the unit over the least cost route."

There is another dimension to the answer, though. Even if the two routes had the same cost, the algorithm would choose to send the entire unit on one route. This is because the constraint set has a very special mathematical structure that guarantees unit flows *always*, so long as a unit is put in and a unit is taken out. The special structure is discussed briefly in Section 5.C.1. It is called *unimodularity*.

The special structure is not common, but when it can be created by special formulation, that formulation should be exploited. If a problem in some other context, say, project scheduling, can be reformulated as a shortest-path problem, and a number of problems can, that form ought to be utilized to guarantee integer solutions.

We return now to a second example of a problem in which the decisions are of a yes or no character, where one indicates "yes" and zero signals "no."

2. The ambulance system of a large city has grown haphazardly with new units being placed as the need appeared—with no shifting of the positions of previously placed units. Now the city manager—hoping to shave the municipal budget—wishes to see if the same coverage of population can be provided with a lesser number of ambulance units, deployed differently in the metropolitan region. A number of possible sites are available, including the current positions of the ambulance units, several city-owned hospitals, and fire stations which have space for an ambulance vehicle. Specifically, the manager would like to find the least number of units needed, as well as their positions, so that all sectors of the city can be reached by an ambulance within 10 minutes or less. The decisions are which sites should be occupied by ambulances; that is, a yes–no decision is needed on ambulance placement for each possible deployment site. This problem is formulated later in the chapter.

3. A long-distance shipper is interested in entering a new market area and plans to open terminals to serve the cities in the region. Each terminal site has an opening cost which will be incurred if the site is developed. For each possible pair of open terminals, market studies have provided the shipper with an estimate of the net revenue that could be earned if both of the two terminals are open. That is, the volume of trade between the terminal pair has been forecast along with cost of transport and the charge levied on the route. The difference between total charge and cost is the net revenue per unit time earned for the particular city pair. The shipper would like to determine which terminals to open in order to maximize the net profit that can be earned in this new market area. Net profit in the area is the difference between the total net revenue from serving open terminal pairs and the cost of opening the terminals. This problem is also formulated later in the chapter.

5.B.3 The Mixed Integer Programming Problem

Some important optimization problems require only some of the variables to be integer valued. Typically, these integer-valued variables must be zero or one. The reasoning behind such models is that a function or operation must be begun or a structure must be built or machine installed before subsequent activities can be undertaken.

For instance, for winter snow removal, before any roads that branch from a rural highway can be plowed, the rural highway itself must be cleared of snow. Plowing that rural highway is a yes-or-no decision, and decisions about plowing spurs off of that highway wait on that yes-or-no choice.

Similarly in a manufacturing operation, products of a particular type may require a specialized machine for fabrication or finishing. The choice of whether or not to purchase the machine is a yes-or-no decision, and decisions about how many units of product to make depend on whether the machine is purchased. If the machine is not purchased, no units of the product can be made. If the machine is purchased, the number of the products that can be made per day is limited by the capacity of the machine and any other resources that go into the product. The machine purchase decision is yes-or-no, zero-or-one; the decision on production is the quantity of the particular item to manufacture.

Decisions on the siting of landfills exhibit similar properties. The choice of whether to build a landfill operation on a particular parcel of land is a yes-or-no decision based on both politics and economics. If politically unacceptable choices are excluded, the decision becomes one only of economics. The neighborhoods or communities which should use a particular landfill follow from the choice of whether or not to build at the site. If the choice at a site is negative, logically no shipments of solid wastes can be made to the site. However if the decision is a "yes," then neighborhoods or communities can transport their solid wastes to the landfill up to the daily limit of the landfill or the daily limit of their own generation of solid wastes. The landfill decision is yes-or-no, zero-or-one; the neighborhood decisions are the quantities to transport to the landfill—if it is open.

A final example is a problem in emergency services, much like the one described earlier, but with interesting twists. In this problem, more than one ambulance can be housed at a particular site, indeed may be needed at a particular site because of high call rates in a dangerous sector of the city. The choice of whether or not to build an ambulance dispatching station at a particular site is yes-or-no, zero-or-one. Once a station has been built, there is a further choice—the number of ambulances to deploy at that station. Interestingly, this number must also be an integer. It need not be confined to zero or one as the station decision is, but may be 0, 1, 2, 3, and so on. In a certain sense, this problem statement belongs elsewhere as all its decisions must be integer. In another sense, it rightly belongs here because of the presence of the zero-one siting variable which dictates the extent of subsequent activity.

5.C NETWORK FORMULATIONS

For problems which can be written as linear programs whose variables can take on any value, the simplex procedure or a variant is a very adequate method of solution. On the other hand, if your linear problem is a zero-one formulation, or an all-integer formulation, or a mixed integer formulation, the possibilities and options become more complex and may even depend on the size of your problem. Some problems can, in fact, still be solved as linear programs with good or even guaranteed success at achieving integers. And for some problems, it may turn out to be impossibly hard to find an efficient method that finds the exact integer optimal solution. In such cases, you may have to settle for a good, and perhaps not optimal, solution. The challenge here is recognizing which tool to reach for when the problem is at hand. Fortunately, for a number of problems, the tool is fairly obvious.

5.C.1 Linear Programming Solves the Integer Program—Exactly

There is a class of integer programming problems that can be solved efficiently and exactly (that is the true optimal will be found with certainty) by linear programming algorithms. It is a happy circumstance if your problem falls in this category. It is a matter only of coding the problem for computer solution using the linear programming algorithm of your choice.

These problems have a special constraint form. Sometimes the constraint form is easy to recognize and sometimes it is not. On all occasions, the right-hand side values of the constraints are integer (including zero)—in addition to the problems having a special constraint form. The special constraint form has the mathematical name *unimodular* because solution of the associated LP always yields integer solutions. In addition, the determinants of all submatrices of the constraint set are all 0, 1, or -1. This property is, in fact, the proof of unimodularity, but as you can surmise, the proof is difficult to achieve in most cases. Unimodularity, the condition of having a unimodular constraint form, is discussed in Nemhauser and Wolsey, *Integer and Combinatorial Optimization* (Wiley-Interscience, 1988).

Unimodularity is, however, a mathematician's construct. It is not necessary to show that a problem has a unimodular constraint matrix to attempt its solution—or to solve it. Our approach is one of sufficiency. If the LP works, if it achieves an all-integer answer, it is sufficient. While it might be nice to prove that we can in all cases achieve all-integer answers, if we find such answers in all cases that we need to solve, what does it matter?

Nonetheless, some problems are known to be unimodular, are known to always provide integer solutions when solved by a linear programming algorithm. The shortest-path problem discussed earlier is an example. We discuss some of these problems next.

The Transportation Problem. One famous problem of this sort is the transportation problem. The name isn't wonderful; really the problem ought to be called the distribution problem, but the name was applied early on and has stuck, so there's no hope to change it. In the transportation problem, each source has a limited amount of goods available each month, and each destination has a monthly requirement for the goods. An array of costs exists; these are the costs to ship the single type of good from each of a number of sources to each of a number of destinations. The sources might be warehouses; the destinations might be retail stores; they are scattered across the map. The objective is to accomplish the monthly distribution of goods at the least total cost.

The problem requires the following notation:

i, m = the index and total number of sources;

j, n = the index and total number of destinations;

a_i = the monthly availability of the good at source i;

b_j = the monthly amount required at destination j; and

c_{ij} = the cost to ship one unit from source i to destination j.

For a problem with three sources and six destinations, the problem looks like this:

$$\begin{aligned}\text{Minimize } Z = \; & c_{11}x_{11} + c_{12}x_{12} + \cdots + c_{16}x_{16} \\ & + c_{21}x_{21} + c_{22}x_{22} + \cdots + c_{26}x_{26} \\ & + c_{31}x_{31} + c_{32}x_{32} + \cdots + c_{36}x_{36}\end{aligned}$$

$$\begin{aligned}\text{Subject to: } & x_{11} + x_{12} + \cdots + x_{16} && \leq a_1 \\ & x_{21} + x_{22} + \cdots + x_{26} && \leq a_2 \\ & x_{31} + x_{32} + \cdots + x_{36} && \leq a_3 \\ & x_{11} + x_{21} + x_{31} && \geq b_1 \\ & x_{12} + x_{22} + x_{32} && \geq b_2 \\ & x_{13} + x_{23} + x_{33} && \geq b_3 \\ & x_{14} + x_{24} + x_{34} && \geq b_4 \\ & x_{15} + x_{25} + x_{34} && \geq b_5 \\ & x_{16} + x_{26} + x_{36} && \geq b_6\end{aligned}$$

and

$$x_{ij} \geq 0, \qquad i = 1, 2, \ldots, m,$$
$$j = 1, 2, \ldots, n.$$

The first three constraints say that the amount shipped out of each of the sources will not exceed the amount available at each of these sources. The last six constraints ensure that the amount shipped from all three sources to each of the six destinations will be at least the quantities required by each of the destinations. In condensed notation, the problem is

$$\text{Minimize } Z = \sum_{i=1}^{m} \sum_{j=1}^{n} c_{ij}x_{ij}$$

$$\text{Subject to: } \sum_{j=1}^{n} x_{ij} \leq a_i, \qquad i = 1, 2, \ldots, m,$$

$$\sum_{i=1}^{m} x_{ij} \geq b_j, \qquad j = 1, 2, \ldots, n,$$

$$x_{ij} \geq 0, \qquad i = 1, 2, \ldots, m,$$
$$j = 1, 2, \ldots, n.$$

Close inspection of the constraint matrix reveals a pattern in which each column of the constraint matrix has exactly two ones and the rest of its elements are 0. The ones descend in a staircaselike pattern in successive columns. This particular form, whenever identified, is an immediate indicator of unimodularity or integer termination on solution by linear programming. If a problem in another context can be put in the form of a transportation problem, you should use this form to ensure integer termination of the variables. Another form of the constraint matrix which provides all-integer solutions to linear programming problems has exactly one one and one negative one in each column. The constraint set of the shortest-path problem that you saw earlier in this chapter is an example of a problem which has this special form.

The shortest-path problem and the transportation problem are in fact closely related. The parameter c_{ij} was described as the cost to ship one unit from source i to destination j. How was it determined? Most likely it is the product of the cost to ship one unit just one mile multiplied by the number of miles between i and j. And the number of miles between i and j? The distance utilized should be the shortest distance between i and j, and this number would come from the solution of a shortest-path problem. Hence, solution of shortest-path problems is needed to solve transportation problems.

The Terminal Selection Problem. The terminal selection problem is also an interesting problem from the point of view of constraint structure. In its simplest setting, the problem begins with a network or even just a line on which there are cities. The cities will trade with each other if there is a transportation terminal at each of the cities. That is, trade will occur between city pair (i, j) if terminals are at both i and j. The volume of trade between each pair of cities is known in advance, and since the path between i and j already exists, the costs of moving goods from i to j and from j to i is only the cost of transportation which is known. Profit margins are also assumed to be known, so the profit of transporting demand between i and j can be calculated. However, the monthly profit from trade between any city pair is received by the carrier if and only if terminals are open at both i and j. Each terminal has a fixed monthly cost to remain open; this cost does not vary with the volume of shipment through the terminal. The network can be shown as a simple line without loss of generality:

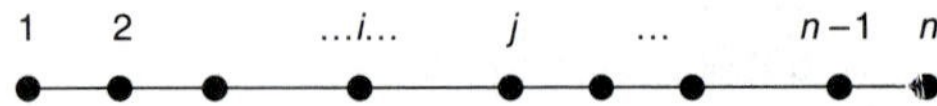

We need to define the following indices, parameters, and decision variables:

i = index of nodes $(1, 2, \ldots, n)$;

p_{ij} = potential monthly profit from the trade between nodes i and j; profit is the sum of profits from shipping goods i to j and j to i;

f_i = monthly cost of a terminal at i;

x_i = 0, 1; it is 1 if a terminal is open at node i, and 0 otherwise; and

y_{ij} = 0, 1; it is 1 if terminals are established at both i and j, and 0 otherwise.

Now the problem can be stated in two ways. One way produces fractions almost always when solved by LP. The other produces all 0, 1 variables when solved by LP—always. We state the integer unfriendly version of the problem first:

$$\text{Maximize } Z = \sum_{i>j}^{n} \sum_{j=1}^{n} p_{ij}y_{ij} - \sum_{i=1}^{n} f_i x_i$$

$$\text{Subject to: } y_{ij} \leq \tfrac{1}{2}(x_i + x_j), \quad \text{for all } (i, j) \text{ pairs,}$$

$$x_i, y_{ij} = 0, 1 \qquad i = 1, 2, \ldots, n; \quad j = 1, 2, \ldots, n.$$

The objective sums the profit over all city pairs. By only summing over i greater than j, we count each city pair just once. The single constraint type is designed to prevent profit from being counted for a city pair (i, j) unless both of the cities have open terminals.

Unfortunately, solution of this problem with relaxed linear programming, that is, with no zero-one requirement placed on any of the variables, would be expected to yield fractional y_{ij} with great frequency even if all x_i are 0-1. This is because even if only one member of an (i, j) pair is one, the corresponding y_{ij} will be one-half in order to gain half of the possible profit. Some form of postsimplex integer programming (as yet undescribed—see Section 5.C) would be required to resolve this problem. This problem formulation is definitely integer unfriendly.

However, by simply doubling the number of constraints, the problem can also be formulated in a fashion that is integer friendly, in this case, even guarantees all 0, 1 variables.

$$\text{Maximize } Z = \sum_{i>j}^{n} \sum_{j=1}^{n} p_{ij}y_{ij} - \sum_{i=1}^{n} f_i x_i$$

$$\text{Subject to: } y_{ij} \leq x_i, \quad \text{for all } (i, j) \text{ pairs,}$$

$$y_{ij} \leq x_j, \quad \text{for all } (i, j) \text{ pairs,}$$

The new constraints again prevent profit from being counted unless terminals are open at both city i and city j.

Rhys (1970) shows that the matrix of this second formulation is, in fact, unimodular. As a consequence, each and every extreme point of the linear programming formulation is all (0, 1). Hence, the LP solution is guaranteed to be all

0, 1. In addition, changes in the coefficients of the objective function will not alter this property. The formulation with the constraints expanded is not simply integer friendly; it is guaranteed to achieve zero-one variables on all occasions.

5.C.2 Linear Programming Problems That Produce All 0-1 Solutions Very Frequently

Certain problems can be formulated in a way that favors zero-one outcomes when the problem is solved by linear programming. The style of formulation does not guarantee all zero-one solutions on all occasions but does increase the frequency of such solutions, sometimes in a dramatic fashion. We call such formulations "integer friendly." Not only do such formulations increase the frequency of all zero-one termination, they also may make the integer resolution of partially fractional solutions relatively easy and swift. We alluded earlier to a postsimplex add-on procedure called branch and bound, which is commonly used to resolve into zero-one form such fractional solutions as may result from application of linear programming. The extent of the need to utilize branch and bound, often decreases substantially if the original linear programming formulation is constructed in an integer friendly fashion.

The Location Set Covering Problem. The ambulance deployment problem (or location set covering problem) described earlier can be formulated here in an integer friendly way. Recall that a number of sites are available for the placement of ambulances. Furthermore, the areas or the neighborhoods of the metropolitan area that require coverage may be represented as nodes or points. In fact, the nodes may be centroids of the neighborhoods, and covering the centroid point effectively assures coverage of the neighborhood because the area of the neighborhood is so small. Formally, each node that demands coverage requires at least one ambulance stationed within a time standard of the node. That is, an ambulance must be able to reach the demand node from its deployment site within the specified time standard.

To formulate this problem, we again make use of the symbol $\in$ that means "contained in." We also define sets N much like those in the shortest-path problem. We define

j, n = index and total number of potential ambulance deployment sites;

i, m = index and total number of demand areas;

t_{ji} = the shortest time from a potential ambulance station at j to demand node i;

S = a time standard, say, 10 minutes; each demand node should have a server that can reach the node within this time;

N_i = {set of ambulance sites eligible by virtue of time to cover demand node i} = $\{j \mid t_{ji} \leq S\}$ = {those j such that the time from j to i is less than or equal to S}; and

x_j = 0, 1; it is 1 if an ambulance is sited at j and 0 otherwise.

The objective of the problem is to site the least number of ambulances that will "do the job." The "job" is coverage within the time standard of each and every demand node. The problem can be stated as a linear integer program as follows.

$$\text{Minimize} \quad Z = \sum_{j=1}^{n} x_j$$

$$\text{Subject to:} \quad \sum_{j \in N_i} x_j \geq 1, \qquad i = 1, 2, \ldots, m,$$

$$x_j \geq 0, 1, \qquad j = 1, 2, \ldots, n.$$

The objective, as desired, is merely a counter of the number of ambulances that have been sited. A coverage constraint is written for each demand node i that says that the sum of the number of ambulances that are stationed within the time standard S (that is, within N_i) is at least one. Finally, ambulances are represented as zero one variables; portions of an ambulance make no sense.

This formulation, based as it is on the geography of a region, is integer friendly to a very high degree. That is, when the zero-one constraints are replaced by the simply nonnegativity constraints of linear programming, the solution is very often all zero-one. The problem that is solved is

$$\text{Minimize} \quad Z = \sum_{j=1}^{n} x_j$$

$$\text{Subject to:} \quad \sum_{j \in N_i} x_j \geq 1, \qquad i = 1, 2, \ldots, m,$$

$$x_j \geq 0, \qquad j = 1, 2, \ldots, n.$$

Experience by Toregas et al. (1971), suggests that on the order of 95% of the linear programming solutions to this problem are all zero-one—without resorting to any add-on procedure such as branch and bound. Of the remaining 5% or so fractional terminations, nearly all are resolvable to all zero-one solutions by a very simple procedure. The remainder, a very few problems, are usually contrived situations and hence of less interest, although they could be resolved by branch and bound.

The simple add-on procedure that resolves most of the fractional solutions that occur is particular to each situation. Almost every fractional solution involves an objective value from solution of the linear programming problem that is some integer number of ambulances plus a fraction.[1] Suppose the fractional solution suggested the need for $8\frac{1}{2}$ ambulances, stationed variously across the network, with pieces of ambulances here and there. No amount of conjuring could possibly reduce the required total number of ambulances to 8. In fact, since partial ambulances make no sense, the total number of ambulances needed in the system must be at least 9. A constraint that states this requirement is added to the original problem, and the augmented problem is solved again by linear programming. That is, we solve

$$\text{Minimize} \quad Z = \sum_{j=1}^{n} x_j$$

$$\text{Subject to:} \quad \sum_{j \in N_i} x_j \geq 1, \qquad i = 1, 2, \ldots, m,$$

$$\sum_{j=1}^{n} x_j \geq 9,$$

$$x_j \geq 0, \qquad j = 1, 2, \ldots, n.$$

Most solutions to noncontrived problems which originally presented fractional answers will now present with all zeros and ones with the addition of this constraint. Cases of nonzero-one termination are rare.

The Plant Location Problem. A problem in the location of manufacturing plants can also be formulated in an integer friendly way. The plant location problem described here is a version of the transportation problem in which the location of demands are known, but the origins that will supply the demands are not as yet chosen. In addition, an opening cost is levied when a plant opens at some point. The opening cost, or fixed charge, is to be counted just once no matter the number of destinations the plant supplies.

We discuss the plant location problem in two versions. The first is a condensed version that nearly always gives fractional solutions on application of linear programming, and the second is in a constraint-expanded version that is integer friendly to a very high degree.

We define

i, I = index and set of eligible plant sites;

j, J = index and set of demand areas;

n = number of plant sites;

m = number of demand areas;

[1] This is, in fact, the most likely fractional outcome, although it is possible and can occur that the objective value is an integer and some siting variables are fractions. To date, most such occurrences seem to be contrived examples.

f_i = cost to open plant i, independent of the volume or number of demands served;

d_j = demand for product at area j;

c_{ij} = cost to transport all of j's demand from plant i;

e_i = cost to manufacture each additional unit at plant i;

$e_i d_j$ = cost to manufacture j's demand at plant i;

x_{ij} = 1, 0; it is one if j's demand is fully supplied by plant i and zero otherwise;

and

y_i = 1, 0; it is one if plant i opens and zero otherwise.

The condensed problem statement is

$$\text{Minimize } Z = \sum_{i=1}^{n} f_i y_i + \sum_{i=1}^{n}\sum_{j=1}^{m} (c_{ij} + e_i d_j)\, x_{ij}$$

$$\text{Subject to: } \sum_{i=1}^{n} x_{ij} = 1, \qquad j = 1, 2, \ldots, m,$$

$$\sum_{j=1}^{m} x_{ij} \le n y_i, \qquad i = 1, 2, \ldots, n,$$

$$x_{ij} = 0, 1, \quad y_i = 0, 1 \;\forall\, i \in I, j \in J.$$

The objective minimizes the sum of opening costs, manufacturing at the plant, and transportation of the goods. The first constraint type says that each demand node j must get its full requirement from the source nodes. The second constraint type ensures that no shipment to any node j will occur from a plant at i unless the plant is open.

Solution via linear programming, without any zero-one requirement, of this version of the plant location problem will almost always result in fractional values of variables. An exhaustive and prohibitive post-LP branching and bounding would be necessary to resolve this problem structure in zero-one variables.

A *constraint expanded version* can also be formulated. The objective remains the same as in the condensed version, as does the first set of constraints:

$$\sum_{i=1}^{n} x_{ij} = 1, \qquad j = 1, 2, \ldots, m.$$

The second set of constraints is replaced by m sets of "tight" constraints

$$x_{ij} \le y_i, \quad i = 1, 2, \ldots, n, \quad j = 1, 2, \ldots, m$$

so that the original m plant constraints are expanded by a factor of n. Summation over j of each of the sets of equations $x_{ij} \leq y_i$ yields the condensed constraints above. These tight constraints say that as soon as any demand node gets its full supply from a plant i, the plant will open. In general, the constraints say that no demand can be supplied from site i unless a plant is open at i.

Morris (1978) showed for 600 randomly generated problems that this version of the plant location problem produced all zero-one solutions 96% of the time. ReVelle and Swain (1970), in experience with a closely related problem, observed that when branch and bound was required to resolve fractional variables produced by linear programming, the extent of branching and bounding needed was very small, always fewer than 6 nodes of a branch-and-bound tree. Constraint expansion thus makes zero-one solutions much more likely. And it appears to make the need for branch-and-bound use in a particular problem far less frequent.

CHAPTER SUMMARY

In this chapter, we discussed the characteristics of integer programming problems emphasizing three situations. In the first situation, only integers were admissible answers for the decisions; the production of 17.3 chairs made no sense.

In the second setting, yes-or-no, zero-one, decision variables were needed exclusively to indicate whether or not a set of actions would be taken. Three specific examples were provided of problems in which yes-or-no decisions were needed. These were (1) a problem of building a highway between two spatially distant cities (the shortest path problem), (2) a problem of deploying ambulance units in a city, and (3) a problem in which shipping terminals are to be established. The first problem, the shortest-path problem, required decisions on which specific links of the highway to build to establish the connection at least cost. The second required variables which indicated where to site ambulance units, and the third called for variables which designated some cities, but not others, as shipping terminals.

In the third situation of possible variable values, some but not all of the variables needed to be zero-one; these problems were termed mixed integer programming problems. These problems originated when some of the variables could be continuous, such as the quantity of solid wastes to be shipped from an origin node to a transfer station, and other variables needed to be zero-one, such as whether or not to establish a transfer station at a particular site on the network. The shipment of solid wastes to the transfer station site obviously could not take place unless the transfer station was established.

After discussing these three situations of variable types, we introduced and defined a class of problems, known as network problems, whose solutions, when linear programming is employed, will always be all-integer. Examples of such problems included the (1) shortest-path problem, discussed earlier; (2) the transportation problem, which we described mathematically and which involved the distribution of goods; and (3) the terminal selection problem, which we also described mathematically.

Discussion of an additional class of problems, which provided all zero-one solutions vary frequently when solved by linear programming, concluded this chapter. The problems included the location set covering problem for siting emergency facilities such as ambulances as well as the plant location problem, which positioned manufacturing plants and assigned demands to each plant's customer set.

EXERCISES

5.1. A MINING PROBLEM

A mining operation has identified an area where the ore is rich enough to excavate. The excavation proceeds in distinct blocks from the surface downward. (Digging a hole is one of the few jobs in the world where you start at the top.) The problem is sketched in its two-dimensional form with numbered blocks in the following figure. In practice, a three-dimensional problem would be solved, but the problem is presented to illustrate a point, and the two-dimensional problem version is sufficient for this purpose.

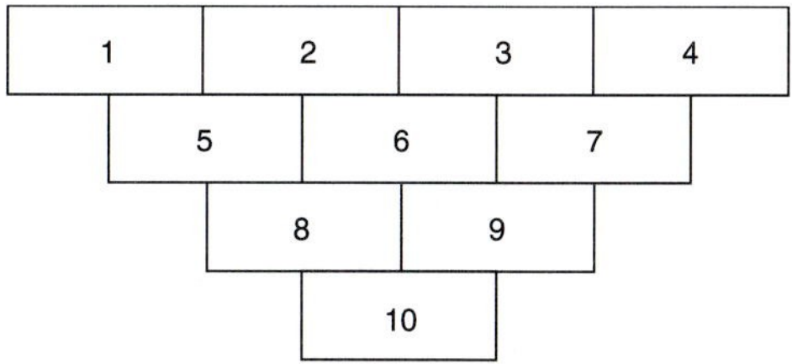

Because of the angle of slip, block 5 cannot be mined unless both blocks 1 and 2 are mined. Similar relations hold for the rest of the blocks in the arrangement. By the use of bore holes, the mining company has estimated the ore content in each of the blocks. From this information, a profit can be associated with each block. The profit is the value of the refined ore less the cost of removal and the cost of refining (beneficiation).

We let p_j = profit associated with block j; and

$x_j = 1, 0$; it is 1 if block j is removed and 0 otherwise.

This is a case where there are two formulations of the problem. One formulation is integer friendly; the other is integer unfriendly. The second formulation, the integer unfriendly one, is obtained by summing pairs of the constraints from the first formulation. The first formulation, it turns out, is more than integer friendly. The matrix of the constraint set is unimodular, but there is no need for you to prove this. Formulate this zero-one programming problem both ways and explain by a simple example why the integer unfriendly version is integer unfriendly. You may leave your formulation in non-standard form to explain integer friendliness/unfriendliness.

5.2. A PERSONNEL SELECTION PROBLEM

A contractor is assembling a crew with six positions. The contractor will select the 6 people from among a dozen of the firm's employees who are willing to travel.

Each of the dozen people who are eligible for the assignment has been ranked according to effectiveness in each of the 6 slots. The selection problem is to choose the 6 crew members from the 12 eligible employees in such a way that the crew's total effectiveness is a maximum. You are given the effectiveness e_{ij} of the ith eligible crew member in the jth position on the crew, in all 72 parameters. The decision variables are

x_{ij} = 1, 0; it is 1 if eligible employee i is chosen for the jth job on the crew, and 0 otherwise.

Formulate this problem as a zero-one programming problem. Suppose this problem was solved as a continuous or relaxed linear programming problem, that is, with the variables only constrained to be positive. What evidence do you have that this problem, solved as a continuous LP, will solve with all variables equal to zero or one?

5.3. THE MAXIMUM HARMONIOUS PARTY

A successful businessperson wishes to give a party for the local clients of the firm that the executive manages. The firm has many clients (n of them), and the executive wishes to have the largest party possible subject to just one kind of constraint, compatibility. All the invitees need to get along, or be compatible, and in this case, a number of pairs of clients do not get along, indeed might even get into noisy arguments if they drink too heavily. The executive knows which pairs of clients do not get along, but wishes still to have the largest possible party that can be arranged. Formulate a zero-one programming problem which, if it could be solved, would select which clients to invite. You are given all the (j, k) pairs that are incompatible, that as, the pairs of clients who cannot get along.

5.4. AN AMBULANCE SITING PROBLEM

The nine communities on the road network shown below are all part of the same rural county. Travel times between pairs of communities are shown on the arcs of the network. They have banded together to form a single ambulance service to serve all of the nine towns with the eligible sites for ambulances restricted to any of the nine communities. The county council desires to site the least number of ambulances such that all of the towns can be reached within 15 minutes. Set up the linear integer program that achieves the county's objective. Do not solve unless you are instructed to do so.

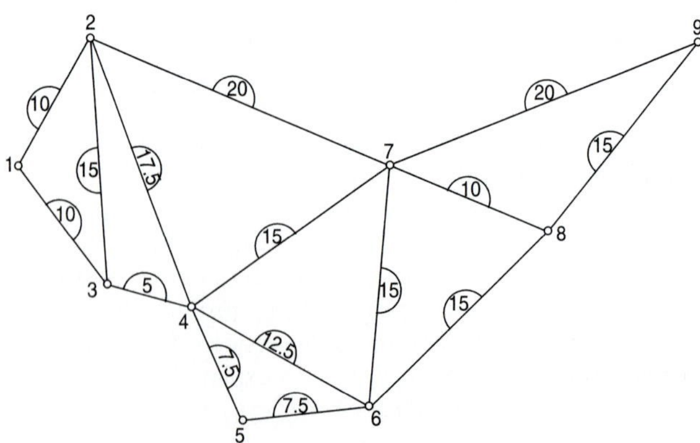

A set of nine communities with times in between them (in minutes).

5.5. SITING ROAD MAINTENANCE DEPOTS (This model was applied in Victoria Province, Australia, by G. Rose, D. Bennett, and D. Chipperfield.)

Historically, the system of road maintenance depots in Victoria was expanded a depot at a time as needs for pavement repair, drainage works upkeep, litter collection, and so on increased on rural roads. Road maintenance patrols, stationed at the depots, traveled to assigned road segments for such management activities.

Vic Roads, Victoria's road authority, wanted to see if the depot system could be consolidated into fewer depots more strategically located without a loss of service efficiency. The authority hired BTA Consulting who built a model to examine possibilities for consolidation.

First, the road system in about one-fifth of the state was divided into segments that were indexed by i. Levels of maintenance activity or need for service trips based on traffic volume were calculated for each segment. Eligible depot locations were then identified; these included 19 existing depots and 15 towns with sufficient population and community structure to support a depot. A 60-minute service time standard was utilized. An integer programming model was constructed to maximize the segments weighted by their needed trips for service that could be covered by various reduced numbers of depots. It was found that crews from 12 well-placed depots could cover all segments within the time standard and that as few as 9 depots achieved 94% coverage.

The model that Rose et al. applied resembled the location set covering problems discussed in the chapter. Its definitions, variables, and parameters include

j, n = index and total number of potential depot sites;

i, m = index and total number of demand segments;

t_{ji} = the shortest time from the potential depot site at j to segment i;

S = a time standard, 60 minutes; each segment should have a depot from which a crew can reach the segment within this time;

N_i = {set of depot sites eligible by virtue of time to cover segment i} = $\{j \mid t_{ji} \leq S\}$ = {those j such that the time from j to i is less than or equal to S}; and

x_j = 0, 1; it is 1 if a depot is placed at j and 0 otherwise.

In addition, you will need three new definitions:

m_i = maintenance need of road segment i (known);

y_i = 1, 0; a decision variable that is 1 if road segment i is covered and 0 otherwise; and

D = maximum number of depots to be sited.

With these definitions and variables, structure a problem that maximizes the activity weighted segments that can be covered in 60 minutes by D depots, as Rose et al. did.

5.6. A PROBLEM IN SOLID WASTE MANAGEMENT

A new and strict state law is forcing most cities and towns in the state to upgrade their solid waste facilities from dumps to lined landfills—unless, of course, their facilities already meet the standards. A group of four towns in one of the counties of the state are investigating cooperative approaches and have formed a compact in which they have agreed to share costs in a regional partnership where the partnership, according to the agreement, must undertake the least cost alternative. That doesn't necessarily mean that one grand solid waste facility will be built. Several may be built and operated, but the total cost will be divided among the towns in proportion to their population. The compact calls in your consulting firm to do the analyses to see what arrangements or groupings produce the least cost alternative. The consulting firm studies all possible physical arrangements of groupings of towns that make sense. Each grouping of two or more towns implies that a landfill site has been identified for that grouping and the arrangement has been "costed out" so that the annual costs of the facility and of hauling are known. Single-town alternatives are also costed out. The map of the towns and the listing of groupings with their costs are provided below.

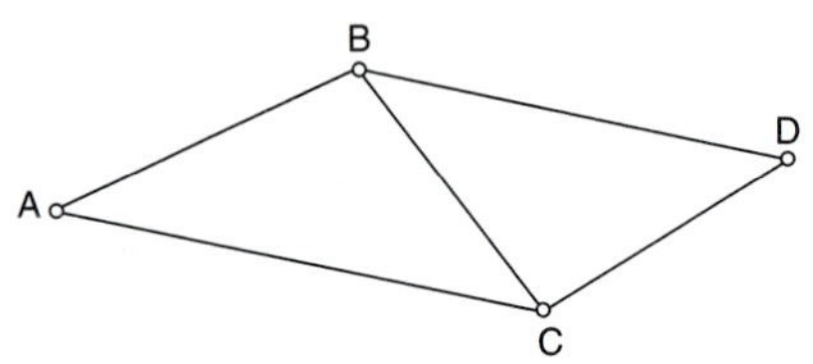

COSTS OF VARIOUS GROUPINGS

	Grouping	*Annual Cost (millions)*
1.	A, B	2.3
2.	A, C	3.0
3.	B, C	3.3
4.	B, D	2.6
5.	C, D	3.2
6.	A, B, C	4.2
7.	A, B, D	3.5
8.	A, C, D	4.0
9.	B, C, D	4.3
10.	A, B, C, D	5.4
11.	A	1.1
12.	B	1.5
13.	C	2.2
14.	D	1.3

Your job as consultant to the compact is to advise on the arrangements that cost the towns, in total, the least money. Perhaps that is (A, B) and (C, D) or perhaps some other arrangement. There are two ways to deal with the problem. One way is to enumerate all alternatives. This will work in this case because the problem is small, but it will be very difficult in larger problems. Hence, enumeration is *not* the methodology you are asked to discuss. What is desired here is a zero-one programming approach. Your programming approach should seek the least cost arrangement of towns subject to the constraint that each town is included in exactly one coalition, even if that coalition is just itself. Here are some variables and definitions to start you off.

j = index of coalitions of which there are 14;

x_j = 1, 0; it is 1 if grouping j (say, A, B) is included in the least cost solution, and 0 otherwise; and

N_i = the set of coalitions to which town i could belong (it can only belong to one though).

More explicitly,

$$N_A = \{1, 2, 6, 7, 8, 10, 11\}$$

$$N_B = \{1, 3, 4, 6, 7, 9, 10, 12\}$$

$$N_C = \{2, 3, 5, 6, 8, 9, 10, 13\}$$

$$N_D = \{4, 5, 7, 8, 9, 10, 14\}.$$

where the numbers in brackets are the numbers identifying coalitions.

Write a zero-one programming problem that minimizes cost subject to constraints that each town belongs to precisely one coalition (which could be a grouping of just itself). This problem is called the set partitioning problem (because of the sets N_A, N_B, etc.) and typically is very integer friendly when solved by linear programming.

REFERENCES

ASHFORD, R. and R. DANIEL, "Some Lessons in Solving Practical Integer Programs," *Journal of the Operational Research Society*, Vol. 43, no. 5, 1992, pp. 425–433.

MORRIS, J., "On the Extent to Which Certain Fixed-Charge Depot Location Problems Can be Solved by LP," *Journal of the Operational Research Society*, Vol. 29, no. 1, 1978, pp. 71–76.

REVELLE, C., "Facility Siting and Integer Friendly Programming," *European Journal of Operational Research*, Vol. 65, 1993, pp. 147–158.

REVELLE, C. and R. SWAIN, "Central Facilities Location," *Geographical Analysis*, Vol. 2, 1970, pp. 30–42.

RHYS, J., "A Selection Problem of Shared Fixed Costs and Network Flows," *Management Science*, Vol. 17, no. 3, 1970, pp. 200–207.

TOREGAS, C., SWAIN, R., REVELLE, C. and L. BERGMAN, 'The Location of Emergency Service Facilities," *Operations Research*, Vol. 19, 1971, pp. 1363-1371.

WILLIAMS, H. P., "Experiments in the Formulation of Integer Programming Problems," *Mathematical Programming Study*, Vol. 2, pp. 180–197, Amsterdam: North Holland, 1974.

NEMHAUSER, G. and L. WOLSEY, *Integer and Combinatorial Optimization*, Wiley-Interscience, 1988, pp. 763.

CHAPTER 6

Integer Programming and Network Problems—Part II

6.A INTRODUCTION

In the previous chapter, we described first of all the importance of integer programming as well as types of integer programs. We then proceeded to develop a number of network-based integer programming problems, all of which had special properties. The properties related to whether or not and how frequently the application of linear programming methodology provided all integer solutions. We discovered that some formulations always give integer answers while others almost always do so.

In this chapter, we take up the methodologies that become necessary when linear programming alone fails to resolve the formulation in integers. The first and most important methodology is *branch and bound*, a procedure that, if run to completion, can guarantee the optimal integer answer. We also discuss the merits of total enumeration, the evaluation of all possible integer answers. We consider, in addition, heuristic procedures, methods which do well in approaching optimal solutions, it is hoped with reduced effort. And we briefly take up multiobjective integer programming.

We then return to the network problems with which we began these two chapters and lay out the strong relationships between the fundamental network problem statements. Last, we consider the famous travelling salesman problem.

6.B SOLVING INTEGER PROGRAMMING PROBLEMS

No problem so preoccupies the researcher in mathematical programming as integer programming. Efficient solution of the general problem, that is, the problem without the special structure of network problems such as the shortest path or transportation problems, seems almost as elusive today as it was in the 1950s. Nonetheless, a relatively good procedure has evolved, the technique of branch and bound. The procedure, however, does not always do the job—often because it exhausts computing resources. In this case, analysts turn to heuristic procedures to approach good solutions. Occasionally, also, a problem is small enough that enumeration of alternatives may be undertaken. And special methods are needed for multiobjective integer programming problems.

6.B.1 Branch and Bound

The discussion of integer programming methodology in the previous chapter focused on ways to formulate problems so that linear programming solutions will always be integer or will frequently be integer. The progression of topics made sense but for one situation—the situation in which linear programming fails to deliver the all zero-one or all integer answer. In this case, the analyst probably will benefit from employing the procedure known as branch and bound.

We delayed the presentation of branch and bound to this chapter, even though we needed to refer to it on several occasions, so that the need for the procedure would be apparent. We are forced to call on branch and bound when the linear programming algorithm delivers a fractional answer. We say "forced" because solution by linear programming is definitely preferable. For problems that do not possess integer friendly structures, branch and bound is a very time-consuming computer procedure which can produce optimal zero-one or all integer answers to only moderately sized problems, and this is when it is used in conjunction with mainframe computers or powerful lab computers. Desktop computers can usually solve modestly sized integer programs only if they possess special constraint structure.

Branch and bound is not perfect, but it does solve problems that we could solve in no other way. In that sense, branch and bound is quite marvelous because solution of the general integer or general zero-one programming problem is extraordinarily difficult.

Mathematicians have, in fact, been unable to develop a general procedure that will in all situations resolve these problems. Branch and bound is the closest anyone has come, and branch and bound dates back in the literature to at least 1960 (Land and Doig).

Branch and Bound Applied to the Zero-One Programming Problem. To understand the power and the cleverness of this algorithm as well as its limitations, we need to examine it in detail. We first provide a description of the procedure as it applies to zero-one programming problems and then sketch how the algorithm is extended to the general integer programming problem.

Our discussion of branch and bound presumes a knowledge of linear programming on the part of the reader. Our application is to a minimizing problem where all variables are to be 0, 1.

The variables are the x_j, and their values are restricted by a set of structural constraints which define feasibility. In a linear integer program the constraints could be of the form

$$\sum_{j=1}^{n} a_{ij}x_j = b_i, \qquad i = 1, 2, \ldots, m,$$

where a_{ij} and b_i are known parameters. Note that the constraints of all linear programs can be brought to this form by adding slack variables or subtracting surplus variables depending on the sense of the constraint. An objective function in a linear integer program would have the form

$$z = \sum_{j=1}^{n} c_j x_j.$$

During the steps in a branch-and-bound algorithm for a zero-one linear program, a solution being examined must always be feasible with respect to the structural constraints shown above; it may, however, be infeasible with respect to achieving all variables at 0 or at 1 during the steps of the algorithm. Branching is accomplished by setting some currently fractional variable first to zero and then to one. Each branching produces a node. The node consists of the problem being solved at this stage with the original structural constraints all enforced and with a number of variables required to be zero or to be one. Also associated with the node is the value of the objective function that is achieved by optimization subject to the structural constraints and subject to the variables that are set equal to zero and the variables that are set equal to one. The nodes pile up, one after another, in a cascading or pine tree–type structure that is called a branch–and–bound tree. (See Figure 6.1).

As the zero-one requirements are successively added to a minimizing problem, the value of the objective will be expected to increase (although in some situations it could remain at the same level). When sufficient zero-one requirements are imposed (this number is, it is hoped, less than the number of zero-one variables) and all decision variables are at zero or one, the solution is then feasible with respect to the integer requirements. A feasible solution at a particular node can then be optimal if no other solution dangling from the branch-and-bound tree exists whose associated objective value is less than or equal to the objective value at this node.

The procedure begins with the space of all solutions which are feasible with respect to the structural constraints. That is, the top node of the branch-and-bound tree is the solution to the original problem without any zero-one require-

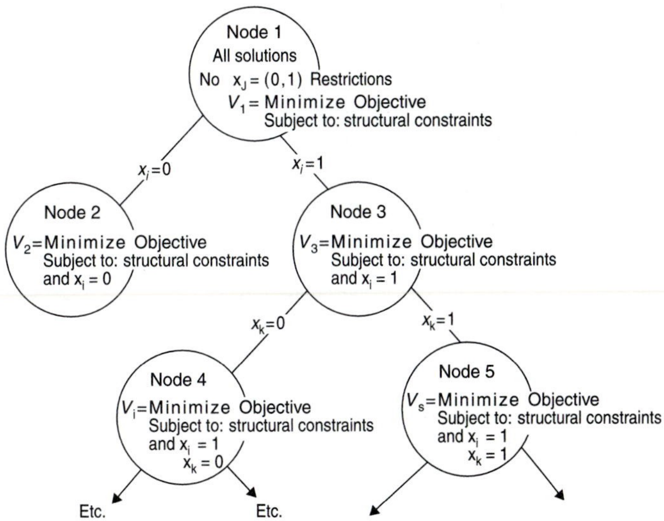

Figure 6.1 A branch-and-bound tree.

ments being imposed. It is the minimization of the objective subject only to the structural constraints. The feasible space of this node, as defined by the structural constraints, includes all the zero-one solutions and all the continuous solutions which honor those structural constraints. An objective value is associated with this node; it is the minimal value of the objective given that all structural constraints are met and that no zero-one requirements are imposed. The value of the minimum cost solution to this problem will be less than or equal to the value of the minimum cost solution subject to both the structural constraints and the zero-one requirements. With great likelihood, of course, the objective value is strictly less because addition of the integer requirements should force the objective value to increase. If the solution with this minimal value at the apex or top node, as shown in the diagram, happens to be all zero or one, even though no zero-one requirements have been imposed, it is optimal. No additional steps are needed. If the solution is not all zero-one, the zero-one requirements are added successively; that is, the branch-and-bound process begins.

A branching takes place from the apex node in which some variable, say, x_i, is fixed first at zero and then at one; then the problem is resolved. Since these two values of x_i are the only possible values that x_i may take on, the two nodes produced include all *possible* solutions to the problem That is, no zero-one solutions have conceptually been excluded; all are still possible. At each of these two nodes, a solution is determined which minimizes the objective given that x_i is zero (at the left node) or alternatively given that x_i is one (at the right node) and that the structural constraints are met. There are four possible outcomes to the branching from the apex node that has produced these two new nodes. In the first outcome, both solutions could be all zero-one. In this case the branch which produced the node with the lowest objective value is optimal. No additional steps are needed.

A second outcome is that for each node, the associated solution has fractional values for at least one variable; that is, neither solution meets the all zero-one requirement. Branching must continue from at least one of these two nodes. The rule is that branching takes place from the node with lowest associated solution value. We will explain why in a moment.

A third outcome is that one solution is all integer while the other is not, and that the solution value associated with the all zero-one solution is greater than the value associated with the fractional solution. In this case, branching (setting some x_k equal to zero and then to one) takes place from the node with the lower value of the objective, that is, the node with one or more fractional variables.

The fourth possibility is that one solution is all zero-one and that its solution value is the lowest value of the two nodes. This solution must be declared optimal since further branching from the fractional solution cannot decrease its value to the level of the zero-one solution.

These objective values associated with the nodes are better termed "lower bounds" since all we really know is that the optimal solutions at the end of these paths must have values greater than or equal to these. If further branching is required (see Figure 6.1), two new nodes are produced from each node branched on. The union of the solutions to these two new nodes with the solutions of the node not branched on continue to make up the space of all possible solutions. That is, still, no zero-one possibility has been excluded.

Branching and determining solutions with the minimal associated objective values continues to produce new dangling nodes, the union of whose solutions make up the entire feasible solution space. A dangling node, as opposed to a node interior to the tree, is either a node which is already all zero and one or a node which has not yet been branched on. *When a node is reached which has all variables zero or one whose minimal associated objective value is less than or equal to the values associated with all other dangling nodes, the solution at that node is optimal.* This is so since the remaining dangling nodes must be in one of two conditions: either a fractional solution is associated with the minimal objective value, in which case further branching required to produce a zero-one feasible solution cannot possibly reduce the objective value to that of the designated node, or the dangling node is already all zero and one and has a higher value of the objective function, in which case it is already worse than the designated node.

If, on the other hand, the node with the smallest objective value is not zero-one feasible—not all zero-one—further branching from the node with the lowest bound, is still required.

Several further notes may heighten appreciation of the branch and bound rules. First, even though branching from the node with the lowest lower bound or objective value is a heuristic rule, that is, a rule that seems to work well, it may be justified by noting that it will be impossible to terminate on *that* branching if it is done from a node which does not have the lowest lower bound. This is because the lower-valued node will always remain to be branched from. (In a maximizing problem, branching takes place from the highest-valued terminal node; see if you can explain why.)

Second, nodes temporarily abandoned, because they do not at the moment have the lowest objective value among the dangling nodes, may be returned to later in the procedure if and when their associated objective value becomes lowest. This leads us to a procedure to prune or pare the branch-and-bound tree so that only relevant branches are retained—that is, branches that can lead eventually to the optimal solution. In this regard, a node may be discarded from further consideration for branching if a feasible zero-one solution is known or found whose objective is lower than this node's bound. The presence of such a zero-one solution means that branching from the node in question cannot possibly lead to the optimal solution. The procedure of discarding is called pruning or paring of the branch-and-bound tree.

Branch and bound may be implemented by hand, but modern computer codes usually include a branch-and-bound option which can be preselected by the user to resolve the problem in zero-one variables.

How large a problem can be handled by a branch-and-bound procedure? The answer would be given in terms of the number of zero-one variables that the procedure could resolve in a reasonable length of time. That number is difficult to estimate because so many different computers and computer codes are now in use. Further, some problems have structure that favor zero-one variables on termination of the linear programming algorithm—as we indicated in an earlier discussion. Other problems, such as the condensed version of the plant location problem, have structures that favor outcomes with *many* fractional variables. If a problem does not fall in the category of being integer friendly, it probably would be wise to steer clear of problems with more than 20 to 30 zero-one variables. The amount of branching and bounding for problems larger than this appears to consume time and storage to such an extent that codes have to terminate far short of the optimum—often without any answer.

Branch-and-Bound and Mixed Integer Programming. A problem that falls in the category of lacking special structure and on which branch and bound is likely to fail to yield an answer for moderately sized problems is the condensed version of the plant location problem that was described earlier in chapter 5. Recall that shipment variables, x_{ij}, were written for each plant (i) demand (j) combination. These variables were the fraction of j's demand supplied by a plant at i. Each demand point required its full demand from some plant or combination of plants. That is, the x_{ij} were not required to be zero-one, although they always come out that way in a situation where there is no limit on plant capacity once the plant is established.

Recall also that a plant opening variable, y_i, was utilized to indicate when the opening cost would be charged *and* when flow could occur from the plant to demand points. The condensed version, the integer unfriendly version, of the problem is

$$\text{Minimize} \quad Z = \sum_{i=1}^{m} f_i y_i + \sum_{i=1}^{m} \sum_{j=1}^{n} (c_{ij} + e_i d_j) x_{ij}$$

$$\text{Subject to:} \quad \sum_{i=1}^{m} x_{ij} = 1, \qquad j = 1, 2, \ldots, n,$$

$$\sum_{j=1}^{n} x_{ij} \le ny_i, \qquad i = 1, 2, \ldots, m,$$

$$x_{ij} \ge 0, \qquad i = 1, 2, \ldots, m, \quad j = 1, 2, \ldots, n,$$

$$y_i = 0, 1, \qquad i = 1, 2, \ldots, m.$$

Notice that the x_{ij} are subject only to the nonnegativity constraints. Only the y_i are required to be zero-one. This set of restrictions makes perfect sense. It doesn't matter if all of *j's* demand is supplied from one plant or two or three, as long as that demand is met. Hence we have no requirement for the x_{ij} variables to be zero-one. We must, however, require the plant to be either open or closed. If we do not, we will not count the costs correctly, and the problem will have no meaning.

With these arguments we have now created a different type of problem. This is not a problem with only continuous variables—as the blending problem is. This is not a problem with only zero-one variables as the terminal selection problem is. It is a problem with some variables that can be allowed to be continuous and some that must be zero or one. This problem type occurs often; it is called a **mixed integer programming problem**.

How can the mixed integer programming problem with its mixture of continuous and zero-one variables be solved? It is most often attacked by an application of branch and bound in which *only* the zero-one variables are declared as branching opportunities. All remaining variables are allowed to fall where they may at each of the nodes in the branch-and-bound tree. The width of the tree and its potential depth are determined by the number of zero-one variables, not the total number of variables.

It is true, however, that the linear programming solution at each node of the tree must consider the entire problem with all the continuous variables, with some of the plant opening variables declared zero or declared one and some not yet declared in status. When a node is found *with all plant opening variables at zero or one*, whose objective value is less than the value at all other dangling nodes, the solution at that node is optimal. That is how branch and bound operates—only on the integer variables—of the mixed integer programming problem.

With the operation of branch and bound on a mixed integer programming problem kept in mind, we can return to the issue of how large a problem can be solved or should be attempted. Referring now to the problem we have called the mixed integer programming problem, the most important measure of size is the number of variables required to be zero or one. The suggested limit on problem size, again if there is no special structure favoring integer outcomes, is *still* 20–30 zero-one variables. If you were to try to solve a plant location problem in its condensed version, you should probably attempt a problem with no more than 30 plant opening variables. (If you were to try to solve the constraint expanded version which is integer friendly the limit would be determined largely by the problem dimensions that the linear programming code could handle and could be much larger.) We conclude that mixed integer programming—which uses branch and bound as its fundamental operation to achieve zero-one variables—is powerful but limited, limited in general to problems of only moderate size.

Branch and Bound Applied to the General Integer Programming Problem. The algorithm we have described is for a zero-one programming problem. It can be extended to the general integer programming problem as well without much conceptual difficulty. The extension is based on a branching using the integers rather than zero-one variables. For instance, suppose the top node has been solved—the minimization subject to structural equality constraints and nonnegativity constraints only.

The variables need to be integers for meaningful solution of the problem, but x_7, has terminated at 6.125. The value 6.125 is not admissible for x_7, but integers less than or equal to 6 and greater than or equal to 7 both are. Branching may be done from the top node on the variable x_7. One branch sets $x_7 \leq 6$ and solves the original problem subject to this constraint; the other branch sets $x_7 \geq 7$ and solves the original problem with this constraint in place. Absolutely no integer possibilities have been excluded by this branching; the space of all integer solutions remains intact.

The rules for which node to branch from and when to terminate and how to pare the tree are precisely the same as in the algorithm for the zero-one problem. In a minimizing problem, branch from the fractional node with the lowest value of the objective. (Again, branching takes place from the highest-valued terminal node in a maximizing problem.) Terminate when the lowest-valued dangling node is all integer—it is the optimal. When an all-integer dangling node (a feasible solution) exists and its associated objective value is less than that at some other dangling node, the branch represented by the second of the dangling nodes may be cut off—since the integer optimum could not be found by following that branch.

The same cautions on problem size exist for the all-integer problem as for the zero-one problem. More than 30 integer variables without special structure in the constraints (such as exists in the transportation problem) means that solution may not be achievable. The reader interested in the algorithm for the general integer programming problem is referred to the early expository article by Dakin (1965).

6.B.2 Enumeration

For moderately sized integer programming problems, branch and bound is most often the method of choice. For larger problems, heuristic methods (see Section 6.B.3) may be called upon to find good solutions. For smaller problems, in contrast, enumeration may do perfectly well. By enumeration we mean the creation and evaluation of *every* feasible solution to the problem. A final ranking yields the best of these solutions.

Enumeration is not theoretically elegant, but just as we would not use a backhoe to plant a petunia, we are well advised not to use branch and bound when the problem can be easily enumerated by hand. To make the advice more plain, we present a problem that appears at first glance to require branch and bound, but that on closer inspection really cries out for analysis by hand.

Our illustrative problem involves a builder/developer who builds tracts of homes. The builder has a choice of five large parcels of land that are for sale in different parts of the county. The parcels are of different sizes, topography, cost, and desirability to potential home buyers. As well, the costs to bring in water, gas, and install sewerage are different. Minimal lot sizes differ among the tracts as well, influencing the price of homes that can be sold. Because of equipment limitations, the builder doesn't feel capable of building on more than two tracts at once. Furthermore, since crews will be shuttling back and forth between the sites, the builder wants the two tracts chosen to be within 4.5 miles of one another. (See Table 6.1.) The builder can handle total up-front costs of only $7.5 million. These up-front costs are the sum of tract purchase costs, utility and road costs, and the costs of building the homes. The limited cost reflects the line of credit the builder has at the bank.

TABLE 6.1 DISTANCES BETWEEN TRACTS UNDER CONSIDERATION

	Distance to Parcel j				
Parcel i	$j = 1$	2	3	4	5
$i = 1$	0	6	5.25	8.25	9
2	6	0	4.5	6.75	4.5
3	5.25	4.5	0	2.25	3.75
4	8.25	6.75	2.25	0	2.25
5	9	4.5	3.75	2.25	0

TABLE 6.2 CHARACTERISTICS OF THE TRACTS (MILLIONS OF DOLLARS)

Index No. of the Tract	Number of Homes	Av. Home Cost	Av. Home Price	Road and Utilities Cost	Tract Purchase Cost	Net Revenue	Up-front Costs
1	7	0.18	0.30	0.03	0.6	0.220	1.890
2	10	0.08	0.20	0.045	0.9	0.255	1.745
3	15	0.12	0.25	0.10	1.5	0.350	3.400
4	12	0.20	0.35	0.07	1.4	0.330	3.870
5	22	0.10	0.22	0.105	2.1	0.435	4.405

The first step is to calculate for each tract the net revenue after all homes in the tract are built and sold. The net revenue is the sixth column in Table 6.2. Up-front costs, the total of home cost, tract cost, and road/utility costs are in column 7.

The problem the builder wants to solve is which two tracts to purchase and develop to maximize net return given a limited ability to absorb up-front costs. Furthermore, the builder wants the two tracts within 4.5 miles of one another. Variables for the problem are $y_j = 0, 1$; it is 1 if tract j is purchased and developed.

The objective is

$$\text{Maximize} \quad 0.22y_1 + 0.255y_2 + 0.350y_3 + 0.330y_4 + 0.435y_5.$$

The limited ability to absorb up-front costs is expressed by

$$1.890y_1 + 1.745y_2 + 3.400y_3 + 3.870y_4 + 4.405y_5 \leq 7.5.$$

The restriction that no more than two sites can be developed is given by

$$y_1 + y_2 + y_3 + y_4 + y_5 \leq 2.$$

Last, tracts must be within 4.5 miles of one another. This requires excluding those pairs of tracts which are farther apart than 4.5 miles. Using Table 6.1, the distances between tracts, we can write

$$\begin{aligned} y_1 + y_2 &\leq 1 \\ y_1 + y_3 &\leq 1 \\ y_1 + y_4 &\leq 1 \\ y_1 + y_5 &\leq 1 \\ y_2 + y_4 &\leq 1. \end{aligned}$$

These constraints assure that all infeasible pairs of tracts are excluded. In addition, all variables must be zero or one.

Isn't this clearly a case for branch and bound? No. It is a case for careful enumeration of feasible alternatives. The 4.5 mile constraint limits the number of feasible pairs drastically—to just five in number. From looking at the distance matrix, we can see that those feasible pairs are (2, 3), (2, 5), (3, 4), (3, 5), and (4, 5). In addition, the individual tracts can be developed alone, but these won't deliver as much revenue as pairs of parcels. The up-front costs of the five feasible pairs are

Pairs of Parcels	Up-front Costs
(2, 3)	5.145
(2, 5)	6.15
(3, 4)	7.27
(3, 5)	7.805
(4, 5)	8.275

Since the total of up-front costs the builder can afford is 7.5 million, only alternatives (2, 3), (2, 5), and (3, 4) remain feasible. These three pairs of parcels can now be compared on the basis of net revenue.

Pairs of Parcels	Net Revenue (millions of $)
(2, 3)	0.255 + 0.350 = 0.605
(2, 5)	0.255 + 0.435 = 0.690
(3, 4)	0.350 + 0.330 = 0.680

It looks as though the optimal pair of parcels is (2, 5) since this pair gives the greatest net revenue, but enumeration has provided an insight that straight integer optimization would not have. The pair (3, 4) is nearly as good and, given the uncertainty in the cost and sales price estimates, may be just as good as (2, 5). Furthermore parcels (3, 4) are only 2.25 miles apart, whereas parcels (2, 5) are 4.5 miles apart. The builder could opt for the second best of the alternatives because it provides other qualities. Can you think of a way (not including enumeration) to solve zero-one programs and then find the second best and third best alternative?

Our conclusion from this exercise is that not all problems need the power of branch and bound, that sometimes common sense can do just as well in finding integer or zero-one answers.

6.B.3 Heuristics

The branch-and-bound algorithm described above is a rather elegant, but sometimes computationally intensive, methodology to resolve linear programming problems in integers. For large problems without special structure, however, beyond the reach of branch and bound, methods which find good or optimal solutions need to be tailor-made for the problem at hand. Often, however, no efficient method can be created which will find the guaranteed optimal solution.

To approach a relatively fast solution of such problems, approximate algorithms may be specially designed. By "approximate" is meant that the solutions found are probably good, that is, not far from the real optimal objective value. The solutions may even be optimal, but in contrast to linear programming where the optimal is guaranteed, there may be no way, short of enumeration and evaluation of *all* possible feasible alternatives, to verify the optimality of the chosen solution. The approximate algorithms are called *heuristics* or *heuristic methods*, a name that describes a sequence of logical steps that lead, in aggregate, to improved values of the objective function. Heuristics are generally custom designed for each problem by a researcher in the field of optimization. When you encounter an heuristic, you should be aware that solution may be fast, but the solution chosen may not be the true optimal answer.

To illustrate what we mean by a heuristic procedure, we describe the famous knapsack problem, a zero-one programming problem that has inspired numerous specialized algorithms. Although the knapsack problem is often a good candidate for branch and bound, we want to explore an approximate and often very good method to "solve" the problem. Our approximate method falls in the category of heuristic procedure because it is not guaranteed to find the optimal solution. The knapsack problem is stated in the following way.

A hiker is planning a trip into the wilderness for a period of days and must decide which items to take along. The hiker can carry no more than 30 pounds of food, clothing, and equipment because of a lack of regular exercise. This 30 pounds is the amount over and above the weight of the pack itself which, of course, must be carried as a given of the problem. The hiker has been able to associate a value with each item under consideration and a weight.

Only one of each item would be carried, for example, one cup, one miniature stove, one extra pair of shoes, one half toothbrush, one flashlight, one emergency blanket, one candle lantern, one pair of sunglasses, one towel, and so on. Assume that the volume and shape of each item do not need to be considered.

Let: n = number of items being considered for the pack;

$x_j = 1, 0$; it is one if the jth item is chosen for the pack, and 0 otherwise;

v_j = value associated with item j; and

w_j = weight of item j (in pounds).

The problem may be stated as

$$\text{Maximize} \quad Z = \sum_{j=1}^{n} v_j$$

$$\text{Subject to:} \quad \sum_{j=1}^{n} w_j x_j \le 30$$

$$x_j = 0, 1, \qquad j = 1, 2, \ldots, n.$$

This problem is one of the simplest imaginable constrained optimization problems—except for the requirement that decisions are yes or no, take along or leave behind. It has only an objective and one constraint. Nevertheless, the yes-or-no decisions make it difficult to solve optimally, especially if the number of items under consideration is large.

First, let us explore what would happen if we solved this zero-one problem as an LP problem with only the zero-one requirements replaced by an upperbound of one and a lower bound of zero. That is,

$$\text{Maximize} \quad Z = \sum_{j=1}^{n} v_j x_j$$

$$\text{Subject to:} \qquad \sum_{j=1}^{n} w_j x_j \leq 30$$

$$\left.\begin{array}{l} x_j \leq 1 \\ x_j \geq 0 \end{array}\right\} j = 1, 2, \ldots, n.$$

The result of applying LP to this problem would generally have the following form. A number of x_j would be precisely equal to one, indicating that those items j are desirable for inclusion in the knapsack. One x_j would be fractional, indicating that although the item is desirable, it weighs too much to be fully included—given the other items in the pack. The remaining x_j would be zero, indicating noninclusion. Occasionally, for reasons you will see in a moment, the solution is all zero-one, but this is an unusual situation. It looks as though the zero-one portion of the LP solution could be very attractive, and it could well be, but LP is not even needed to derive that solution. The heuristic we describe next will find that answer en-route to its own proposed solution.

The heuristic methodology proceeds as follows. Each item j has two parameters associated with it, a value and a weight. The value relative to the weight for each item can be calculated. It is the ratio (v_j/w_j). This value-to-weight ratio is calculated for all items and the items are ranked from the item with greatest ratio to the item with the lowest ratio. One at a time these items are loaded into the pack (if there is a tie in the ratio, arbitrarily load the one with the lowest index first). When the latest item added either just fills the pack (in terms of the weight limit) or exceeds the weight limit, stop. Now the odds on being able to load, in order of the value-to-weight ratio, right up to the precise weight limit are small, but the situation could occur. The zero-one solution that occurs would be optimal; take it. It is the situation referred to in the discussion of the LP solution to the knapsack that is all zero-one. More likely, however, the latest item will only partly fit in the knapsack. What should be done next?

Remember that this is a heuristic solution procedure that we are discussing. We are not seeking an optimal solution, only a good solution. We began with an excellent rule—load in order of the value-to-weight ratio. This rule leads, in fact, to the exact same solution that LP would give; so the solution is already probably quite good even if we simply decide to take only those items recommended by the rule—without the last and fractional item. That's correct; we could stop with the items recommended and brought in fully according to their value-to-weight ratios. That solution is probably quite near the optimal value. We have a heuristic solution; we will call it solution A.

We could do more, however. We could probably get an even better value of the objective by looking at the weight capacity remaining after the last full item is loaded in. Now it may be that only 1/2 pound is left to fill and that the next item in order of the ratio is the 3/4-pound 36-function Swiss army knife with a miniature telescope, chronograph, and sextant that you got for Christmas. It violates your weight limit by 1/4-pound. Maybe the knife is worth the extra 1/4-pound. For a 1/4-pound violation in the weight limit, you will be able to achieve the optimal solution—albeit to a slightly different problem. Think about it. It may not be so bad, if your weight limit isn't really tight.

Alternatively, the weight limit may be precise—suppose the knapsack is a satellite and the items are experiments that would be placed on board. Still, there may be something you can do to improve the solution over and above solution A. Again, look at the weight capacity remaining in solution A. Does any item in the list of items not loaded have a weight at that level or less? Create a group of these items. Find the item (in this group) with the largest value. Load it in, and the solution is improved without violating the weight limit. It still may not be optimal, but you probably have a solution worth implementing.

The knapsack heuristic is a good example of a procedure with attractive rules of thumb that are easy to implement and that produces good solutions. For each zero-one problem that standard algorithms do not solve or do not solve in reasonable time, the challenge is to find those characteristics that can be exploited in logical sequence to provide a good answer. Heuristics are very much particular to the problem at hand.

6.B.4 Multiple-Objective Integer Programming

A fascinating connection exists between multiobjective programming, which was discussed in a preceding chapter, and integer programming, the subject of this chapter. You remember from Chapter 4 the argument that ran something like "It is important to display the trade-off curve between objective values to decision makers. It assists them in arriving at an informed choice and in entering into the analysis itself." This argument led us to develop methods to derive the trade-off curve, one of which was the weighting method.

In a two-objective problem, the weighting method combined both objectives, using weights that summed to one, into a single objective for optimization. For purposes of discussion, let us assume that we are dealing with a maximizing problem—that is, both objectives are to be maximized. With two objectives, Z_{A} and Z_{B}, the combined objective is given by

$$Z_{\text{C}} = \lambda Z_{\text{A}} + (1 - \lambda)\, Z_{\text{B}}, \qquad 0 \leq \lambda \leq 1.$$

For a sequence of values of λ within its prescribed limits, the objective Z_{C} is optimized subject to the problem's constraints. For each value of λ, a value of Z_{C} results as well as component values of Z_{A} and Z_{B}. New values of Z_{A} and Z_{B} do not necessarily occur for every increment of λ, but in general as λ increases, Z_{A} increases (because it is weighted more heavily) and Z_{B} decreases (because it is weighted less heavily). The tradeoff curve between Z_{A} and Z_{B} is generated as λ moves from large to small or small to large. The weighting method is both relatively easy to implement and not hard to understand—and it is often ideal for solving multiple-objective zero-one programming problems.

In what way is it ideal? In a problem which has an integer friendly structure except for a goal that is being parametrically varied in the constraint set, the weighting method, by shifting that goal out of the constraint set and into the objective function, maintains the integer friendly character of the constraint set. Put another way, if the fundamental set of constraints in a problem is either integer friendly or produces a unimodular constraint matrix such as in the transportation problem, an additional goal placed in the constraint set is likely to destroy the integer friendly or unimodular character. That same goal placed with a weight alongside the goal already in the objective function leaves the constraint set intact —still integer friendly or unimodular. Optimization of the weighted combined objective subject to the original constraints leaves the integer friendly problem still prone to zero-one solutions and the unimodular problem still unimodular.

Probably this lucky characteristic of the weighting method is best illustrated by a problem. We will use the terminal selection problem to suggest how the weighting method of multiobjective programming functions effectively in an otherwise unimodular problem. The terminal selection problem maximized profit from the trade between cities by opening terminals. Trade between any pair of cities could not take place unless terminals were open at each of the cities. In mathematics, the problem is to

$$\text{Maximize} \quad Z_{\text{A}} = \sum_{i>j}^{n} \sum_{j=1}^{n} p_{ij} y_{ij}$$

$$\text{Subject to:} \quad \left. \begin{aligned} y_{ij} &\leq x_i \\ y_{ij} &\leq x_j \end{aligned} \right\} \forall\ i, j.$$

where p_{ij} is the profit to the shipper from the trade between city pair (i, j); x_i is 1 if a terminal is open at i, 0 otherwise; and y_{ij} is 1 if terminals are open at both cities i and j, 0 otherwise. This formulation, you remember, guarantees all 0, 1 solutions because of its special structure.

Now the shipper is interested in more than today's profit figures. The shipper wants to build a presence in high population markets in order to foster future business growth. If a_i is the population in the city corresponding to potential terminal site i, the shipper would also like to

$$\text{Maximize} \quad Z_{\text{B}} = \sum_{i=1}^{n} a_i x_i$$

that is, to maximize the population served by the firm's terminals.

The two most fundamental ways to approach the derivation/generation of the trade-off curve are the weighting method and the constraint method. If the constraint method is used and population served is placed in the constraint set, the problem looks like

$$\text{Maximize} \quad Z_A = \sum_{i>j}^{n} \sum_{j=1}^{n} p_{ij} y_{ij}$$

$$\text{Subject to:} \quad \left.\begin{array}{l} y_{ij} \le x_i, \\ y_{ij} \le x_j, \end{array}\right\} \forall\ i, j,$$

$$\sum_{i=1}^{n} a_i x_i \ge A,$$

$$x_{ij}, y_i = 0, 1, \qquad \forall\ i, j,$$

where A is a population level for service that will be varied from small to large.

This formulation, if one attempts to solve it by relaxed linear programming, will not even be integer friendly; fractional variables are to be expected, and branch and bound will be needed to resolve each of the runs of this problem as A is varied from small to large.

If, on the other hand, the weighting method is used and population served is placed in the objective function alongside profit, the problem looks like

$$\text{Maximize} \quad Z_C = \lambda \sum_{i>j}^{n} \sum_{j=1}^{n} p_{ij} y_{ij} + (1 - \lambda) \sum_{i=1}^{n} a_i x_i$$

$$\text{Subject to:} \quad \left.\begin{array}{l} y_{ij} \le x_i \\ y_{ij} \le x_i \end{array}\right\} \forall\ i, j,$$

$$x_{ij}, y_i = 0, 1, \qquad \forall\ i, j,$$

where λ is varied from small (emphasis on population served) to large (emphasis on profit).

This formulation will not only be integer friendly; zero-one solutions are guaranteed when relaxed LP is used to solve the problem—because the unimodularity of the constraint matrix is maintained. Thus, the trade-off curve for two-objective zero-one programs can be efficiently generated by the weighting method when the original constraint matrix is either integer friendly or unimodular.

6.C ADDITIONAL INTEGER NETWORK MODELS

Network-based problems constitute one of the widest classes of problems to which linear programming and linear integer programming have been applied. To say that such problems are numerous is understatement; they are virtually ubiquitous. Later in the chapter we discuss a number of application areas in civil and environmental engineering where network concepts find important use. In addition, the ideas of network modeling that we develop here are significant in many other settings as well.

We will treat just two problems in this section of the chapter: the transshipment problem and the traveling salesman problem. The transshipment problem is discussed because it illustrates mass balance/input-output relations common in so many problems. The traveling salesman is discussed to demonstrate mutual exclusivity constraints in the context of a network problem. These two additional problems really are only the "tip of the iceberg" of a class of problems that attract the attention of many researchers and practitioners from industry to academia to the military and the public sector.

6.C.1 The Transshipment Problem

Formulation of the Transshipment Problem. The network shown in Figure 6.2 illustrates a system, perhaps a highway system, where the least cost distribution of a good is the goal. The nodes at which the goods are available have an entering heavy arrow. Perhaps these are the sites of manufacturing plants where goods are manufactured up to a certain quantity per month. The nodes at which goods are demanded have an exiting dark arrow. These sites may be warehouses which supply the goods to local customers, the retail stores, in known quantities per month.

Other nodes have roads (arcs) entering or exiting with no goods available and none demanded and, even further, no requirement that the goods even pass through these nodes. These are merely the junctions of the road network. The costs per unit of flow moving on each arc or road would be indicated in the half circle on each arc. Only names of the costs are indicated in this example rather than the cost itself. The costs are assumed to be symmetric; that is, $c_{ij} = c_{ji}$, but they need not be. The question is how to supply the demand nodes at the least total distribution cost.

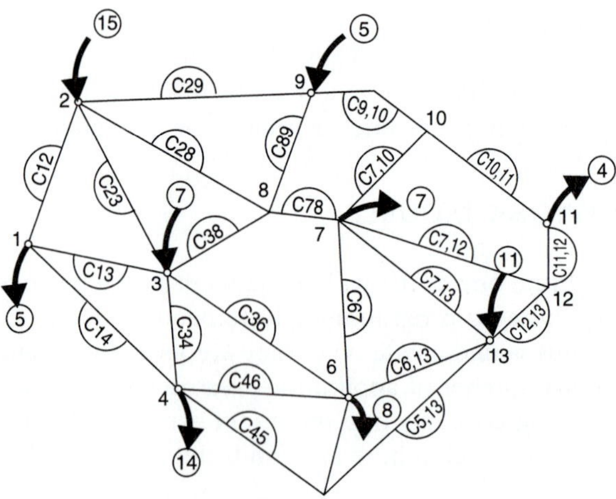

Figure 6.2 The network of a transshipment problem.

The formulation makes use of the following variables and definitions.

x_{ij} = flow from node i to node j;

a_i = the amount available to enter the system at node i;

b_j = the amount demanded to leave the system at node j; and

N_j = {the set of nodes directly connected to node j}.

For this version of the problem, we assume that the monthly total of amounts available, that is, entering the system through manufacture, is equal to the monthly total of amounts demanded, that is, required to exit the nodes to customers.

At every junction at which goods enter the system from outside, for instance, node 3, the sum of the flow of goods into the node from all other nodes plus the amount entering the system at the node must be equal to the sum of the flow of goods out of the node.

For node 3, this constraint is

$$x_{13} + x_{23} + x_{43} + x_{63} + x_{83} + 7 = x_{31} + x_{32} + x_{34} + x_{36} + x_{38}.$$

For node i, this constraint is

$$\sum_{j \in N_i} x_{ji} + a_i = \sum_{j \in N_i} x_{ij}.$$

Of course, these constraints would be put in standard form in any final linear programming problem statement.

At every junction at which goods must exit the system, the total flow of goods into the node must be equal to the flow out of the system plus the total of flow to other nodes. For node 6, this constraint is

$$x_{36} + x_{46} + x_{56} + x_{76} + x_{13,6} = x_{63} + x_{64} + x_{65} + x_{67} + x_{6,13} + 8.$$

For node j, this constraint is

$$\sum_{i \in N_j} x_{ij} = b_j + \sum_{i \in N_j} x_{ji}.$$

Of course, these constraints would be put in standard form in any final linear programming problem statement. For those nodes j with no availability and no requirement, the equations are simply input equals output, which in standard form, is

$$\sum_{i \in N_j} x_{ij} - \sum_{i \in N_j} x_{ji} = 0.$$

Thus an equation of input equals output plus or minus requirement or availability if any is written for each node. In addition, variables must be nonnegative. These three types of flow equations all put in standard form plus the nonnegativity constraints constitute the constraint set of the transshipment problem. The objective, of course, is to minimize costs of distribution; that is,

$$\text{Minimize } Z = \sum_{i \in N_j} \sum_{j} c_{ij} x_{ij}$$

If you were to look closely at the constraint matrix to this problem, you would see that each flow variable x_{ij} appears in exactly two equations, once as an input to node j and once as an output from node i. Thus, the column for variable x_{ij} has one 1 and one -1; the remaining elements are zeros. This condition indicates a unimodular constraint set, and along with integer right-hand sides, is sufficient to guarantee all integer-valued decision variables in the solution to a linear program.

Application of Transshipment Concepts. The transshipment problem is more than an interesting network flow problem whose solution variables terminate integer on LP solution. The input-output equations written at each node illustrate a type of relationship among variables that occurs in many settings in civil and environmental engineering.

In water distribution design problems, water input to a pipe junction equals water output. In structures, the forces incident on the stable junction of members equate to zero. In highway transportation networks, the hourly vehicle input to an intersection equals the hourly vehicle output.

In the multipurpose operation of reservoirs, the natural inflow to the reservoir plus the preceding period's storage less intentional releases less evaporation and seepage is equal to the storage at the beginning of the next period. In a factory that manufactures modular homes or building structures such as trusses, the inventory of units at the end of the current month is the inventory at the end of the preceding month plus the number of units manufactured less the number of units shipped during the month.

In a sewer system, wastewater flowing into a junction equals wastewater flowing out of the junction. In a segment of a river in which the concentration of organic pollution is stable (unchanging), the flow of pollutant into the segment minus the removal of the pollutant by natural biodegradation and settling less the flow of the pollutant out of the segment equals zero, yielding the stable concentration condition. In a solid waste management system, transfer stations may be utilized. At these stations, small trucks offload into larger trucks which then make the trip to the regional landfill. The daily tonnage into these nodes in small trucks must equal the daily tonnage out of these nodes in larger trucks. As you can see, the input output or mass balance relationships illustrated by the nodal equations of the transshipment problem are very frequently encountered and very important in modeling engineering systems.

Relation of the Transshipment, Transportation, and Shortest-Path Problems. A significant relation exists between the three problems that occupy the heading of this section. They are, in a sense, close relatives and, in another sense, all one problem.

The transshipment problem, when you look at it closely, consists, in addition to the network of connections, of a set of nodes at which flow enters the system, a set of nodes through which flow can pass, but is not required to pass, and a set of nodes where flow must exit the system. If the problem could be somehow reduced to only the nodes where flow enters and the nodes where flow exits, the setting would be that of the transportation problem that was discussed earlier in chapter 5.

In fact, the problem can be reduced to this setting. Consider any pair of input and output nodes, the one with flow entering, the other with flow exiting the system. The input is specified, and the output is given. Flow need not move from this input node to this output node, but it could. If it did, what would be the cost? The cost per unit flow would consist of the cost per unit flow on each of the arcs that are chosen for the route the flow follows from the input node to output node. What route will that be? It will be the shortest route, the least costly route, between the input node and the output node. The route and its cost can be determined by the shortest-path formulation given in chapter 5.

We can do this for every input-output pair of nodes—now we can say—for every origin-destination pair of nodes. When this is done, we have the c_{ij} of the transportation problem, the cost to ship one unit from each origin node to each destination node. With the a_i, the flow available at each origin node, and the b_j, the flow required at each destination node, we have all the information necessary to formulate and solve a transportation problem. The solution to the transportation problem will specify the optimal movement of goods between the nodes with flow available and the nodes with a flow requirement. That movement, between any pair of origin-destination nodes will always take place on the shortest (least cost) path between those nodes.

Thus, preprocessing by the shortest-path problem of the network costs of the transshipment problem converts the transshipment problem to a transportation problem. We said, also, that these problems, in addition to being close relatives, could also be considered just one problem, and they can.

We have already shown that the transshipment problem can be converted to the transportation problem. It remains for us to show the equivalence of the transshipment and shortest-path problem. Recall that the shortest-path problem has a unit of flow entering at the single origin node, a unit exiting at the single destination node, and a series of input-output relations at all intermediate nodes. Flow may pass through these intermediate nodes or it may not, but none of the intermediate nodes have any flow available to enter from outside the system or present any demand or flow requirement for removal from the system. All that would be necessary to convert the formulation of the shortest-path problem to that of the transshipment problem is to allow flow availability in any specified quantity, including zero at the origin, or any intermediate node, and flow demand in any specified quantity, including zero at any of the intermediate nodes or the destination.

With these small changes, the shortest-path problem becomes the transshipment problem which is the transportation problem.

6.C.2 The Traveling Salesman Problem

This problem is so famous and difficult to solve that a book has been written about it.[1] Smaller versions of the problem have been solved by the technique we will describe, but you will see that our procedure is too hard to implement for large problems. Nonetheless, the procedure to be described is of interest because it illustrates a network problem with mutual exclusivity constraints.

The problem assumes that a salesman is to visit each of n cities where his product is to be sold. The salesman is to begin and end at the same city which, is presumably his home. The actual starting and ending point doesn't matter, however, since this is a tour—a complete traversal of all of the nodes, starting and ending at the same node—and any of the cities could be the start/end city. The tour is to be achieved with the least total distance. The formulation we present is an early approach but remains educational nonetheless.

Let: $x_{ij} = 1, 0$; it is 1 if the tour proceeds from city i to city j, and 0 otherwise, and

c_{ij} = cost (distance) for the salesman to travel from city i to city j.

We begin by noting that the tour must fulfill three elementary but required conditions. First, a step of the tour must enter each city exactly once. That is, a tour must leave some city i among the $(n - 1)$ cities and enter city j.

$$\sum_{\substack{i=1 \\ i \neq j}}^{n} x_{ij} = 1.$$

Second, a step of the tour must enter each city exactly once. That is, the tour must leave city i and go to one among the remaining $(n - 1)$ cities.

$$\sum_{\substack{j=1 \\ j \neq i}}^{n} x_{ij} = 1.$$

So far, the problem looks exactly like a version of the transportation problem with unit right-hand sides.

Third, the tour must visit all cities in a continuous sweep. That is, the tour cannot visit a subset of cities in one loop and another subset in a second loop. This is the most difficult condition to enforce and the essential challenge of the traveling salesman problem. The challenge, however, only occurs if the solution to the problem violates the continuous tour requirement.

[1] *The Travelling Salesman: A Guided Tour of Combinatorial Optimization*, E. Lawler, J. Lenstra, A. Rinnooy Kan, and D. Shmoys, eds. New York: John Wiley, 1985, 465 pp.

The problem to be solved first is

$$\text{Minimize} \quad Z = \sum_{i=1}^{n} \sum_{j=1}^{n} c_{ij} x_{ij}$$

$$\text{Subject to:} \quad \sum_{j=1, j \neq i}^{n} x_{ij} = 1, \qquad i = 1, 2, \ldots, n,$$

$$\sum_{i=1, i \neq j}^{n} x_{ij} = 1, \qquad j = 1, 2, \ldots, n,$$

$$x_{ij} = 0, 1, \qquad i = 1, 2, \ldots, n, \quad j = 1, 2, \ldots, n, \quad i \neq j.$$

If this problem is solved as a relaxed linear program, that is, with the variables only constrained to be nonnegative, it will yield all variables equal to zero or one since the problem has precisely the form of the transportation problem which is known to always yield integer valued variables. That much is fine, but Figure 6.3 illustrates that the solution, while all zero-and-one, might not achieve the continuous tour requirement, but could contain subtours.

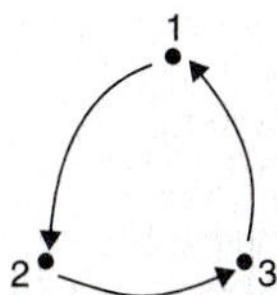

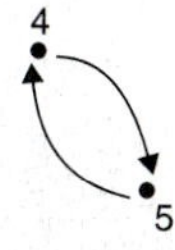

Figure 6.3 Subtours in a traveling salesman problem.

To break subtours, constraints are added which make the subtours infeasible. One of the subtours, according to the figure, is

$$x_{45} = 1, x_{54} = 1.$$

This subtour can be forced out by resolving the original problem with the following additional constraint:

$$x_{54} + x_{45} \leq 1.$$

That is, the tour cannot proceed from node 5 to node 4 and back to node 5. Probably, all such subtours that occur should be forced out at once to increase the likelihood of achieving a continuous tour on the next solution of the linear program.

To eliminate the three-city tour requires two constraints.

$$x_{12} + x_{23} + x_{31} \leq 2$$

and

$$x_{13} + x_{32} + x_{21} \leq 2.$$

As you can see, larger problems may require many constraints and a number of repeated solutions of successively larger linear programs.

The traveling salesman problem could apply to a building inspector who must visit a number of construction sites on a given day. The ideas of the traveling salesman have been applied to the routing of a hydrographer who must read stream gauging station data at a number of sites. Similarly, a school bus routing might benefit from the ideas of the traveling salesman. Finally, a progressive party that proceeds from appetizer at one house to second course at a second house, and so on, until it reaches liqueurs and mints at the last house might be planned with the traveling salesman problem in mind.

CHAPTER SUMMARY

In this chapter, we described the multiple-step methodology for integer programming known as branch and bound. We indicated that branch and bound is utilized for resolving a problem–which had been solved by linear programming and had yielded fractional solutions–into integer variables. Three situations were investigated: (1) solution of a problem with all zero-one variables, (2) solution of a problem in which some of the variables are required to be zero-one (the mixed integer programming problem), and (3) solution of a problem in which only the integers (0, 1, 2, 3, ...) are admissible. Branch and bound was discussed because it is the best available procedure for integer programming, but it is, nonetheless, very labor intensive in that much computer time can be expended in the production of the optimal integer answer.

We also pointed out that cases exist in which branch and bound can be replaced by a simple enumeration of alternatives. An illustrative example of this situation was offered. And we discussed heuristic methodologies briefly to indicate how "good rules" can be used to approach good solutions to problems. The well-known knapsack problem was used as an example to illustrate the application of a heuristic methodology to an integer programming problem. We concluded discussion of solution methodologies with a consideration of multiobjective integer programming problems, noting that the weighting method of multiple objective programming is an excellent choice to preserve the integer friendly character of a particular problem.

We returned then to a further discussion of network problems, considering the trans-shipment problem for the distribution of goods on a network. This problem will always terminate with integer flows–as long as inputs and outputs are integers. We pointed out the wide applicability of transshipment concepts to practical problems of a civil or environmental engineering character. We also noted the strong theoretical/formulational relations of the shortest path problem, the transportation problem and the transshipment problem. Our discussion of integer programming network problems concluded with a formulation of the famous traveling salesman problem. We used that problem to illustrate further how logical relations can be built into integer programming problems.

EXERCISES

6.1. EPA'S CONSTRUCTION GRANT PROGRAM

Currently the U.S. Environmental Protection Agency funds community water quality improvement projects via a priority-ranking scheme using what has been termed "The Funding Line Approach." In this approach, each project is assigned a dimensionless priority number based on the water quality improvement it achieves. Projects with higher priority numbers are better than lower-priority-number projects. A cost to fund is associated with each project, and a total amount of money is available for all the projects chosen in a particular funding cycle. The procedure of funding is take the first-, second-, third-, and so on, priority projects in order, adding their costs as they are placed on the list, until the sum of costs of the projects chosen exceeds the amount of the budget. Then the last project added is removed from the list, and the remaining projects from the top of the priority list down to the funding line are those chosen for funding in that cycle.

This procedure may not yield the most cost-effective collection of projects. Probably you could increase the total priority points achieved within the budget limit by taking the individual project cost into account in the choice of which project to choose next for the list. Construct a method that does this, that maximizes priority points achieved subject to a budget limit. You will need the following parameters:

c_j = cost of the jth project;

p_j = priority points associated with the jth project;

B = budget available for this funding cycle; and

n = number of projects.

Your method should consist first of a zero-one programming formulation and second of a procedure to solve that formulation. Describe the procedure briefly.

6.2. WATER SUPPLY ALTERNATIVES

This problem is a simplification of a water supply planning problem faced by the adjacent cities of Beijing and Tianjin, China. These cities are experiencing a combination of growth in water demand and overdrafts of their groundwater supplies and so are reaching out for new supplies. Three fundamental types of alternatives are available: new reservoirs on rivers yet untapped, a diversion canal from the Yellow River, and water conservation/demand reduction. Demand reduction has a number of dimensions, but in the case of these cities, the largest reduction can come about by changing local crop irrigation to a trickle irrigation system. This problem has all three of these elements, but is obviously reduced in scope from the planning problem that the two cities face. Each alternative has a cost associated with it, and each has a volumetric yield in cubic kilometers per year. Even this system is made complicated, however, by the interaction of the components. The three components are shown below.

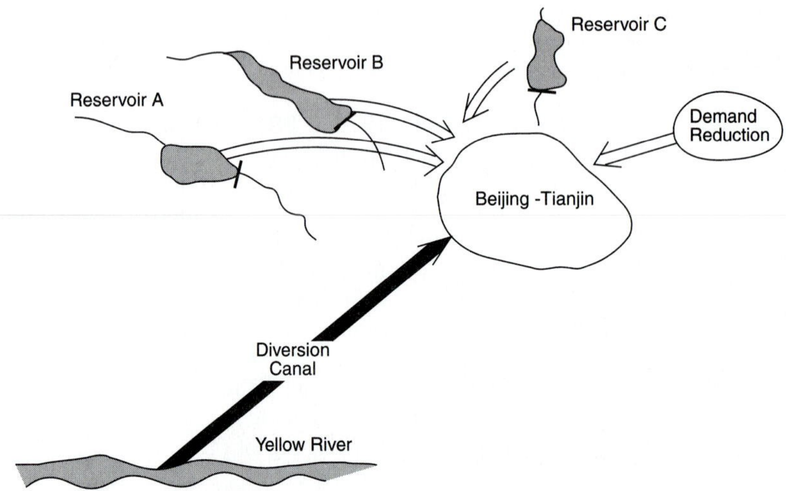

In the accompanying table, the water supply alternatives are numbered, and their cost and yield are listed. Interacting groups of the alternatives are also numbered, and their costs and yields are listed as well. Note that alternatives 6, 7, 8, and 9, the interacting groups, consist of combinations of reservoirs. The costs of these combinations are nothing but the sum of the costs of the components, but the yields are *not* the sum of yields. By methods of water resources systems, an engineer can calculate the firm yield of a pair or triplet of reservoirs, and this yield is typically larger than the sum of the individual yields, often by a good margin. Hence, the yield q_7, for instance, is greater than $q_1 + q_3$, the sum of the yields of the component alternatives.

Number	Alternative	Cost	Yield
1	Reservoir A	c_1	q_1
2	Reservoir B	c_2	q_2
3	Reservoir C	c_3	q_3
4	Diversion canal	c_4	q_4
5	Conservation/demand reduction	c_5	q_5
6	Reservoirs A and B	$c_6 = c_1 + c_2$	q_6
7	Reservoirs A and C	$c_7 = c_1 + c_3$	q_7
8	Reservoirs B and C	$c_8 = c_2 + c_3$	q_8
9	Reservoirs A, B, and C	$c_9 = c_1 + c_2 + c_3$	q_9

You are given a budget B that can be expended on all projects. What is the largest firm yield that can be obtained under this budget limitation? Structure a zero-one programming problem to show how to solve this problem.

Note that if reservoirs A and B are both built, it is equivalent to building alternative 6, and you must not count the individual yields, only the joint yield. Similarly, if A, B, and C are all built, the project/alternative 9 has been undertaken and the joint yield of 9 has been achieved, not the sum of individual yields. That is, you cannot undertake alternative 9 and any of alternatives 1, 2, or 3 at the same time. You are given a zero-one variable, x_j, to represent whether alternative j is undertaken or not undertaken.

6.3. UPGRADING A HIGHWAY SYSTEM IN A DEVELOPING COUNTRY

The highway system in many developing countries may be a tree-shaped structure only. By tree we mean a network with no loops, or one in which only one route exists between any two cities of the network. We begin by assuming that such a tree network connects the major cities in such a country. The highways are currently in very poor condition, but a modest deposit of oil has been discovered offshore and a highway upgrading is desired.

The Transport Ministry has been allotted B units of money for highway improvement and wishes to spend it in a way that connects the maximum population with the improved highway system, which radiates out from tho coastal city of P. The population is assumed to be concentrated at the junction points of the tree network. Each link j that is improved connects a new increment of population a_j to the system. Further, no link can be constructed unless the link that immediately precedes it enroute to the coastal city of P is also built.

Structure this problem as a zero-one programming problem with the objective of achieving the largest possible population on the continuous improved network subject to a budget constraint. By continuous, we mean that no gaps with improved highway on one link and unimproved highway on an adjacent link and then improved highway again are allowed. Use the diagram to guide your constraints. Let c_j = costs of link j, and a_j = the population connected when link j is built.

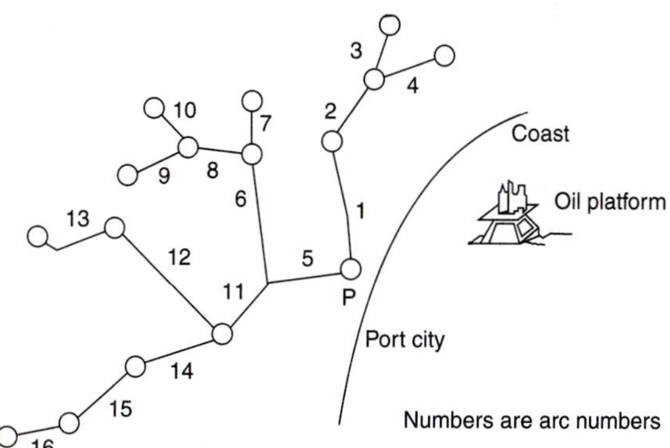

6.4. THE LEANEST EFFECTIVE EXPEDITION TEAM

An expedition is planned across the wilderness of Nepal to ascend Mount Everest via the treacherous rear ascent route. The expedition leader has identified a large number of compatible climbers, set J, all with the endurance to survive the trip and conquer the mountain but has not yet selected those that would be asked. This is because the leader has also listed a set of skills, I, which must be present for the expedition to be successful. For purposes of logistics, the expedition must be kept as lean as possible; hence the selection must lead to the smallest number of climbers. Structure a zero-one program to pinpoint climbers for selection in such a way that all skills are included somewhere in the expedition team and the team itself has the fewest total members.

Let: N_i = {j that possess skill i}, and

x_j = 1, 0; it is 1 if climber j is selected for the expedition and 0 otherwise.

6.5. A DELIVERY PROBLEM

The local distributor of Blahs Beer has set up a warehouse in an industrial park on the south side of Sudsville. The distributor is interested in increasing the effectiveness of the delivery process and has had the office staff identify all the demand points for Blahs, all the bars, restaurants, and liquor stores that are willing to distribute a product of such dubious value. In addition, the staff has been instructed to create on paper many feasible routes. Each route begins and ends at the warehouse and will fit a four-hour period, so that the trucker can load, drive, and distribute the entire contents of the truck and return to the warehouse in the time between check-in and lunch or between the end of lunch and quitting time. Each route involves a truck whose total volume of delivery does not exceed its carrying capacity. Also, when the delivery truck stops at one of the drop-off points, it fills the entire demand at that drop-off point. Each delivery route is assumed to have approximately the same cost as any other. The distributor calls the local university to see if he can get a systems engineering class to select the least number of delivery routes that will serve all the firm's customers. This is your task—structure the appropriate 0, 1 program for the distributor's problem.

Let: i, I = the index and set of demand points;

j, J = index and set of delivery routes; and

N_i = {the set of delivery routes j that include delivery drop-off point i}.

Suppose the optimal solution produced by the method you suggested included two routes which both had delivery to the same point. How would you tell the distributor to treat this result?

6.6. MATRIX CHARACTERISTICS OF INTEGER-GUARANTEED AND INTEGER-FRIENDLY FORMULATIONS

You have read about linear programming formulations that guarantee integer or all zero-one solutions, and you have read about formulations that "favor" integer or zero-one solutions. The latter are the formulations we called integer friendly. Look back at the formulations in these two categories and see if you can find features in common in them. Specifically, look at the coefficients (elements) of the constraint matrix and observe their values. Look as well at right-hand sides of the constraints and observe their values. (You are unlikely to find commonality in the number of elements in the columns.) What do the two kinds of problems have in common?

REFERENCES

CAMM, J., RATURI, and S. TSUBAKITANI, "Cutting Big M Down to Size," *Interfaces*, Vol. 20, no. 5, 1990, pp. 61–66.

DAKIN, R., "A Tree Search Algorithm for Mixed Integer Programming Problems," *Computer Science Journal*, Vol. 8, 1965, pp. 280–285.

DANTZIG, G., *Linear Programming and Extensions*. Princeton, NJ: Princeton University Press, 1963, 632 pages.

GLOVER, F., In *Handbook of Operations Research*, Moder, J., and S. Elmaghraby, eds. New York: Van Nostrand Reinhold, 1978.

HADLEY, G., *Linear Programming*, Reading, MA: Addison-Wesley, 1962, 520 pages.

LAND, A. and A. DOIG, "An Automatic Method of Solving Discrete Programming Problems," *Econometrica*, Vol. 28, 1960, pp. 497–520.

WILLIAMS, H., *Model Building in Mathematical Programming*, 2nd ed. New York: John Wiley and Sons, 1985, 349 pages.

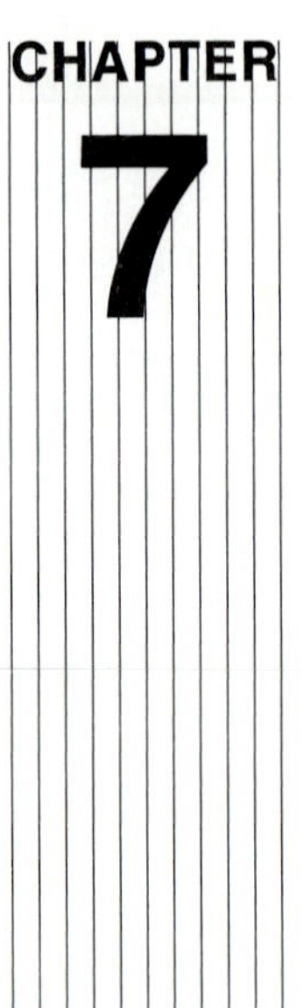

Scheduling Models: Critical Path Method

7.A INTRODUCTION

Construction and design projects involve a wide variety of individual activities, each making use of particular labor and equipment categories. The activities must be identified, sequenced in time, and coordinated with respect to resource requirements so that the project proceeds without any unnecessary delay.

The *critical path method (CPM)* is a family of analysis techniques aimed at scheduling activities on a project. Large construction projects are typically analyzed using the critical path method both before and after contracts are awarded, and CPM is required on most public projects. Unlike the techniques described in previous chapters, CPM is not usually structured as a mathematical programming problem, although it can be. Rather, a project is depicted in a series of visual displays that help the engineer to coordinate the design and/or construction activities.

Project activities are sequenced and displayed in either an *arrow diagram* or *precedence diagram*, while the economics of each project activity is summarized in its *cost-duration curve*. Using this information, a time table, or *activity schedule*, can be derived. From the activity schedule, another graphic, a *bar chart*, or *Gantt chart*, is developed (usually with alternative forms of presentation), which can be used to conduct *resource leveling*, and/or *project compression*. The bar chart can also be used to derive *progress curves* for the project, which are useful in financial analysis and project-control. Figure 7.1 is a flow chart of the major topics included in the critical path method. Each is discussed in this chapter. Additional treatment can be found in Antill and Woodhead (1990), Halpin and Woodhead (1980), Harris (1978), and O'Brien (1971).

7.B ARROW DIAGRAM

7.B.1 History

In late 1956, the E. I. du Pont de Nemours & Co. initiated a study with the Remington Rand UNIVAC division of Sperry Rand Corporation to determine whether recently developed computers could help in managing engineering projects. By mid-1957, the essential theory of the critical path method had been developed, and by March 1959, the method had been successfully applied to three construction projects, including a new chemical plant facility and the shutdown, maintenance, and overhaul of chemical processing units (Kelley and Walker, 1959). In the latter case, a reduction in the average shutdown time of over 25% was enough to repay the development cost of CPM many times over. The method received national publicity and was rapidly introduced in the construction industry (Davis, 1974).

7.B.2 Arrow Format

An *arrow diagram* can be drawn to represent the flow of activities on a design or construction project. Each arrow in the diagram represents a specific activity, and each node represents an event, or point in time. Figure 7.2a indicates a single activity, which can be denoted either "activity A" or "activity 1-2." Node 1 represents the event, or point in time, when activity A begins, and node 2 represents

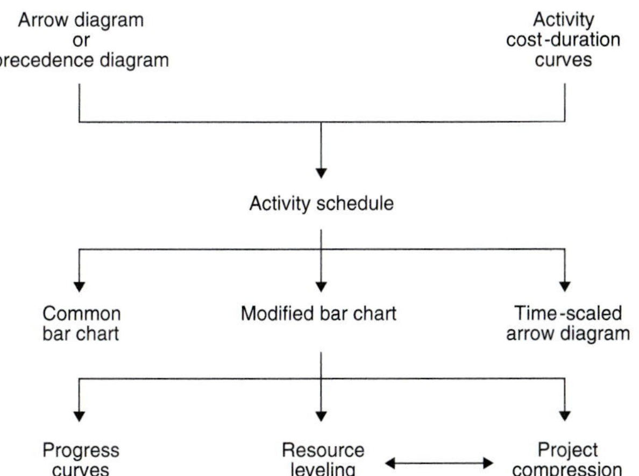

Figure 7.1 Common stages in the critical path method.

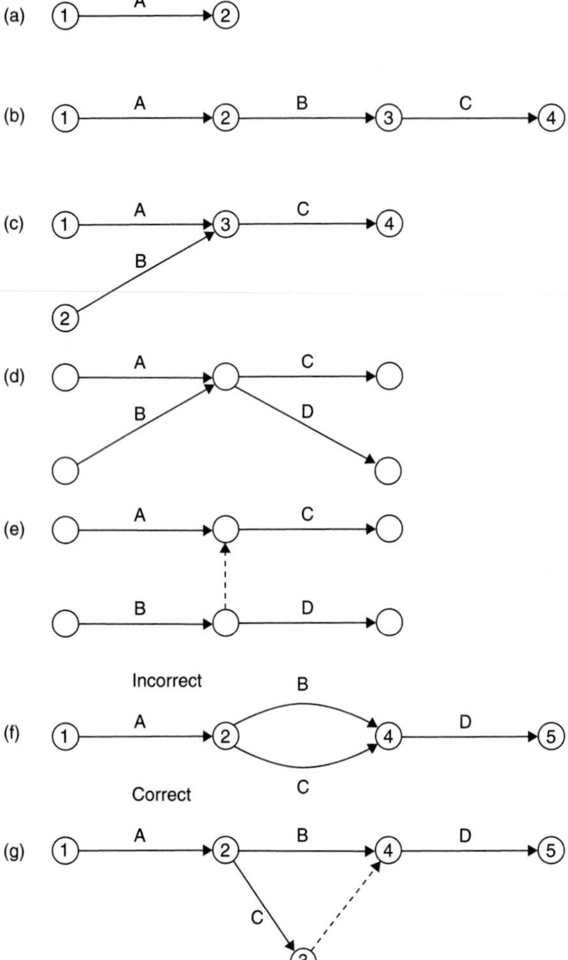

Figure 7.2 Examples of segments of an arrow diagram.

the event, or point in time, when activity A is completed. Figure 7.2b represents a chain of activities which is interpreted to mean that "activity A must precede activity B, and activity B must precede activity C." Figure 7.2c, on the other hand, represents the statement that "both activities A and B must precede activity C." This means that both activities A and B must be *finished* before activity C can begin. Node 3 represents the event, or point in time, when activities A and B are *both* complete, and activity C can begin.

Use of Dummy Activities. There are occasions when a dotted arrow, or "dummy activity," is required in an arrow diagram. For example, consider the statements activity A precedes activity C;

activity B precedes activity C; and

activity B precedes activity D.

Verify that Figure 7.2d does *not* capture the full meaning of the above statements, since it implies that "activity A precedes activity D," which is not a condition. To avoid imposing a condition that does not exist, Figure 7.2e is required. The dummy activity ties activity B into activity C ("activity B precedes activity C"), but avoids saying that activity A precedes activity D.

Use of dummy activities is also required in order to avoid ambiguities in denoting activities. Consider the statements

activity A precedes activity B, and B precedes D, and

activity A precedes activity C, and C precedes D.

Figure 7.2f is an incorrect representation of the above statements. In addition to causing nonlinear "arrows," an ambiguity arises when speaking of "activity 2-4." To avoid this, a dummy activity is added, as in Figure 7.2g. Each activity now has its own unique name, and the original sequence of dependencies is maintained.

One Starting Point, One Ending Point. Convention requires that an arrow diagram have one starting point (node) and one ending point (node). In addition to adding clarity to the diagram, computation of the activity schedule (to be discussed later) assumes this format. Figure 7.3 shows the conversion of an unfinished arrow diagram to one with one starting point and one ending point.

Diagram Sparsity and Flow. An arrow diagram should not have any unnecessary arrows, nodes, or dummy activities. Also, arrows should flow to the right as much as possible. The conversion in Figure 7.3 shows these characteristics.

Node Numbering. Convention also requires that the nodes be numbered so that in the (i, j) node notation for activities, the j-index is always larger than the i-index. Figure 7.3 illustrates this format. A convenient method to achieve this for any arrow diagram is the following (Fulkerson, 1962):

1. Number the origin node 1, and erase all arrows eminating from this node.
2. Find all nodes that have only outward-pointing arrows (no arrows coming into the node).
3. Number all such nodes consecutively (in any order) then erase all arrows eminating from these nodes.
4. Repeat steps 2 and 3 until the final node (end of project) is numbered.

Verify this procedure for the diagram in Figure 7.3. Only one alternative numbering scheme is possible in the diagram (the numbering of nodes 2 and 3 can be

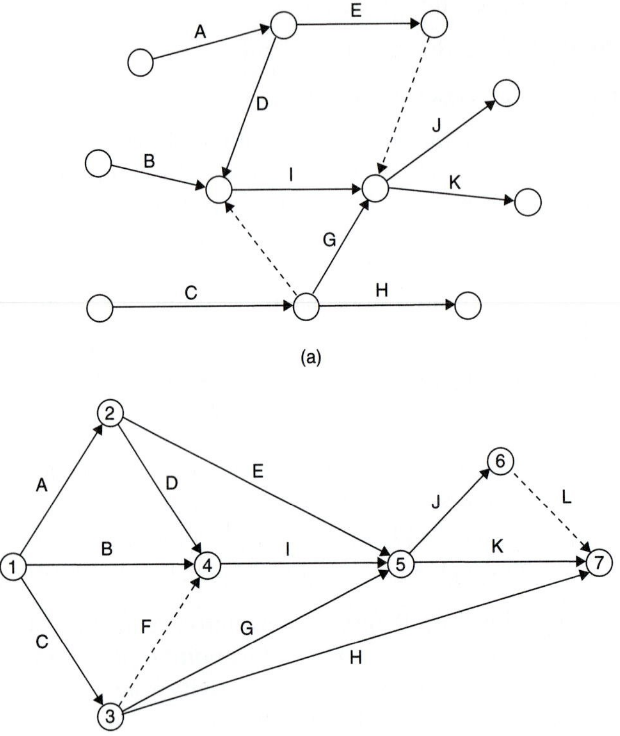

Figure 7.3 Conversion of an unfinished arrow diagram (a) to final form (b).

reversed). Essentially the same approach can be used to label the arrows themselves so that lower-letter activities flow into higher-letter activities. This is also shown in Figure 7.3.

Example 7-1

Arrow Diagram for Foundation Construction. As part of a larger arrow diagram, a contractor wants to include the following activities related to construction of a foundation for a light commercial building:

place steel sideforms;

pour foundation;

excavate for foundation;

order, wait, and receive steel sideforms;

order, wait, and receive steel rebars; and

place steel rebars.

It is assumed that a sufficient number of employees are available to simultaneously order sideforms, order rebars, and excavate for the foundation. Sufficient labor

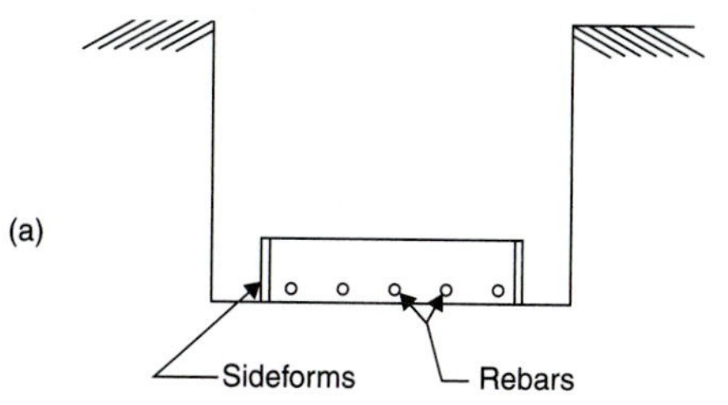

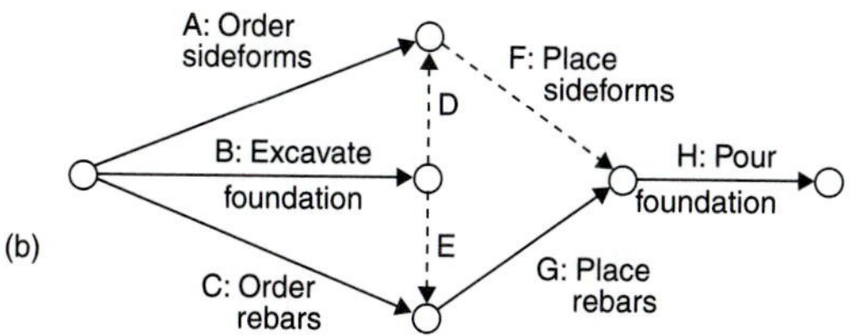

Figure 7.4 Cross-sectional view of foundation (a) and corresponding arrow diagram (b)

is also available to simultaneously place sideforms and rebars. Pouring of the foundation is not to be done in stages. Figure 7.4 illustrates the cross-sectional view of the foundation and the corresponding arrow diagram. Note the use of dummy activities to clarify physical dependencies. The activities should now be lettered as shown in the diagram.

7.B.3 Cost-Duration Curves

To minimize the cost of completing a design or construction project, each activity must be done as economically as possible. Through experience, project managers come to know what type and combination of labor skills and equipment results in the least cost to complete an activity. The time needed to complete the activity at least cost is the *normal time* for the activity, expressed in work-units (work-days, etc.). However, considering the project as a whole, it is sometimes necessary to complete an activity in a time shorter than its normal time. Figure 7.5 shows the

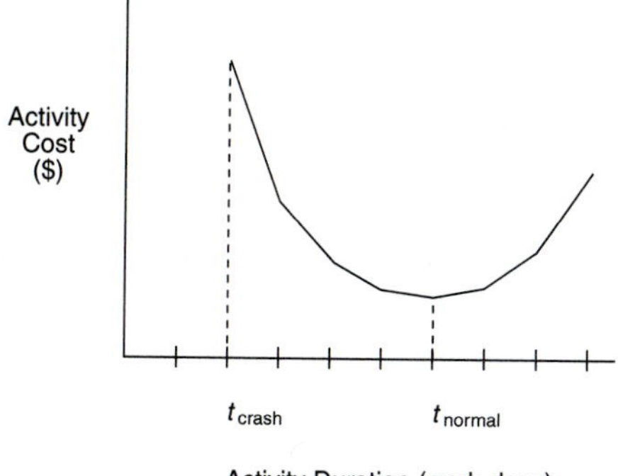

Figure 7.5 Activity cost-duration curve.

cost-duration curve for an activity. The cost is a minimum at the normal time, and increases as the activity is shortened, or *compressed* in time. Compression can be achieved in various ways: (1) with the same crew size by using overtime, (2) increasing the crew size by outside hiring, (3) buying or renting more productive equipment, (4) increasing the quality and/or level of supervision, or (5) other project-specific means. All these approaches directly increase cost, since it is assumed that the least cost way of completing the activity has already been determined. The cost of compression is also increased through more frequent accidents, inefficiencies caused by crowding in the work space, more frequent errors that must be corrected, and decreased quality of the design or constructed product, which may influence future marketing success. Figure 7.5 shows that a point is reached eventually where further time compression becomes impossible, either physically or economically. This shortest completion time is called the *crash time* for the activity.

It is common to assume a convex cost-duration curve and to approximate the smooth curve with a series of straight-line segments, as shown. Cost also increases to the right of the normal time, due to higher direct labor and equipment costs when employed for a longer period of time, lost interest due to delayed payments for work completed, lost bonus payments for early project completion, and any penalties for late project completion.

Each activity in an arrow diagram can be labeled with its normal completion time. These times, along with the logic of the arrow diagram, determine the total duration of the project.

7.B.4 Critical Path

Figure 7.6 shows an arrow diagram using node numbering to designate activities. In the middle of each arrow is listed the normal time for the activity (ignore all other numbers for now). How long does the total project take to complete? The project is not complete until all activities are complete, and an activity cannot begin until all activities leading up to it are completed. Simple inspection and enumeration of the seven possible paths from the starting node to the ending node results in finding the *longest path* from start to finish: 1-3, 3-4, 4-6, 6-9, 9-10. Its length is 25 work-days. The longest path (sequence of jobs) from start to finish in an arrow diagram is called the *critical path* and is denoted by cross-hatched lines on the *critical path activities*. The length of the critical path is the project duration, also called the *critical time*, t_{cr}.

In large arrow diagrams, the critical path is not always obvious from simple inspection. In addition, a project manager is interested in many other scheduling questions that can only be answered by analyzing the arrow diagram more systematically.

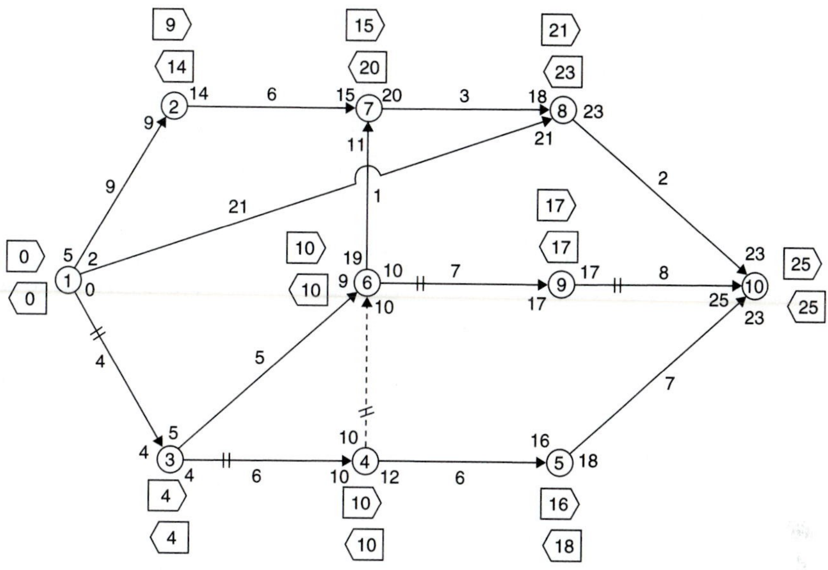

Figure 7.6 Arrow diagram showing foreword pass/backward pass calculations

7.C ACTIVITY SCHEDULE

7.C.1 Basic Definitions

To carry out detailed project planning and control, the following definitions of time are needed:

d_{ij} = duration of activity i-j (normal time);

EST_{ij} = earliest time at which activity i-j can start;

EFT_{ij} = earliest time at which activity i-j can finish;

LST_{ij} = latest time at which activity i-j can start without causing an increase in the total project duration, and;

LFT_{ij} = latest time at which activity i-j can finish without causing an increase in the total project duration.

Clearly,

$$EFT_{ij} = EST_{ij} + d_{ij} \tag{7.1}$$

$$LFT_{ij} = LST_{ij} + d_{ij} \tag{7.2}$$

7.C.2 Forward Pass/Backward Pass

If the earliest and latest start times of each activity were known, equations (7.1) and (7.2) could be used to find the earliest and latest finish times. A common approach is to find the earliest start time on a *forward pass*, and the latest start time on a *backward pass* through the arrow diagram. Figure 7.6 shows these computations.

The method begins by assigning an earliest start time of "zero" to all activities emanating from the start node. This should be interpreted to mean "end of day zero" (or "beginning of day one"). The duration of each beginning activity is then added to its earliest start time of zero to give the earliest finish time, which becomes the earliest start time for the next activity in a single path. However, care must be taken at junction nodes such as 6, 7, 8, and 10; when two or more activities converge on a node, the *maximum* of their earliest finish times becomes the earliest start time for all activities leaving the node. This is the case since all activities converging on the node must be completed before any of the succeeding activities can begin. The longest path, therefore, controls the earliest start time for succeeding activities. The earliest finish time for each activity is recorded at the *tip* of its arrow, and the largest of these, the controlling earliest start time for succeeding activities, is recorded in the forward-pointing box beside each node. Note that computations cannot proceed until all paths into a node have been evaluated.

Proceeding through the diagram, the final node is reached. The maximum earliest finish time of activities converging on the terminal node is "end of day 25." The project takes 25 work-days to complete ($t_{cr} = 25$) and the critical path can be traced. The backward pass can now begin.

The project length of 25 days can be used to calculate the *latest start time* of activities leading into the terminal node, as $LST_{ij} = 25 - d_{ij}$. The LST of preceding activities can be found in the same way by recognizing that they must not delay the LST of succeeding activities. If activity *i-j* precedes activity *j-k*, then

$$LST_{ij} = LST_{jk} - d_{ij} \tag{7.3}$$

The LST of each activity is recorded at the *base* of its arrow. In tracing backward in the diagram, care must be taken at juncture nodes such as 6, 4, 3, and 1; when two or more arrows originate at a node, the controlling LST that must be used to calculate the LST of preceding activities is the *minimum* of the latest start times recorded at the base of the arrows originating at the node. This is the case since, at the latest, preceding activities must start in time to be finished at the same time as the smallest of the latest start times of succeeding activities, or the total project will be delayed. This "controlling latest start time" is recorded in the backward-pointing box beside each node. It is important to note, however, that this controlling latest start time is *not* the LST of all activities leading out of the node; the actual LST of each activity is recorded at the base of its arrow. This is in contrast to the controlling EST values found on the forward pass; in that case, the control-

TABLE 7.1 ACTIVITY SCHEDULE FOR ARROW DIAGRAM OF FIGURE 7.6

Act.	Crit?	d_{ij}	EST	EFT	LST	LFT	TF	IF	FF	IndF	SF
1-2		9	0	9	5	14	5	5	0	0	5
1-3	Yes	4	0	4	0	4	0	0	0	0	0
1-8		21	0	21	2	23	2	2	0	0	2
2-7		6	9	15	14	20	5	5	0	0	0
3-4	Yes	6	4	10	4	10	0	0	0	0	0
3-6		5	4	9	5	10	1	0	1	1	1
4-5		6	10	16	12	18	2	2	0	0	2
4-6	Yes	0	10	10	10	10	0	0	0	0	0
5-10		7	16	23	18	25	2	0	2	0	0
6-7		1	10	11	19	20	9	5	4	4	9
6-9	Yes	7	10	17	10	17	0	0	0	0	0
7-8		3	15	18	20	23	5	2	3	0	0
8-10		2	21	23	23	25	2	0	2	0	0
9-10	Yes	8	17	25	17	25	0	0	0	0	0

EST = Earliest Start Time
EFT = Earliest Finish Time
LST = Latest Start Time
LFT = Latest Finish Time
TF = Total Float
IF = Interfering Float
FF = Free Float
IndF = Independent Float
SF = Safety Float

ling *EST* applied to all activities emanating from the node. As a check on the calculations, when the initial node is reached, the controlling *LST* should be zero.

The *EST* and *LST* values so calculated for each activity can be transferred to a tabular listing, the activity schedule.

7.C.3 Activity Schedule Format

Table 7.1 shows the activity schedule for the arrow diagram of Figure 7.6. A common format is to list each activity by its *i-j* order, first on the *i*-index and then, within that order, on the *j*-index. Basic to the schedule are the *EST* and *LST* values of each activity, which are transferred from the arrow diagram calculations. The *EFT* of each activity is also available from the arrow diagram, or can be calculated simply by adding the activity duration to the *EST*. Similarly, adding activity duration to the *LST* produces the *LFT*.

7.C.4 Categories of Float Time

Float time represents a safety factor relative to completing a project on time. Float permits flexibility in scheduling activities, thereby controlling the timing and use of labor and equipment, and serves as a buffer against such unforeseen circumstances as inclement weather, labor difficulties, equipment breakdowns, and inaccuracies in activity work quantities or production rates, all of which may lengthen the time to finish an activity. In scheduling activities, project managers have found it useful to define various categories of float time.

Total Float. The most common measure of float time is total float, defined as

TF_{ij} = total float of activity *i-j*; the maximum amount that activity *i-j* can be delayed without causing an increase in the total project duration, assuming that predecessor activities have started as early as possible.

Total float can be calculated directly in the activity schedule as

$$TF_{ij} = LST_{ij} - EST_{ij} \tag{7.4}$$

$$= LFT_{ij} - EFT_{ij} \tag{7.5}$$

Equation (7.4) is recommended, since primary calculations on the arrow diagram find the *EST* and *LST*. Table 7.1 indicates that all noncritical activities have total float.

Free Float, Interfering Float. Total float is an important measure of activity scheduling flexibility. However, delaying an activity may consume the float time of succeeding activities. If so, the float of the activity "interferes" with the float of succeeding activities. The amount of interference is called the *interfering float.* Conversely, in some cases an activity can be delayed without interfering with the float of succeeding activities. The degree to which this can be done is called the *free float* of the activity, defined as

FF_{ij} = free float of activity *i-j*; the maximum amount that the activity can be delayed without delaying the earliest start time of any succeeding activity, assuming that predecessor activities have been started as early as possible.

Free float is best calculated using the arrow diagram and the relationship

$$FF_{ij} = EST_{jk} - EFT_{ij} \tag{7.6}$$

where activity *i-j* is assumed to immediately precede activity *j-k*. The last activity in a chain of activities with total float will have free float equal to its total float if it converges into the critical path (activities 3-6, 5-10, and 8-10 in Figure 7.6). Also, free float exists for activities having higher total float values than the activities into which they converge (activities 6-7 and 7-8 in Figure 7.6). The amount of free float in both these cases can be found using the *EST* of succeeding activities and the *EFT* of the converging activity. These values have been recorded on the arrow diagram.

Conceptually, free float is the first to be consumed when an activity is delayed; any further delay is interfering float (e.g., activity 7-8 can consume 3 days of free float before using 2 days of interfering float from its total float of 5 days). The free float of each activity is recorded in the activity schedule. Interfering float is the difference between total float and free float

$$IF_{ij} = TF_{ij} - FF_{ij} \tag{7.7}$$

and is recorded next to total float in Table 7.1.

Finally, it must be emphasized that for free float and total float to exist to their fullest, previous activities must not have consumed more than their free float, since any use of interfering float consumes an equivalent amount of the free float and total float of succeeding activities (e.g., in Figure 7.6, activity 7-8 cannot be delayed more than 3 days, or it begins to consume the free float and total float of activity 8-10).

To summarize, a free-float activity tends to be the last activity in a chain of activities that converges either into the critical path or into a noncritical path that has less total float at the juncture than does the activity itself. The calculation of free float can be done directly on the arrow diagram. Interfering float is the difference between total float and free float.

Independent Float. To determine the effect of delays in predecessor activities on the float of a successor activity, *independent float* has been defined:

$IndF_{ij}$ = independent float of activity *i-j*; the maximum amount that activity *i-j* can be delayed without delaying the earliest start time of successor activities, assuming that predecessor activities have been completed as late as possible.

Assuming the sequence of activities *h-i*, *i-j*, and *j-k*, independent float can be calculated as (Harris, 1978)

$$IndF_{ij} = EST_{jk} - (LFT_{hi} + d_{ij}) \tag{7.8}$$

where zero is recorded if $IndF_{ij}$ is less than zero. The *LFT* values are in the activity schedule, while the rest of the data for the calculation have been recorded on the arrow diagram.

For example, the calculation for activity 6-7 is

$$IndF_{6-7} = 15 - (10 + 1) = 4$$

and for activity 7-8 is

$$IndF_{7-8} = 21 - (20 + 3) = -2 \rightarrow 0$$

Note that independent float is always less than or equal to free float, so that if free float is zero, the calculation for independent float need not be made.

Independent float activities tend to be associated with activities emanating from the critical path (including the origin), and the amount indicates the degree to which the activity is "independent" of late starts by preceding activities.

Safety Float. A measure of float that is a bit less stringent than independent float is *safety float*, defined as

SF_{ij} = safety float of activity *i-j*; the maximum amount that activity *i-j* can be delayed without causing an increase in the total project duration, assuming that predecessor activities have been completed as late as possible (Thomas, 1969).

Assuming the sequence of activities *h-i*, *i-j*, safety float can be calculated as

$$SF_{ij} = LST_{ij} - LFT_{hi} \tag{7.9}$$

Values for both LST and LFT must be found from the activity schedule. For example, for activity 6-7,

$$SF_{6-7} = 19 - 10 = 9$$

and for activity 7-8,

$$SF_{7-8} = 20 - 20 = 0$$

Activities with safety float tend to be those that emanate from the critical path (including the origin).

Relationship Between Float Times. It is useful to keep in mind the general relationship between the relative number of activities that have float time in different categories. Activities with total float are, by far, the most numerous (nine activities in Figure 7.6), usually followed by those with interfering float (six activities). Those with free float and safety float are usually the next most common (five each). Activities with independent float, being a subset of the activities with free float, are rather rare (only two). The relationships vary somewhat, of course, depending upon the nature of the arrow diagram.

Table 7.2 summarizes definitions of the four basic float times based on predecessor and successor start times.

TABLE 7.2 CLASSIFICATION OF FOUR FLOAT TIMES

		Successors completed	
		Early	Late
Predecessors completed	Early	Free Float	Total Float
	Late	Independent Float	Safety Float

Source: Thomas (1969).

7.D BAR CHART

7.D.1 Common Bar Chart

The activity schedule contains all the essential information concerning timing of activities. If used alone, however, it would be rather difficult to picture the flow of

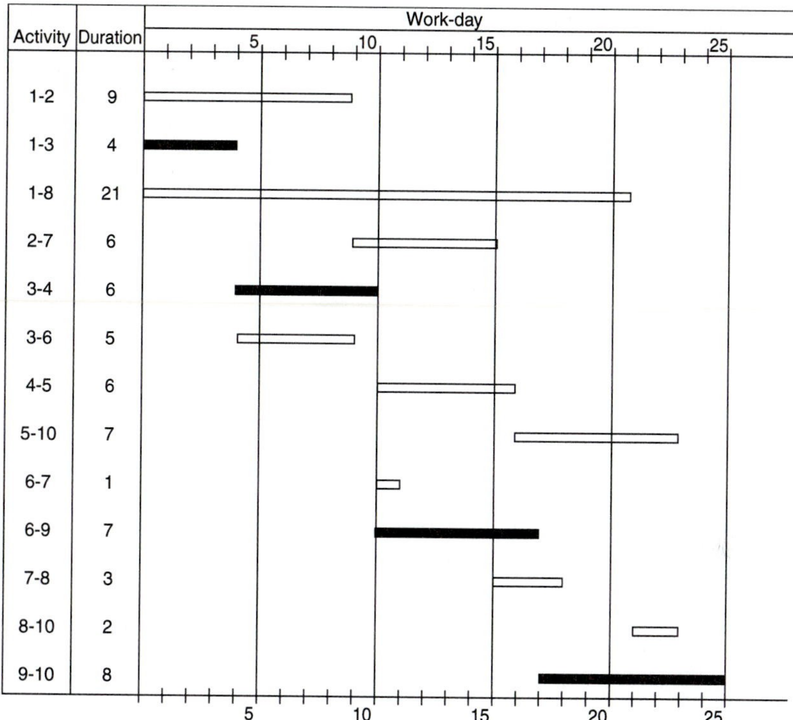

Figure 7.7 Common bar chart: node-notation ordering.

activities through time, since the table format is rather difficult to read. As a visual aid, the *common bar chart*, also called the *Gantt chart*, excels. Figure 7.7 shows the common bar chart for the arrow diagram of Figure 7.6, using information from the activity schedule of Table 7.1. Activities are listed in the same node-notation order as in the activity schedule, and each activity is assumed to start at its *EST* and to finish at its *EFT*. A bar drawn between these two times represents when each activity is planned to take place. For example, on work-day 16, activities 1-8, 4-5, 6-9, and 7-8 are expected to be active. It is common to denote critical path activities with a shaded bar, to draw attention to keeping these activities on schedule.

Figure 7.8 shows the same bar chart, but with activities listed first in order of their *EST*, then within this order, in order of their *EFT*. The resulting bar chart is even easier to read than that of Figure 7.7, in terms of the flow of activities through time. Either format, and others, may be used.

The shortcoming of the common bar chart is that it does not provide enough information on which to modify the scheduling of activities by utilizing float times. The common bar chart has been extended to make it more useful in project planning and control.

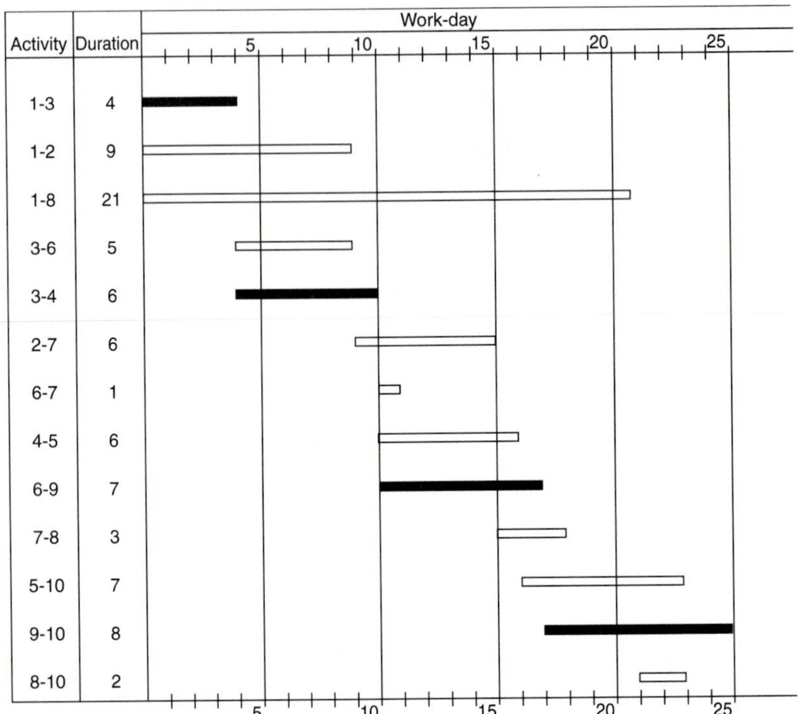

Figure 7.8 Common bar chart: EST-EFT ordering.

7.D.2 Modified Bar Chart

Figure 7.9 displays the *modified bar chart* corresponding to the arrow diagram of Figure 7.6 and activity schedule in Table 7.1. Node-notation ordering of the activities is recommended. Just as with the common bar chart, it is assumed that activities will start at their *EST* and finish at their *EFT*. This is shown by the short line in the pair of lines opposite each activity. However, the fact that the activity could be delayed, started, and completed anywhere between its *EST* and *LFT* is indicated by the long line in the pair of lines. The distance between the end of the short line and the end of the long line is the *total float* for the activity.

Vertical lines are drawn to show activity precedence relationships. Most vertical lines indicate that the *EST* of activities lower in the diagram depend on the *EFT* of an activity above and to the left (preceding them). Activities on the critical path are joined in this manner, as are 1-2, 2-7, and 7-8; 1-3 and 3-6; 1-8 and 8-10; 3-4 and 4-5; 3-4 and 6-7; and 4-5 and 5-10. All activities will have a vertical line determining their *EST*. As long as the node-notation ordering system is used to list activities, these vertical lines will give the modified bar chart a "stepped-down" appearance.

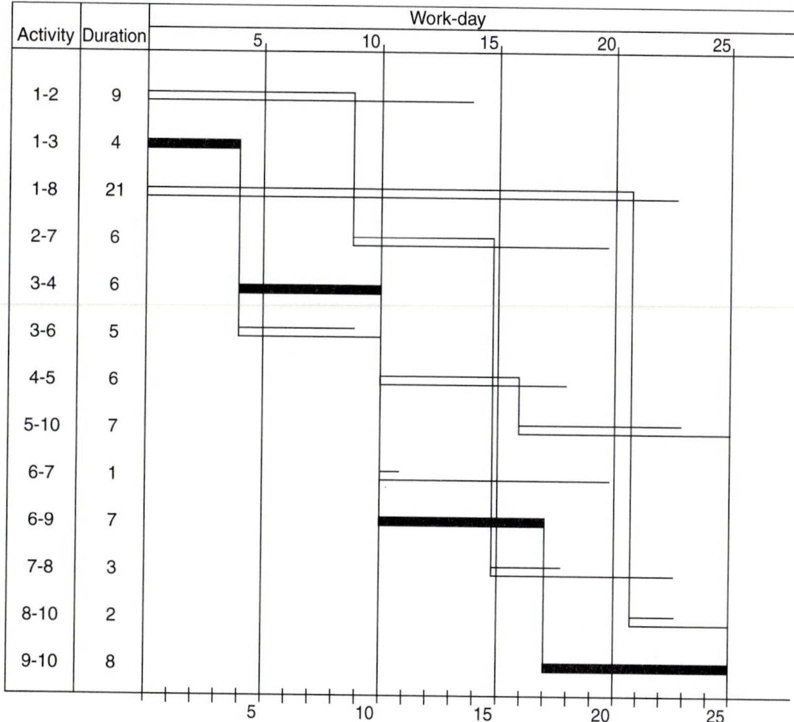

Figure 7.9 Modified bar chart.

Occasionally, a vertical line will appear at the end of the long line in the pair of lines for an activity instead of at the end of the short line. In this case the *LFT* of the activity is controlled by the *LST* (and *EST*) of a successor critical activity. Activities 3-6, 5-10, and 8-10 fall in this category. In a sense, the dependency is "upward" in the diagram, from the critical activity to the activity in question.

Finally, some activities will be joined to others by only one vertical line, not two. Activities 6-7 and 7-8 are typical. Neither activity's *EFT* value determines the *EST* of the succeeding activity, and neither intersects the critical path, so the *LFT* value is not set by the *LST* (*EST*) of a critical activity. Rather, the *LFT* is determined by the *LST* of the immediately succeeding noncritical activity. The *LST* of such an activity does not appear on the bar chart, however, so no vertical line can be drawn. The apparent "decoupling" of dependent activities in such cases is a shortcoming of the modified bar chart that must be kept in mind when making schedule changes.

The modified bar chart is an excellent visual display of much of the information found in the arrow diagram and activity schedule. In particular, total float and most activity dependencies are clearly indicated, allowing activities to be rescheduled should the need arise. Also, drawing vertical lines for dependencies serves as a check on the calculations made to obtain the activity schedule. The activities shown in the bar chart should flow together vertically just as indicated in the arrow diagram.

7.E RESOURCE LEVELING

7.E.1 General Problem

Resources are required to complete every activity in a design or construction project, and multiple activities take place simultaneously. The project manager must be concerned that sufficient resources are available at all times. Personnel and equipment are usually of greatest concern, but other resources must also be considered. Storage space can become a limiting factor if material arrival is not staged properly, and large expenditures during short periods of time may cause a significant cash flow problem for the contractor. Additionally, supervisory personnel may be in short supply, making it desirable to avoid a large number of employees at one time. The activity schedule and bar chart derived from the arrow diagram ignore possible resource limitations, and sometimes result in large fluctuations in required resources. Every project must be examined to see whether a leveling of resources is needed to comply with resource limitations.

Figure 7.10 derives the resources needed on each work-day to complete the project portrayed in the arrow diagram of Figure 7.6. The number of personnel required to complete each activity is listed beside the activity. Assuming that each activity starts at its *EST* and finishes at its *EFT*, the total resources needed on a given day are obtained by summing the resources associated with each activity under way on that day. The totals are listed at the bottom of the modified bar chart. It can be seen that there is a high demand for personnel on days 10 and 11 of the project. If the contractor had a labor force of only 14 people, the project would not be feasible as scheduled. A revised schedule must be found that reduces the number of required people on any day to 14 or fewer.

A common additional goal is to reduce the day-to-day fluctuation in required labor, that is, to "level" the resource requirements on the project. For example, on day 16, it would be beneficial to raise the resource requirement from 10 people to something much closer to 14, the available labor force. As a measure of fluctuation, the sum of squared deviations from a desired level is often used. A single, large deviation is therefore penalized much more than are several small deviations. If the desired level in Figure 7.10 is taken to be 14, the deviation on day 16 incurs a penalty of $(14 - 10)^2 = 4^2 = 16$.

There are practical reasons why leveling of personnel is desirable. There are costs associated with a fluctuating labor force: (1) new arrivals on the project are not as familiar with site layout and procedures, and therefore are not initially as productive as those continuously employed; (2) temporary labor may be quite a bit more expensive than permanent employees; and (3) the permanent labor force understandably prefers steady employment rather than frequent layoffs.

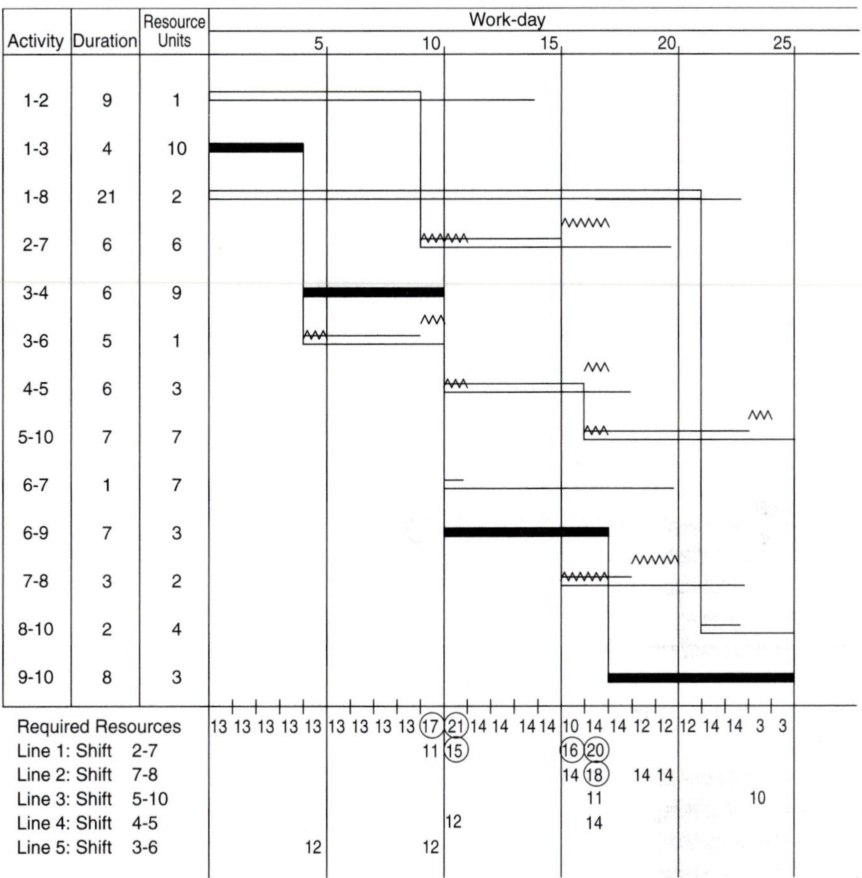

Figure 7.10 Scheduling adjustments for resource leveling.

7.E.2 Heuristic Procedure

In rescheduling activities to avoid the resource violation and to level resources, many rules-of-thumb have been developed to guide the process toward acceptable final results (Harris, 1990). There is no one best procedure, since the degree of success of any heuristic approach will depend on the characteristics of the individual project. What works well in one case may not be as successful in another. Most of the heuristic approaches are very detailed. Nevertheless, to illustrate some common considerations in resource leveling, the following generalized rules are given:

1. Derive the modified bar chart for the project, sum resources for each work-day, identify resource violations, and sum the squares of resource deviations for each resource category. Rank the resources in order of priority.

2. For the most severe resource violation period for the currently highest-priority resource, identify activities using the resource that have free float. Rank these activities giving highest priority to those activities with the greatest free float and next highest priority to those with the greatest total float.
3. Shift the fewest possible high-priority activities to the right (beyond their *EST*) only as far as required to remove the resource violation and to not cause another resource violation. Revise the resource requirements accordingly, identify the currently greatest resource violation, and return to step 2. If none, select the next-highest-priority resource with a resource violation and return to step 2. If all resource violations have been removed for all resources, repeat steps 2 and 3 for squared resource deviations for the highest-priority resource until no further improvement is possible. Consider the next highest-priority resource by returning to step 2. If none, stop. At any point, if resource violations cannot be removed or if further improvement in minimizing squared resource deviations is desired, consider activity interruption in step 4.
4. For the greatest constraint violation period (or greatest squared deviation), rank the interruptible activities in terms of float (as in step 2) and resource use, and interrupt the fewest possible high-priority activities that will remove the resource violation (or reduce the squared deviation), shifting a portion of each activity to the right only as far as necessary. Repeat with the next greatest resource violation (or squared deviation) until all are removed (or resources are leveled as much as possible). Consider the next-highest-priority resource by returning to step 2. If none, stop. If violations remain, or further improvement in minimizing squared resource deviations is desired, consider activity compression in step 5.
5. For the greatest constraint violation period (or squared deviation), determine the least cost means of compressing and possibly interrupting those activities that utilize the resource so as to remove the resource violation (or minimize the squared deviation). Consider the next-highest-priority resource by returning to step 2. If none, stop.

In summary, the strategy is to first shift resources within their free float, and then shift within their interfering float. If this does not solve the problem, activities can be interrupted and, as a last resort, compressed in time. The procedure is carried out for the highest-priority resource and then repeated for subsequent resources. It is emphasized, however, that the method is only a guide—modifications can be made to take advantage of problem circumstances in manual computations.

The procedure is applied in Figure 7.10, where wavy lines indicate schedule modifications. The worst resource violation is on day 11, and the next worst is on day 10; consider handling these together. The only activity under way on both days that can be shifted is 2-7, and it uses 6 units of the resource: shift two days and revise the resource requirements as shown (line 1). Shifting 2-7 causes activity

7-8 to shift also, as indicated by the resource changes of line 2. Day 17 now has the greatest resource violation (18 units, as circled). The only activity on that day that can be shifted is 5-10, so it is moved one day to the right (line 3). The largest resource violation is now on day 11 (15 units). Activity 4-5 or 6-7 can be shifted, but while 6-7 has the greatest free float, its shift would cause too great a change in resource requirements. Shifting 4-5 not only solves the problem on day 11, but also serves to level the resources needed on day 17, as shown (line 4). All resource violations are now removed, and leveling of resources can be taken as the primary objective. The greatest deviation of resources (except for the last two days of the project) occurs on day 10 (11 units). By shifting activity 3-6 within its free float of one day, the sum of squared deviations can be reduced. At this point, inspection indicates that no further resource leveling is possible, if allowance is made for a phase-out of resources on the final two days of the project (alternatively, activities 1-8 and 8-10 could be shifted two days to achieve greater leveling in the final two days).

Note the heuristic nature of the above moves. In particular, in the first step it was decided to handle both days 10 and 11 simultaneously. If the rules had been adhered to exactly, only day 11 would have been considered and activity 6-7, with a large amount of free float, would have been shifted (to day 16). However, since activity 6-7 carries a large amount of the resource (7 units), this would have led (in just two steps) to an unsolvable resource violation on day 16 (for practice, verify this).

On an actual project, many other considerations also would be taken into account when rescheduling. The above procedure illustrates many of the activity interactions, however.

7.E.3 Effect on Float

Figure 7.11 shows the resulting resource-constrained, modified bar chart. When the labor constraint of 14 people is considered, three more activities become critical. They cannot be delayed or the resource constraint would be violated. Also, the total float is generally reduced for other activities, as shown in the figure. Additionally, it is important to note that the activity chains are no longer independent in terms of using float. For example, activities 1-8 and 8-10 could be delayed two days without causing a resource violation. However, it would then be questionable whether activities 4-5 and 5-10 could be shifted one day to the right as is indicated in the bar chart. In this case, it is still possible to shift 4-5 and 5-10 by one day without causing a resource violation on the last day of the project. In another situation, it may not. Such dependencies therefore need to be considered separately in working with the bar chart. The resource-constrained modified bar chart can be looked upon as representing one-at-a-time activity chain shifts that are possible. When any change is actually made, the bar chart must be updated to reflect the current distribution of resources.

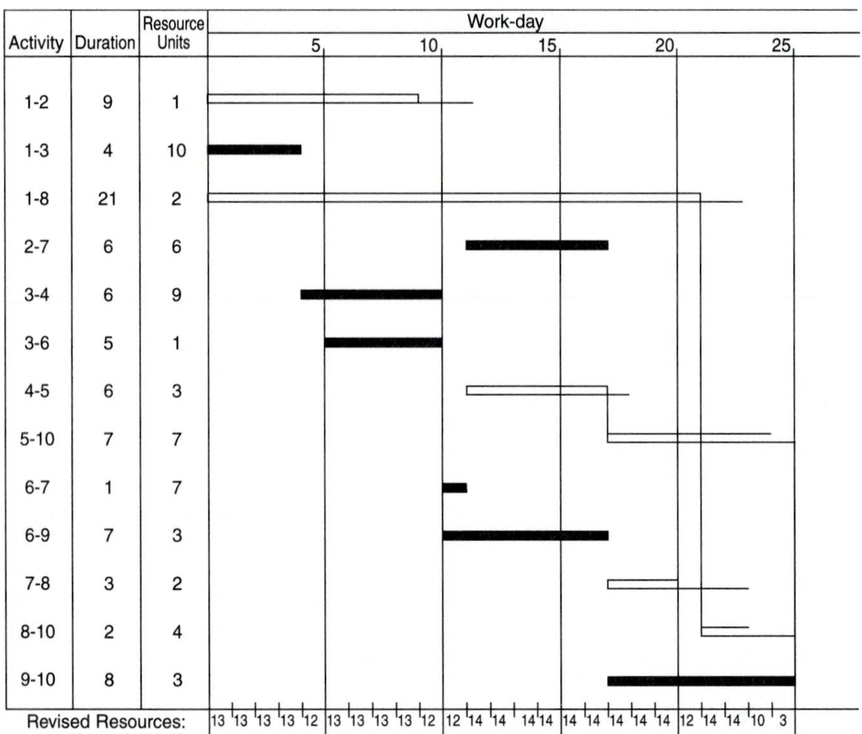

Activity
Duration
Resource Units
Work-day
5
10
15
20
25
1-2 9 1
1-3 4 10
1-8 21 2
2-7 6 6
3-4 6 9
3-6 5 1
4-5 6 3
5-10 7 7
6-7 1 7
6-9 7 3
7-8 3 2
8-10 2 4
9-10 8 3
Revised Resources:
13 13 13 13 12 13 13 13 13 12 12 14 14 14 14 14 14 14 14 14 12 14 14 10 3

Figure 7.11 Resource-constrained modified bar chart.

7.F PROJECT COMPRESSION

Once again consider the arrow diagram of Figure 7.6, for which the project duration is 25 work-days. If the owner wanted the project finished in 24 days, attention would have to focus on the cost-duration curves for the critical path activities as shown in Figure 7.12. Which activity should be shortened by one day? Key to answering this question is to first note that the cost curves represent total cost and that the current cost incurred for each activity is given at the low point of each curve, since the activity is scheduled to be done in its normal time. If an activity is shortened by one day, the total cost would increase by the incremental height change in moving from the normal time, t_n, to $(t_n - 1)$. The minimum increase in cost is desired, which is equivalent to finding the activity cost-duration curve that has the least *slope* to the left of the current cost. Activity 6-9 satisfies this condition, and would be shortened by 1 day to give a project length of 24 days. The critical path activities remain the same.

For a desired project length of 23 days, the cost curves for the critical activities are again examined, and activity 3-4 is found to have the least slope to the left of the current cost position (note that the cost curve for activity 6-9 becomes much

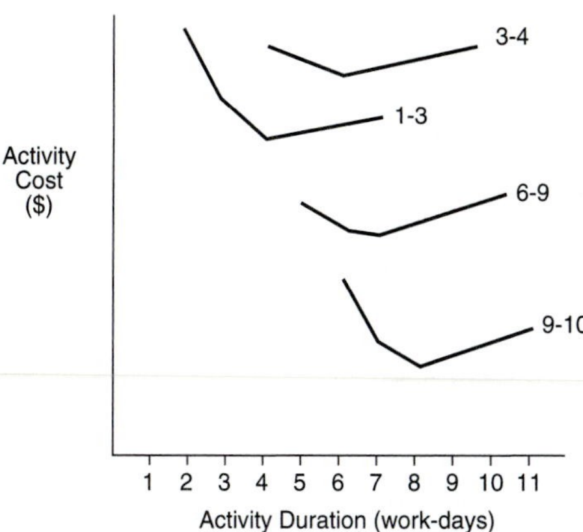

Figure 7.12 Activity cost-duration curves for Figure 7.6.

steeper after it is shortened by 1 day). Activity 3-4 is shortened to 5 days, and the project length becomes 23 days. Now, however, there are *three* critical paths with a length of 23 days. If the project were to be shortened by another day, the cost-duration curves for activities on all *three* paths would have to be considered. Since there are two completely independent paths (path 1-8, 8-10, and any one of the other *two* critical paths), at least two critical activities would have to be shortened simultaneously (one on each independent path) to achieve a project length of 22 days. It is obvious that if the process were continued, eventually *all* paths and activities would become critical.

Note that the project compression example was analyzed ignoring resource constraints and the desire to level resources. Inclusion of all three factors causes a large increase in problem complexity and is beyond the scope of the present treatment.

7.G TIME-SCALED ARROW DIAGRAM

Knowing the format of the modified bar chart allows it to be read almost as an arrow diagram. Some have suggested that the arrow diagram itself be modified to resemble a bar chart, resulting in the *time-scaled arrow diagram*. Various formats have evolved; a simple form is shown in Figure 7.13 for the arrow diagram of Figure 7.6, using data derived in the activity schedule of Table 7.1. The selected format preserves the basic arrangement of the original arrow diagram, but for greater clarity, the activities could be grouped by project site location (foundation, first floor, second floor, etc.), type of work (structural frame, ductwork, electrical, etc.), subcontractor, or any other classification that would enhance clarity and the planning and control function. For time-scale accuracy, nodes have been omitted and activities labeled by their *i-j* index. If the time scale were coarser (e.g., weeks or months), nodes could be used. The time-scaled arrow diagram displays dependencies more clearly than the modified bar chart, but is not as convenient to use when making scheduling changes for resource leveling.

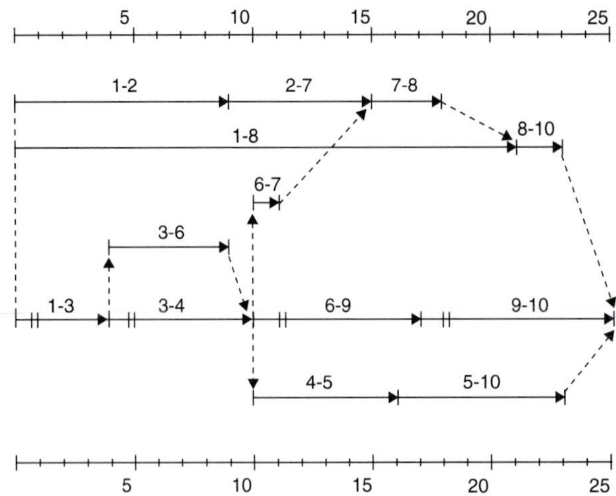

Figure 7.13 Time-scaled arrow diagram for Figure 7.6.

7.H PROGRESS CURVE

7.H.1 Alternative Formats

The activity schedule and modified bar chart depict when activities on a project are to be carried out. Associated with each activity are the estimated cost to the contractor and the estimated value that the contractor is to receive from the owner for completing the activity. Using the activity schedule or modified bar chart, it is a simple matter to determine the cumulative cost and value of a project over its duration by summing the cost and value of activities and portions of activities completed to date. Figures 7.14a and c show typical *progress curves*, or *S-curves* for project cost and project value, respectively. It is emphasized that these curves are *estimated* before the project is undertaken, and are usually based on each activity starting at its earliest start time. The typical S-shape is due principally to sequential, low-cost and low-value start-up and shut-down activities, and to the presence of many parallel, high-cost and high-value activities during the middle part of a project. The x-y axes for the progress curve in this format represent time and dollars, respectively.

Alternatively, the progress curve can be shown nondimensionally as in Figures 7.14b and d. Both axes in this format represent percent, and each axis is the same length. This allows typical S-curves to be derived for projects of a particular class, even though the individual projects may be of significantly different size. The Corps of Engineers, U.S. Army, has found that the following equation is a good generalized predictor (Perry, 1970):

$$y = \text{Sin}^2 (90x) \tag{7.10}$$

where

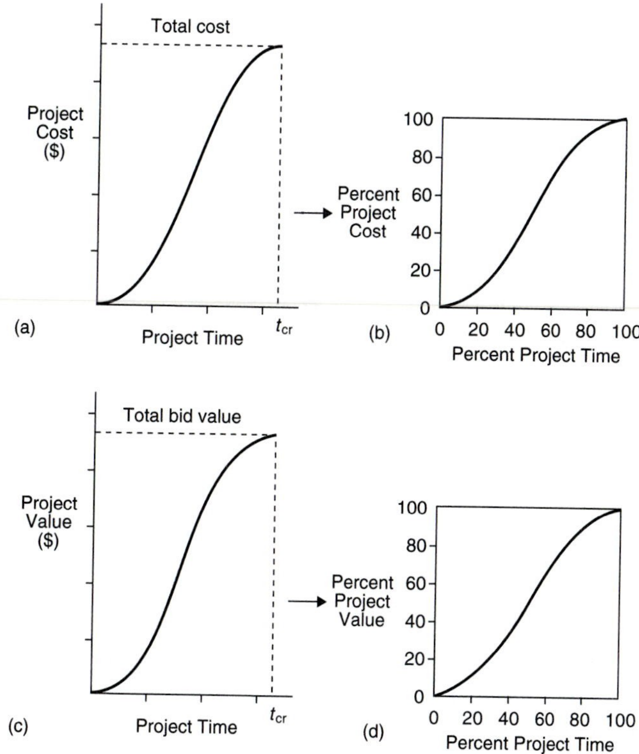

Figure 7.14 Alternative formats for progress curves.

y = decimal fraction of project value completed, and

x = decimal fraction of project time completed.

7.H.2 Control Function

One goal in estimating progress curves for a project is to help in project control. If actual progress curves differ significantly from those planned, the reason should be determined immediately. In judging whether a deviation is significant or not, it is helpful to have not just one estimated progress curve but two—one based on an early start (ES) of each activity and one based on a late start (LS). The LS curve will lie to the right of the ES curve, and the actual progress curve should fall between, typically at the average of the two, unless something unusual occurs.

If the actual progress curve for project value falls below the estimated curve based on an activity late start, it would indicate that, if the problem is not corrected, the project will not be completed in its planned duration. Conversely, if the actual progress curve for project value falls above that estimated based on an activity early start, it would indicate that the project will be finished ahead of schedule, due to higher productivity than estimated or to some other reason.

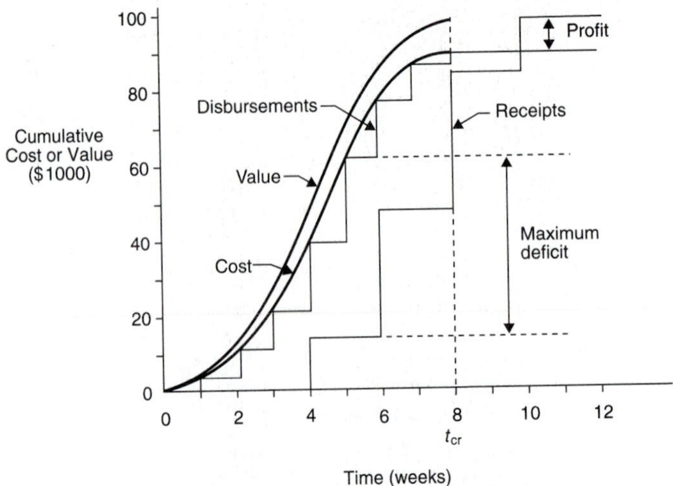

Figure 7.15 Cash flow forecast.

Direct conclusions are not as straightforward from a comparison of the estimated and actual progress curves for project cost. For example, if actual cost falls below the LS curve, it may be due to increased productivity or lower unit costs, but it may also be due to scheduling delays that will lengthen the project duration. Referral to the progress curve for project value is necessary before reaching a conclusion.

7.H.3 Cash Flow Forecast

Another important use of the progress curve is in forecasting cash flow. Figure 7.15 shows the progress curves for cost and value as estimated for an eight-week project. It is assumed that disbursements are made weekly to cover costs incurred during the week. The staircase function next to the cost progress curve shows cumulative outlays. By agreement, billings to the owner for value added are made every two weeks and are paid two weeks later (two-week lag in receipt). The cumulative value of receipts is represented by the rightmost staircase function. This payment-receipt pattern results in a continuous cash deficit over the period of the project and for two weeks thereafter. At the end of week 10, the final receipt creates a positive net cash flow (profit) of about $9000, or 9.1% of project value. The maximum difference between cumulative disbursements and receipts is almost $49,000 and represents the amount of cash that the contractor must have available to draw on to meet payroll and other expenses.

If the contractor has sufficient finances, the cash flow projection allows the greatest possible interest to be earned on assets. Of the \$49,000, only \$4500 is needed at the end of week 1, \$7000 more is needed at the end of week 2, and so on. Long-term investments that earn a higher interest rate than short-term accounts can be timed to mature when needed on the project. If the contractor must borrow money to meet expenses, the cash flow calculation can be used to determine the smallest loan amount that would be adequate.

7.H.4 Activity Progress Curve

In estimating project progress curves, it is usually assumed that cost and value are produced uniformly throughout the duration of an activity. This need not be the case, however. Activities may have start-up and shut-down periods similar to projects, and other characteristics may cause nonlinear progression of cost and value. Actual project data are shown in Figure 7.16 for the activity of placing concrete formwork during the construction of nine reinforced concrete structures (Carr et al., 1974). The duration required to place the formwork ranged from 6 months to 29 months, and the quantity varied from 103,000 ft^2 to 942,000 ft^2. The plot illustrates that activity progress, expressed in percent activity value, followed an S-curve on each project, and that there is a high degree of consistency between the curves. The example also demonstrates that, for specific activities, progress can be measured in physical units.

Nonlinear progress on an activity can be accounted for in a bar chart by writing the anticipated percentage progress at various points directly above the bar for

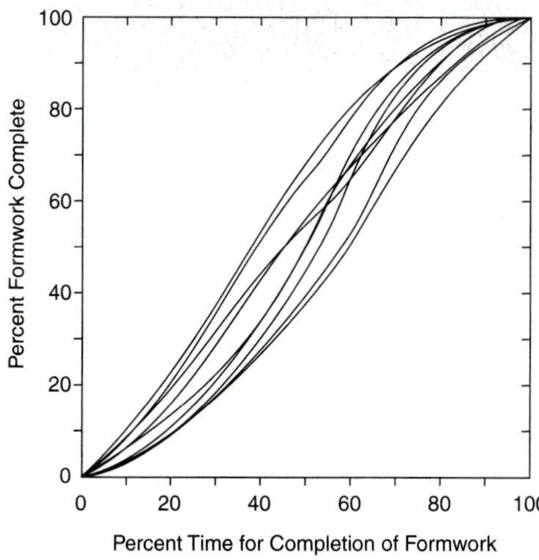

Figure 7.16 Progress curves for formwork activity on nine projects (Carr et al., 1974).

the activity. The day-by-day running sum of cost or value can then be found in the usual way for the project.

7.I MATHEMATICAL PROGRAMMING MODELS

Mathematical programming models can be developed for resource leveling, project compression, and resource scheduling. However, there are several reasons why a less formal approach is adequate at this time. For large projects with hundreds or even thousands of activities, a mathematical programming model becomes very large and rather cumbersome. More important, scheduling problems are inherently multiobjective in nature (Andreu and Corominas, 1989). Project engineers must satisfy a range of considerations when revising the project schedule; removing a resource constraint and minimizing of the sum of squared resource deviations from an established value are only two such concerns. Other, less tangible factors are often difficult to incorporate into an optimization model. Third, some activities can be interrupted, that is, stopped and then restarted again, sometimes multiple times. Permitting this option increases the size of the optimization model, and leads to the incorporation of heuristic rules to speed computations (Andreu and Corominas, 1989; Burgess and Killebrew, 1962).

7.J IMPORTANCE AND USE OF CPM

Results of a 1993 survey of 11,200 general contractors, specialty contractors, and architectural/engineering design firms point out the importance of CPM-related topics (Deloitte & Touche, 1993). The survey confirmed the widely held view that the construction industry is highly competitive and that profit margins are small. Only 20% of survey respondents had pretax net income as a percentage of revenue (profitability) of more than 5%, while 30% of general and specialty contractors and 24% of architects/engineers had pretax losses of 1% or more. The survey found that highly profitable firms focused more on project management (budgeting/control and scheduling) than high-loss firms and concluded that greater emphasis on project management may increase profits. Further, all three categories of respondents ranked project management either first or second among key drivers of their business success.

The same survey reported that 67% of general contractors, 35% of specialty contractors, and 81% of architectural/engineering firms were using project scheduling information systems. Previously, a 1974 survey of the largest 400 construction firms in the United States found that about 80% of such firms were using CPM methods (Davis, 1974). Considering that large firms tend to use the technique more than small firms, the percentages in the two surveys compare quite well. An important finding of the 1974 survey, however, was that only about one-half of CPM users employed the technique for project control, whereas almost all used it for detailed planning of construction activities prior to start of construction. This was probably due to the inconvenience of having to constantly update activity schedules on mainframe computers as changes occurred on the project. Today, the availability of desktop computers makes this task much easier, but it is safe to say that CPM is still used more for project planning than for project control. The 1974 survey also reported that 76% of the construction firms stated that they were either moderately or very successful in achieving the advantages attributed to use of CPM. Further, it is important to note that the successful use of CPM was strongly linked to good top management support for use of the technique.

CHAPTER SUMMARY

CPM is commonly used in design and construction projects. In many cases, bid specifications require construction contractors to submit an arrow diagram or bar chart with their bids, and the successful bidder must work with the owner and engineer to prepare a network that is mutually acceptable (Gleason and Ranieri, 1964). Depending on project conditions, CPM takes different forms: simple activity planning and scheduling, resource leveling, project compression, estimation of cash flow, project control, and others. The increased availability of inexpensive, fast microcomputers and convenient commercial software should enhance implementation.

EXERCISES

7.1. Draw the arrow diagram that represents each set of the following conditions. Note that the symbol "<" means "precedes," but not necessarily "immediately precedes." (Hint: Only one solution is possible for each arrow diagram; activity letters are not in standard form.)

(a) A < E
D < C
B < A, D

(b) G < B, E, D
A < B, E, D
C < G, A
F < G, A

(c) A < B
C < B
D < C
G < A
E < D, G, F

(d) A < D
D < B
F < C
C < E
G < F
H < I
I < B
F < I

(e) C < B, E
A < F, E
B < D
F < E

7.2. Using the forward pass/backward pass technique, derive the activity schedule and modified bar chart for the arrow diagram shown. What is the critical path? Project duration?

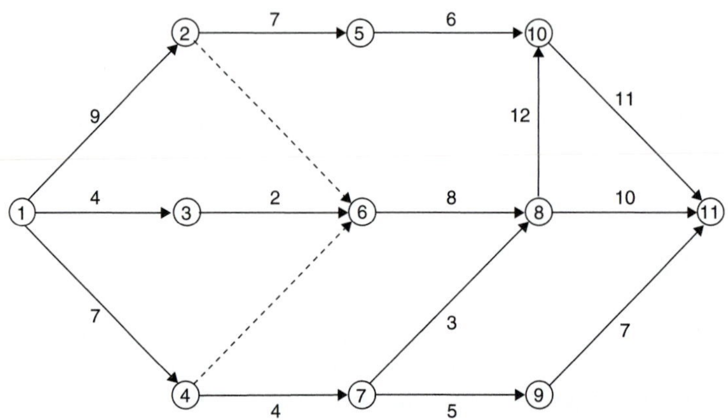

7.3. The activities in Exercise 7.2 require the resources shown in the table.

(a) Show the modified bar chart for Exercise 7.2 and list the number of resources of each type required on each work-day.

(b) Resources A and B are such that the maximum number of each resource needed on any day must be employed throughout the project. Shift the activities and interrupt activities where necessary to minimize the number of resource units of A and B constantly employed. Give first priority to resource A. Assume that all activities can be interrupted. How many units of each resource will be constantly employed? (Note: Do not attempt to minimize the sum of squared deviations).

	Resource			*Resource*	
Activity	A	B	Activity	A	B
1-2	0	2	6-8	3	3
1-3	2	1	7-8	0	0
1-4	3	0	7-9	1	1
2-5	1	1	8-10	2	1
3-6	2	2	8-11	0	0
4-7	0	3	9-11	2	0
5-10	2	0	10-11	1	3

7.4. Figure 7.10 shows the resource requirements (unleveled) for the arrow diagram of Figure 7.6, assuming that each activity will start at its EST.

(a) Draw the modified bar chart assuming that each activity will start at its LST, and show the (unleveled) resource requirements corresponding to this format.

(b) Assume that resource units active on a day are linearly proportional to project value added on that day. Plot the S-curve for the project under the assumption of an early start for each activity, and, on the same diagram, the S-curve under the assumption of a late start for each activity. Let the *x-y* axes of the plot be expressed in percent project time completed and percent project value completed, respectively.

(c) Actual progress on the project has been calculated at the end of days 5, 10, and 13. The percent of total project value in place at these times is 10, 20, and 40.

(1) Plot the S-curve for actual progress on the diagram of part (b). At the end of day 13, what would be a fair estimate of the time to complete the project?

(2) What is your subjective assessment of the chances that the project will be completed on time?

REFERENCES

ANDREU, RAFAEL, AND ALBERT COROMINAS, "SUCCES92: A DSS for Scheduling the Olympic Games," *Interfaces*, Vol. 19, no. 5, September–October 1989, pp. 1–12.

ANTILL, JAMES M., and RONALD W. WOODHEAD, *Critical Path Methods in Construction Practice*, 4th ed. New York: John Wiley and Sons, 1990, 422 pp.

BURGESS, A. R., and JAMES B. KILLEBREW, "Variation in Activity Level on a Cyclical Arrow Diagram," *The Journal of Industrial Engineering*, Vol. 13, no. 2, March-April 1962, pp. 76–83.

CARR, ROBERT I., BRIGHTMAN, THOMAS O., and FRANKLIN B. JOHNSON, "Progress Model for Construction Activity," *Journal of the Construction Division*, Vol. 100, no.C01, March 1974, pp. 59–64.

DAVIS, EDWARD W., "CPM Use in Top 400 Construction Firms," *Journal of the Construction Division*, Vol. 100, no. C01, March 1974, pp. 39–49.

DELOITTE & TOUCHE, *Insights in Construction: Survey Report on the Nation's Leading Construction Companies and Design Firms*. Washington, D.C.: The Associated General Contractors of America, 1993, 27 pp.

FULKERSON, D. R., "Expected Critical Path Lengths in PERT Networks," *Operations Research*, Vol. 10, no. 6, November–December 1962, pp. 808–817.

GLEASON, WILLIAM J., and JOSEPH J. RANIERI, "First Five Years of the Critical Path Method," *Journal of the Construction Division, Proceedings of the American Society of Civil Engineers*, Vol. 90, no. C01, March 1964, pp. 27–35.

HALPIN, DANIEL W., and RONALD W. WOODHEAD, *Construction Management*. New York: John Wiley and Sons, 1980, 483 pp.

HARRIS, ROBERT B., "Packing Method for Resource Leveling (PACK)," *Journal of Construction Engineering and Management*, Vol. 116, no. 2, June 1990, pp. 331–350.

HARRIS, ROBERT B., *Precedence and Arrow Networking Techniques for Construction*. New York: John Wiley and Sons, 1978, 429 pp.

KELLEY, JAMES E., JR. and MORGAN R. WALKER, "Critical-Path Planning and Scheduling," *Proceedings of the Eastern Joint Computer Conference*, Boston, December 1–3, 1959, pp. 160–173.

O'BRIEN, JAMES J., *CPM in Construction Management*, 2nd ed. New York: McGraw-Hill, 1971, 321 pp.

PERRY, W. W., "Automation in Estimating Contractor Earnings," *The Military Engineer*, Vol. 62, no. 410, November–December 1970, pp. 393–395.

THOMAS, WARREN, "Four Float Measures for Critical Path Scheduling," *Industrial Engineering*, Vol. 1, October 1969, pp. 19–23.

CHAPTER

8 Decision Theory

8.A RISK AND UNCERTAINTY

Most decisions made on large civil and environmental engineering projects involve elements of risk and/or uncertainty. *Risk* is defined as the situation wherein objective data exist upon which to estimate the probability of an event. Preferably, the probabilities would be based upon the results of experimental tests and historical data applicable to the case in question, but, lacking such "hard" data, *subjective probabilities* may be assigned through the subjective judgment of the decision maker or experts who have had experience in similar situations. Examples of project conditions involving risk are (1) the soil strength for a foundation design is not known exactly due to soil spatial heterogeneity and sampling errors of tests, (2) the yield of a full-scale water supply well or well field must be estimated using small-scale and spatially limited pump tests, and (3) the probability of a design or construction bid being awarded to a firm depends on many outside factors. In all these cases, reasonable probability estimates can be made about likely conditions or outcomes based on experimental data or past experience on similar projects. *Bayesian decision theory* is that branch of decision theory which makes use of these probability estimates, and structures the problem in such a way as to help in the decision-making process. Sections 8.B and 8.C are devoted to this case.

Uncertainty, by contrast, has been defined as the situation wherein no objective data exist upon which to base an estimate of the probability of an event; that is, complete ignorance exists, and no probabilities can be assigned. The descriptor "complete" uncertainty might be more appropriate, however, since some authors have used the term uncertainty to include "risky" situations. Fortunately, complete uncertainty is rare in engineering situations, but when it does occur it raises significant questions as to how to proceed. Decision making in the absence of probabilities is discussed in Section 8.D.

8.B SIMPLE DECISION TREE ANALYSIS

8.B.1 Example 8-1. Construction Bidding

Perhaps the best way to introduce the basic approach of Bayesian decision theory is by considering an example—similar to that discussed by Adrian (1973)—in which a contractor is faced with the decision of whether or not to bid on one of two heavy construction projects, a dam or a highway. The contractor is limited to choosing, at most, only one of the projects for bidding due to limited manpower and equipment available to him. He may also choose not to bid on either of the projects, in which case it is assumed that he would neither gain nor lose anything.

The contractor has collected data and performed some preliminary analyses on the two projects as summarized in Table 8.1. For either the dam or highway project, he has the option of bidding either high or low. From past records of bids made on similar types of projects, he is able to obtain estimates of the probability of being awarded the contract. For dam projects similar to the one in question, he determines that when a low bid was submitted, he received the contract 4 out of 10 times and lost it 6 of 10. On the other hand, when he submitted what he considered to be a high bid, he received the contract award only 2 out of 10 times and lost it 8 of 10. For jobs similar to the highway project, he finds that when low bids were submitted, he was awarded the contract 2 out of 10 times and lost it 8 of 10. Conversely, for high bids, the same figures were 1 of 10 awarded and 9 of 10 lost.

Also, profits or losses from the projects under consideration incur risk due to the unpredictable nature of weather, equipment breakdowns, labor strikes, and unknown project conditions such as the amount of rock encountered in excavations, subsoil strength, and so on. Moreover, estimated quantities of work items and estimated unit prices often do not match those found when the project is actually built. Again, using records of past experience on similar types of projects, the contractor is able to make estimates of his profit under high and low bid conditions if awarded the contract as shown in Table 8.1 (figures include cost of drawing up the bid). Note how profits vary for a given level of bid. Costs incurred in drawing up bids on the projects are \$50,000 and \$100,000 for the dam and highway projects, respectively.

The problem can be summarized graphically in a *decision tree* as shown in Figure 8.1. A decision tree depicts the choices open to a decision maker at any point in time, the chance events that might result later in time if these choices were

TABLE 8.1 DATA FOR EXAMPLE 8-1. CONSTRUCTION BIDDING

Project	Type of Bid	Historical Profit ($)	Probability of Winning Bid
Dam	High	800,000	0.2
	High	400,000	0.5
	High	−200,000	0.3
	Low	500,000	0.3
	Low	100,000	0.5
	Low	−400,000	0.2
Highway	High	2,000,000	0.3
	High	1,000,000	0.6
	High	−400,000	0.1
	Low	800,000	0.2
	Low	400,000	0.6
	Low	−400,000	0.2

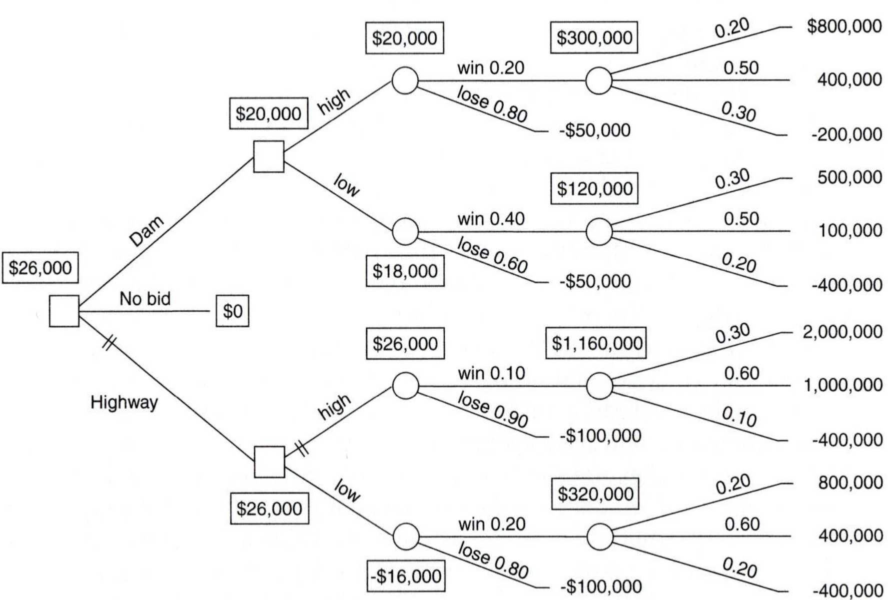

Figure 8.1 Decision tree for Example 8-1. Construction bidding.

actually made, and the final payoff from having traveled along a specific path. A decision tree has been described as depicting the moves in a game of chance, wherein the two players are the decision maker and "chance." In drawing the tree, circles indicate chance nodes and squares represent decision nodes.

For example, at time zero, the contractor could choose to bid on either the dam project, the highway project, or neither. If a project is chosen, the contractor then must choose whether it is best to bid high or low. If a high or low bid is actually made, then chance will determine whether the contractor wins or loses the bid. If the bid is lost, the game is over, and the contractor has lost the cost of drawing up the bid. On the other hand, if the bid is won, the game can continue, with chance then deciding the level of profit obtained from actually completing the project. The payoff from taking a specific path is shown at the "top" of the tree.

Having constructed the tree from the bottom (time zero) to the top, the analysis can then proceed in the opposite direction—from the top of the tree to the bottom. The expected value of profit can be computed at the series of chance nodes at the top of the tree. These nodes represent the point in time when the contractor has won the bid, but has not yet started the actual construction process. The decision maker can evaluate his or her financial position *assuming that the game could be played a large number of times.* If this were the case, the expected value would represent the long-run average profit to be expected from being at this point in time. The numbers recorded in the boxes next to the chance nodes represent expected values.

The decision maker is now ready to step down one level in the tree. This lower level is also a series of chance nodes, wherein the bid is either won or lost with the corresponding probabilities. Again, the expected value of being at this point in time can be found and recorded in the box next to the node (the calculation amounts to finding the expected value of expected values).

The next lower level in the tree corresponds to the decision maker's choice. A high or low bid can be made on either project. Naturally the contractor would choose the path that would result in the highest expected profit. This highest value is therefore recorded in the box at this level. Only one lower level remains (time zero), and it also corresponds to the decision maker's choice. Again, the choice is clear; choose the path yielding the highest expected profit, and record that profit at time zero. An expected profit of $26,000 results from choosing to bid high on the highway project. This is the contractor's optimal decision and is indicated in the decision tree with double bars on the optimal action branches.

Note that in any one "play" of the game in Example 8-1, the contractor will not receive the expected profit amount. In fact, the payoffs actually received vary quite widely. Use of the expected value criterion weights the actual payoffs by their relative frequency of occurrence, producing a decision which is best in the long-run, average sense. In the next section, the payoffs are considered from a short-run viewpoint.

8.B.2 Utility Considerations

In the construction bidding problem, profits were expressed in dollar terms at the top of the tree, and expected values were calculated. The assumption in taking an expected value in this way is that a linear relationship exists between profit and true value, or worth. There are cases where this assumption is not valid, however. For example, if the contractor were on the brink of bankruptcy, and a payment of $500,000 had to be made to a bank to avoid foreclosure, any profit less than this amount would have very little value. A linear relationship would not be representative of the real situation.

There are many other cases where it would be more appropriate to derive a *utility function* to represent the true value of the payoffs in one particular play of the game. A utility function transforms dollars into units more representative of psychological value to the decision maker. Figure 8.2a illustrates the shape of a utility function that might represent someone who is sensitive to a debt, D. Figure 8.2b might represent someone intent on becoming rich; small payoffs are not seen as being very valuable, while large payoffs are viewed much more highly. On the other hand, Figure 8.2c represents a situation of declining marginal utility of money to the decision maker as more and more needs are met.

Procedures to develop utility functions are beyond the scope of this text, but are available elsewhere (de Neufville, 1990; Schlaifer, 1969; Von Neumann and Morgenstern, 1944). It will be assumed in this chapter that either dollar or utility units are used at the top of the tree to represent payoffs. The expected value criterion can then be used to derive a decision appropriate in either the long run or short run.

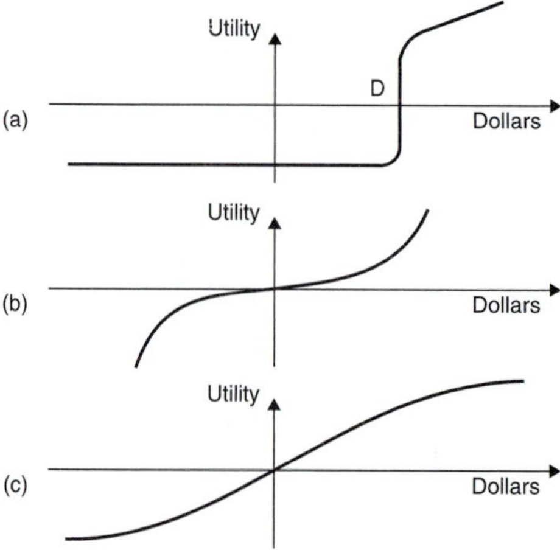

Figure 8.2 Alternative utility functions for individuals.

8.B.3 Summary

A decision tree serves as a convenient way to organize the analysis of staged, probabilistic engineering problems. The tree is built from the bottom up, starting at time zero and proceeding sequentially through time, as would the actual engineering problem. Payoffs, either in dollar or utility terms, are shown at the top of the tree and represent all financial transactions occurring along the specific path to the topmost branch. Analysis of the tree proceeds from the top down, taking expected values at chance nodes and optimal decisions at decision nodes. When the origin is reached, the optimal decision is known.

In many situations, the engineer has the option of refining estimates of the probability of certain outcomes by conducting experimental tests. This option expands the decision tree and complicates the calculation of relevant probabilities. Both these aspects are discussed in the next section.

8.C GENERAL DECISION TREE ANALYSIS: EXPERIMENTATION

8.C.1 Problem Setting

For most civil and environmental engineering problems, it is possible to perform some type of experimental test to find out more about the actual "state of nature." For example, in addition to known geologic conditions, more information about the soil strength for a building foundation can be found by taking core samples and testing these in a laboratory. However, it must also be recognized that any test has some degree of error associated with it: samples cannot be taken everywhere, the number of tests run in the laboratory is usually small, and the tests themselves are imperfect. Thus, even after the test is run, the state of nature is not known with certainty. Nevertheless, the test results do serve to change and improve the estimates of the probabilities of finding the true soil strength (state of nature) within certain ranges. Is the experimental test worth the extra cost involved, or should the design be based on the probability of finding different states of nature as estimated without the experiments? A decision tree can be constructed to answer this question.

8.C.2 General Structure Decision Tree

Define:

e_i = experiment type i;

z_j = experimental result j;

a_i = action i;

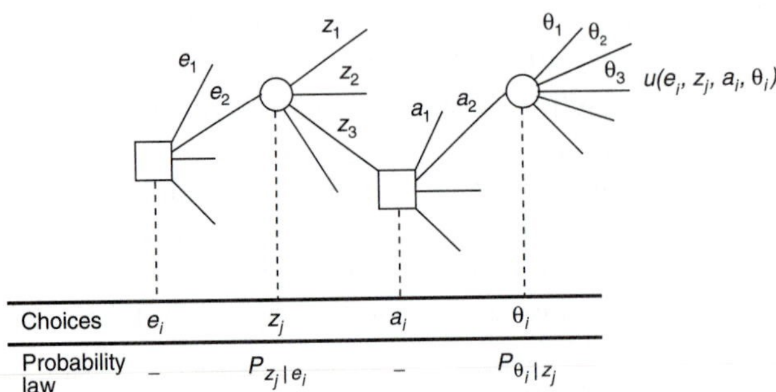

Figure 8.3 General structure of a decision tree: Experimentation.

$$\theta_i = \text{state of nature } i\text{; and}$$

$$u(e_i, z_j, a_i, \theta_i) = \text{utility, or payoff.}$$

The sequence of events in the general engineering decision problem is (1) selection of an experiment, e_i, to perform; (2) observation of an experimental result, z_j; (3) selection of a certain action, a_i; (4) observation of a particular true state of nature, θ_i; and (5) receipt of a payoff or utility, $u(e_i, z_j, a_i, \theta_i)$, depending on the path followed.

Figure 8.3 illustrates the general structure decision tree. The decision maker chooses the type of experiment to perform, chance determines the experimental outcome, the decision maker selects an action, and chance determines the state of nature. Together, this sequence results in the final payoff. Probabilities must be assigned to the chance nodes.

Define:

$P_{z_j|e_i}$ = marginal (prior) probability of an experimental outcome z_j given that experiment e_i has been selected; and

$P_{\theta_i|z_j}$ = conditional (posterior) probability of having a state of nature θ_i given that an experimental result z_j has been observed.

Use of the term "prior" implies that a probability estimate is made before any experimentation takes place. A "posterior" probability, on the other hand, is one calculated after having seen an experimental result. The location of these probabilities in the general structure decision tree is indicated in Figure 8.3, and their calculation will be described in Section 8.C.4. Assuming for now that they are known, the solution process for the general structure decision tree can be specified.

8.C.3 Solution Process

The general structure tree can be treated as a game of chance that is played a large number of times. The solution procedure is

Step 1. At the top of the tree, calculate the value of being at each chance node by taking the expected value of the payoff using the conditional (posterior) probabilities of the states of nature that branch from each chance node;

Step 2. Drop down one level in the tree to the decision nodes, and record for each decision node the optimal expected payoff resulting from selection of the best action possible from all the actions that branch from each decision node;

Step 3. Drop down one level in the tree to the chance nodes representing the experimental results, and calculate the value of being at each chance node by taking the expected value of the payoff using the marginal (prior) probabilities of the experimental results that branch from each chance node;

Step 4. Drop down one level to the decision node at the origin of the tree, and record the optimal expected payoff resulting from selection of the best action possible from all the actions that branch from the decision node. This is the overall optimal expected payoff, and the associated action is the overall optimal course of action. Then;

Step 5. (Implementation) Conduct the optimal experiment, observe the actual experimental outcome, and based on this outcome, implement the optimal design action from the possible design actions at the corresponding decision node, observe the true state of nature and receive the associated payoff.

Just as with the simple decision tree, the general structure decision tree is constructed from the bottom up, analyzed from the top down, and finally implemented from the bottom up. Note that the general structure decision tree can be combined with a simple decision tree at the origin to represent the options of experimentation or direct action. Another important point is that the branches of the decision tree proceed forward in time exactly as would steps in the real decision process. Keeping this time sequence in mind helps greatly when drawing the tree for actual problems.

The solution process is one of backward induction, sometimes referred to as "averaging out" (at chance junctures) and "folding back" (at decision junctures) (Raiffa, 1968, p. 23). As such, it resembles the backward induction approach of dynamic programming (see Chapter 12), wherein once the optimal policy cost is known, the calculation steps must be retraced to find the actual decisions yielding that cost.

8.C.4 Probability Calculations

Joint Probabilities. Consider the situation wherein, historically, experimental tests have been conducted and engineering designs have been implemented (facilities constructed). For example, a core sample has been tested for soil strength, a foundation design has been selected, and the structure has been built. The result is a paired observation of actual soil strength, θ_i, as demonstrated by performance of the structure, and predicted soil strength, z_j, as indicated by the experimental result. If a large number of projects are available in which the joint occurrence of θ_i ad z_j have been observed, the relative frequency, or probability, of each pair can be calculated as:

$$P\{\theta_i, z_j | e_i\} \equiv P_{\theta_i, z_j|e_i} = \frac{n_{ij}}{N} \tag{8.1}$$

where

$P_{\theta_i,z_j|e_i}$ = joint probability of θ_i and z_j occurring simultaneously, given that experiment type i has been performed;

n_{ij} = the number of times that θ_i has been observed simultaneously with z_j; and

N = the total number of observations, given by

$$N = \sum_i \sum_j n_{ij}.$$

Table 8.2 provides example data and probability calculations for the soil strength of a foundation. The total number of historical observations is $N = 100$. For example, on 24 occasions the actual soil strength was found to be low, θ_1, when the experimental result indicated a low strength, z_1.

However, there were also five occasions when the experiment predicted a medium soil strength, z_2, but the actual strength was found to be low, θ_1. On one occasion, the experiment indicated a high soil strength when the actual was low. By examining Table 8.2, it can be seen that the experimental results are highly correlated with actual strength, but some error still remains. The experimental test also seems to be biased toward low strength predictions. Triaxial soil tests exhibit this tendency due to disturbance of the soil in obtaining the sample.

TABLE 8.2 CALCULATION OF JOINT AND MARGINAL PROBABILITIES FOR FOUNDATION SOIL STRENGTH (N = 100)

	Experimental Results			Marginal Probability of the State of Nature
State of Nature	z_1 (low)	z_2 (medium)	z_3 (high)	
θ_1 (low)	24/100 = 0.24	5/100 = 0.05	1/100 = 0.01	30/100 = 0.30
θ_2 (medium)	15/100 = 0.15	20/100 = 0.20	5/100 = 0.05	40/100 = 0.40
θ_3 (high)	5/100 = 0.05	15/100 = 0.15	10/100 = 0.10	30/100 = 0.30
Marginal probability of the experimental result	44/100 = 0.44	40/100 = 0.40	16/100 = 0.16	100/100 = 1.0

Prior probabilities. Along with the joint probabilities, two other types of probabilities can be calculated directly in Table 8.2. By adding across a row, the relative frequency of seeing a particular state of nature can be found. Historically, a low soil strength has been observed 30 times out of 100. For the current project, and prior to any experimentation, the probability of having a low soil strength can be estimated as 0.30. This is the prior (marginal) probability of seeing a state of nature θ_1.

Similarly, before an experiment is run, an estimate of the probability of seeing a particular experimental result can be made using the historical data. For example, by summing down the z_1 column, the prior (marginal) probability of z_1 occurring is 44 times out of 100, or 0.44. Prior probabilities can thus be found as the sum of the joint probabilities along a row or column.

Posterior probabilities. After an experimental result has been observed, a revised and improved estimate can be made of the probability of seeing a particular state of nature. This conditional probability can be found using basic probability relationships.

Let x and y be events with probabilities $P\{x\}$ and $P\{y\}$. The conditional probabilities associated with these events are $P\{x|y\}$ and $P\{y|x\}$. It is well known that the probability that both events occur is given by

$$P\{x, y\} = P\{x|y\}\, P\{y\}. \tag{8.2}$$

Rearranging,

$$P\{x|y\} = \frac{P\{x, y\}}{P\{y\}}. \tag{8.3}$$

Interpreting (x, y) as (θ, z), the desired result is obtained:

$$P_{\theta_i|z_j} = \frac{P_{\theta_i, z_j|e_i}}{P_{z_j|e_i}} = \frac{P_{\theta_i, z_j}}{P_{z_j}}. \tag{8.4}$$

Equation 8.4 is an equivalent statement of Bayes formula, which is fundamental to bayesian decision theory and can be used to calculate the posterior probability of the state of nature, θ_i, given the experimental result, z_j.

Based on the joint and marginal probabilities in Table 8.2, posterior probabilities can be calculated as shown in Table 8.3. As a check on the calculations, note that for a given experimental outcome, z_j, the posterior probabilities of the different states of nature must sum to one. Also, as a memory aid, note that the conditional probability for a row (θ_i) based on a column (z_j) is given by the joint probability for the row-column intersection divided by the column (z_j) marginal probability. The data in Tables 8.2 and 8.3 are utilized in Example 8.2: foundation design.

TABLE 8.3 CALCULATION OF POSTERIOR PROBABILITIES FOR FOUNDATION SOIL STRENGTH (BASED ON DATA IN TABLE 8.2)

$P_{\theta_1 \mid z_1} = \dfrac{0.24}{0.44} = 0.545$	$P_{\theta_1 \mid z_2} = \dfrac{0.05}{0.40} = 0.125$	$P_{\theta_1 \mid z_3} = \dfrac{0.01}{0.16} = 0.630$
$P_{\theta_2 \mid z_1} = \dfrac{0.15}{0.44} = 0.341$	$P_{\theta_2 \mid z_2} = \dfrac{0.20}{0.40} = 0.500$	$P_{\theta_2 \mid z_3} = \dfrac{0.50}{0.16} = 0.313$
$P_{\theta_3 \mid z_1} = \dfrac{0.05}{0.44} = 0.114$	$P_{\theta_3 \mid z_2} = \dfrac{0.15}{0.40} = 0.375$	$P_{\theta_3 \mid z_3} = \dfrac{0.10}{0.16} = 0.625$
Sum: 1.0	1.0	1.0

8.C.5 Example 8-2: Foundation Design

An engineer must decide between spread footings or piles for the foundation to support each column of a building. The soil strength is not known for certain, but could be low (θ_1), medium (θ_2), or high (θ_3). Spread footings for the building have a total cost of \$150,000, while piles cost \$250,000. The piles are considered "safe" in that if the actual soil strength turns out to be low, no damage will result to the building. The spread footings are less expensive but would result in some building damage if the soil strength is low or medium, and they would have to be replaced with piles (under difficult construction conditions) if the soil strength is low. Data are provided in the Table 8.4.

For \$10,000, field studies and a triaxial strength test can be conducted. Assume that the joint, prior, and posterior probabilities given in Tables 8.2 and 8.3 apply to this site. Determine the optimal strategy and cost.

Solution. The decision tree for this analysis is shown in Figure 8.4, along with all payoffs, probabilities, and expected values. The optimal action at time zero is to use pile foundations at an expected cost of \$250,000. Experimentation is not economically justified in this case. If experimentation were conducted, the expected overall cost would be \$257,618. Note that this cost is less than the simple sum of the optimal cost without experimentation of \$250,000 for pile foundations and \$10,000 for the test. Performing the test allows more accurate future decisions based on posterior probabilities. For example, if the test were conduct-

TABLE 8.4 COST DATA FOR EXAMPLE 8-2: FOUNDATION DESIGN

	Spread footings			Piles		
State of Nature	Foundation Cost	Damage Cost	Replacement Cost	Foundation Cost	Damage Cost	Replacement Cost
θ_1	\$150,000	\$100,000	\$1,000,000	\$250,000	—	—
θ_2	\$150,000	\$50,000	—	\$250,000	—	—
θ_3	\$150,000	—	—	\$250,000	—	—

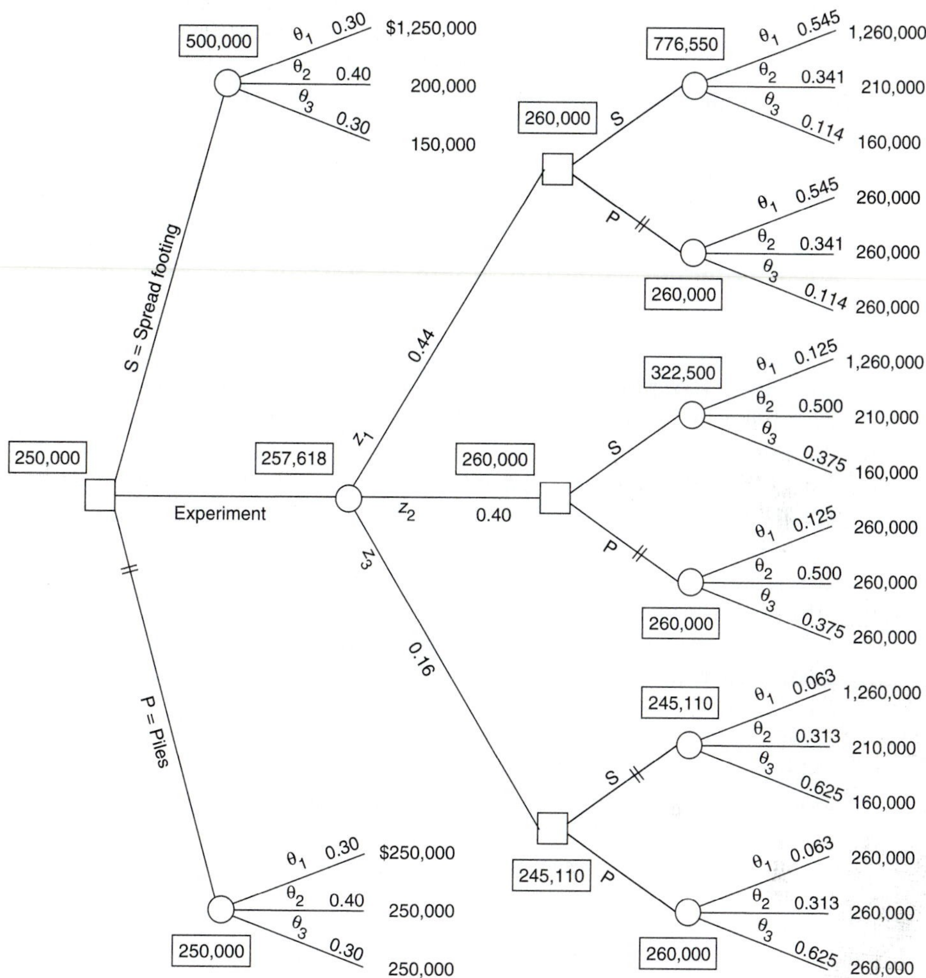

Figure 8.4 Decision tree for example: Foundation design.

ed and resulted in a high strength experimental result, z_3, the engineer would select the spread footing at an expected cost of $245,110, which is less than the cost of choosing pile foundations without experimentation. However, because the probability of such an experimental outcome (0.16) is low, the experimental test is not economically justified, as shown by its $257,618 overall expected cost. Conclusion: Use pile foundations at an expected cost of $250,000.

8.C.6 Expected Monetary Value of Information

As seen in Example 8-2, information gained from an experimental test, in the form of revised probabilities of the states of nature, allows better decisions to be made subsequently. However, in this particular example, the experimental test recaptures only $2,382 [= ($250,000 + $10,000) − $257,618] of its $10,000 cost. The $2,382 is the *expected monetary value of information* (EMVI) derived from the experiment, or its expected monetary *benefit*. The benefit-cost ratio for the experiment is $2,382/$10,000 = 0.2382, or 0.24, showing it not to be worthwhile. In many other situations, the EMVI will more than cover the cost of the experiment, making experimentation advisable.

8.C.7 Expected Monetary Value of Perfect Information

Assume that there existed an experiment or source of information that would predict with certainty what the true state of nature will be. How much would a decision maker be willing to pay to obtain this perfect information? In Example 8-2, this amount can be calculated from the decision tree shown in Figure 8-5. If perfect information said that a particular state existed, then the foundation option that minimizes cost for that state would be chosen as shown in the decision tree. However, the decision maker has formed a judgment (prior probabilities) as to the relative frequency with which the various states will be predicted using the perfect information experiment. The prior probabilities are applied to yield an expected value of cost under perfect information of $200,000, which is $50,000 less than that associated with the optimal action without any experimentation. The *expected monetary value of perfect information* (EMVPI) is therefore $50,000. Note that the decision maker surely will not pay more than the EMVPI for any real experiment because it would yield less than perfect information. Any experiment that costs more than this amount, therefore, could be immediately eliminated from consideration. From this perspective, it was worthwhile considering the $10,000 experiment in Example 8-2, even though it was ultimately rejected.

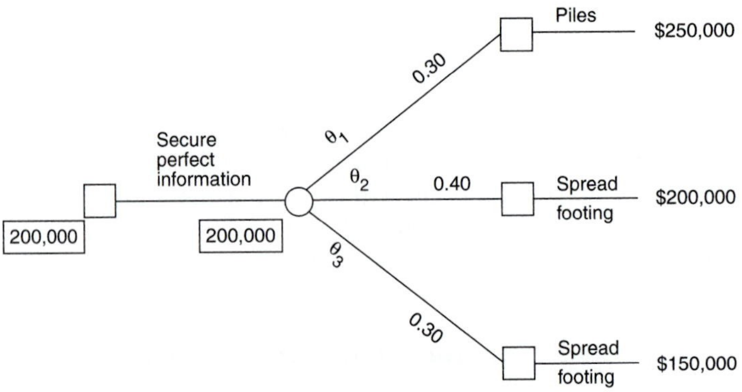

Figure 8.5 Decision tree using perfect information.

8.C.8 Summary

The option of experimentation can be handled by expanding the concept of the simple decision tree and by introducing the calculation of conditional probabilities based upon joint and marginal probabilities. Results from analysis of the expanded decision tree can be used to compute the expected monetary value of information gained from experimentation.

8.D DECISION MAKING IN THE ABSENCE OF PROBABILITIES

8.D.1 Introduction

In certain situations, the decision maker may be so unfamiliar with the problem setting that he or she is either unable or unwilling to assign probabilities to the various possible true states of nature. An example might be an engineer for a large international consulting firm who has just been put in charge of a project in a country in which he has had no prior experience. The engineer has just learned that an important construction permit must be obtained from the country's ministry of public works. The engineer may be entirely unable or unwilling to assign probabilities to the possible outcomes of a permit application (approved in one month, two months, etc., disapproved in one month, two months, etc). However, other project decisions must be made by the engineer despite the complete uncertainty surrounding the outcome of the permit application. Useful decision criteria have been proposed to aid in making decisions in the absence of probabilities, but it must be said that all of the proposals have serious theoretical shortcomings, as illustrated by Luce and Raiffa (1957) and Milnor (1951).

8.D.2 Dominance

Consider the *payoff* or *utility matrix* in Table 8.5. The numbers represent either the savings or utility to be gained by the decision maker under different actions and

TABLE 8.5 PAYOFF OR UTILITY MATRIX REDUCED BY DOMINANCE

	θ_1	θ_2	θ_3	θ_4			θ_1	θ_2	θ_3	θ_4
a_1	6	3	5	2		a_1	6	3	5	2
a_2	3	9	0	5	→	a_2	3	9	0	5
a_3	4	7	8	1		a_3	4	7	8	1
a_4	3	2	0	3						
	Initial						Reduced			

states of nature. The matrix should first be inspected to see if any action is dominated by another. If so, the dominated action can be eliminated from further consideration. An action is subject to *dominance* if another action can be found that is preferable under some states of nature and at least equal under all others. Making pairwise comparisons, it can be seen that action a_4 is dominated by action a_2. Action a_4 can therefore be eliminated from the matrix. Note that actions a_1 and a_3 do not dominate a_4 since if state θ_4 occurred, action a_4 would be preferable. The reduced matrix in Table 8.5 is now considered further.

8.D.3 Laplace Criterion

The Laplace criterion is based on the "principle of insufficient reason," first set forth by Jacob Bernoulli (1654–1705). The criterion states that if one is truly completely ignorant of the relative likelihood of the different states of nature, then they should all be treated as equally likely, being assigned the probability $1/n$, where n is the number of states. Doing so transforms the problem to one of risk, and the expected value criterion can be used as before. Applied to the reduced matrix of Table 8.5, the criterion ranks the actions as

$$a_3(5.00) > a_2(4.25) > a_1(4.00)$$

where ">" can be read as "is preferred to," and numbers in parentheses are the expected payoffs. Action a_3 would be taken.

8.D.4 Maximin Criterion

The decision maker could take the pessimistic viewpoint that, if a particular action is chosen, the worst possible state of nature will occur. Under this assumption, it would be best to choose that action which maximizes the minimum payoff among all possible actions. This is the maximin criterion and results in the following ranking for the actions:

$$a_1(2) > a_3(1) > a_2(0)$$

where numbers in parentheses represent the minimum payoff for each action. Action a_1 would be taken.

8.D.5 Maximax Criterion

Alternatively, the decision maker could take the optimistic viewpoint that, if a particular action is chosen, the best possible state of nature would occur. Under this assumption, it would be best to choose that action which maximizes the maximum payoff among all possible actions. This is the maximax criterion, and results in the following ranking:

$$a_2(9) > a_3(8) > a_1(6)$$

where numbers in parentheses represent the maximum payoff for each action. Action a_2 would be taken. Both the maximin and maximax criteria are attributed to von Neumann (1944).

8.D.6 Hurwicz Criterion

It is unlikely that the decision maker is a complete pessimist or optimist. The Hurwicz criterion (Hurwicz, 1951a, b) allows the decision maker to lie anywhere between these two extremes by computing an α-index for each alternative as

$$\alpha\text{-index of } a_i = \alpha M_i + (1 - \alpha)m_i$$

where M_i is the maximum and m_i is the minimum payoff or utility for action a_i over all possible states of nature. For example, for action a_1, the maximum payoff is 6, and the minimum is 2. The equation for the α-index becomes

$$\alpha\text{-index of } a_1 = \alpha(6) + (1 - \alpha)(2) = 2 + 4\alpha$$

Figure 8.6 is a plot of the α-index for each action as α is varied from zero (complete pessimism) to one (complete optimism). As indicated in the figure, the ranking of the actions varies with α, the degree of optimism chosen. However, the plot should help the decision maker by indicating the sensitivity of the rankings to α. For example, any degree of optimism above $\alpha = 0.50$ would result in the ranking $a_2 > a_3 > a_1$. For $\alpha = 0.0$, the Hurwicz criterion is the same as the maximin criterion, and for $\alpha = 1.0$ it is the maximax criterion.

8.D.7 Minimax Regret Criterion

Often, decision makers have taken a certain action only to regret it later when the true state of nature is known; the payoff was not as great as it could have been if a different action had been taken. The difference between what could have been obtained and what was actually obtained has been defined as the "regret". For ex-

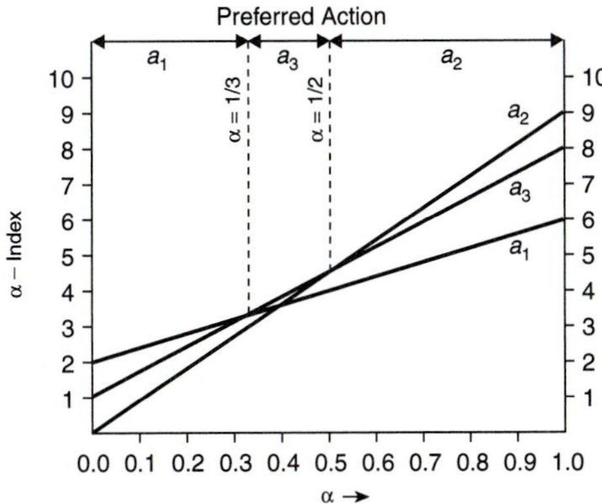

Figure 8.6 Hurwicz criterion.

TABLE 8.6 REGRET MATRIX

	θ_1	θ_2	θ_3	θ_4	Maximum Regret
a_1	0	6	3	3	6
a_2	3	0	8	0	8
a_3	2	2	0	4	4

ample, in Table 8.5 if action a_3 had been taken only to find out that θ_1 was the true state of nature, a payoff of 4 would have been received. The decision maker might regret that he had not chosen, a_1, where he would have received the maximum payoff in the θ_1 column of 6. The degree of his "regret" for taking a_3 is the difference, or 2. Decision makers may well decide to take that action which will minimize the maximum regret that they may later experience. This is the minimax regret criterion, attributed to Savage (1951).

The first step in the method is to transform the payoff or utility matrix to a "regret matrix." This is accomplished by subtracting all of the elements in a column of the payoff matrix from the maximum payoff in that column. This is done for each column. Table 8.6 is the regret matrix associated with the reduced payoff matrix of Table 8.5. The maximum regret possible from taking each action is shown in the last column. The minimax regret criterion seeks to minimize the maximum regret and would rank the actions as:

$$a_3(4) > a_1(6) > a_2(8)$$

where the numbers in parentheses represent the maximum regret possible from taking the action. Action a_3 would be taken.

8.D.8 Summary

Various criteria have been proposed for decision making in the absence of probabilities. Since it is unlikely that they will result in the same recommendation, the decision maker must study the results of all the approaches and decide on the action with which he will be most comfortable in a given situation.

CHAPTER SUMMARY

One of the primary advantages of decision theory is the structure that it provides in the form of a decision tree. Decisions, experimental outcomes, states of nature, and payoffs are organized for analysis. Often, construction of the tree itself is enough to improve the decision process, even if a complete probability analysis is not carried out. Assignment of probabilities can be done through subjective (expert) judgment or analysis of historical experimental results. Also, decision criteria have been developed to aid in the rare event that probabilities cannot be assigned.

EXERCISES

8.1. A young engineer is planning ahead for her retirement. She has saved $10,000 and has three investment choices open to her: (1) a corporate bond with a 10-year maturity, (2) common stock in a corporation that she has been following, and (3) a stock mutual fund. Her estimates of the compound annual rates of return and their associated probabilities are shown in the table below. In each case she has decided that the investment will not be touched for 10 years.

Investment	Rate of Return (%)	Probability
Corporate bond	7	1.0
Corporate stock	15	0.4
	5	0.3
	−5	0.2
	−15	0.1
Mutual fund	10	0.3
	5	0.4
	−5	0.3

(a) Determine the best investment strategy. Draw the complete decision tree, showing all probabilities, payoffs, expected values, and so on. What is the expected return on the optimal investment?

(b) The engineer learns that management is about to change in her closely watched individual corporation. The new management is known to favor larger dividend payouts to holders of common stock. If this occurs, it will not only increase the annual rate of return on the stock, but will also make the stock more attractive to investors, driving up the price in the long run, further increasing the compound rate of return. The engineer revises her probability estimates for the corporate stock investment as shown in the table below. Draw the complete decision tree for this new situation and determine the optimal strategy and associated expected annual rate of return.

Dividend Action	Probability	Probability of Achieving the Rate of Return Shown			
		15%	5%	−5%	−15%
Raise	0.7	0.50	0.35	0.10	0.05
Same	0.2	0.40	0.30	0.20	0.10
Lower	0.1	0.30	0.20	0.30	0.20

8.2. A student just entering college needs a used car so that he can earn part of his college expenses. He has narrowed it down to two choices—a "cheap" used car or a "good" used car—and he estimates the performance probabilities for both as shown in the table. If he buys the cheap car and it is "totaled" through excessive repair costs, he will not buy another cheap car. Rather, he will move up to a good used car like the one he is currently considering. Similarly, if he buys a good used car and it is totaled through excessive repair costs, he will not buy another "good" car, but will move down to a "cheap" car like the one currently under consideration. "Repairs above normal" are \$300 per year, while "major repairs" are \$600 per year for each of the student's four years of college. Note that normal repair costs represent a zero baseline.

Car	Cost (\$)	Performance Probability			
		Normal Repairs	Repairs Above Normal	Major Repairs	"Totaled"
Cheap	4000	0.20	0.30	0.30	0.20
Good	6000	0.50	0.30	0.10	0.10

Assume a salvage value of \$500 for each car if totaled and no more than one such "total disaster" occurring. Further assume that the replacement cost is simply the purchase cost of the next used car; that is, neglect future expected costs for the replacement car.

Which car should be purchased? What is the optimal, four-year expected cost? If the student expects to be in college for five years instead of four, is the decision changed? What is the new expected cost?

8.3. Use all the data for the construction bidding problem in Example 8-1, but substitute utility for the monetary payoffs at the top of the tree. The utility function for the contractor is provided in the figure.

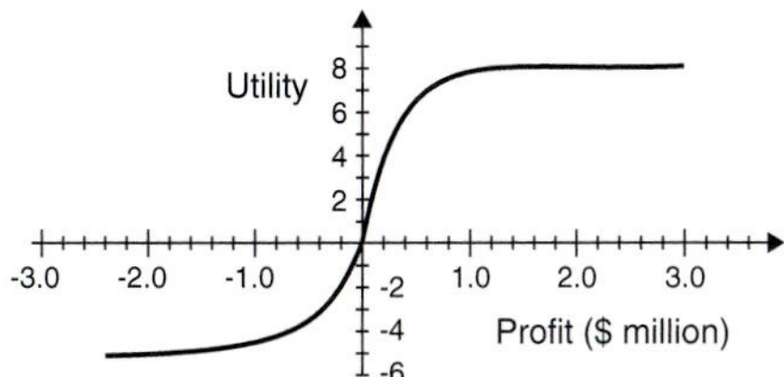

Draw the decision tree and show all computations necessary to determine the optimal bidding strategy. What is the best strategy and associated expected utility? Is this different from the strategy found in Example 8-1? If so, why?

8.4. A contractor is planning to pour concrete on Monday morning, but he has heard a weather forecast that rain is expected. The probabilities of rain are 0.8 on Monday, 0.6 on Tuesday, 0.3 on Wednesday, and 0.1 on Thursday and each succeeding day for five days. If it rains on the day that concrete is poured, there will be an immediate $10,000 damage cost and a delay of five days in the construction schedule. For every day of delay, $4000 is lost. The delay cost applies even if it is the contractor who decides to delay the pour.

Determine on which day the concrete should be poured. What is the associated expected cost?

8.5. A suburb with a population of 20,000 is beginning to undergo rapid housing development, and is experiencing failure of its current individual household septic tank disposal system. The suburb has decided to build a sewage treatment plant to meet its needs for the next 20 years. The cost of the plant depends on the hydraulic flow, which is directly related to the population served. The suburb estimates its population 20 years from now as shown below:

Population	Probability
20,000 ≤ population ≤ 25,000	0.30
25,000 ≤ population ≤ 30,000	0.30
30,000 ≤ population ≤ 35,000	0.40

Estimates are obtained on the cost of a wastewater treatment plant built today to meet the needs of three possible population levels as shown below:

Plant Capacity (persons)	Cost ($)
25,000	1,200,000
30,000	1,400,000
35,000	1,550,000

It has been suggested, however, that money might be saved by building to a small capacity now and expanding the plant after ten years if population begins to exceed capacity at that time. Cost estimates, expressed in present value terms, are obtained for upgrading the size of the plant ten years from now. These are provided in a third table:

Capacity (persons)	Upgraded Capacity (persons)	Present Worth Cost ($)	Probability
25,000	30,000	190,000	0.20
		230,000	0.50
		270,000	0.30
25,000	35,000	320,000	0.25
		390,000	0.55
		410,000	0.20
30,000	35,000	130,000	0.30
		190,000	0.50
		210,000	0.20

Determine the size plant that the suburb should build today. Draw and label the decision tree and state the optimal expected cost.

8.6. A water resources engineer has been called in to help with the design of a temporary bridge for a construction project. The bridge will be used for one year during construction and then removed. The two available options are shown in the figure below. One is an earth fill with a pipe culvert of undetermined diameter, the other is a wooden truss bridge. The cost for the wooden truss structure (including demolition)

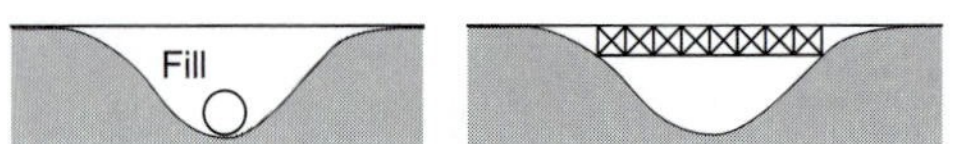

is $65,000. The cost of the earth-fill bridge (including demolition) is given in the table as a function of culvert diameter. There are additional costs for the earth-fill bridge associated with overtopping by flood water unable to be passed through the pipe culvert. They are (1) reconstruction cost of pipe and fill if this type of bridge is destroyed by overtopping and (2) damage costs downstream if the earth-fill bridge is destroyed. Both types of costs are shown in the following table:

Diameter of Pipe Culvert (feet)	Cost of Earth-fill Bridge ($)	Reconstruction Cost If Pipe and Fill Destroyed ($)	Damage Downstream If Pipe and Fill Destroyed ($)
4	20,000	10,000	40,000
6	25,000	15,000	40,000
8	30,000	20,000	40,000
10	40,000	25,000	50,000
12	60,000	30,000	50,000

It is assumed that the earth-fill bridge will not be overtopped more than once in the year and that the wooden bridge will not be damaged regardless of the flood flow level for the one year period.

If the earth-fill bridge is overtopped, it will not always be destroyed. There is a 0.20 probability that it will survive and a 0.8 probability that it will be destroyed. The project engineer has developed flood flow and probability relationships shown in the following table for annual peak flows.

Pipe diameter (ft)	Allowable flow, Q, (cfs $\times$ 1000)	Probability that $Q_{peak} < Q$
4	1.0	0.06
6	1.8	0.20
8	2.8	0.53
10	4.1	0.82
12	7.0	0.98

Using these cost data along with the flow and probability relationships, determine the best bridge type to build. If this is the earth-filled bridge, state the optimal pipe diameter to use (to the closest even diameter pipe, in feet, e.g., 4, 6, 8). Draw the decision tree associated with the engineer's problem noting on it all probabilities, terminal payoffs, and expected values, and state the optimal strategy and associated expected cost.

8.7. In Example 8-2, the engineer learns that it is virtually certain that gravel lenses will be encountered when driving piles at the building site. This will double the cost of pile foundations to $500,000. Rework Example 8-2 under this assumption. Note that if the spread footings need to be replaced with piles, the replacement cost will increase by $250,000.

(a) Draw the complete decision tree.

(b) Determine the optimal strategy and cost.
(c) Calculate the EMVI and the benefit-cost ratio for the experiment.
(d) Calculate the EMVPI.

8.8. *The following problem is motivated by an oil well exploration problem presented by Raiffa (1968).* A water resources engineer is considering the economic feasibility of a centralized well field source for supplying a small community in an arid region. Surface water supplies are available, but at such a distance that they would be expensive to secure. The engineer must decide whether to construct (act a_1) or not to construct (act a_2) the well field. It is uncertain as to whether the well field will provide (1) only an insignificant amount of groundwater (state θ_1), (2) an amount sufficient to supply part of the city's needs (state θ_2), or (3) an amount sufficient to completely supply the community needs (state θ_3).

If a well water supply is found, part or all of the cost of the surface water source can be avoided, but the cost of the well field must be considered. The net returns (savings) under the different states of nature are calculated in the first table below. The cost for installing the well field is \$700,000, and the cost of the remote reservoir source is \$2,700,000.

	Act	
State	a_1	a_2
θ_1	\$ = 700,000	0
θ_2	500,000	0
θ_3	2,000,000	0

In addition, at a cost of \$100,000, the engineer could drill a series of small test wells to determine whether they produced a low yield, a medium yield, or a high yield. Based upon historical records and experience, the engineer estimates the joint probabilities for the states of nature and the results of the test wells as shown in a second table:

	Test Well Yield			
State	Low z_1	Medium z_2	High z_3	Marginal Probability of State
Low yield, θ_1	0.30	0.15	0.05	0.50
Medium yield, θ_2	0.09	0.12	0.09	0.30
High yeld, θ_3	0.02	0.08	0.10	0.20
Marginal probability of test yeld	0.41	0.35	0.24	1.00

The engineer wants to know (1) the optimal action if experimentation (test wells) is not conducted, (2) the optimal strategy for experimentation and action, (3) the best overall strategy, (4) the expected monetary value of information in this case, (5) the benefit-cost ratio for the experiment, and (6) the EMVPI.

8.9. Private utilities must obtain many federal, state, and local permits before constructing a coal-fired electric generating plant. In most cases, the key permit is from the federal agency responsible for writing the environmental impact statement for the plant. Assume that a utility estimates the probability of the permit being denied as 0.2 and of being approved 0.8.

As part of the decision-making process, the utility must determine whether it is worthwhile purchasing land for the plant, and if purchased, whether "limited site work" is to occur, all before the permit decision is made by the federal agency. If the land is purchased, it is usually done anonymously through a real estate representative before any permits are applied for, in order to avoid possibly having to buy at a premium from sellers who may be reluctant to sell their land for a power plant, and to avoid eminent domain proceedings that might be necessary later. If the land is purchased, a further decision must be made during the permitting process as to whether "limited site work" (clearing, grubbing, road work, etc.) is to be carried out, all at the utility's risk. The incentive to begin site work is to advance any subsequent construction process, thereby saving interest costs and avoiding inflated future construction costs. The danger in buying the land and doing site work is that if the permit is later denied, the land would have to be sold, usually at a loss to the utility.

For a particular case, assume that the utility can immediately buy the necessary 1000 acres of land for $1000 per acre, or that it could invest this money and have $1,200,000 accumulated at the time of the agency's decision on the permit (the latter amount is thus the opportunity cost of the land purchase to the utility). If the land were purchased, the utility could also clear the site for an additional $100,000, again expressed as the opportunity cost at the time of the agency's permit decision. If the agency's decision were to deny the permit, the utility could sell the uncleared land for $1000 per acre (zero appreciation in real terms), or the cleared land for $500 per acre. By clearing the land, the utility could advance its construction schedule and save $300,000 in interest and differential inflation costs if the permit were approved.

Alternatively, the utility could wait and purchase the land only if and when the agency approved the key permit. In this case the land would cost $1400 per acre. Also, since the utility would have to negotiate directly with the landowners (or possibly go through eminent domain proceedings), the construction schedule would be delayed, causing a loss of $300,000 in interest and inflation costs. In any case, if the permit is denied, the utility would have to do something else to secure the necessary power, but it is assumed that there is no cost or benefit in so doing.

(a) Should the utility immediately buy the land or not, and if purchased, should 'limited site work' be carried out or not? Draw the complete decision tree and determine the optimal action and associated expected savings.

(b) Increase the probability of permit denial by tenths (to 0.3, 0.4, etc.) and determine the probability levels at which it would first become best not to clear the land, and best not to buy the land immediately.

8.10. A company is to carry out a rain-sensitive component of a construction project. One option is to hire a subcontractor (sub) to do the work at a fixed cost, c, regardless of weather conditions. Another option is for the company to rent equipment and do the job itself. If it does the job itself and it does not rain during the construction period, a savings of 25% will result. However, if it does rain, the company is not prepared for all the modifications that would be necessary, and the final cost would be twice that of hiring the work done. Since this situation comes up frequently, the company has analyzed weather data for the construction period as shown in the table below. It is now 30 days from the beginning of the construction period, and the company has four options: (1) hire the subcontractor immediately, (2) rent (reserve) the equipment immediately in order to do the work itself, (3) wait until the the seven-day weather report to decide, or (4) wait until the one-day weather report to decide. No equipment rental penalties are incurred for waiting, but if the company waits until one day before the construction period, the cost of hiring a subcontractor will be 25% higher.

	Joint Probabilities			
	Seven-Day Forecast		One-Day Forecast	
Actual Outcome	Rain	No rain	Rain	No rain
Rain	0.08	0.12	0.12	0.08
No rain	0.17	0.63	0.18	0.62

(a) Are there biases in the weather forecast? If so, what are they?

(b) Draw the complete decision tree for the contractor and determine the optimal action and associated cost (carry probability calculations out to the third decimal place when necessary). What is the expected percentage cost savings?

(c) Assuming that the two forecasts are independent, if the seven-day forecast is for rain, should the contractor wait for the one-day forecast before deciding what to do, or should he take immediate action?

(d) What is the EMVI gained from knowing only the marginal probabilities of rain and no rain for the construction period? What is the EMVI gained from knowing the joint probabilities for the seven-day and one-day weather forecasts?

(e) What is the EMVPI for this problem, assuming that the company knows the marginal probabilities for rain and no rain?

8.11. Reduce the payoff matrix given in the table, eliminating any action that is dominated by another.

	θ_1	θ_2	θ_3	θ_4	θ_5
a_1	8	7	9	4	10
a_2	12	9	6	3	8
a_3	6	2	8	1	8
a_4	4	8	7	6	5

Then determine the best action using each of the following criteria:

(a) Laplace
(b) maximin
(c) maximax
(d) Hurwicz (show graph)
(e) minimax regret

Do any actions seem preferable overall?

8.12. A young engineer's consulting firm has done well, and the owner wants to purchase new office space for the firm and as a real estate investment. Shopping around, the engineer has found three financing options for the mortgage. A variable-rate mortgage is available that uses a market-sensitive "S"-index. The index closely follows current market interest rates. If interest rates go up, the engineer would have to pay higher interest on the mortgage, based on the S-index. Of course, if interest rates go down, the engineer would pay less interest, again based on the S-index.

Another variable-rate mortgage is available that uses a lagged "L"-index. Since this index includes a multiperiod lag, it is not as sensitive to current market interest rates, and changes more slowly.

Finally, a fixed-rate mortgage is available which sets the interest rate at the current market fixed rate. Fixed rates are usually initially higher than variable rates. The engineer estimates the payoff matrix shown in the table below, where numbers represent annual interest rate savings.

	Future Interest Rates		
	Up	Down	Same
S-index	−3	3	1
L-index	−2	2	1
Fixed	0	0	0

Check for dominance and determine the best action to take using the following criteria:

(a) Laplace
(b) maximin
(c) maximax
(d) Hurwicz (show graph)
(e) minimax regret

8.13. An engineer is planning an investment with a time horizon of from six to seven years. He is considering five different investments, as shown in the following table and has estimated their annual percentage rate of return under four possible future states of the economy. He wants to undertake only one of the investments.

Growth: Inflation:	High High	High Low	Low Low	Low High
a_1: Two 3-year CD's	−2	1	1	−2
a_2: Mutual fund	4	7	0	−3
a_3: Growth stock	9	12	−3	−6
a_4: 7-year bond	−1	2	2	−1
a_5: 30-year bond w/resale	−2	3	3	−2

Check for dominance, and determine the ranking of the remaining actions on the basis of the following criteria:

(a) Laplace
(b) maximin
(c) maximax
(d) Hurwicz (show graph)
(e) minimax Regret

Discuss your results; Which would you, yourself choose, and why?

8.14. A project engineer has assumed responsibility for a construction project in a country unfamiliar to her. She has just learned that an important permit must be obtained from one of the country's federal agencies before the project can proceed. She has no idea whether this permit will be approved or not, but she is assured that the decision will be made in three months or less. She is also aware that the sponsor will pay a \$5 million bonus if the project is finished one month early, \$7 million if two months early, and \$8 million if three months early. The normal start date is three months away, and the first activity is to order materials and equipment which will take one month to arrive, at the earliest.

The engineer decides to investigate whether it would be worthwhile ordering the materials and equipment before the permit is obtained, so that the project could be finished early. Her buyers inform her that if the materials arrive at the project site, she will have to pay for them at that time, and if the materials are not used right away, interest charges of \$1 million per month will be lost. Further, if the order must be canceled after arrival there will be a penalty of \$5 million. If an order is canceled one month before arrival, the penalty is \$2 million, and if two months before arrival, the penalty is \$1 million. To speed up a previous order, an additional cost of \$1 million is incurred.

Order Material to Arrive at the End of Month:	Permit Approval at the End of Month			Permit Disapproved at the End of Month		
	1	2	3	1	2	3
1	8	6	3	−5	−6	−7
2	7	7	4	−2	−5	−6
3	6	5	5	−1	−2	−5
Wait to order	7	5	0	0	0	0

Verify the payoff matrix, check for dominance, and determine the best ranking of actions using the following criteria:

(a) Laplace
(b) maximin
(c) maximax
(d) Hurwicz (show graph)
(e) minimax regret

Based on your results, what do you suggest that the engineer do?

REFERENCES

ADRIAN, JAMES J., *Quantitative Methods in Construction Management.* New York: American Elsevier, 1973.

DE NEUFVILLE, RICHARD, *Applied Systems Analysis.* New York: McGraw-Hill, 1990.

HURWICZ, LEONID, "The Generalized Bayes-Minimax Principle: A Criterion for Decision-Making Under Uncertainty," *Discussion Paper: Statistics, No. 355.* Cowles Foundation. Cowles Commission, New Haven, Conn., February 1951a.

HURWICZ, LEONID, "Optimality Criteria for Decision Making Under Ignorance." *Discussion Paper: Statistics, No. 370*, Cowles Foundation, Cowles Commission. New Haven. Conn., December 1951b.

LUCE, R. DUNCAN and HOWARD RAIFFA, *Games and Decisions: Introduction and Critical Survey.* New York: John Wiley, 1957.

MILNOR, JOHN, "Games Against Nature," *Research Memorandum RM-679.* The RAND Corporation, Santa Monica, September 1951. Also reprinted in Shubik, Martin ed., *Game Theory and Related Approaches to Social Behavior.* New York: John Wiley. 1964.

RAIFFA, HOWARD, *Decision Analysis: Introductory Lectures on Choices Under Uncertainty.* Reading, MA: Addison-Wesley, 1968.

RAIFFA, HOWARD and ROBERT SCHLAIFER, *Applied Statistical Decision Theory.* Cambridge, MA: The M.I.T. Press, 1961.

SAVAGE, L. J., "The Theory of Statistical Decision." *Journal of the American Statistical Association.* Vol. 46, no. 10, 1951, pp. 55–67.

SCHLAIFER, ROBERT, *Analysis of Decisions Under Uncertainty.* New York: McGraw-Hill, 1969.

VON NEUMANN, JOHN and OSKAR MORGENSTERN, *Theory of Games and Economic Behavior.* Princeton, NJ: Princeton University Press, 1944.

CHAPTER

9 Lessons in Context: Simulation and the Statistics of Prediction

9.A INTRODUCTION

In this chapter, we explore the ideas of regression and simulation. These two concepts are investigated in the context of a practical situation in the area of water resources management. That is, we will first describe a water resources planning problem and indicate two specific problem questions that need to be addressed. We will then address the questions individually introducing the statistics of prediction to answer the first question and the methodology of simulation to answer the second.

A town is considering building reservoirs for water supply on two parallel and unconnected streams. The water supplies drawn from these two reservoirs are to be combined to form the water supply for the town.

A number of issues make the problem nontrivial. First, since the town is only thinking about acquiring the watersheds and building the dams, the volumetric capacities of the two reservoirs have yet to be determined. Furthermore, a long record of stream flows (60 years) exists on only one of the streams. The other stream has had its flows measured and recorded for only the last 20 years. Finally, even if both of the reservoirs were built, it is not clear how they would be operated together to yield the maximum supply.

Obviously, these issues of capacity, sufficient record, and joint operation are all related, and all need resolution in some manner. The two techniques we will utilize in this chapter to deal with these issues are statistics and simulation. Statistics will allow us to fill in the record of stream flow for the stream with the shorter record. Simulation will provide us a means to test various operating policies to determine a good strategy for operations. Once a good strategy for joint operations is found, we can test the impact of varying the capacities of the reservoirs.

At this point, probably half of the readers who were with us on the last paragraph have fallen away because of the mention of statistics. Now, with one of our true objects revealed—the teaching of statistics—many readers instinctively recoil. Once again, devious textbook authors have lured innocent students into a subject that has been known to induce immediate sleep in previously lucid and alert people. Are you still with us? Good! There is hope. We want to assure you that we will be brief—for as long as it takes in our discussion of statistics.

This is our complete problem. We want to determine the sizes and operating policies for the two reservoirs. To do this, we need a long record for each reservoir, not just one long record for one of them.

To obtain long records for both streams, we will need to utilize the statistical procedure known as regression, which we now describe in detail in the context of our stated problem.

9.B THE METHOD OF REGRESSION

The subject of statistics has many facets. Most often, statistical procedures are used to characterize and summarize sets of data. You have undoubtedly already been exposed to the ideas of a mean value and standard deviation as well as to other related descriptors of the characteristics of a data set. Mean value and standard deviation are typically applied to data that are unchanging in time but whose realized values are often different. The number of hours required to complete a routine and standard task such as welding a particular type of joint is an example.

Statistical procedures are also used for purposes of prediction. Sometimes, the values of one set of numbers "run with" the values of another set of numbers. For instance, higher per capita amounts of electricity usage in summer are associated with higher summertime temperatures because air conditioners are run longer during periods of elevated temperatures. Retail sales in a community shopping mall are associated with the percentage of workers employed in the nearby area as well as with mean incomes (setting aside inflationary effects). Our interest here is in the use of statistics for prediction of the stream flow values in the stream with the shorter record when the one stream with the longer record was monitored and the other was not.

The two streams, which are to form the water supply for the town, run through regions of similar soils and topography, although the watersheds that feed the two streams are of different areas. In addition, one of the streams lies to one side of a ridge of hills, and the other is on the opposite side of the ridge.

The average wind direction is the same across the two watersheds, but because of the line of hills, slightly more rain per unit area falls on one basin than on the other. That is, the clouds drop more of their moisture on the upwind side of the hills than on the downwind side. Stream 1 whose record of flows is longer is on the upwind (wetter) side of the ridge of hills. Stream 2 whose record is shorter is on the downwind (drier) side of the ridge of hills. Stream 1 has been gauged for the last 60 years, stream 2 for only the last 20 years.

The records of flows from stream 1 is to be used to extend the record of flows on stream 2 backward in time so that we have 60 years of flow record for both streams. Our assumption is that in any particular month of the year, the two flows, though different in value, will "run together." Said another way, higher values of flow in stream 2 will be associated with higher values of flow in stream 1.

To extend the record of stream 2 backward in time, we build a simple linear model of stream flow prediction.[1] Our linear model associates the flow in stream 2 with the flow in stream 1 for each month in the following way:

$$Y = aX + b$$

where

Y = flow in the particular month (e.g., February) being studied in stream 2 in billion gallons;

X = flow in stream 1 in the same month (February) being studied in billion gallons; and

a, b = constants, to be determined, used to predict the flow in stream 2 given the flow in stream 1.

A model just like this one is built for each month, associating the flow in stream 2 with the flow in stream 1 in that same month. It will be used for prediction of the stream 2 flow value when only the record of flow for stream 1 is available.

The task is to find the values of a and b in this simple predictive model. We know, however, that the values of Y for various values of X are only predictions of flows in stream 2. The relationship is not a perfect one.

Consider the hypothetical graph (Figure 9.1), of the flow in stream 2 versus the flow in stream 1 in February. The points on the graph consist of pairs of values (y_i, x_i). A particular (y_i, x_i) pair constitute measurements or observations of the flow in stream 2 (y_i) and the simultaneous (same month) flow in stream 1 (x_i) in year i. Note that the points do not fall on a straight line nor on any simple smooth curve. The points do cluster into a band and suggest a relationship between the flows, but the relationship is not precise. For instance, point A associates a relatively low flow in stream 2 with a high flow in stream 1. In general, we would have expected a higher flow in stream 2, based on the recognition that, in general, flow in stream 2 increases with the flow in stream 1. As you can see from

[1] By simple, we mean that we are going to ignore the correlation that exists between flows from one month to the next.

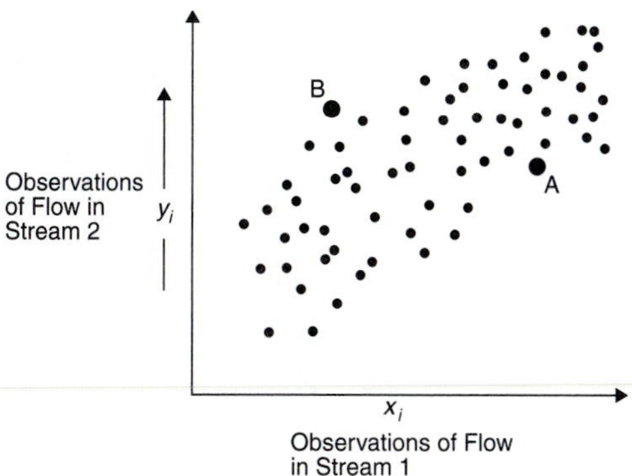

Figure 9-1 Observations of flows in the parallel streams.

the graph, however, this is a general as opposed to a specific conclusion that we have drawn. Likewise, at point B, we have a relatively high flow in stream 2, associated with a not-so-high flow in stream 1. Again, this illustrates that our relation may be good but is not perfect. The fact that statistical predictions are not perfect should not set us against them but only make us more cautious in their use.

We next investigate how to determine values for a and b in the linear predictive model suggested above. We observe that the equation $Y = aX + b$ is a line with slope a and y-axis intercept of b. As these values are being "tuned" to their best values the line they represent shifts in position. Larger values of b shift the curve upward. Larger values of a increase the angle the line makes with the x axis. One intuitive measure of goodness of the predictive line is the vertical distance of the line from the observation at a particular value of flow in stream 1.

Consider the graph in Figure 9.2. Only four data points (1, 2, 3, 4) are shown.

The data points, indicated by the dots, give the clear impression that the flow in stream 2 increases with the flow in stream 1. We show three possible lines as candidates for the best-fitting line. The reader can verify that the sum of the vertical distances from each of the four observations to line A is the same as the sum of distances for line B and is the same as the sum for line C. That is, all three lines seem to do equally well at reducing the sum of the distances from observations to

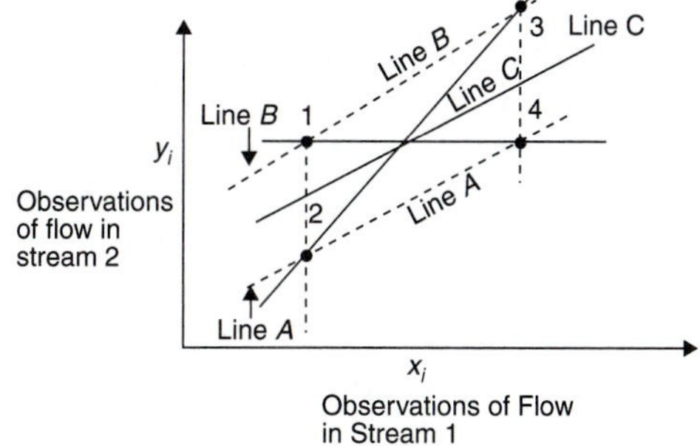

Figure 9-2 Three possible lines that predict the flow in stream 2.

the line. At the same time, line C seems to be most appealing, suggesting that some other factor is at work in our judgment of what is and is not "good."

Apparently when the sum of the vertical distances is used as the measure of goodness of the fit of the line to the data, the resulting line need not be located centrally with respect to the observations. The goal is to select a measure of goodness that will indicate that line C is a better candidate than lines A and B for fitting the data. A widely used and easily implemented measure that fills this need is the so-called "least squares" measure of goodness. By least squares is meant that the sum of the square of the vertical distances from the data points to the predictive line is to be a minimum. This measure, because it focuses on the squares of the distances, brings down the value of the largest distance of any observation to the line, and thus makes the line more central with respect to the data points. In fact, this measure performs beautifully at centering the line with respect to a scattered set of data and reducing the distance between "outliers" and the predictive line.

The equation for the predictive line for the particular month under study is

$$Y = aX + b$$

where Y is the flow predicted in stream 2 and X is the flow recorded in stream 1. The coefficients a and b are constants to be determined. The data points, or observations, are pairs of stream flow values (y_i, x_i), where the subscript i is the indicator of the year in which the pair of flows was recorded. The value y_i is the flow recorded in stream 2 in year i. The value x_i is the flow recorded in stream 1 in year i.

The objective to be minimized is the sum over all observations of the squared vertical distance from point to line. The squared vertical distance from the ith point to the line is

$$[y_i - (ax_i + b)]^2$$

where $(ax_i + b)$ is the predicted value of the flow in stream 2 given the flow in stream 1, and y_i is the value of flow in stream 2 that is actually observed. More precisely, we seek to minimize the sum of squared deviation (SSD):

$$\text{Minimize} \quad Z = \sum_{i=1}^{n} [y_i - (ax_i + b)]^2$$

where n is the number of years during which flow measurements were taken in both of the streams. Remember that we are using those years in which measurements were made of flows in both streams in order to establish a relationship which we will use to predict the flow in stream 2 in those years when only the flow in stream 1 is available.

To determine the values of a and b which minimize the above unconstrained objective function we simply take two partial derivatives, one with respect to the unknown a (where b is assumed a constant) and the other with respect to the unknown b where a is assumed a constant. These partial derivatives are simply the sum of the partial derivatives of the individual terms. The summations are set equal to zero to find the minimum individual terms.

$$\frac{\delta z}{\delta a} = \sum_{i=1}^{n} 2[y_i - (ax_i + b)]\,(-x_i) = 0$$

and

$$\frac{\delta z}{\delta b} = \sum_{i=1}^{n} 2[y_i - (ax_i + b)]\,(-1) = 0.$$

Manipulation of these two equations gives

$$\sum_{i=1}^{n} (y_i x_i - a x_i^2 - b x_i) = 0$$

and

$$\sum_{i=1}^{n} (y_i - a x_i - b) = 0.$$

Applying the summation to individual terms results in two equations that are linear in a and b

$$\sum_{i=1}^{n} y_i x_i - a \sum_{i=1}^{n} x_i^2 - b \sum_{i=1}^{n} x_i = 0$$

and

$$\sum_{i=1}^{n} y_i - a \sum_{i=1}^{n} x_i - bn = 0,$$

which can be set up for solution as

$$a \sum_{i=1}^{n} x_i^2 + b \sum_{i=1}^{n} x_i = \sum_{i=1}^{n} y_i x_i$$

and

$$a \sum_{i=1}^{n} x_i + bn = \sum_{i=1}^{n} y_i.$$

Solution by ordinary methods yields

$$a = \frac{\sum_{i=1}^{n} x_i y_i - (1/n)(\sum_{i=1}^{n} x_i)(\sum_{i=1}^{n} y_i)}{\sum_{i=1}^{n} x_i^2 - (1/n)(\sum_{i=1}^{n} x_i)^2}$$

and

$$b = \left(\frac{1}{n}\right)\sum_{i=1}^{n} y_i - a\left(\frac{1}{n}\right)\sum_{i=1}^{n} x_i.$$

These are the two unknown constants which are needed in the predictive equation $Y = aX + b$. If these constants are determined for each month, the missing data for stream 2 can be "filled in," resulting in two "completed" long records for the parallel streams.

9.C COMPUTER SIMULATION

Now we have two long records. The flows are

I_{1t} = recorded flow in stream 1 in month t, in billion gallons, and

I_{2t} = flow in stream 2 in month t, either recorded or "filled in" in billion gallons.

These flow values are available for all months t ($t = 1, 2, \ldots, n$). The flows will be utilized in a computer simulation of the operation of two reservoirs. One of the reservoirs is on stream 1 and the other on stream 2. We need to establish capacities for each of the reservoirs, and we need to find rules for the operation of the two reservoirs given the capacities that are assigned.

What is computer simulation? Is it a new technique or an old idea? It turns out that simulation is an old idea dressed up in new clothing. The old idea can now be implemented on a scale not thought possible just 50 years ago; modern computing resources have made the difference.

The old idea is "organized trial." We can illustrate organized trial by an example. Suppose that you have a reasonably complex equation in x that is equal to zero. Suppose further that you don't know what value of x solves this equation (makes it equal to zero) and that you can't find a technique to provide the value of x in a single step. Under the circumstances, you might try in succession values of x starting with an initial value of x_0. Each value would be tested in the equation to see if it makes it equal to zero. The value x_0 would be followed by $x_0 + a$, $x_0 + 2a$, $x_0 + 3a$, and so on, where a is some relatively small constant. Each value would be tried in the equation in turn to see if it "solves" the equation (i.e., makes the equation equal to zero).

It may be that two successive values of x that are tried cause the numerical value of the equation to change from positive to negative. In general, this suggests that the appropriate value of x, the one that makes the numerical value equal to zero, is somewhere between the two values of x that caused the sign to change. Exploring this region between these two values of x in an even more precise search with smaller constants should reveal an even narrower range for the correct value of x. The process continues to focus ever more tightly until the range within which x is known to lie is of sufficiently small size to be reported as the answer.

The process of locating a range for the value of x can usually be conducted by hand, but a computer and an appropriate computer program can speed the completion of the steps. This "organized trial" to find x is a lot like simulation and, if a computer program is used, very much like simulation. Simulation is really organized trial with multiple measures of goodness and many variables and perhaps no clear-cut way to evaluate and compare the outputs.

We describe a computer simulation in the context of sizing and operating together the reservoir placed on stream 1 and the reservoir placed on stream 2. To begin the problem, we need temporarily to establish the capacities of the two reservoirs. We will do this symbolically as opposed to numerically.

Let:

C_1 = tentative capacity of reservoir 1, in billion gallons, a numerical value, and

C_2 = tentative capacity of reservoir 2, in billion gallons, a numerical value.

It will be assumed throughout the remaining discussion that the water supply requirement is constant month to month and is equal to D, where D is the amount of water needed for municipal and industrial use. This simplifying assumption is made only to make our discussion easier. In a real situation, water demands do vary by month, and in fact, simulation handles such situations easily. In this case, however, to make the introduction to simulation easier to understand, we avoid this modest complication.

Our next step is to create operating rules for each of the two reservoirs. The first question that arises is how to divide the water supply contribution between the two reservoirs. We will propose three rules for the division between these reservoirs. All three should be evaluated and critiqued. A number of other rules are also reasonable to suggest. You should suggest other rules to test your understanding and intuition of what constitutes "goodness" in water supply reservoir operating rules.

Rule I: The contribution of a particular reservoir toward the water requirement D should be in the same proportion as the ratio of the average annual inflow of that reservoir to the total of the average annual inflows to both reservoirs.

Let:

Q_1 = the average of the annual inflow to reservoir 1;

Q_2 = the average of the annual inflow to reservoir 2;

α = fractional contribution of reservoir 1 toward the water requirement; and

β = fractional contribution of reservoir 2;

where

$$\alpha = \frac{Q_1}{Q_1 + Q_2} \quad \text{and} \quad \beta = \frac{Q_2}{Q_1 + Q_2}$$

Thus, in any month, reservoir 1 contributes αD and reservoir 2 contributes βD. In some months, however, it may not be possible to deliver these quantities because of insufficient inflow and storage, and in some months, so much water may be accumulated that water may be wasted to the stream rather than to water supply. These rules may be further developed and refined using the following notation:

S_{1t} = storage in reservoir 1 at the end of month t;

S_{2t} = storage in reservoir 2 at the end of month t;

W_{1t}, W_{2t} = the amounts spilled or wasted from reservoirs 1 and 2 in month t;

X_{1t}, X_{2t} = the amounts *actually* contributed toward water supply D by reservoir 1 and by 2 in month t; and

I_{1t}, I_{2t} = the inflows to reservoir 1 and to reservoir 2 in month t.

We can describe three basic storage and flow conditions for the reservoirs depending on their relative fullness and the extent of their inflows. These conditions will dictate actual releases and spills. The first condition is an insufficiency of flow and storage to meet release needs. The second is a sufficiency of flow and storage to meet release needs. The third is an overabundance of flow and storage leading to the wasting of water to the stream. These three conditions lead to different release and spill strategies which are described in the following equations. The releases from the two reservoirs are described first.

$$X_{1t} = \begin{cases} S_{1,t-1} + I_{1t} & \dots \quad S_{1,t-1} + I_{1t} < \alpha D, \\ \alpha D & \dots \quad S_{1,t-1} + I_{1t} \geq \alpha D. \end{cases}$$

$$X_{2t} = \begin{cases} S_{2,t-1} + I_{2t} & \dots \quad S_{2,t-1} + I_{2t} < \beta D, \\ \beta D & \dots \quad S_{2,t-1} + I_{2t} \geq \beta D. \end{cases}$$

These equations say that each reservoir behaves relative to water supply releases in the following way. First, if the sum of the previous end-of-period storage and this month's inflow are less than the designated contribution of the particular reservoir, the reservoir will release all the water that it has available toward its water supply target. Nonetheless, a shortfall from its target contribution will occur. Second, if the sum of the previous end-of-period storage and this month's inflow equals or exceeds the designated water supply contribution of the reservoir, the reservoir will release its full target contribution. Using similar conditional equations, we can define when spill or waste occurs. By spill, we do not mean water flowing in the spillway. Such a flow is a last resort in flood situations, and it could actually damage the spillway. Instead, we mean controlled releases to the stream via ordinary outlet works which are over and above planned releases. These releases are made to prevent the need for ever having to use the spillway.

Equations are also needed to ensure that each reservoir behaves in terms of controlled spill to the stream in the following way.

$$W_{1t} = \begin{cases} 0 & \dots \quad S_{1,t-1} + I_{1t} - \alpha D \leq C_1, \\ S_{1,t-1} + I_{1t} - \alpha D - C_1 & \dots \quad S_{1,t-1} + I_{1t} - \alpha D > C_1. \end{cases}$$

$$W_{2t} = \begin{cases} 0 & \ldots \quad S_{2,t-1} + I_{2t} - \beta D \leq C_2, \\ S_{2,t-1} + I_{2t} - \beta D - C_2 & \ldots \quad S_{2,t-1} + I_{2t} - \beta D > C_2. \end{cases}$$

If the sum of the previous end-of-month storage plus current inflow less the reservoir's designated water supply contribution exceeds the storage capacity of the reservoir, the reservoir will spill that excess volume to the stream. Next we describe reservoir storages through time.

$$S_{1t} = S_{1,t-1} + I_{1t} - X_{1t} - W_{1t}$$

$$S_{2t} = S_{2,t-1} + I_{2t} - X_{2t} - W_{2t}$$

where X_{1t}, X_{2t}, W_{1t}, and W_{2t} are given by previous equations. These last equations update the contents of the reservoirs to their values at the end of the current month t. The equations utilize the values of the releases and the spills as they were calculated in the equations that immediately preceded them.

Beginning with full initial storages for the two reservoirs, the above six simulation equations are implemented again and again, each time with the new inflow for the month at hand. The releases from each reservoir are noted for each period as well as any spills. Special note is taken of the months in which a shortage occurs, that is, when, according to the rules, the full value of the water requirement D cannot be delivered. This occurs whenever either reservoir is unable, according to the allocation rule, to release its suggested contribution. The reservoir storages are traced through time with notations for the months in which a reservoir is empty at the end of the month and the months when each of the reservoirs is full.

Statistics are then tabulated on the following events or conditions:

1. The number of months when the full requirement D was not delivered;
2. The number of months when the full requirement D was not delivered and sufficient water to deliver D was actually available when the contents and inflows of both reservoirs were summed;
3. The number of months when one or both reservoirs were empty;
4. The number of months when one reservoir was full; and
5. The number of months when both reservoirs were full.

The analyst looks across the statistics to observe the behavior of the system and in the process notes that three controllable sets of numbers or rules might make a difference in the reservoir system's behavior. These three items are (1) the water requirement D, (2) the rules for joint operation of the reservoirs, and (3) the capacities chosen for the reservoirs. For simplicity, let us assume that the water requirement is a firm figure that must be used for planning. Water conservation efforts or higher prices might, in fact, alter it, but for this problem we will take it as a given.

The second point at which a change might be made is in the rules that govern the relative contribution of each reservoir to the jointly furnished supply. The rule we began with—that the relative contribution of each reservoir is based on its fraction of the average annual flow of both reservoirs together—had some appeal. The rule, however, did not look very closely at individual months and so may have been unduly restrictive. In a moment, we will examine the system performance under a second, less restrictive rule. For now, however, we note that a third control knob is the capacities of the individual reservoirs. We did not say how the initial capacities were chosen; we might have some freedom to increase one of the capacities and decrease another without much change in cost. Once we have settled on a rule of operation, it would make sense to come back and modify one or more capacities incrementally to see what might be accomplished and at what cost. At this moment in time, though, we return to the task of selecting reservoir release allocation rules.

Rule II: The contribution of a particular reservoir toward the water requirement D for each month i of the 12 months of the year should be in the same proportion as the ratio of that reservoir's average inflow in months $i-1$ and i (the sum of these flows) to the total of the average inflows to both reservoirs in months $i-1$ and i ($i = 1, 2, \ldots, 12$).

Let:

q_{1i} = average total inflow to reservoir 1 in months $i-1$ and i;

q_{2i} = average total inflow to reservoir 2 in months $i-1$ and i;

α_i = fractional contribution of reservoir 1 toward the water supply requirement D in month i; and

β_i = fractional contribution of reservoir 2 toward the water supply requirement D in month i;

where

$$\alpha_i = \frac{q_{1i}}{(q_{1i} + q_{2i})}$$

$$\beta_i = \frac{q_{2i}}{(q_{1i} + q_{2i})}$$

Thus, in any month t (say, month $t = 38$, which is month $i = 2$), the contribution of reservoir 1 toward the requirement D is $\alpha_2 D$ and the contribution of reservoir 2 toward the requirement D is $\beta_2 D$. In some months, however, it may not be possible to deliver these quantities because of insufficient inflow to and storage in one or both of the reservoirs.

Using the same notation as for rule I, the releases, spills, and storages can now be described using a new set of equations. Before writing the equations, we need one piece of new notation to deal with the issue of what month i of the 12-month year corresponds to each month t of the historical record. To derive the month i of the 12-month year from a month t of the record, we use the following equation,

$$i = t - 12\left[\frac{t-1}{12}\right]$$

where $[v]$ = the integer part of the largest integer inside of v.

To see how this works, try $t = 53$:

$$i = 53 - 12\left[\frac{52}{12}\right] = 53 - 12\,(4) = 5$$

Now we can write the release, spill, and storage equations:

$$X_{1t} = \begin{cases} S_{1,t-1} + I_{1t} & \ldots \quad S_{1,t-1} + I_{1t} < \alpha_i D, \qquad i = t - 12\left[\dfrac{t-1}{12}\right], \\ \alpha_i D & \ldots \quad S_{1,t-1} + I_{1t} \geq \alpha_i D, \qquad i = t - 12\left[\dfrac{t-1}{12}\right]. \end{cases}$$

$$X_{2t} = \begin{cases} S_{2,t-1} + I_{2t} & \ldots \quad S_{2,t-1} + I_{2t} < \beta_i D, \qquad i = t - 12\left[\dfrac{t-1}{12}\right], \\ \beta_i D & \ldots \quad S_{2,t-1} + I_{2t} \geq \beta_i D, \qquad i = t - 12\left[\dfrac{t-1}{12}\right]. \end{cases}$$

$$W_{1t} = \begin{cases} 0 & \ldots \quad S_{1,t-1} + I_{1t} - \alpha_i D \leq C_1, \qquad i = t - 12\left[\dfrac{t-1}{12}\right], \\ S_{1,t-1} + I_{1t} - \alpha_i D - C_1 & \ldots \quad S_{1,t-1} + I_{1t} - \alpha_i D > C_1, \qquad i = t - 12\left[\dfrac{t-1}{12}\right]. \end{cases}$$

$$W_{2t} = \begin{cases} 0 & \ldots \quad S_{2,t-1} + I_{2t} - \beta_i D \leq C_2, \qquad i = t - 12\left[\dfrac{t-1}{12}\right], \\ S_{2,t-1} + I_{2t} - \beta_i D - C_2 & \ldots \quad S_{2,t-1} + I_{2t} - \beta_i D > C_2, \qquad i = t - 12\left[\dfrac{t-1}{12}\right]. \end{cases}$$

$$S_{1t} = S_{1,t-1} + I_{1t} - X_{1t} - W_{1t},$$

$$S_{2t} = S_{2,t-1} + I_{2t} - X_{2t} - W_{2t}.$$

These six simulation equations, slightly more complex than the first set, are executed for each month of the record. Once again, statistics on reservoir performance are tabulated. In these new equations, account is taken of the historical average reservoir inflows at each time of the year so that releases are keyed to inflow properties which are closer in time than in the previous model. Hopefully, shortages will occur less frequently with this new and more realistic model, but simulation gives us the power to test the hypothesis that these new rules are indeed better. Nonetheless, our exploration of reservoir operating rules can proceed still further. We propose yet another operating rule.

Rule III: The contribution of a particular reservoir toward a water requirement D for each month i should be in proportion to the storage in that reservoir at the end of the previous month over the sum of the storage in both reservoirs at the end of the previous month. Thus,

$$\alpha_t = \frac{S_{1,t-1}}{S_{1,t-1} + S_{2,t-1}}$$

and

$$\beta_t = \frac{S_{2,t-1}}{S_{1,t-1} + S_{2,t-1}}$$

Once again, the six simulation equations can be written and then executed for each month in the record. Once again, statistics on system performance can be tabulated. These new equations, which make relative contributions dependent on *current* reservoir conditions, should do even better at reliably meeting the requirement D; that is, even fewer shortages should occur.

Other rules can be suggested. One such rule is asked for in the exercise that follows the text of this chapter. Our focus now shifts, however, to the issue of capacities. It is generally true, but not always, that reservoirs are proposed to be built at the maximum feasible size for a given site. This is because it is usually cheaper to build just once than to add on in stages. Still, when the maximum feasible sizes deliver far more water than is projected to be needed by the end of the planning horizon, smaller reservoirs are likely to be proposed.

Suppose reservoir 1 is nearer to the city being supplied and hence is proposed at its maximum feasible size. Reservoir 2, being more distant and requiring pumping to obtain its product, is proposed at less than its maximum feasible size. The requirement D may be the amount of water projected to be needed by the city 30 years hence. Analysis of the occurrence of shortages with the best of the three rules, probably rule III, suggests that, with the current proposed capacities, the periods of shortage are too frequent. If the rule for allocation of the relative contribution is the best we can think of, it is time now to experiment with the capacity of reservoir 2 to see how large an increment of capacity is needed to bring the frequency of shortage to an acceptable level. The analysis is straightforward and will not be described further here.

CHAPTER SUMMARY

This chapter has focused on two widely used techniques in environmental systems engineering: regression and simulation. In the first section, we taught the method of "least squares" as a means to predict flows which had not been observed from other flows which had been recorded in a parallel stream. We used the estimates to "fill out" the record of the partially observed stream, giving us two long records. These long records formed the input to a computer simulation.

A computer simulation was suggested for the operation of reservoirs that would provide the city's water supply. The reservoirs would be placed, one on each stream, and would contribute jointly toward the water supply requirement of the city. Three different rules were explored via computer simulation for the relative contributions of the reservoirs. Simulation is really organized trial on a very large scale made possible by use of the digital computer. The present application utilized steps of one month in which the reservoirs' condition were updated by the rules of operation. The consequences of using each rule was investigated by the creation of statistics on the frequency of shortages and other system characteristics.

EXERCISE

9.1. The operating rules that we described allocated the contribution of each of the two reservoirs to current needs according to historical flow records. Still other ways exist to allocate the proportional contribution of the two reservoirs toward the water requirement of the city.

One way (rule III) is to allocate the relative contribution of each reservoir in proportion to the fraction of total system storage which exists in each reservoir at the end of the previous period. That is, the fraction of the water requirement from each reservoir matches the fraction of the total storage in that reservoir, and so on. This allocation is followed for a particular reservoir unless the previous storage stage plus current inflow in the reservoir is smaller than the portion of the requirement that the reservoir is supposed to fill. In this case both storage and current inflow are completely contributed toward the requirement. This rule, which is keyed to current reservoir conditions, is likely to perform even better than the previously discussed rules (I and II). Rule III was also discussed in the text.

Still another way (rule IV) to allocate the relative contributions between reservoirs is even more complicated but could improve performance still further. Let us now consider the fraction of total water in the system projected to be in a particular reservoir at the end of the current period if no release is made from that or any other reservoir. That is, the fraction of the water requirement furnished from each reservoir is to match the fraction of projected system storage in that reservoir at the end of the current month if no releases are made. The projected storage in a particular reservoir at the end of the month is calculated as the sum of the previous storage (end of last month) plus the *projected* inflow without any release toward the water requirement. The projected system storage is the sum of these projected reservoir storages over all reservoirs. This allocation rule would be followed unless insufficient water is available from a particular reservoir to meet its designated contribution—in which case only available water (storage plus current inflow) would be contributed. This rule is tuned even more tightly than previous rules to current conditions.

(a) Using the notation in the chapter, write the six simulation equations for rule III. Define α_t and β_t both verbally and with mathematics.

(b) Writing the simulation equations for rule IV requires more machinery yet. First, it will be necessary to establish the projected inflow for each month in the long record. Hydrologists have observed that flows in a given basin are serially correlated month to month. For instance, the flow in a particular month seems to be well predicted by the immediately preceding flow. Often, the best predictive equation for the flow in a particular month is based only on the flow in the single preceding month. An equation of the form

$$Y = aX + b$$

where

Y = this month's flow (a prediction);

X = last month's flow (an observation); and

a, b = constants

can be used to establish the projected flow in the current month.

(1) This predictive equation needs to be developed for every pair of months of the year, beginning with the January–December pair and ending with the December–November pair. By reference to a particular pair of months (and hence to any pair of months), explain how you would use the long record available to you on each stream to establish the predictive equation for each pair of months.

(2) Having established 12 predictive equations, you can now proceed to develop the simulation equations. For purposes of these equations we will distinguish two different flows in month t for each reservoir. We illustrate for reservoir 1

I_{1t} = the recorded or "filled-in" flow in month t reservoir 1; these flows constitute the historical record used in the simulation; and

$\hat{I}_{1t}$ = the predicted as opposed to actual flow in month t into reservoir 1 as established by the equation that predicts flow in month t from flow in month $t-1$.

With this notation, you can write α_t and β_t as well as the six simulation equations.

REFERENCES

Benjamin, J. and C. Cornell, 1970; *Probability, Statistics, and Decisions for Civil Engineers*, New York: McGraw-Hill, 684 pages.

Clark, A. B. and R. L. Disney, 1985. *Probability and Random Processes: A First Course with Applications*, New York: Wiley, 324 pages.

Law, A. M. and W. D. Kelton, 1982. *Simulation Modeling and Analysis*, New York: McGraw-Hill, 400 pages.

Loucks, D., J. Stedinger and D. Haith, 1981, *Water Resource Systems Planning and Analysis*, Englewood Cliffs, NJ: Prentice-Hall, 559 pages.

Mays, L. and Tung, Y-K, 1992. *Hydrosystems Engineering and Management*, McGraw-Hill, 530 pages.

Soong, T. T., 1981. *Probabilistic Modeling and Analysis in Science and Engineering*, New York: Wiley, 384 pages.

Tocher, K., 1974. *The Art of Simulation*, English University Press, 184 pages.

CHAPTER

10 Lessons in Context: A Multigoal Water Resources Problem Utilizing Multiple Techniques

10.A THE PROBLEM SETTING

The storage volume in a water reservoir evolves through time in response to random inputs and planned outputs. Reservoir operation is keyed to meeting goals with the quantities released as well as meeting goals that involve the storage in the reservoir. The mathematical model of reservoir storage has rather universal appeal since so many human-made systems involve storage, inputs, and releases. For instance, manufacturing inventory systems involves manufacture, storage, and sales. Blood banking involves acquisition, storage, and distribution. The traffic intersection involves vehicle arrivals, storage in the street, and vehicle departures through the intersection. Oil depot operation includes refining, storage, and sale/shipment. Military parts inventory systems acquire parts, store them, and distribute them at times and places of demand. In fact, the water reservoir can serve as a model for virtually any system that involves inventory. The differences between the systems we have cited have primarily to do with whether decisions are made at the input side (how much to manufacture or purchase) or at the output side (how much to release). But the notions of inventory, operation, and meeting goals are typical to nearly all of the situations described.

The lessons we will cover in this chapter are the following

- Target hitting/goal programming (minimizing the sum of absolute values);
- Minimizing the maximum deviation from a target;

- Minimizing the sum of squared deviations from a target (convex and concave programming by piecewise approximation);
- Minimizing the range (minimizing the maximum, maximizing the minimum); and
- Recursive programming.

The system we will discuss is shown in Figure 10.1. It consists of a reservoir which will have multiple functions, all of which are important. The capacity of the reservoir is given; it is already in place. One of its functions is to supply irrigation water to an area of large farms. A target irrigation flow is furnished for each month. This is the allocation of water that results in the best crop yields for the mix of crops being grown. Another function is, to the extent possible, to maintain the flow in the stream below the reservoir within some desirable bounds. Flows larger than the upper bound can cause streambank erosion. Flows less than the lower bound do not provide sufficient flow for fish survival and migration.

A third function of the reservoir is recreation. To meet this need most effectively requires that the contents of the reservoir change as little as possible through the year. That is, the range of the storage values, the difference between the largest and smallest storage over the year, should be as small as possible. This prevents, to a degree, the exposure of the unsightly mounds of earth that exist beneath the water surface. The last function is the production of hydroelectric energy which is generated when water is released through and turns a turbine. A minimum amount of hydroelectric energy is required each month. The same water that is directed through the turbines provides the down-reservoir stream flow, which, as noted earlier, is desired to be within stated bounds.

The operation to be determined is for the period of the upcoming 12 months. The current storage in the reservoir is given, and monthly stream flows are projected for each of the next 12 months, using previous inflows and historical correlations. Obviously, the flows in later months are much less certain than are flows in the upcoming months. Nonetheless, a plan with first steps to be taken is desired for the next 12 months. It can and will be modified as conditions evolve. That is, after the first month's actions are taken and the first month's real inflow has en-

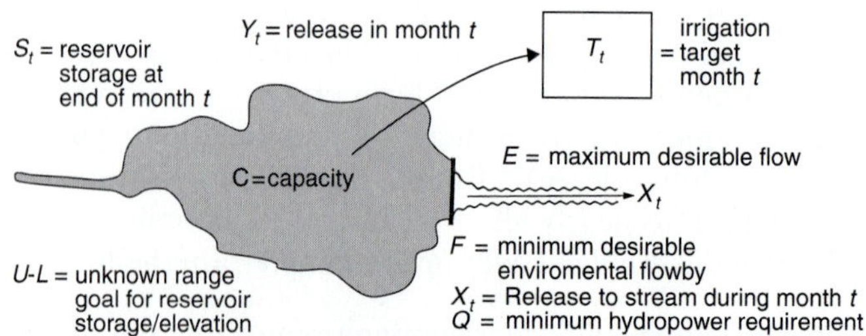

Figure 10.1 Multipurpose reservoir.

tered, the system is in a new state. An additional 12 months of flow are then projected, 12 months of operation are determined, the first month's operation or action is taken, a real inflow occurs, and a new reservoir contents is calculated at the end of the new first month. The process then repeats.

This process of optimization over a future scenario, followed by the actual events of the first interval along with the first action in sequence, followed by recalculation of a new scenario and a subsequent optimization, with the steps repeated over and over, is called *recursive programming*. This programming procedure brings the power of optimization to a process that evolves in time in an uncertain way. When the complete reservoir model is built, we will return to our discussion of recursive programming and elaborate on the sequence of steps. For now, what we have established is that we are determining for a 12-month period of operation with 12 projected monthly inflows, the specific release actions that will guide the reservoir manager for the entire period. Only the first action will be taken, however, before updating predictions and recalculating decisions for another 12 months.

The variables and parameters needed for this problem are

C = capacity of the reservoir, known;

S_t = storage in the reservoir *at the end* of month t, unknown;

X_t = release to the stream during the month t (all of which passes through the turbines);

Y_t = release to the irrigation area during month t (none of which passes through the turbines);

T_t = the target level for the irrigation release in month t;

U = the maximum reservoir storage that occurs over the 12 months, unknown;

L = the lowest level of reservoir storage that occurs over the 12 months, unknown;

E = the largest desirable stream flow; known;

F = the smallest desirable stream flow, known;

Q = the smallest level of the hydropower production rate over the 12-month period in kilowatts, unknown;

I_t = projected inflow to the reservoir in month t, known; and

S_o = initial storage in the reservoir, known.

For this problem, we are ignoring evaporation and seepage from the reservoir. These processes are not particularly difficult to build into the model, but would obscure the programming issues we need to discuss.

We will consider the four functions of the reservoir in turn; these are (1) minimizing the variation in storage, (2) hitting the irrigation target closely, (3) achieving stream flows within desirable bounds, and (4) producing a steady and specified hydropower production level.

10.B THE BASIC MODEL WITHOUT GOALS IMPOSED

There is an underlying model here that needs to be explained before we can focus on the operational objectives or purposes of the reservoir. The model is that of inventory or mass balance, a basic relation in hydrology, that turns out to be a linear relation. As a consequence of this linearity, the inventory or balance equation, although it is a strict equality, can be an integral part of a linear program. This may be your first exposure to a constraint that *defines* a relationship, rather than enforces a limit on a function, but this logic of linear definition works just exactly the same as any other linear constraint—as you will see shortly.

The mass balance equation for the reservoir says that the storage at the end of the current period is equal to the storage at the end of the previous period $(t - 1)$ less any releases during the current period plus inflow during the current period. That is,

$$S_t = S_{t-1} - X_t - Y_t + I_t$$

where Y_t and X_t represent releases toward irrigation and toward down-reservoir stream flow. In standard form, written for each of the 12 months, the balance equation is

$$S_t - S_{t-1} + X_t + Y_t = I_t, \qquad t = 1, 2, \ldots, 12.$$

This *equation* will constitute a basic constraint of the reservoir model. As a constraint of the model, it will always be enforced. That is, the appropriate mass balance equation will always be found to be honored in any postoptimization analysis of the storage-input-output sequence.

In addition to the inventory equation/constraint, the basic model also limits the storage or reservoir contents to the capacity of the reservoir:

$$S_t \leq C, \qquad t = 1, 2, \ldots, 12.$$

Water in excess of capacity will be directed either to the irrigation area or released through the turbines to down-reservoir stream flow.

Finally, it is convention in the building of reservoir models—which of necessity operate through the cycle time of one year—to ensure that water is not borrowed from initial storage and squandered over the year. If it were, operation in the subsequent year could be adversely affected. To prevent "borrowing water," a single constraint is appended to the model that forces the storage at the end of 12 months of operation to equal or exceed the known initial storage. In this case, the constraint is

$$S_{12} \geq S_o.$$

The mass balance equation and these two additional types of equations complete the basic model of reservoir storage and operation. We now proceed to explore how to structure the goals for reservoir operation and how to structure the constraints that dictate reservoir function. Both involve adding new constraints to the basic model. The basic model is summarized as

$$S_t = S_{t-1} - X_t - Y_t + I_t, \qquad t = 1, 2, \ldots, 12,$$

$$S_t \leq C, \qquad t = 1, 2, \ldots, 12,$$

$$S_{12} \geq S_o,$$

or in standard form

$$S_t - S_{t-1} + X_t + Y_t = I_t, \qquad t = 1, 2, \ldots, 12,$$

$$S_t \leq C, \qquad t = 1, 2, \ldots, 12,$$

$$S_{12} \geq S_o,$$

and

$$S_t, X_t, Y_t \geq 0, \qquad t = 1, 2, \ldots, 12.$$

10.C DECREASING THE FLUCTUATION IN STORAGE

The variability in reservoir storage can be approached in a number of ways. In fact, by the time you finish this chapter, you will be able to suggest at least several approaches in addition to the one which we present here. Our suggestion for providing a stable pool level in the reservoir is to minimize the range of storage values which occur through the 12-month period. By the range of storage values, we mean the difference between the largest storage that occurs in the 12 months and the smallest storage that occurs in the 12 months. If this difference is pushed to its smallest value, we have successfully minimized the range of values.

Another way to decrease fluctuations is to limit the month-to-month change in storage. This can be done, as we will show, but it does not limit effectively the range over the 12 months—unless the month to month change is very tightly controlled. Nonetheless, we could add a constraint that forces any individual month-to-month change in storage to be less than or equal to, say, 25% of the range. What we are preventing here is an excursion of storage values that causes the entire range value to occur in just one month-to-month transition.

To limit the range, we operate first on the largest storage to occur over the 12 months. The largest storage is defined by the smallest value of storage which all end-of-period storages are less than. This value is called U, for upper limit. It is precisely equal to the largest of all 12 storages. That is,

$$S_t \leq U, \qquad t = 1, 2, \ldots, 12,$$

or

$$S_t - U \leq 0, \qquad t = 1, 2, \ldots, 12,$$

where some minimization of the unknown U is implied.

We next operate on the unknown smallest storage to occur over the 12 months. The smallest storage is defined by L where

$$S_t \geq L, \qquad t = 1, 2, \ldots, 12,$$

or

$$S_t - L \geq 0, \qquad t = 1, 2, \ldots, 12,$$

and L is precisely equal to the smallest of all 12 storages. Further, a maximization of L is implied.

To minimize the range of storage values requires both sets of constraints to be written together:

$$S_t \leq U, \qquad t = 1, 2, \ldots, 12,$$
$$S_t \geq L, \qquad t = 1, 2, \ldots, 12,$$

which in standard form are

$$S_t - U \leq 0 \qquad t = 1, 2, \ldots, 12,$$
$$S_t - L \geq 0 \qquad t = 1, 2, \ldots, 12,$$
$$U \geq 0, L \geq 0.$$

The minimization is of

$$z = U - L.$$

Effectively, this minimization simultaneously pushes down on the maximum value and up on the lowest value, thus minimizing the range of storage values.

We mentioned earlier that it might be of interest to control month-to-month changes. Storages can, of course, go up or down, forcing us to write not only constraints on $(S_t - S_{t-1})$ but also on $(S_{t-1} - S_t)$. If the limit of a month-to-month change is 25% of the range, the constraints are

$$S_t - S_{t-1} \leq 0.25\ (U - L), \qquad t = 1, 2, \ldots, 12,$$
$$S_{t-1} - S_t \leq 0.25\ (U - L), \qquad t = 1, 2, \ldots, 12.$$

And in standard form this is finally

$$S_t - S_{t-1} - 0.25U + 0.25L \leq 0, \qquad t = 1, 2, \ldots, 12,$$
$$S_{t-1} - S_t - 0.25U + 0.25L \leq 0, \qquad t = 1, 2, \ldots, 12.$$

10.D HITTING THE IRRIGATION TARGET

The water provided from the reservoir for irrigation does not pass through the turbines nor contribute to down-reservoir stream flow. It goes only toward application to the soil on the area of large farms to stimulate the growth of crops. The agency that operates the reservoir has been given a target release for the irrigation area for month t; it is T_t. Releases may exceed T_t or be less than T_t, but T_t is the optimal release, the release that will provide the best growth characteristics.

In this section, we offer three criteria for target hitting, three measures of how closely releases cluster around the monthly targets. These are (1) minimization of the sum of absolute deviations from targets, (2) minimization of the maximum deviation from targets, and (3) minimization of the sum of squared deviations from targets. The structuring of all three criteria require that additional specialized constraints be appended to the basic model.

10.D.1 Minimizing the Sum of Absolute Deviations

This criterion suggests that the absolute value of the difference between the irrigation release and the irrigation target measures the "match" between the two. Larger values of difference imply less than optimal crop yields. Smaller values of difference imply crop yields that are closer to the maximum. The measure suggests that the decline in the value of the crop yield is one-to-one linear with the shortage or excess relative to the target. The sum of absolute deviations from the target is not a perfect measure, but a rough measure with some appeal.

A loss function that indicates actual economic losses due to deviations from targets would be better—if it could be produced. Unfortunately the loss function is generally only a gleam in an economist's eye; it is rare to exist and rarer still to be reliable. Among other issues, the loss incurred in month $t + 1$ by a particular value of deviation depends on the deviation in the previous month t. If the corn crop was already lost by a shortage in the month of June, a July shortage would be of little consequence. Thus, utilizing a loss function is a grand idea, but one with little possibility of achievement. The absolute value of the deviation then acts as a stand-in or surrogate for the unknown or unattainable loss function.

When we say that we seek the set of irrigation releases that minimizes the sum of absolute deviations from targets, we mean that we desire to

$$\text{Minimize } Z = \sum_{t=1}^{12} |Y_t - T_t|$$

where the absolute value symbol encloses a function that could be positive or negative. That is, release could exceed or be less than the target. Unfortunately, operating directly on the absolute value of a function with linear programming is not possible. Instead, the modeling of absolute deviations from a target is accomplished by introducing the difference between two new nonnegative variables as follows:

$$Y_t - T_t = W_t^+ - W_t^-$$

where

$$W_t^+ = \text{a positive number if } Y_t \text{ exceeds } T_t, \text{ and}$$
$$W_t^- = \text{a positive number if } T_t \text{ exceeds } Y_t,$$

or, in standard form,

$$Y_t - W_t^+ + W_t^- = T_t.$$

This equation is written for all time periods, $t = 1, 2, \ldots, 12$.

If Y_t exceeds T_t, we want W_t^+ to be positive and W_t^- to be zero. And if T_t exceeds Y_t, we want W_t^- to be positive and W_t^+ to be zero. Also, if Y_t equals T_t, both W_t^+ and W_t^- should be zero. Obviously, there are many possible ways in which these requirements could be violated. For instance, suppose $Y_t = 34$ and $T_t = 40$. The following are a few of the infinite number of pairs of W_t^+ and W_t^- that satisfy the definitional constraint above.

$$(W_t^+, W_t^-) = (0, 6) = (2, 8) = (10, 16) \text{ and so on.}$$

It would seem like magic if we could arrange the special condition that we desire in which always one of the two variables is positive and the other is zero, but in fact we can—with no difficulty. The device was created by Charnes and is used in his invention of goal programming. In fact, hitting targets with the deviation measured by the absolute value of the difference between release and target is precisely the form of goal programming that Charnes invented.

How do we do it? How do we achieve no more than one of the two deviational variables positive at a time? Before answering the question, it is useful to set up the objective of minimizing the sum of absolute deviations. We wish to

$$\text{Minimize } Z = \sum_{t=1}^{12} W_t^+ + \sum_{t=1}^{12} W_t^-$$

The W_t^+ are to be the positive deviations; the W_t^- are to be the negative deviations. Together they represent the sum of deviations—if we can ensure that when W_t^+ is positive, W_t^- is zero, and vice versa.

Remarkably, simply minimizing the above objective ensures this relationship! Here's why. Suppose we again had the situation in which

$$Y_t = 34 \quad \text{and} \quad T_t = 40.$$

We display once again the three examples we gave that satisfied the definitional equation for W_t^+ and W_t^-; this time, their sum of values is indicated:

W_t^+, W_t^-	0, 6	2, 8	10, 16,
$W_t^+ + W_t^-$	6	10	26.

The set of values for W_t^+ and W_t^- that gives the smallest sum of W_t^+ and W_t^- is always the set in which at least one of the two variables is zero. Thus, by simply placing the deviational variables in the objective, we have automatically ensured

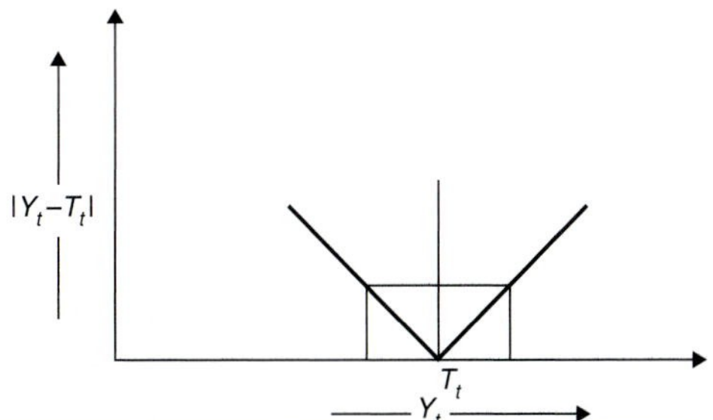

Figure 10.2 Shape of the objective in a minimization of the sum of absolute deviations.

that the variables will properly represent the difference between release and target. It is certainly a coincidence that structuring the objective this way makes the variables take on the desired values, but it is a coincidence worth exploiting.

It is informative to look at the shape of this objective function in this problem (Figure 10.2). When Y_t and T_t are equal, this objective has a value of 0. For every unit that Y_t is below T_t, the objective increases by one unit. For every unit that Y_t is above T_t, the objective also increases by one unit. Thus, the objective of minimizing the sum of absolute deviations has a distinctive *V* or notched shape as in Figure 10.2.

To summarize this goal relative to the irrigation targets, we superimpose the following objective on the basic model (see Section 10.B for the basic model) along with the constraints that define the positive and negative deviations. That is,

$$\text{Minimize} \quad Z = \sum_{t=1}^{12} W_t^+ + \sum_{t=1}^{12} W_t^-$$

$$\text{Subject to:} \quad Y_t - W_t^+ + W_t^- = T_t, \qquad t = 1, 2, \ldots, 12,$$

$$W_t^+ \geq 0$$

$$W_t^- \geq 0$$

$$+ \text{ constraints of the basic model (Section 10.B)}$$

10.D.2 Minimizing the Maximum Deviation from a Target

This criterion, minimizing the maximum deviation from a target, suggests that the losses that occur due to shortages or excesses relative to the irrigation target T_t climb quite steeply with distance from the target. By minimizing the maximum deviation from the target, the largest losses are thus avoided. As with the objective of minimizing the sum of the absolute value of deviations, minimizing the maximum deviation may be viewed as a stand-in for the underlying objective of minimizing losses.

To minimize the maximum deviation from the targets, we need to introduce two new constraints per time period, in addition to an objective. That is,

$$\text{Minimize} \quad Z = M$$

$$\begin{aligned} \text{Subject to:} \quad & Y_t - T_t \le M, && t = 1, 2, \ldots, 12, \\ & T_t - Y_t \le M, && t = 1, 2, \ldots, 12. \end{aligned}$$

where

M = the maximum deviation from any of the 12 targets.

The first of these inequations deals with the case in which the release Y_t is greater than the target. This constraint limits the positive difference $(Y_t - T_t)$ between the release and the target to no more than the unknown M. The second of the inequations is written for the situation in which the target is greater than the release. This constraint limits the positive difference $(T_t - Y_t)$ to less than or equal to the unknown M. For any period t, one of these two conditions must occur. If release exceeds target, the first constraint defines the deviation or excess, and the second constraint is not binding. If target exceeds release, the second constraint defines the shortage, and the first constraint is not binding. In minimizing M, we are pushing down on the maximum difference, whether it be release less target or target minus release. In standard form, the constraints are

$$\begin{aligned} & Y_t - M \le T_t, && t = 1, 2, \ldots, 12, \\ & Y_t + M \ge T_t, && t = 1, 2, \ldots, 12, \end{aligned}$$

but it is most useful to look at the constraints in their initial form to understand their meaning.

The solution to this problem will compress over all months the largest release shortage or overage to its smallest possible value. The solution would be obtained by minimizing the objective of M subject to the constraints that define M and the constraints of the basic model (see Section 10.B). That is,

$$\text{Minimize} \quad Z = M$$

$$\begin{aligned} \text{Subject to:} \quad & Y_t - M \le T_t, && t = 1, 2, \ldots, 12, \\ & Y_t + M \ge T_t, && t = 1, 2, \ldots, 12. \end{aligned}$$

+ constraints of the basic model (Section 10.B)

10.D.3 Minimizing the Sum of Squared Deviations from Targets

The goal of minimizing the sum of squared deviations of irrigation releases from targets can be easily written

$$\text{Minimize} \quad Z = \sum_{t=1}^{12} (Y_t - T_t)^2$$

where all variables are as defined earlier. Unfortunately, the goal cannot be minimized in this form but requires some manipulation. In fact it requires extensive transformation, since up to now we have dealt only with the optimization of linear functions.

We will deal with transforming the objective in steps, but before we begin the process of transformation, we note again that minimizing this goal is also a stand-in for minimizing losses. As with the goal of minimizing the sum of absolute deviations from targets, this goal suggests that losses are symmetric about the monthly target value. And, as with the goal of minimizing the maximum deviation from the targets, this goal suggests that losses rise more steeply as the release "distance" from target increases on either side of the target (Figure 10.3).

The objective can be expanded into three individual component sums as follows:

$$\text{Minimize} \quad Z = \sum_{t=1}^{12} (Y_t - T_t)^2 = \sum_{t=1}^{12} (Y_t^2 - 2T_tY_t + T_t^2)$$

$$= \sum_{t=1}^{12} Y_t^2 - \sum_{t=1}^{12} 2T_tY_t + \sum_{t=1}^{12} T_t^2.$$

Of the three components, the last one is recognized as a sum of constants and, hence, nonoptimizable. This last term does not need to be written again in this problem because nothing can be accomplished to change its value. The second term consists of 12 individual linear functions in Y_t each with a slope of negative $2T_t$. This set of terms is in the correct form for proper application of linear programming. Hence, for these terms, ordinary linear programming may be applied without any further steps.

The first term, on the other hand, the sum of the squares of the 12 releases is an entirely new form for us to deal with. Linear programming cannot handle such terms unless each of the functions is specially decomposed or separated into its subcomponent terms. Once this is done, however, linear programming *will work*. We have to introduce a new mathematical technology at this point to deal with these 12 terms—**piecewise approximation**— a procedure sometimes referred to as separable programming. You will see that the method of piecewise approximation of these convex functions produces multiple linear functions which can closely describe the actual shape of the convex functions. Piecewise approxima-

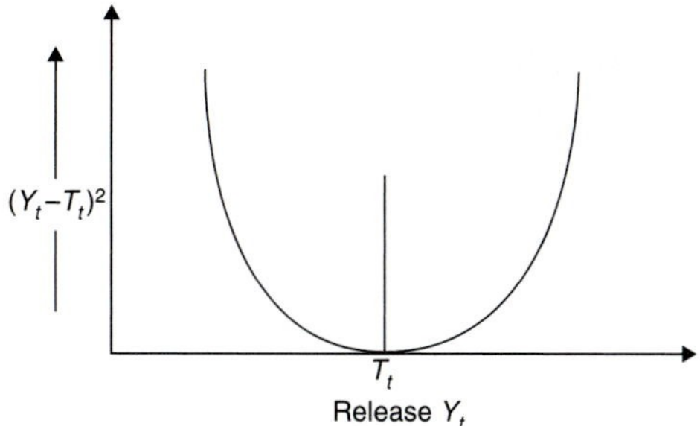

Figure 10.3 Shape of a single term in the objective of minimization of the sum of squared deviations.

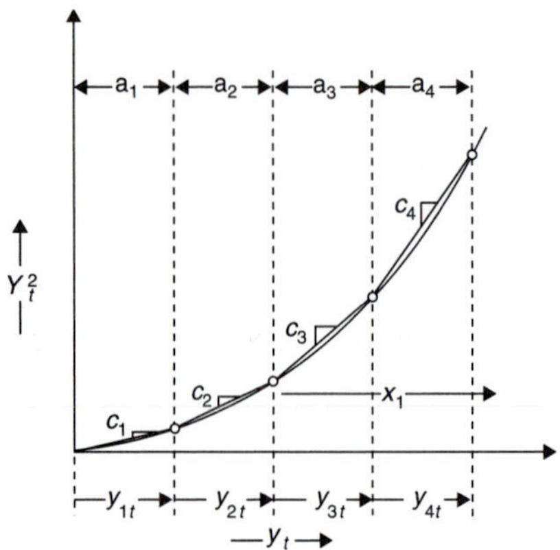

Figure 10.4 Piecewise approximation of Y_t^2.

tion is presented in the box following Figure 10.4. You should read that boxed discussion at this point and then return to the text here to find out how it is applied.

With this new technology, we can break up each of the 12 quadratic functions into individual segments which total to the entire anticipated range of the variables Y_t. Note that the quadratic functions, Y_t^2, are convex and that we are minimizing the sum of these convex functions. For this reason, the segments will enter in the proper order.

The linear approximation of Y_t^2 is

$$Y_t^2 = c_0 + c_1 y_{1t} + c_2 y_{2t} + c_3 y_{3t} + c_4 y_{4t}$$

where

y_{it} = the extent or length of the ith component segment that is occupied;

a_i = the full or maximum length of the ith component segment; and

c_1, c_2, c_3, and c_4 are positive numbers.

These terms plus 11 other sets of terms for the 11 other monthly releases are placed in the objective function. Ordinarily, if the Y_t^2 functions had been different shapes, they would be approximated differently so that both the component segment lengths and slopes would be different, but in this case, since they are all the same function, only one division into segments and slopes is needed.

Of course, the constraints are expanded as well to include for every month t

$$y_{1t} \leq a_1,\; y_{2t} \leq a_2,\; y_{3t} \leq a_3,\; y_{4t} \leq a_4$$

as well as

$$Y_t = y_{1t} + y_{2t} + y_{3t} + y_{4t}$$

or in standard form

$$Y_t - \sum_{i=1}^{4} y_{it} = A, \qquad t = 1, 2, \ldots, 12.$$

This constraint need not be written formally if $(A + \sum_{i=1}^{6} y_{it})$ had been substituted for Y_t on every occasion in every constraint that it occurred.

Piecewise Approximation of Nonlinear Functions

In this discussion we will only refer to functions of a single variable. Functions of multiple variables that cannot be separated into the sum of functions of single variables will not be considered. We can classify functions of a single variable in four ways: linear, convex, concave, and neither convex nor concave. Convex functions are bowl-shaped upward; a line joining any two points on a convex function lies entirely above the function except at its endpoints (Figure 10.5a). Concave functions are bowl-shaped downward; a line joining any two points on a concave function lies entirely below the function except for its endpoints (Figure 10.5b). The last type of function cannot be classified as convex or concave because it has one or more portions that are of one type and one or more of the other (Figure 10.5c).

What we will show here is how to minimize, via linear programming and piecewise approximation, the sum of convex functions. We will find in the process that we *can minimize* such a sum, but that we *cannot maximize* such a sum using this technique. We will also explain why we cannot maximize such a sum using piecewise approximation. Finally, we will ask you to consider how to deal with optimizing a sum of concave functions.

Suppose we had an objective that consisted of the sum of two convex functions, $f_1(x_1)$ and $f_2(x_2)$, that we wished to minimize subject to a set of linear resource constraints and the usual nonnegativity constraints. The two functions are shown in Figure 10.6 with short approximating segments drawn between adjacent points on the curves. The distance between adjacent points is kept relatively short to produce good approximations of the curves.

The actual spacing of points depends on the curvature of the functions and how closely we want to approximate their values. The parameters and the variables of the two approximations are listed in Table 10.1.

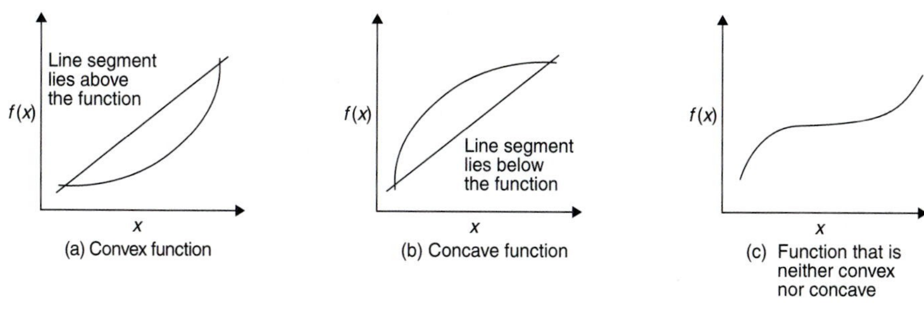

Figure 10.5 Types of function of a single variable.

TABLE 10.1 APPROXIMATION OF TWO CONVEX FUNCTIONS

Function	Number of the Segment	Length of Segments (max value of the variable)	Name of Variable in That Segment	Slope in the Segment
1	1	a_1	y_1	c_1
	2	a_2	y_2	c_2
	3	a_3	y_3	c_3
	4	a_4	y_4	c_4
2	1	b_1	z_1	d_1
	2	b_2	z_2	d_2
	3	b_3	z_3	d_3

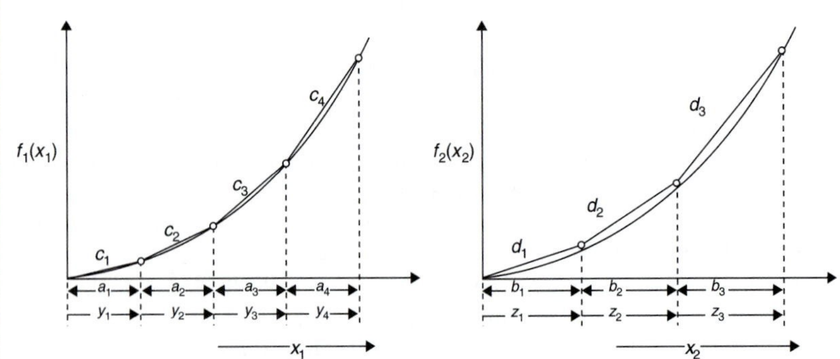

Figure 10.6 Two convex functions whose sum is to be maximized.

The approximation produces two broken-line curves each of which reproduces the original functions within a margin of error dictated by the analyst—who performs the approximation.

The function $f_1(x_1)$ is approximated by

$$f_1(x_1) = c_1y_1 + c_2y_2 + c_3y_3 + c_4y_4$$

which can be placed in the objective instead of $f_1(x_1)$. In the constraints, we need to add

$$y_1 \leq a_1, y_2 \leq a_2, y_3 \leq a_3, y_4 \leq a_4$$

where the variables y_1, y_2, y_3, y_4 represent the extent to which each of the numbered segments of function 1 is filled. In addition, we need to write an identity equation (in standard form), namely, that x_1 is the same as the sum of y_1 to y_4; that is,

$$x_1 - (y_1 + y_2 + y_3 + y_4) = 0$$

Similarly, the function $f_2(x_2)$ is approximated by

$$f_2(x_2) = d_1z_1 + d_2z_2 + d_3z_3$$

which can be placed in the objective function in place of $f_2(x_2)$. In the constraints we need to add

$$z_1 \leq b_1, z_2 \leq b_2, z_3 \leq b_3$$

where the variables z_1, z_2, z_3 represent the extent to which each of the numbered segments of function 2 is filled. In addition, we need to write an identity equation (in standard form) that equates x_2 with the sum of z_1, z_2, and z_3. That is,

$$x_2 - (z_1 + z_2 + z_3) = 0.$$

Of course, the y_j and z_j are constrained to be nonnegative variables.

The approximation procedure we have described has been rather mechanical; we have used segments and slopes and partial sums. We know the approximation is good, but will it work? That is, will the new, replaced objective and the new constraints operate together to produce a continuous curve? For instance, for $f_1(x_1)$, it is conceivable that y_1 and y_3 would be positive and $y_2 = 0$. The approximation would be meaningless if such a situation occurred. Can it be prevented?

It turns out that, so long as the function is convex and the objective is to minimize, such meaningless situations will never occur and variables will always enter in the proper order. That is, if y_3 were positive but did not fill its segment, y_1 would equal a_1, y_2 would equal a_2, and y_4 would be precisely zero. Similarly, if y_2 were positive but did not fill its segment, y_1 would equal a_1, but y_3 and y_4 would be zero. The combination of objective minimization and the function's convexity ensure that this will always be the case because it is always least expensive to take the earlier segments first—their slopes are less. And the variables will always fill these earlier segments in order before later variables fill later segments.

This rather surprising phenomenon, our ability to piecewise approximate a convex function and have the segments enter in the correct sequence, is limited, however. If either of the two conditions, minimization or convexity, were to change, the optimization would not work. If we were to try to maximize the sum of convex functions, the last segment would enter first, then the second last, and so on—a meaningless situation. If we were to try to minimize the sum of concave functions—using piecewise approximation—again the last segment would enter first. Why? As a consequence, minimization via piecewise approximation can be accomplished only when the function being cut into segments is strictly convex.

On the other hand, suppose the objective was to be maximized and the functions were concave. Would the piecewise approximation method work? That is, would the segments enter in the proper order, first segment filled first, second next, and so on? What are the relative slopes of the first segment, second segment, and so on? Which segment gives the greatest "bang for the buck"? You probably have the answer by now. You can maximize a sum of convex functions by piecewise approximation—just as you can minimize the sum of convex functions by this method. You should make a chart of what you can and can't do with piecewise approximation and why. It will help you think about the technique—which incidentally is another contribution of Charnes.

A related topic in piecewise approximation is what shapes work in greater than or equal constraints and what shapes work in less than or equal constraints. Using the same "bang for the buck" argument as above, a constraint consisting of the sum of concave functions which is greater than or equal to the right-hand side will have the piecewise variables enter in the proper order. This will not work if the constraint is a less than or equal to constraint. And, using a like but inverse argument, a constraint consisting of the sum of *convex* functions which is less than or equal to the right-hand side will also have the piecewise variables enter in the proper order. The variables will not enter in the proper order if the constraint is of the greater than or equal form. You should create a chart of these relationships as well.

10.E PLACING THE RELEASE TO STREAM FLOW IN A TARGET INTERVAL

In Section 10.D, we showed how linear programming could be used to cluster irrigation releases around a target. In contrast, in this section, we will be attempting to insert the releases to streamflow into a target interval or desirable range. The upper limit of the interval is E, the flow that damages stream banks through erosion. The lower bound of the interval is F, the least rate of release that provides adequate flow for fish survival and migration. A release within these bounds is desirable; a release outside of these limits is permissable but less desirable.

The goal will be to minimize the sum of absolute deviations of stream flow release from the target interval. This goal is very much like the minimization of absolute deviations from a target.

This criterion and the means to achieve it was also suggested by Charnes. To achieve this goal we need to define four new variables; all are deviational variables. One pair of variables is used to define the deviations around the upper stream flow target E. The other pair of variables is used to define the deviations around the lower stream flow target F.

$$X_t - E = U_t^+ - U_t^-, \qquad t = 1, 2, \ldots, 12,$$

$$X_t - F = V_t^+ - V_t^-, \qquad t = 1, 2, \ldots, 12,$$

where

U_t^+ = the amount by which the release exceeds E;

U_t^- = the amount by which the release falls short of E;

V_t^+ = the amount by which the release exceeds F; and

V_t^- = the amount by which the release falls short of F.

Of the four deviational variables, we are only concerned with U_t^+ and V_t^-; these variables represent deviations outside the target interval. The variables U_t^-

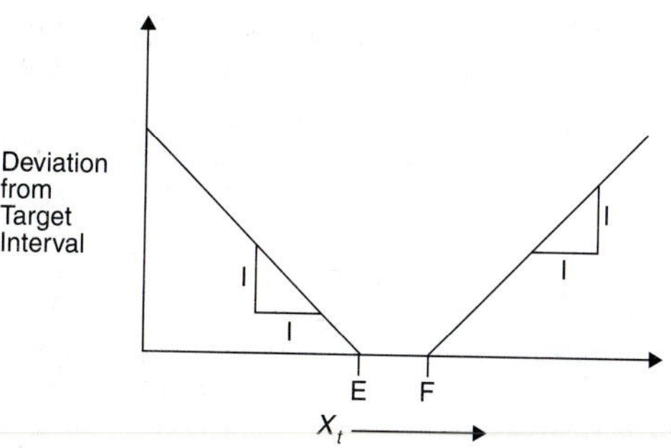

Figure 10.7 A single term of the objective of minimizing absolute deviations from a target interval.

and V_t^+ generally, though not always, represent deviations within the target interval. In any event, only U_t^+ and V_t^- are of importance here, and it is the sum of these variables that is minimized. If the sum of U_t^+ and V_t^- over all 12 months can be driven to zero, then every release to stream flow will have fallen within the target interval.

The objective function term for a single month looks like Figure 10.7.

The problem of directing stream flow releases into the target interval can then be written as

$$\text{Minimize} \quad Z = \sum_{t=1}^{12} U_t^+ + \sum_{t=1}^{12} V_t^-$$

Subject to:

$$X_t - U_t^+ + U_t^- = E, \qquad t = 1, 2, \ldots, 12,$$

$$X_t - V_t^+ + V_t^- = F, \qquad t = 1, 2, \ldots, 12,$$

$$+ \text{ basic model constraints (Section 10.B).}$$

It may be that exceedances of E are of greater importance than shortfalls less than F. A weight greater than one could then be placed on U_t^+. And if shortfalls were of greater concern than excedances, the weight greater than one could be placed on V_t^-.

10.F DELIVERING A HYDROPOWER REQUIREMENT (NONLINEAR PROGRAMMING TECHNIQUES)

Hydropower is a common purpose of reservoirs, and it is useful for illustrating nonlinear programming techniques because it involves the product of decision variables. We will show two nonlinear programming techniques to optimize hydropower production. The first is iterative approximation, and the second is logarithmic decomposition plus piecewise approximation of concave constraints. The hydropower equation indicates that the power generated is proportional to the product of head and the release through the turbines, with specific parameters determined by turbine design. That is, the power generated in month t is given by

$$P_t = k\bar{H}_t X_t$$

where

P_t = rate of power output in kilowatts in month t;

$\bar{H}_t$ = average head (height through which water falls) in feet during month t;

X_t = release during month t; and

k = a constant.

The head in the reservoir at any moment can be related via a linear approximation to the storage in the reservoir

$$H = mS + n$$

where m and n are chosen to give the best possible fit of a straight line to the concave curve that describes head as a function of storage. The average head during month t is estimated by the head that occurs at the average storage in month t, or

$$\bar{H}_t = m\bar{S}_t + n$$

where

$$\bar{S}_t = \frac{1}{2}(S_t + S_{t-1})$$

so that

$$\bar{H}_t = \frac{m}{2}(S_t + S_{t-1}) + n$$

Recall that S_t and S_{t-1} are defined by the inventory or mass balance equations of the basic model, and the basic model is retained as the foundation of this hydropower model.

We pointed out earlier in this chapter that linear programming handles definitional equality constraints as well as the more common inequality constraints. The inventory or mass balance equations referred to above are examples of such equality relations that we leave linear programming to enforce. To avoid algebraic manipulation, we do not replace $\bar{H}_t$ in the power equation with its equivalent in terms of S_t and S_{t-1}. Instead, we add a definitional constraint for average head in every month. In this definitional constraint, average head is expressed in terms of beginning and ending storage. In standard form this is

$$\bar{H}_t - \frac{m}{2}S_t - \frac{m}{2}S_{t-1} = n, \qquad t = 1, 2, \ldots, 12.$$

This constraint type defines average head in terms of variables already defined in the basic model. It must, as a consequence, be written *with* the constraints of the basic model in order for it to have meaning.

With this definition of head now included along with the base model, we can rewrite the power equation, which has a multiplicative term, as a constraint

$$k\bar{H}_t X_t \geq Q, \qquad t = 1, 2, \ldots, 12,$$

where Q is the minimal monthly level of hydropower to be produced over the 12-month period. This constraint is clearly nonlinear and cannot be handled in its displayed form by linear programming.

10.F.1 The Method of Iterative Approximation for Nonlinear Programming

The name "iterative approximation" is a nice way of saying: guess values of decision variables you cannot calculate; insert these values in the model to make the model linear and solvable; solve the model, finding new values for the decision variables you guessed; and insert those new values. Repeat the process until the variable values that are produced by model solution are within some small fraction of the variable values you inserted. The process is surprisingly powerful in a number of instances, often stabilizing in six or seven steps or iterations and producing what appear to be "good" solutions to otherwise relatively intractable problems. We say "good" solutions since we have no provably optimal solutions to compare them with.

How does iterative approximation work in the hydropower model case that we are considering here? The constraints of the model for hydropower production can be summarized as

$$\bar{H}_t - \frac{m}{2} S_t - \frac{m}{2} S_{t-1} = n,$$

$$k\bar{H}_t X_t \geq Q,$$

plus the basic model constraints (Section 10.B)
and all variables greater than or equal to 0

The hydropower constraints are the only constraints with nonlinearities, and it will be these constraints where variable estimates are inserted to linearize the inequalities.

For this model, with $F \leq X_t \leq E$, a fair guess of X_t might be $(E + F)/2$, that is, at the midpoint of the desired interval. All 12 values of X_t would initially be set at this level in the hydropower constraints, and only in the hydropower constraints. The values of X_t in all other constraints are left free—free to assume their best values. Each hydropower constraint has now become linear in a single variable, $\bar{H}_t$, yielding an entirely linear model. The linear programming model is then solved under some objective—perhaps one of the ones we have discussed. Values of all of the decision variables, the X_t among them, are determined by solution of the linear programming model. The new values of X_t are inserted in the hydropower constraints and the process is repeated—until the values of the X_t which the model produces are sufficiently close to the values of the X_t from the previous iteration that were inserted. This final run is taken as the solution to the problem.

10.F.2 Logarithmic Decomposition and Piecewise Approximation

This methodology, like iterative approximation, is useful for handling constraints with products of variables. It is more subtle, however.

First, the hydropower constraint is decomposed by taking logarithms. Thus,

$$k\bar{H}_t X_t \geq Q, \qquad t = 1, 2, \ldots, 12,$$

becomes

$$\log \bar{H}_t + \log X_t \geq \log\left(\frac{Q}{k}\right), \qquad t = 1, 2, \ldots, 12.$$

These constraints, although they have separated the variables, are still nonlinear, but they do consist of the sum of two concave functions, and we do know how to deal with such functions (see the box in Section 10.D.3).

We noted earlier in the chapter that the concave functions in a constraint that forces a sum of concave functions greater than some right-hand side could be piecewise approximated and would yield variables that enter in the proper order. That is precisely what we have here—a sum of concave functions greater than or equal to a number. In this case piecewise approximation is a bit more subtle, because the functions are logarithmic, but still possible.

In Figure 10.8, we display the logarithmic function of X_t. The value of the lowest conceivable release to stream flow is G and is shown on the graph. The

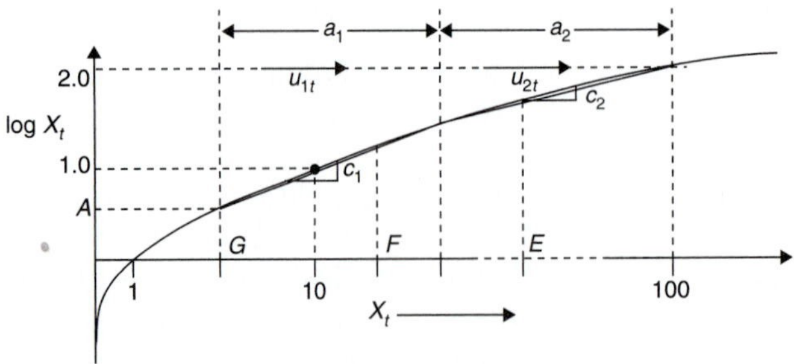

Figure 10.8 Piecewise approximation of the logarithm of X_t.

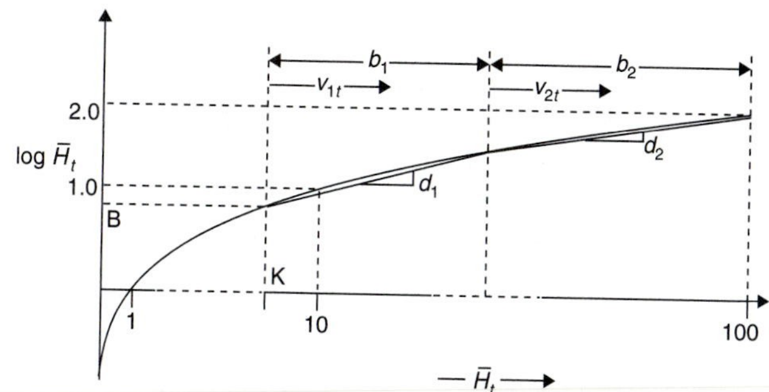

Figure 10.9 Piecewise approximation of the logarithm of $\bar{H}_t$.

piecewise approximation begins at G and uses two segments with lengths a_1 and a_2, costs c_1 and c_2, and initial value A, where A is the value of the logarithm at $X_t = G$. The approximation equations of the logarithm of X_t are then

$$\log X_t = A + c_1 u_{1t} + c_2 u_{2t}$$

$$u_{1t} \leq a_1, \; u_{2t} \leq a_2$$

and

$$X_t = G + u_{1t} + u_{2t}$$

In Figure 10.9, we display the logarithmic function of $\bar{H}_t$. The value of the lowest conceivable average monthly head is K and is shown on the graph.

The piecewise approximation begins at K and uses two segments with lengths b_1 and b_2, cost d_1 and d_2, and initial value B, where B is the value of the logarithm at $\bar{H}_t = K$. The approximation equations of the logarithm of $\bar{H}_t$ are then

$$\log \bar{H}_t = B + d_1 v_{1t} + d_2 v_{2t}$$

$$v_{1t} \leq b_1, \; v_{2t} \leq b_2,$$

$$\bar{H}_t = K + v_{1t} + v_{2t}.$$

These two sets of approximation equations are now combined into the two-component logarithmic constraint on power production. That is,

$$\log \bar{H}_t + \log X_t \geq \log \left(\frac{Q}{k} \right)$$

is replaced by

$$\overbrace{B + d_1 v_{1t} + d_2 v_{2t}}^{\log \bar{H}_t} + \overbrace{A + c_1 u_{1t} + c_2 u_{2t}}^{\log X_t} \geq \log \left(\frac{Q}{k} \right)$$

and

$$\bar{H}_t = K + v_{1t} + v_{2t}$$

$$X_t = G + u_{1t} + u_{2t}$$

plus

$$v_{1t} \leq b_1, \qquad v_{2t} \leq b_2,$$

$$u_{1t} \leq a_1, \qquad u_{2t} \leq a_2,$$

where all constraints are finally written in standard form and for all $t = 1, 2, \ldots, 12$. Of course, this set of constraints would be used in addition to the basic set of model constraints as well as the definitional constraints of $\bar{H}_t$ in terms of storage.

10.G RECURSIVE PROGRAMMING

Early in the chapter we described recursive programming, a technique for optimization through time when the magnitude of future events are uncertain. We return to recursive programming now to solidify understanding of the concept. In our case, the uncertain events are stream flow inputs to the reservoir. The stream flows can be estimated, however, with early stream flow predictions being likely much better than later predicted values. As an example, soil moisture or snow pack in the present month may give strong clues to next month's flow but less information about the second, third, and later months. Nonetheless, a set of decisions are needed for the upcoming periods which optimize over the entire cycle—not just over the next period.

The predicted inflows are thus used in an optimization model that spans the entire cycle, not just the next period of operation. This is done despite the recognized poorer quality of predictions for later periods. The reason we are willing to optimize over the entire cycle with such data is that the decisions that result from the optimization will not be used in their entirety. In fact, only the initial decisions—and these are the decisions based on the best data—are utilized. Once the first period's decisions have been taken, that is, releases have been made to down-reservoir stream flow as well as to irrigation, and the first month's actual inflow has arrived in storage, the optimization is conducted again.

By reconducting the optimization, we mean that (1) a new set of 12 monthly inflows is estimated for the next 12-month period and (2) the programming model is resolved to provide values of the decisions X_t and Y_t for those 12 months. As a consequence of this procedure focusing on operation in the here-and-now, the importance of the more uncertain future monthly flows is diminished, while the importance of the more certain prediction of the upcoming month's inflow is emphasized. Once the actual inflow has materialized, we *adapt* to it by reestimating future inflows and recalculating decisions. For this reason, the procedure may also be referred to in the literature as *adaptive programming.* The procedure is a cunning approach to the need for here-and-now decisions in the face of an uncertain future.

CHAPTER SUMMARY

Our goal in this chapter was to introduce new programming ideas in the context of an actual problem setting. We first described a fundamental model of inventory, the water reservoir, a model useful in a number of practical modeling situations. Then we introduced a number of purposes which the reservoir was to fulfill. Probably there is a reservoir somewhere in the United States, maybe a number of reservoirs, that have the functions that we have suggested and goals that resemble the goals we described. Although we described multiple objectives for the reservoir operation, we did not presume to trade off one objective against another, because we did not really prescribe which objective would be chosen for a given purpose.

As we described in turn each objective in words, we introduced programming ideas to implement the optimization of that objective. For instance, to reduce the range of storage fluctuations over the year, we introduced the notion of the difference between the values of the unknown upper and lower bounds on variables, a function that is precisely the range. To hit irrigation targets closely, we introduced three new programming ideas. The first was minimizing the sum of the absolute values of deviations from the target, a technique that used an equation that defined the difference between two nonnegative variables. The second idea was to minimize the maximum deviation from targets, a deviation that could occur on either side of the target. The third idea was to minimize the squared deviations from targets. This last idea required us to explain the techniques of convex and concave programming by piecewise approximation.

We wanted down-reservoir stream flow to be within a specified range. To accomplish this, we showed how to minimize the sum of deviations from the target interval. To achieve a particular hydropower production level, we had to introduce a function that included the product of two variables. To handle this nonlinear function, we described two techniques. The first—iterative approximation—involved estimating values of variables to linearize the constraints. This was followed by solution by linear programming which produced corrected values of those same variables. The new values were inserted and the process was repeated until the values of the variables stabilized. The second technique was to take logarithms of the product function thereby decomposing each of the constraints into the sum of two concave functions. Since the hydropower constraints were of the greater than-or-equal to form, piecewise approximation was appropriate to utilize to approximate the concave functions, and the component variables were expected to enter in the correct order.

Finally, we discussed recursive programming, a sequence of optimizations, each of which uses currently updated data and produces a set of decisions over a cyclic time horizon. Only the most near-term decisions are ever implemented before the process of updating data is repeated followed by another stage of optimization. Recursive programming is designed to produce decisions in an environment made uncertain by imperfect prediction of the magnitudes of future events.

EXERCISES

10.1. Show how to minimize the maximum month-to-month change in storage. Use standard form and indicate non-negativity for any new variable(s).

10.2. We gave the desirable interval for releases to streamflow as F to E. Show how to minimize the maximum deviation from the target interval. Use standard form and indicate new variables as nonnegative.

10.3. Suppose that all the releases to stream flow can be placed in the interval between F and E. Show how to minimize the range within which the releases to stream flow fall, given that all are between F and E.

10.4. In determining irrigation releases to hit a target closely, we suggested that the "distance" measure be the difference between the release and the target. This would be an especially meaningful objective if the targets were not far in value from one another. It may be, however, that the targets may be very different from say the very early part of the growing season to a late month of fast growth prior to harvest. Suppose that the early month target is 5000 and the late month target is 10,000. If the release Y_t is 4000 in the early month and 9000 in the late month, we count these as equivalent failures in several of the models that we built. Yet the first is a 20% shortage, and the second is a 10% shortage. Perhaps they are not equivalent. We need to put the two measures of difference on a common scale so that we are calculating impacts correctly. Our suggestion for the two models described below is that the distance be calculated as the difference between release and the target, divided by the target. In this way, deviations are "normalized"; that is, a difference of 1000 when the target is 5000 is equivalent to a difference of 2000 when the target is 10,000.

(a) Set up in standard form the constraints and objective for minimizing the sum of the normalized absolute deviations from targets. That is, minimize the sum of the absolute values of the fractional deviations from targets. Assume the constraints of the basic model are already in place. Be sure to note that your variables are nonnegative.

(b) Set up the objective for minimizing the sum of the squared fractional deviations from targets. Again, the constraints of the basic model are already in place. Indicate how this problem should be solved; name the technique.

10.5. Suppose that all irrigation releases are either equal to or short of the target. Show how to minimize the maximum month-to-month change in the fractional delivery of the irrigation target. Use standard form and indicate nonnegativity appropriately.

10.6. Describe another method of iterative approximation for the hydropower problem of Section 10.F.1. That is, do not guess X_t in the hydropower constraints.

CHAPTER

11 Lessons in Context: Transportation Systems

This is an advanced chapter. To comprehend the models in this chapter, the reader should understand the transportation problem as described in section C of Chapter 5, Integer Programming and Network Problems—Part I. The reader should also be familiar with the water resources chapter, Chapter 10, especially Sections 10.B and 10.D, where a reservoir is modeled through time and irrigation releases are programmed to hit target values as closely as possible. Without these background materials, the first, third and fourth sections of this chapter may not be penetrable.

This chapter, unlike other chapters of the book, consists of a number of separate and distinct problem settings. The common thread in the four settings we present is that all problems arise in the arena of transportation system planning, a subdiscipline of civil engineering that is growing steadily in importance and that draws heavily upon the methodologies discussed in this book.

The first problem treats a classical civil engineering problem in highway design, the optimal vertical alignment of a highway, in which cutting and filling are to be balanced to yield a highway segment of minimum cost. The second problem examines the hiring and shift assignment by day of a crew of bus drivers with the objective of hiring the fewest number of drivers. It is presented in simplified form to be sure, but in its complicated form is a problem of major importance in the passenger transport industry. The third problem utilizes the transportation model of linear programming as a statistical technique to estimate the origin-destination flows on an urban expressway. The fourth and last problem deals with goods movement, specifically supplying at least cost the empty containers needed for loading at each of a number of sites. The empty container distribution problem is still being structured and understood by researchers as they gradually come to grips with its remarkable complexity. Our version is, of necessity, basic, but complicated enough.

11.A OPTIMAL VERTICAL ALIGNMENT OF A HIGHWAY

In this problem, you will learn about free variables, about the absolute values of deviations from a target, and about curve fitting using linear programming.

A highway is to be constructed between points A and B through rolling terrain. The highway is replacing an existing road that winds and turns sharply and which has frequent and unacceptable elevation changes by today's standards. Thus, the horizontal alignment of the highway is already known because it is dictated by the position of the current road as well as by land ownership patterns along the road. The new highway is planned in order to make travel safer and faster between A and B. Accordingly, the sharp bends are corrected in the new horizontal alignment. Further, the rapid elevation changes also need to be corrected, in this case, by the vertical alignment which you are to determine. In addition, the rate of change of grade needs to be constrained, influencing such factors as the ability to see a specified distance and the reach of headlights.

The vertical alignment should follow a smooth curve that is anchored in height at the two ends of the highway. In between the curve will have a limited grade in both directions as well as a limited rate of change of grade. The problem is to determine the elevation of the road as a function of distance from the initial point A in the direction of point B. Instead of determining the elevation point by point, it is easier simply to create a smooth mathematical function which meets the constraints we have indicated on grade and on the rate of change of grade.

The topography is too rough for the new road to follow without some cutting and filling to lessen the slope of the road. The cuts are not so deep as to require blasting or tunneling. Nor are the fill sections deep enough to require bridges. It is a matter of shaving a piece here and padding a spot there. The decisions needed are how deep to cut and how much to fill so that the total cost of cutting and filling is minimum. The road must be a smooth curve or function; it cannot have sudden changes in slope.

The slope or "grade" is limited to s_R feet of rise per foot of distance and s_F feet of fall per foot of distance. Finally, the rate of change of slope is limited to q to allow for sufficient distance vision on the highway. The slope of the highway must also "feather in" to match the slope of the road in the two areas in which it joins with the old road.

To begin the problem, we will show how to estimate the cost of cutting and filling based on the depth of cutting and filling. From Figure 11.1, the following definitions are apparent:

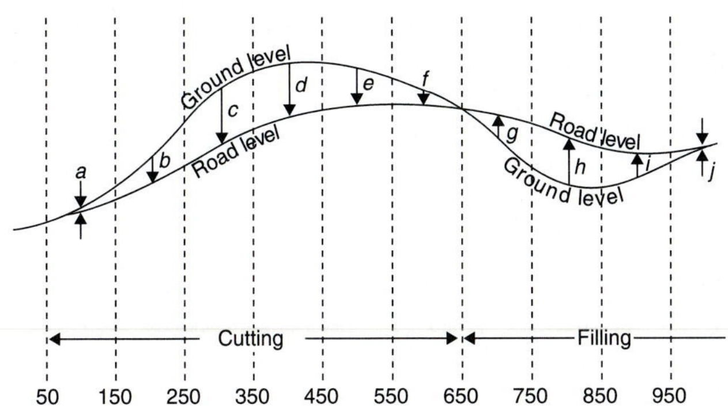

Figure 11.1 Cutting and filling to achieve an acceptable road grade curvature.

a =	average depth of the cut between	50 and 150 feet	measured on the x axis
b =	"	150 and 250 feet	"
c =	"	250 and 350 feet	"
d =	"	350 and 450 feet	"
e =	"	450 and 550 feet	"
f =	"	550 and 650 feet	"
g =	average depth of the fill between	650 and 750 feet	"
h =	"	750 and 850 feet	"
i =	"	850 and 950 feet	"
j =	"	950 and 1050 feet	"

cost/volume of cut = p dollars/cubic foot

cost/volume of fill = q dollars/cubic foot

The road is to be built w feet in width.

The volume of the cut between 50 and 150 feet is calculated as the product of the depth of the cut times the road width times the segment length or $(a)(w)(100)$. The volume of the next cut is $(b)(w)(100)$, the volume of the third cut is $(c)(w)(100)$, and so on. The total volume of cutting between 50 and 650 feet is

$$(100w)(a + b + c + d + e + f)$$

and the total cost of cutting is

$$(p)(100w)(a + b + c + d + e + f).$$

The volume of filling between 650 and 750 feet is calculated as the product of the depth of filling times the road width times segment length or $(g)(w)(100)$.

The volume of the next three segments of filling are $(h)(w)(100) + (i)(w)(100) + (j)(w)(100)$. The total volume of filling between 650 and 1050 feet is

$$(100w)(g + h + i + j)$$

and the total cost of filling is

$$(q)(100w)(g + h + i + j).$$

The total cost of cutting and filling is

$$100w[p(a + b + c + d + e + f) + q(g + h + i + j)]$$

or

$$Z = 100w[p(\text{total depth of cuts}) + q(\text{total depth of fill})].$$

To minimize this objective, it is sufficient to discard $100w$ as it is a constant factor, and

$$\text{Minimize } Z = p \textstyle\sum \text{cut depths} + q \textstyle\sum \text{fill depths}.$$

Assume that a polynomial function of degree 5 is chosen for the profile of the road. Higher orders are found to give no significant decrease in cost. That is, we wish to find a curve of the form

$$y = k_5x^5 + k_4x^4 + k_3x^3 + k_2x^2 + k_1x + k_0$$

where y is the elevation of the road at distance x from the origin. The constants k_1, k_2, k_3, k_4, k_5, and k_0 are unknowns which are to be determined. The actual elevations of the current surface are noted at 100-foot intervals:

$$r_1 = \text{elevation at 100 feet (known);}$$

$$r_2 = \text{elevation at 200 feet (known);}$$

$$\vdots$$

$$r_i = \text{elevation at } 100i \text{ feet (known);}$$

$$\vdots$$

$$r_n = \text{elevation at } 100n \text{ feet (known).}$$

There are assumed to be n increments of 100 feet.

The problem is then stated this way:

Find the constants of the fifth-order polynomial curve which constants minimize the cost of cutting and filling subject to grade and rate of change of grade constraints. Remember k_0 through k_5 are free to be positive or negative.

We define the following parameters and variables:

k_j = the unknown coefficient of the variable x raised to the jth power, the coefficient must be free to be positive or negative;

k_j^+ = the unknown value of k_j if it is a positive coefficient;

k_j^- = the unknown value of k_j if it is a negative coefficient;

x_i = the value of distance at the ith point on the horizontal axis, known;

r_i = the actual height of the ground prior to excavation or filling at the ith point on the horizontal axis, known;

y_i = the height of the highway surface after cutting and filling as defined by the function of x;

u_i = the depth of filling at the ith point on the horizontal axis; u_i represents the positive difference between value of the function evaluated at the point i and the height of the actual ground surface when the value of the function exceeds the height of the ground; it is unknown; and

v_i = the depth of the cutting at the ith point on the horizontal axis; it represents the difference between the value of the function at i and the ground surface when the height of the ground surface is larger than the value of the function; it is unknown.

We first write a definitional equation of the height of the highway surface after the cutting and/or filling has taken place.

$$y_i = x_i^5 k_5 + x_i^4 k_4 + x_i^3 k_3 + x_i^2 k_2 + x_i k_1 + k_0, \qquad i = 1, 2, \ldots, n.$$

Note that the unknowns in this equation are the k values and y_i. The values of x_i are known. In the problem summary, we convert this equality to standard form.

To model the depth of cut and/or fill, we write an equation defining the difference between the height of the highway surface at point i as given by the function of x and the actual surface prior to cutting or filling. That is,

$$y_i - r_i = u_i - v_i, \qquad i = 1, 2, \ldots, n.$$

The equation will be written in standard form in the problem summary. It can be seen that u_i represents the depth of filling. It is positive when the new highway surface exceeds in height the height of the prior surface. It can also be seen that v_i represents the depth of cutting as it is positive when the height of the old ground surface exceeds the height defined by the functional value for the height of the new highway. The variables u_i and v_i will appear in the objective with the values of u_i multiplied by the cost of filling and values of v_i by the cost of cutting. It is important that these two variables should never both be positive at any value of distance. If the variables are handled properly in the constraints, as we shall explain, we can be confident that this condition will be met.

Grade constraints require us to differentiate the function that gives the elevation of the new highway. The grade of the highway, or the rate of change of its elevation, is given by

$$\frac{dy}{dx} = 5x_i^4 k_5 + 4x_i^3 k_4 + 3x_i^2 k_3 + 2x_i k_2 + k_1$$

The value of this function evaluated at each point i cannot exceed s_R to allow cars to climb at an adequate rate of speed. The value of this function must exceed $-s_F$ so that the highway does not fall so quickly that brake use will be excessive. The constraints have the form

$$5x_i^4k_5 + 4x_i^3k_4 + 3x_i^2k_3 + 2x_ik_2 + k_1 \le s_R, \qquad i = 1, 2, \ldots, n$$

and

$$5x_i^4k_5 + 4x_i^3k_4 + 3x_i^2k_3 + 2x_ik_2 + k_1 \ge -s_F, \qquad i = 1, 2, \ldots, n.$$

Recall that the unknowns are the k_j ($j = 1, 2, \ldots, 5$) and the known coefficients in the above constraints are $5x_i^4$, $4x_i^3$, $3x_i^2$, $2x_i$, and 1.

The constraints on the rate of change of grade require us to take the second derivative of the function that gives the elevation of the new highway. The rate of change of grade is given by

$$\frac{d^2y}{dx^2} = 20x_i^3k_5 + 12x_i^2k_4 + 6x_ik_3 + 2k_2$$

The value of this function evaluated at each point cannot exceed q. The constraints have the form

$$20x_i^3k_5 + 12x_i^2k_4 + 6x_ik_3 + 2k_2 \le q, \qquad i = 1, 2, \ldots, n.$$

The problem's objective is to minimize the cost of cutting and filling or as shown conceptually early in the discussion of the problem

$$\text{Minimize} \quad Z = p\sum_{i=1}^{n} v_i + q\sum_{i=1}^{n} u_i.$$

This completes most of the guts of formulation, but there are two issues to clear up before we summarize the problem. The issues are related in concept.

The first concern centers on the variables u_i and v_i which are defined as the depth of fill and the depth of cut in the equation

$$y_i - r_i = u_i - v_i, \qquad i = 1, 2, \ldots, n,$$

where y_i is the elevation of the new highway surface as defined by the fifth-order polynomial in x. For the problem to have meaning, the two variables must not simultaneously be positive. If they were both positive at some distance x_i from the origin, the costs would be counted wrong. We would be paying for both cutting and filling at the same moment, an impossible situation.

For two reasons, the condition of simultaneously positive cutting and filling will not occur. The first reason is easy to explain. We noted that if u_i and v_i were both positive, the costs would be counted wrong. That is one answer in a nutshell. Because we are minimizing the costs of cutting and filling, both variables will not be positive together—else the solution is not minimum. The other answer requires a dose of linear programming and linear algebra theory that we did not provide in the early chapters of the book and which was not treated as a prerequisite for reading the book. As a consequence, we can describe the condition that prevents both variables from being positive at the same moment, but we cannot justify it.

In the linear programming theory chapter, we used the notion of a basis, a collection of vectors that defined which variables had the potential to be positive at a particular extreme point. If the ith vector were in the basis, the variable x_i associated with that vector could be positive in the basis. What we did not say was that two vectors that were a simple multiple of one another would never be present together in the basis. That is, the variables associated with the two vectors would never both be eligible to be positive at the same time. The vectors associated with a (u_i, v_i) pair are just the negative of each other. One vector is all zeros and a single one; the other is all zeros and a single negative one in the same position as the positive one in the first vector. That is, the two vectors associated with a (u_i, v_i) pair are related to one another by a simple multiple. One vector is -1 times the other. Hence the two vectors will not both appear in a basis at the same time, and hence, u_i and v_i will not both be positive at the same time. Thus, we have two explanations for the result that the simultaneous appearance of a positive u_i and positive v_i will not occur.

We have another result to explain, however. The coefficients, the k_j, are to be free to be positive or negative. To achieve this freedom, we add definitional constraints to the problem of the form

$$k_j = k_j^+ - k_j^-, \qquad j = 1, 2, 3, 4, 5.$$

These constraints can be written separately as we have done here, or they can be incorporated into the functional equations that define the elevation of the new highway. With the equation written as we have immediately above, it is easy to see that the pair of vectors associated with k_j^+ and k_j^- are the negative of one another, that is, a simple multiple of one another. As a consequence, the associated vectors will not be in the basis at the same time, and k_j^+ and k_j^- will not both be positive at the same time. Thus, the representation using the above equation achieves a variable that is free to be positive (when $k_j^+ > 0$) or negative (when $k_j^- > 0$).

The mathematical problem statement can be summarized in the following form:

$$\text{Minimize} \quad Z = p \sum_{i=1}^{n} v_i + q \sum_{i=1}^{n} u_i$$

Subject to:

$$y_i - x_i^5 k_5 - x_i^4 k_4 - x_i^3 k_3 - x_i^2 k_2 - x_i k_1 - k_0 = 0, \qquad i = 1, 2, \ldots, n,$$

$$y_i - u_i + v_i = r_i, \qquad i = 1, 2, \ldots, n,$$

$$5x_i^4 k_5 + 4x_i^3 k_4 + 3x_i^2 k_3 + 2x_i k_2 + k_1 \leq s_R, \qquad i = 1, 2, \ldots, n,$$

$$5x_i^4 k_5 + 4x_i^3 k_4 + 3x_i^2 k_3 + 2x_i k_2 + k_1 \geq -s_F, \qquad i = 1, 2, \ldots, n,$$

$$20x_i^3k_5 + 12x_i^2k_4 + 6x_ik_3 + 2k_2 \le q, \qquad i = 1, 2, \ldots, n,$$
$$k_j - k_j^+ + k_j^- = 0, \qquad j = 1, 2, 3, 4, 5,$$
$$y_i, u_i, v_i \ge 0, \qquad i = 1, 2, \ldots, n,$$
$$k_j^+, k_j^- \ge 0, \qquad j = 1, 2, 3, 4, 5.$$

Several points about the practicality of this model should be made. It may be desirable to maintain relatively constant grades over some distances. This would prevent continual shifting of gears. Such a goal, as reflected by permissible changes in grade at a number of success points of measurement could be accomplished by constraining the difference in slopes (as given by the derivative) between a series of points.

Another issue that emerges occasionally during road repairs is traffic flow that shifts sides. If this situation is of concern, the minimum of the absolute value of the rate of rise or rate of fall would govern and a single number of grade would be used in both directions.

11.B HIRING/SCHEDULING A CREW OF BUS DRIVERS

In this problem, you will see how decision variables can be generated which are not at all obvious from an initial assessment of the problem.

A small town is starting up a bus system with a single route to alleviate automobile traffic through the downtown shopping and commercial area. The bus system is to run for 16 hours each day, from 5 A.M. until 9 P.M. Drivers will work on two shifts, the first from 5 A.M. until 1 P.M., the second from 1 P.M. until 9 P.M. The route from first stop in the morning to the last stop in the outbound direction plus the return trip back to the first stop requires approximately 120 minutes, including the time for the stops. Thereafter, the route repeats throughout the day, seven more times during the day.

Throughout each weekday, Monday through Friday, the bus should visit each stop every 15 minutes. If there were only one bus, a stop would not be revisited for 120 minutes. If eight buses were operating, one bus would be able to start from the first stop every 15 minutes, and hence each stop would be visited every 15 minutes. The requirement of eight buses translates into eight drivers for each shift on each weekday.

On the weekend, each stop will be visited every 30 minutes. Since the round trip takes 120 minutes, four buses will be required to operate at a time on the weekend. This translates into a need for four drivers on each shift on Saturday and Sunday.

The public agency that has been created to operate the bus system has to hire drivers for the buses and wants to hire the least number of drivers possible that fulfill the service requirements. Each driver will be hired/paid for five days, eight hours per day, each week. A driver's workweek has to be five consecutive days

TABLE 11.1 PATTERNS OF WORK

	(i = workday)						
	S	M	T	W	Th	F	S
j = Pattern Number	(1)	(2)	(3)	(4)	(5)	(6)	(7)
1	1	1	1	1	1	0	0
2	0	1	1	1	1	1	0
3	0	0	1	1	1	1	1
4	1	0	0	1	1	1	1
5	1	1	0	0	1	1	1
6	1	1	1	0	0	1	1
7	1	1	1	1	0	0	1
	4	8	8	8	8	8	4

followed by two days off. The first of the five days could occur on any day of the week. If more drivers are hired than are usually needed at any moment in time, spare drivers can provide backup or sick leave service to a driver not currently available. Of course, there are two shifts, but the answers for each of the two shifts are the same since each shift has the same driver requirements. Since each shift has the same requirements, we want to solve for the minimum number of drivers on each shift.

Table 11.1 shows that there are seven ways to arrange a consecutive five-day workweek. This is accomplished simply by beginning each five-day week on a different one of the seven days of the week. On the vertical axis is listed the number of the workweek pattern, actually the number of the day that begins that workweek. In the horizontal dimension is the day of the week. A "one" appears at the intersection of the work week pattern and day of the week—if that day is a workday in the workweek pattern. Otherwise, a zero appears. The bottom line of the table displays the number of drivers needed on each particular day.

The step of formulation of bus crew scheduling is described next. We unintentionally make the formulation look easy by presenting a method that we know by experience will work. But this is not an easy formulation; it is, in fact, quite subtle because the variables are not "natural." In Chapter 2, we described a problem cutting various sizes of smaller boards from large 4 foot × 8 foot plywood sheets. It was Example 2-7, the Thumbsmasher Lumber and Home Center, Part II—Cutting Plywood: An Integer Programming Problem with Hidden Variables. We indicated that the board cutting problem did not have natural variables either, but that first it was necessary to generate feasible cutting patterns. That is, different feasible and efficient patterns of cutting the plywood sheets were imagined. Then the decisions were how many sheets to cut according to each of the feasible patterns that had been generated. The variables are said to be columns, columns of the linear programming problem, and the methodology is referred to as "column generation."

The bus crew scheduling problem we have described here is also formulated by "column generation." Natural variables we might think of would be the number of drivers assigned to each shift and day of the week. Then, however, we would need to figure out how to write constraints that ensured that each driver worked five consecutive days with two straight days off. In fact, if natural variables are used, we don't know how to do that. The "natural variable" approach really doesn't seem to work here. Instead, we resort to column generation as in the plywood cutting problem. We first generate feasible patterns of work, and here we are very fortunate. The number of work patterns is very small—especially because we require two consecutive days off. There are only seven five-day patterns, one beginning on each day of the week. If we allowed the two days off to be any two days of the week, the number of variables (the number of feasible work schedules) would be many more.

In fact, the crew scheduling problem—as it is structured in the airline industry and in the mass transit administrations of our major cities—is a vastly more complicated problem than the problem we deal with here. It is more complicated because workdays and workweeks can be broken in many and various ways and the number of feasible work patterns is very, very large. Thousands of work patterns may be considered. Nonetheless, the column generation approach remains the only available way to attack the problem. Because there are so many "columns" (or feasible work patterns), it becomes necessary to generate them by computer. Once the problem is "solved," the analyst sees the work patterns generated and then chosen for the very first time.

The following parameters, definitions, and decision variables are needed to structure the problem:

i = day number (1 = Sunday, 2 = Monday, etc.);

j = pattern number;

r_i = number of drivers required on day i;

x_j = number of drivers assigned to pattern j, an integer valued variable; and

N_i = the patterns j that include i as a workday.

The values of r_i are

$$r_1 = 4, r_2 = 8, r_3 = 8, \ldots, r_6 = 8, r_7 = 4.$$

The patterns N_i are

$$N_1 = (1, 4, 5, 6, 7),$$

$$N_2 = (1, 2, 5, 6, 7),$$

$$N_3 = (1, 2, 3, 6, 7),$$

$$N_4 = (1, 2, 3, 4, 7)$$

$$N_5 = (1, 2, 3, 4, 5),$$

$$N_6 = (2, 3, 4, 5, 6), \text{ and}$$

$$N_7 = (3, 4, 5, 6, 7).$$

The driver requirement for day 1 constraints may be written as follows:

$$x_1 + x_4 + x_5 + x_6 + x_7 \geq 4.$$

This constraint says that any driver hired for pattern 1 (Sunday through Thursday) + any driver hired for pattern 4 (Wednesday through Sunday) + any driver hired for pattern 5 (Thursday through Monday) + any driver hired for pattern 6 (Friday through Tuesday) + any driver hired for pattern 7 (Saturday through Wednesday) can provide one of the four drivers needed on Sunday.

Similar constraints can be written for day 2 to day 7, namely,

$$x_1 + x_2 + x_5 + x_6 + x_7 \geq 8, \quad \text{day 2,}$$

$$x_1 + x_2 + x_3 + x_6 + x_7 \geq 8, \quad \text{day 3,}$$

$$x_1 + x_2 + x_3 + x_4 + x_7 \geq 8, \quad \text{day 4,}$$

$$x_1 + x_2 + x_3 + x_4 + x_5 \geq 8, \quad \text{day 5,}$$

$$x_2 + x_3 + x_4 + x_5 + x_6 \geq 8, \quad \text{day 6,}$$

$$x_3 + x_4 + x_5 + x_6 + x_7 \geq 4, \quad \text{day 7.}$$

The objective is to

$$\text{Minimize } Z = x_1 + x_2 + x_3 + x_4 + x_5 + x_6 + x_7.$$

The constraints and objective can be written in condensed forms. The constraint for day 1 can be written

$$\sum_{j \in N_1} x_j \geq 4.$$

The constraint says that any of those drivers hired according to the patterns in the set N_1 (the patterns that provide service on day 1) can be utilized to meet the driver requirement for day 1.

In general, the constraint for day i can be written

$$\sum_{j \in N_i} x_j \geq r_i.$$

This constraint says that any of those drivers hired according to the patterns in the set N_i (the patterns that provide service on day i) can be utilized to meet the driver requirements for that day.

The problem can be summarized as

$$\text{Minimize} \quad Z = \sum_{j=1}^{7} x_j$$

$$\text{Subject to:} \quad \sum_{j \in N_1} x_j \geq 4$$

$$\sum_{j \in N_i} x_j \geq 8, \qquad i = 2, 3, \ldots, 6$$

$$\sum_{j \in N_7} x_j \geq 4$$

$$x_j = \text{integer}, \quad j = 1, 2, \ldots, 7.$$

The answer to this problem is to hire a total of ten drivers, according to the following rules:

x_1 = the number hired Sunday through Thursday = 1

x_2 = the number hired Monday through Friday = 4

x_3 = the number hired Tuesday through Saturday = 1

x_4 = the number hired Wednesday through Sunday = 1

x_5 = the number hired Thursday through Monday = 1

x_6 = the number hired Friday through Tuesday = 1

x_7 = the number hired Saturday through Wednesday = 1

With these decision variables all weekday constraints are met as precise equalities with no surplus drivers on any of those days. The weekend constraints are exceeded by just one; that is, five drivers are available on each of Saturday and Sunday rather than the needed four. The extra driver on these days can be assigned duties at the bus terminal, and the individual so assigned can be rotated among the five available.

The problem we have just structured and solved is a small and simplified version of an immensely important and very complicated problem in transportation planning. Crew scheduling must be achieved by the railroad companies, by the airline companies, and by the transit companies that serve major and minor cities. Work rules including maximum workday length, broken into pieces of work, with relief at fixed geographic points, and restart only after allowable rest give rise to enormous integer programming problems. These problems are regularly being structured and solved and restructured and solved again by the major airlines, railroads, and transit companies. A full-scale problem was beyond the scope of this text, but it is hoped that the small problem we introduced provides a flavor of the problems being solved.

11.C ESTIMATING ORIGIN-DESTINATION FLOWS ON AN EXPRESSWAY

An urban expressway is under study by the city traffic department. Delays and backups have been occurring on the expressway with increasing frequency. The city council is considering authorizing construction of an increased number of lanes in selected sections of the highway. They are also considering a set of traffic lights which would limit access to the highway to predetermined rates that do not cause backups. And they may make changes in local roads to speed up nonexpressway traffic and draw travelers from the expressway to the local roads. The actual decisions taken will depend on the vehicle flows between the various sections of the city, flows that are presently unknown.

To aid the council in its deliberations, the traffic department has undertaken to count vehicles entering and exiting the expressway at all the junctions of the expressway with local streets. They will also be counting the flow of vehicles on all segments of the expressway itself, that is, between all points of access/egress to the expressway. These counts of vehicle rates (vehicles per hour) are being performed in the hope of estimating the point-to-point flows of vehicles.

The diagram of the expressway and its entry-exit points is shown in Figure 11.2. By organized counting of vehicles, the following hourly rates of vehicle access/egress were determined during the two-hour period of traffic flow in the direction of the central business district. Note that the entrance-exit points are numbered sequentially from the origin of the expressway to its termination in the central business district. We have

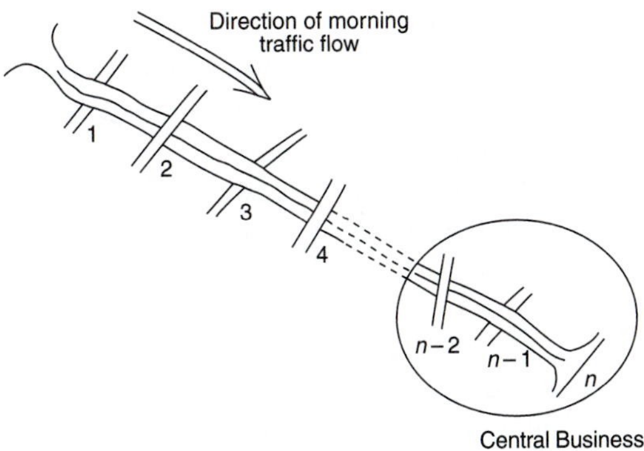

Figure 11.2 The expressway and its exit/entrance points.

a_i = the vehicles per hour entering the expressway at the ith access point;

b_j = vehicles per hour exiting the expressway at the jth exit point; and

n = the number of entry-exit points.

In addition to measuring rates of vehicle entry and exit, the traffic department sets up counters on each segment of the expressway itself so that the total flow of vehicles toward the central business district is known. Thus, we have

$q_{i,\,i+1}$ = vehicles per hour traveling on the segment of the expressway between exit-entrance i and $i + 1$.

The data obtained by the measurement exercise are certainly suggestive of the underlying situation. They are not, however, the product desired. Even though the traffic department now knows flows entering, flows on, and flows exiting the expressway, they do not know the flow that enters the expressway at the ith point bound for exit at the jth point. It is this origin-destination (O-D) information that is critical for decision making, especially for improvement of local options and light timing.

A long-time member of the traffic department suggests that the vehicle count be supplemented by a survey of motorists who are entering the expressway. As the vehicles enter the highway, police would indicate motorists to park on the shoulder, and people with clipboards would ask the driver at which local road they would exit. The survey would require considerable personnel at each of the entry ramps all at the same time. It would also disrupt traffic and subject people to delays that could cause lateness, and definitely the project would cause aggravation.

A young recent hire in the traffic department suggests the possibility of using linear programming to derive the needed origin-destination flows from the measurements of flows entering, flows on, and flows exiting the expressway. The management of the traffic department is skeptical that such information can be derived from the existing data, but is willing to discuss the issue. Their thoughtful approach is that whatever this young person can do must "explain" the data that are already in hand, and they tell the employee what the criteria are that the O-D data must meet.

"The derived origin-destination flows that are present on each segment of the expressway must sum to a number reasonably near the actual flow measured on that segment. Furthermore, if all the derived O-D flows originating at an entry point are summed over all of their destinations, that sum should be the rate of entry that was actually measured at the entry point. Likewise, if all the derived O-D flows that terminate at a particular exit point are summed over all of their origins, that sum should be the rate of exit that was actually measured at the exit point."

The response of the young employee is enthusiastic. The employee tells the director of the traffic department, "You've described the model that I can build." Since the traffic survey will be costly and potentially painful and since the employee, working alone, promises the needed origin-destination flows in a week, the director of the department says to go ahead.

This is the problem that the young traffic engineer structured and solved. Let:

x_{ij} = the unknown hourly flow of vehicles between origin i and destination j.

As the management of the traffic department described, the sum over all destinations of the flow of vehicles originating at i must be the entry rate measured at point i. That is,

$$\sum_{j=i+1}^{n} x_{ij} = a_i, \qquad i = 1, 2, \ldots, n.$$

Note the limits of the summation. It is assumed that the flows that enter at i can only exit "downstream" of point i, that is, at points $i + 1, i + 2, \ldots, n$. In a similar fashion, the sum over all origins of the flow of vehicles exiting at j must be the exit rate measured at point j. That is,

$$\sum_{i=1}^{j-1} x_{ij} = b_j, \qquad j = 2, 3, \ldots, n.$$

Again, the limits of the summation should be noted. Although flows to j can originate starting at the first entry point, the last point of origination of flows that can discharge at j is the point just preceding j, that is, $j - 1$. Note also that we have no discharge at the first point so we have no explaining to do.

It remains for us to structure an objective that pushes these origin destination flows in the direction of explaining the flows observed on the expressway itself. To do this, we consider the flow on the segment between the ith and $(i + 1)$st exit-entry points. The flows on this segment could have originated at any of the entry points from 1 up to the ith entry point. In addition, these flows must have destinations at points from the $(i + 1)$st point to point n, the last point of the expressway in the central business district. The sum of these flows is

$$\sum_{i=1}^{i} \sum_{j=i+1}^{n} x_{ij}$$

This sum is to be as reasonably near to the actual flow, $q_{i,\, i+1}$, as possible, but it is not to be greater than $q_{i,\, i+1}$ because such a set of flows would not make physical sense. Thus,

$$\sum_{i=1}^{i} \sum_{j=i+1}^{n} x_{ij} \le q_{i,\, i+1}, \qquad i = 1, 2, \ldots, n - 1,$$

Introducing the slack variable gives

$$\sum_{i=1}^{i} \sum_{j=i+1}^{n} x_{ij} + u_{i,\, i+1} = q_{i,\, i+1}, \qquad i = 1, 2, \ldots, n - 1.$$

With this equation, with its slack variable, we can now structure at least three objectives, all of which are meaningful. First, we can seek those origin-destination flows which minimize the sum of absolute deviations from below the flows on all of the segments. That is,

$$\text{Minimize} \quad Z = \sum_{i=1}^{n-1} u_{i,\,i+1}$$

Subject to:

$$\sum_{i=1}^{i} \sum_{j=i+1}^{n} x_{ij} + u_{i,\,i+1} = q_{i,\,i+1}, \qquad i = 1, 2, \ldots, n-1,$$

$$\sum_{j=i+1}^{n} x_{ij} = a_i, \qquad i = 1, 2, \ldots, n,$$

$$\sum_{i=1}^{j-1} x_{ij} = b_j, \qquad j = 2, 3, \ldots, n,$$

$$x_{ij} \geq 0,\ i = 1, 2, \ldots, n, \quad j > i.$$

Second, we can minimize the squared deviation from below as follows:

$$\text{Minimize} \quad Z = \sum_{i=1}^{n-1} u_{i,\,i+1}^2$$

subject to the same constraints as the formulation above. Actual implementation of minimizing the sum of squares of variables requires the use of piecewise approximation of convex functions, a procedure we described in Chapter 10 in the context of closely hitting irrigation targets with the releases from a reservoir.

Finally, we can minimize the maximum deviation from below from among the $u_{i,\,i+1}$ in the following way:

$$\text{Minimize} \quad Z = m$$

Subject to:

$$u_{i,\,i+1} \leq m, \qquad i = 1, 2, \ldots, n-1,$$

$$\sum_{i=1}^{i} \sum_{j=i+1}^{n} x_{ij} + u_{i,\,i+1} = q_{i,\,i+1}, \qquad i = 1, 2, \ldots, n-1,$$

$$\sum_{j=i+1}^{n} x_{ij} = a_i, \qquad i = 1, 2, \ldots, n,$$

$$\sum_{i=1}^{j-1} x_{ij} = b_j, \qquad j = 2, 3, \ldots, n.$$

And all variables are nonnegative.

11.D THE EMPTY CONTAINER DISTRIBUTION PROBLEM: THE OTHER SIDE OF THE SHIPPING COIN

In this problem you will see how the transportation problem can be structured through time and how storages can be cumulated at origins and destinations.

In Chapter 5, you read about the transportation problem, a problem in which goods are shipped from origins to destinations (factories to warehouses is one example). Often bulky goods are moved in large metal containers that can serve either as a truck or railcar payload. These containers can shift from rail to truck to ocean vessel and back. The question naturally arises as to how to best secure the empty containers that are to be loaded at each point of disembarkation. The empty containers that are needed for loading must be moved to the shipping site from sites of supply where the empty containers have accumulated from prior movements. These points of supply where empty containers are warehoused are at seaports and inland depots. The optimal distribution of containers from points of supply where they sit empty to points of demand where they will be loaded is called the empty container distribution problem.

This problem is exceedingly complicated to formulate in its most realistic form. Thus, to make the problem accessible for this text, we have to simplify its features extensively. Even so, the problem remains challenging to state neatly and in an understandable fashion. Nonetheless, because the problem is of major importance in the goods movement/shipping industry, we do take the time and effort to state it here. One realistic feature we will omit is randomness or uncertainty in both the supply of and demand for empty containers. Remember there is more to this problem, but this introduction as well as references at the end of the chapter should give the careful student entry to the growing literature on the empty container problem.

To simplify the problem, we consider, first, that there is only one kind/size of container, and second, that there are only five months of supply and demand. We could use more months, but estimates of supply and demand degrade as the planning horizon is extended. The number actually chosen is arbitrary but convenient for purposes of presentation. We also assume that supplies and demands are known with certainty through the planning horizon. The five months of supply and demand are offset by just one month. That is, the demand at the beginning of the first month cannot be met by the supply that is available at the beginning of the first month because of the time it takes to ship the empty containers to each of the possible destinations. The demand for empty containers at the beginning of the first period then needs to be met by some prior plan outside of the present scheduling model. Furthermore, we assume that all empty container shipments can reach their destination within one month so that a shipment from a supply point at the beginning of one period is fully available at the point of demand to which it was shipped by the beginning of the next period. The delivery window, the time of arrival for the quantity of empty containers required at a destination at the beginning of time period t, is any time period prior to t. That is, storage at the sites of demand is possible.

The unknown decision variables for this problem are, first, flow variables between supply and demand points, which occur during each month, and second, storage variables, which reflect the inventory of empty containers at sites of availability and demand. The known parameters are (1) the number of empty containers that become newly available at each site of supply at the beginning of each month, (2) the number of empty containers that are needed at each site of demand at the beginning of each month, (3) the cost of shipment between sites of supply and demand, and (4) the costs of container storage at sites of availability and demand. The variables and parameters are defined as follows:

x_{ijt} = the number of empty containers shipped from supply point i to demand point j during month t. These containers are assumed to be shipped at the beginning of t to arrive at j by the end of period t so that they may be used for loaded shipment from j in the next month $(t + 1)$. The containers cannot be loaded until the period after the month of shipment;

a_{it} = the number of empty containers that newly arrive at supply point i throughout the month prior to month t and become fully available at the beginning of month t;

S_{it} = the number of empty containers in storage at supply point i at the end of month t. This is calculated from the storage at the end of the prior month plus new arrivals available for shipment in the current month (a_{it}) less all shipments from i in the tth month;

b_{jt} = the number of empty containers required for loading at demand point j at the beginning of month t. This is the number of containers that need to be loaded for shipment at j;

Z_{jt} = the number of empty containers in storage at demand point j at the end of month t. This is composed of the prior month's end-of-period storage plus new empty container arrivals at the beginning of t less withdrawals (b_{jt}) at the beginning of month t;

c_{ij} = the cost to ship one container from supply point i to demand point j;

d_i = the cost to store an empty container at supply point i for one month;

e_j = the cost to store an empty container at demand point j for one month;

m = the number of supply points; and

n = the number of demand points.

The availabilities at supply points, the requirements at demand points, and the storages at both supply and demand points are summarized in terms of time of measurement in Table 11.2.

TABLE 11.2 SUPPLY AND DEMAND FOR EMPTY CONTAINERS AT TWO SITES. SITES i AND j ARE ASSUMED CLOSE ENOUGH IN SHIPMENT TIME THAT CONTAINERS LEAVING SITE i AT THE BEGINNING OF ONE TIME PERIOD WILL ARRIVE AT SITE j BY THE BEGINNING OF THE NEXT PERIOD. THEY WILL NOT BE AVAILABLE UNTIL THAT PERIOD AS WELL

	Arrivals and Storage at Supply Point i		Storage and Shipments at Demand Point j	
	Newly Arrived Empty Containers Available For the Upcoming Period	Empty Containers Stored At the End of the Month	Empty Containers Stored at the End of the Month	Empty Containers Needed For Loading
		S_{i0}	Z_{j0}	
Beginning of period 1 1st Time period End of period 1	a_{i1}	S_{i1}	Z_{j1}	b_{j1}*
Beginning of period 2 2nd Time period End of period 2	a_{i2}	S_{i2}	Z_{j2}	b_{j2}
Beginning of period 3 3rd Time period End of period 3	a_{i3}	S_{i3}	Z_{j3}	b_{j3}
Beginning of period 4 4th Time period End of period 4	a_{i4}	S_{i4}	Z_{j4}	b_{j4}
Beginning of period 5 5th Time period End of period 5	a_{i5}	S_{i5}	Z_{j5}	b_{j5}
Beginning of period 6	a_{i6}		Z_{j6}	b_{j6}
				b_{j7}

* Demand being met by a prior plan.

We begin the model by developing the storage equations: The storage at supply point i at the end of month t is given by

$$S_{it} = S_{i,\,t-1} + a_{it} - \sum_{j=1}^{n} x_{ijt}, \qquad t = 1, 2, 3, 4, 5.$$

At supply point i, the storage at the end of t is equal to the storage at i at the end of $t - 1$ plus any new arrivals of empty containers that became available for use at the beginning of month t minus all shipments out of i during t. The newly available empty containers actually accumulated during the previous month $(t - 1)$ but are now considered available for shipment during t.

At demand point j, the storage at the end of month t is given by

$$Z_{jt} = Z_{j,\,t-1} + \sum_{i=1}^{m} x_{ij,t-1} - b_{jt}, \qquad t = 1, 2, 3, 4, 5.$$

The storage at the jth demand point at the end of month t is equal to the storage at the end of $t - 1$ plus inputs from supply points that were shipped at the beginning of month $t - 1$ arriving at j at the beginning of t, minus the number of empty containers taken for loading at the beginning of month t. Note that the shipments from supply points originated during month $t - 1$ (at the beginning), but they are becoming available for loading for the first time at the beginning of month t. This arrangement prevents us from meeting current demand using containers that were shipped to j during the current month—our assumption is that they are not available until the next month. It is assumed that all demand for empty containers is met in each time period so that no unmet demand accumulates through the planning period. These storage equations will be converted to the standard form of equality constraints in a moment.

In addition to the storage equations, we can write constraints on shipment—from supply points and to demand points. The amount shipped from supply point i to all demand points j is limited to the number of empty containers available at supply point i. That is,

$$\sum_{j=1}^{n} x_{ijt} \leq S_{i,\,t-1} + a_{it}, \qquad i = 1, 2, \ldots, m, \quad t = 1, 2, \ldots, 6,$$

and the number of containers available to demand point j for loading at the beginning of t must exceed the quantity required

$$\sum_{i=1}^{m} x_{ij,\,t-1} + Z_{j,\,t-1} \geq b_{jt}, \qquad j = 1, 2, \ldots, n, \quad t = 2, 3, \ldots, 5.$$

In fact, however, while these last two constraints may seem appropriate, they are unnecessary. The storage equations already enforce these conditions in the following way. Note that each storage equation defines the storage at the end of a month. In the equation for storage at each supply point, it is seen that storage is diminished by the sum of shipments to demand points. Since the storage can never be negative (it is a linear programming variable), the sum of shipments out of the supply point automatically is limited to that storage value plus new arrivals as an upper bound.

In the equation for storage at each demand point, it is seen that storage at the end of t is diminished by the current (month t) demand for empty containers and incremented by the inflow of empty containers during t. Again, since the storage at the end of t can never be negative, the storage at the end of $t-1$ plus the inflow of containers during t must exceed the number of containers that were loaded at the beginning of month t. That is, the storage definition equation already enforces the logic of the constraint that requires an adequate number of empty containers to be available at the beginning of each month.

The objective is

$$\text{Minimize} \sum_{i=1}^{m}\sum_{j=1}^{n}\sum_{t=1}^{6} c_{ij}x_{ijt} + \sum_{i=1}^{m}\sum_{t=1}^{6} d_i S_{it} + \sum_{j=1}^{n}\sum_{t=1}^{6} e_j Z_{jt}$$

Subject to:

$$S_{it} - S_{i,\,t-1} + \sum_{j=1}^{n} x_{ijt} = a_{it}, \qquad i = 1, 2, \ldots, m,$$

$$t = 1, 2, \ldots, 5,$$

$$-Z_{jt} + Z_{j,\,t-1} + \sum_{i=1}^{m} x_{ij,\,t-1} = b_{jt} \qquad j = 1, 2, \ldots, n,$$

$$t = 2, 3, \ldots, 6,$$

$$S_{it}, Z_{jt}, x_{ijt} \geq 0, \qquad i = 1, 2, \ldots, m,$$

$$j = 1, 2, \ldots, n,$$

$$t = 1, 2, \ldots, 6.$$

The reader will appreciate our leaving the empty container distribution problem at this point and may wish to take an aspirin or non-aspirin headache medication as the authors did after preparation of this section of the chapter. You could, if you wish, attempt to extend the problem statement to the situation in which not all demands are met, but the authors recommend rest, food, a walk in the sunlight, or other nonmathematical diversion.

EXERCISES

An Airplane Pilot Buying Fuel

11.1. Puddlejumper Airlines (PA, for short) has a regular route from city 1 through six more cities with return to city 1. The number of passengers on board each leg of the trip is known fairly well in advance. PA is just getting off the ground, so to speak, and has to count every dollar. As a consequence, the systems department (Jeff) wants the pilot to purchase fuel at stops along the way rather than try to carry the entire flight's needs from city 1. Prices along the route vary so the decision of where and how much to buy is a complicated one. The systems department knows linear programming and has an idea on how it may be applied.

The fuel consumed in flying the kth leg of the journey depends not only on the number of passengers on board (really their weight plus the weight of their luggage) but also on the weight of fuel being carried. That is, it costs money to carry excess fuel from one city to the next. The rule is that the plane must carry at least the amount of fuel that is needed to carry its load from city k to city $k + 1$ (leg k) plus some reserve amount.

Let:

b_k = amount of fuel consumed on leg k if no extra fuel is carried on board,

and

r_k = the reserve amount of fuel to be carried on leg k.

By extra fuel is meant any fuel in excess of $(b_k + r_k)$, that is, in excess of the minimum requirement on leg k. In addition to b_k and r_k, we have the additional parameters:

u_k = maximum amount of fuel that the plane can safely carry on leg k (actually at the start of leg k);

f_k = fraction of extra fuel on leg k burned on leg k; and

c_k = cost of fuel purchased at k before beginning the kth leg (the trip from k to $k + 1$).

The decision variables for the problem are

y_k = the amount of fuel in the plane's tanks when it lands at city k, and

x_k = the amount of fuel purchased at city k.

To calculate the fuel consumed on leg k requires that we know the extra fuel on board at takeoff from k. That number is

$$y_k + x_k - (b_k + r_k)$$

so that the extra fuel burned on leg k, in addition to b_k, is

$$f_k(y_k + x_k - b_k - r_k).$$

The total fuel burned is

$$b_k + f_k(y_k + x_k - b_k - r_k).$$

so that you can calculate the fuel in the tanks when the plane arrives at city $k + 1$.

The PA systems department needs to determine fuel amounts to buy at each city so that the total cost of fuel purchases is as small as possible. Write the standard form linear program. (This problem in another situation is the creation of Norman Waite of the IBM Corporation.)

A Detail in the Vertical Highway Alignment Problem

11.2. A pair of constraints were intentionally left out of the formulation of this problem. The constraints were described verbally but were not written in mathematics. Specifically, the highway segment under plan has to be "feathered" into the two prior segments that were already built to needed specifications. That is, at each of the two points of attachment to previously built portions of the highway, the slope of the segment under plan as well as the highway elevation have to match the slope and elevation of the highway portion to which it is attaching. Write these two pairs of constraints. The slope at zero distance is s_0 and at the end of the segment, s_E. The elevation at zero distance is y_0 and at the end of the segment y_E.

Light Timing at a Road Intersection With a Steady Rate of Vehicle Arrivals

11.3. A light is to be timed at an intersection of two one-way roads for the period of the morning rush hour. The intersection is relatively isolated from other decisions being made in the highway network. The length of a light cycle has already been set at 3 minutes. There are 40 light cycles in the morning rush hour. Of this cycle time, a certain amount is lost to the movement of vehicles. That is, when the light turns from red to green in each of the two directions, there is a delay in vehicles entering the intersection. Generally, this total loss time is approximated by the length of the amber phase. We will use 20 seconds as the loss time. A survey of traffic on the two roads during the morning rush hour has given arrival rates at the intersection through time. (See the accompanying figure.)

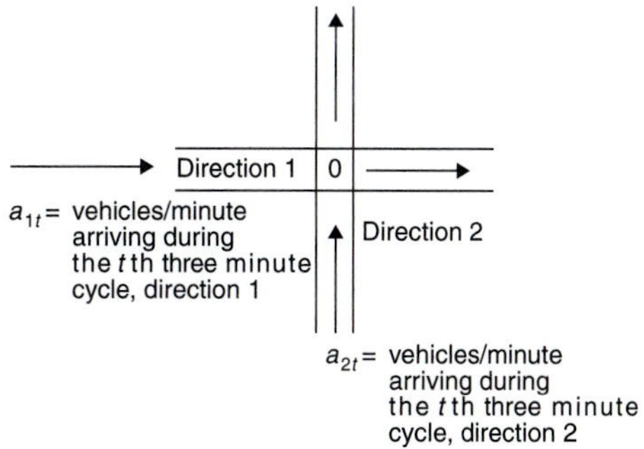

The light cycle is the standard green-red-amber sequence. The decision variables for light timing are

$$g_t = \text{the minutes of green time in direction 1, and}$$

$$\overline{g_t} = \text{the minutes of green time in direction 2.}$$

The diagram of light timing shows that green time, direction 1, corresponds closely to red time, direction 2, and green time, direction 2, corresponds closely to red time, direction 1.

Amber ↓ Amber ↓

Direction 1	G		R		G		R	
Direction 2			G		R		G	

↑ Amber ↑ Amber

Thus, the sum of green time in the two directions is the total usable time in the light cycle. That is,

$$g_t + \overline{g_t} = 3 - 0.33 = 2.67 \text{ minutes}, \qquad t = 1, 2, \ldots, 40.$$

Not all the green time in each direction is always needed. Sometimes, all vehicles in line in a particular direction can be cleared in less than the available green time, leaving a slack green time when the light is green when no vehicles are available to clear. Thus, we need four new variables:

x_t = used green time, direction 1;

u_t = slack green time, direction 1;

$\overline{x_t}$ = used green time, direction 2;

$\overline{u_t}$ = slack green time, direction 2.

The sum of these variables is equal to the usable time in the light cycle.

$$x_t + u_t + \overline{x_t} + \overline{u_t} = 2.67 \text{ minutes}$$

and this equation replaces the previous equation.

An intersection has a clearance rate in each direction, the maximum rate at which vehicles move through the intersection in each direction, once flow "gets up to speed."

Let

r_1 = vehicles per minute that can clear in direction 1, and

r_2 = vehicles per minute that can clear in direction 2.

The vehicle movement through the intersection in direction 1 is the product of the clearance rate and used green time $r_1 x_t$. The movement in direction 2 is by similar convention $r_2 \overline{x_t}$. Assume that both directions start with no vehicles in line waiting at the time of the first green phase.

Let:

S_{1t} = vehicle storage at the end of the tth light cycle in direction 1, unknown,

and

S_{2t} = vehicle storage at the end of the tth light cycle in direction 2, unknown.

Write the storage/inventory equations for directions 1 and 2 that calculate S_{1t} and S_{2t} from the data given and the decisions made on green time in each direction.

Using the storage equations, show how to minimize the maximum line length (storage of vehicles) in any direction through the morning rush period.

Applying Recursive Programming to the Empty Container Distribution Problem

11.4. How would you apply the concept recursive programming, discussed in the water resources chapter, to improve the solution of the empty container distribution problem?

REFERENCES

Optimal Vertical Alignment of a Highway

GOH, C., CHOW, E., and T. FWA, "Discrete and Continuous Models for Computation of Optimal Vertical Highway Alignment," *Transportation Research B*, Vol. 22B, pp. 399–409, 1988.

PARKER, N., "A Systems Analysis of Route Location," doctoral thesis, Cornell University, Ithaca, NY, 1971.

Empty Container Distribution

DEJAX, P. and T. CRAINIC, "A Review of Empty Flows and Fleet Management Models in Freight Transportation," *Transportation Science*, Vol. 21, no. 4, pp. 227–247, November, 1987.

MENDIRATTA, V. and M. TURNQUIST, "Model for the Management of Empty Freight Cars," *Transportation Research Record*, Vol. 838, pp. 50–55, 1982.

RATICK, S., OSLEEB, J., KUBY, M., and K. LEE, "Interperiod Network Storage Location-Allocation (INSLA) Models." In *Spatial Analysis and Location-Allocation Models*, A. Ghosh, and G. Rushton, eds. New York: Van Nostrand Rheinhold, 1987.

WHITE, W. and BOMBERAULT, A., "A Network Algorithm for Empty Freight Car Allocation," *IBM Systems Journal*, Vol. 8, pp. 147–169, 1969.

Transportation and Crew Scheduling

BLAIS, J.-Y., LAMONT, J., and J.-M. ROUSSEAU, "The HASTUS Vehicle and Manpower Scheduling System at the Société de Transport de la Communauté urbaine de Montréal," *Interfaces*, Vol. 20, no. 1, pp. 26–42, January-February, 1990.

BROWN, G., GRAVES, G., and D. RONEN, "Scheduling Ocean Transportation of Crude Oil," *Management Science*, Vol. 33, no. 3, p. 335–346, March 1987.

GERSHKOFF, I., "Optimizing Flight Crew Schedules," *Interfaces*, Vol. 19, no. 4, pp. 29–43, 1989.

HOFFMAN, K. and M. PADBERG, "Solving Airline Crew Scheduling Problems by Branch and Cut," *Management Science*, Vol. 39, pp. 657–682, 1993.

ROUSSEAU, J.-M., ed., *Computer Scheduling of Public Transport*, Vol. 2. Amsterdam: North-Holland, 1985.

SMITH, B. and A. WREN, "A Bus Crew Scheduling System Using a Set Covering Formulation," *Transportation Research A*, Vol. 22A, no. 2, pp. 97–108, 1988.

WREN, A., ed., *Computer Scheduling of Public Transport*. Amsterdam: North-Holland, 1981.

CHAPTER 12 Dynamic Programming and Nonlinear Programming

12.A INTRODUCTION

We are concerned in this chapter with five methods, all of which are useful for finding the optimum solution for problems with nonlinear objective functions. These methods are dynamic programming, unconstrained optimization, calculus with substitution, Lagrange multipliers, and gradient search. We discuss the mathematical aspects of these topics and provide examples of their application, particularly in environmental, hydraulic, and structural engineering. None of these nonlinear methods utilizes the mathematical technology developed for linear and integer programming. They are different orders of ideas from the simplex-based optimization we have been discussing up to now.

The techniques presented in this chapter do not exhaust the realm of nonlinear optimization. Many other techniques exist, and some are simplex based. The simplex-based methods are discussed earlier in this text in Chapter 10, which covers an extended problem in water resources management. The water resources problem is used as a vehicle to introduce piecewise approximation of convex and concave functions. A self-contained discussion of piecewise approximation is presented there. Piecewise approximation can be applied both to certain nonlinear objective functions and certain nonlinear constraints. In addition, Chapter 10 contains a treatment of a nonlinear constraint in which two decision variables multiply one another. In that treatment, the nonlinear constraint is converted, by approximation, to a linear constraint. This chapter, in contrast, focuses on methodologies which are specific for nonlinear objective functions with no constraints, with one constraint, or perhaps with a few constraints. The first topic taken up is dynamic programming.

12.B DYNAMIC PROGRAMMING

12.B.1 Theory

Recall that linear programming means the determination of "programs" of action for linear problems. Dynamic programming has a parallel meaning; it is the determination of "programs" of action for problems whose decisions are made at successive stages, often time stages. A better, but perhaps less classy, name for dynamic programming might be "staged optimization," implying decision making at each of a number of successive stages. Whereas linear programming utilizes a directed search of the feasible extreme points defined by the constraint set, dynamic programming utilizes the "principle of optimality." Dynamic programming and the principle of optimality which underlies it are the creation of the applied mathematician Richard Bellman and date from the 1950s. Bellman's original statement of the principal of optimality was a rather lengthy and convoluted sentence which students and researchers read and promptly forgot, even though the mathematical underpinnings were retained. In the 1960s, Kaufmann, a French operations research analyst, restated the principle in a simplified yet mathematically elegant way. We utilize his restatement to introduce you to dynamic programming. Kaufmann put the principle this way:

An optimal policy contains only optimal subpolicies.

This clean restatement affords us the opportunity of some nonmathematical discussion.

12.B.2 Application

Siting Solid Waste Recycling Centers. Imagine that we can open a number of recycling centers in half a dozen counties of a state. Each additional recycling center positioned in each county increases access of the people in that county to the recycling centers. A county can have 0, 1, 2, up to at most 25 recycling centers, and 25 centers is the limit of the number that can be located in total in all the counties. That is, all the centers could be allocated to a single county if that is optimal. Access of the people in a county is measured by the number of people within 15 minutes by car of a recycling center in that county. We assume that when a recycling center is sited in a county, it is positioned optimally for the number of centers in the county; that is, if p recycling centers are allocated to a county, they are presumed to be sited in such a way that the maximum number of people are within 15 minutes of a center.

Now assume that the recycling centers have been allocated to the six counties by a method that provides a globally optimal solution to the allocation. Suppose that solution looks like the following:

$$(3, 6, 2, 4, 4, 6)$$

where the numbers in parenthesis represent, in order, the number of centers allocated to county 1, county 2, and so on, up to county 6. Now suppose we had an additional requirement on our allocation, namely, that exactly 10 centers must be allocated in total to county 3, county 4, and county 5. This happens to be the exact amount already allocated to counties 3, 4, and 5 and distributed (2, 4, 4) to those counties. Is this allocation of 10 to counties 3, 4, and 5 optimal? It must be, else the original optimal allocation of 25 units to all six counties would be wrong. In fact, the allocation of (3, 6, 2) to the first three counties must be the optimal way to distribute 11 to these counties and (4, 4, 6) must be the optimal way to distribute 14 to the last three counties. That is, an optimal policy must contain only optimal subpolicies. Dynamic programming utilizes this principle of subsystem optimization to implement a methodology for finding globally optimal solutions.

In general, a dynamic programming problem finds the allocation $(x_1, x_2, \ldots, x_j, \ldots, x_n)$ to n stages which optimizes an objective function subject to one or a few constraints. That is, the dynamic programming problem may be stated as

$$\text{Maximize or Minimize } f_1(x_1) + f_2(x_2) + \cdots + f_j(x_j) + \cdots + f_n(x_n) \quad (12.1)$$

$$\text{Subject to: } x_1 + x_2 + \ldots + x_j + \ldots + x_n = B$$

where the return or loss function is $f_j(x_j)$, a monotonically increasing or decreasing function of x_j, the amount of resource allocated to stage j. The methodology for deriving the optimal allocation is created in the following way.

Let us say that 25 units of resources are to be allocated among all the n stages of the optimization. The subsystem optimization begins at the first stage and proceeds stage by stage with allocations. Let us say we have reached the fifth stage and are examining possible allocations. At the fifth stage we are developing not the optimal allocation to the first five stages, but all the optimal ways to allocate 0, 1, 2, ..., 25 units to those five stages. For example, if 10 units were to be allocated to the first five stages, with the remaining 15 to the last $n - 5$ stages, these 10 units might be distributed optimally as

$$(2, 3, 0, 4, 1)$$

with an objective function value of 470,000 for these five stages. If 11 units were to be allocated, the distribution might be

$$(2, 3, 1, 4, 1)$$

with a return of 475,000. If 12 units were to be allocated, the distribution is

TABLE 12.1 OPTIMAL ALLOCATIONS TO THE FIRST FIVE STAGES

Amount (Number of Centers) Allocated to Stages 1 to 5	Optimal Allocation of Centers by Stage	$F_{1...5}(x_1, x_2, ..., x_5)$ Objective Value Associated with the Optimal Allocation
⋮	⋮	⋮
9	(2, 2, 0, 4, 1)	469,000
10	(2, 3, 0, 4, 1)	470,000
11	(2, 3, 1, 4, 1)	475,000
12	(2, 4, 1, 4, 1)	479,500
13	(2, 4, 2, 4, 1)	483,500
14	(3, 4, 2, 4, 1)	487,000
⋮	⋮	⋮

$$(2, 4, 1, 4, 1)$$

with a return of 479,500 and so on. We can construct a table for allocation to the first five stages which shows the objective value and the sequence of allocated numbers for each possible allocation to the five stages. A portion of this table appears as Table 12.1. Even though, we don't show allocation of 25 units, this will be part of the completed table, as will allocation of 24, 23, and so on, down to zero units.

We now move on to the sixth stage where we will develop a table with similar information to that developed for the fifth stage. The development of the table makes use of a combination of a simple enumeration and Bellman's principle of optimality. To build the new table, we need a listing of the payoffs or returns from allocation of 0, 1, 2, ..., 25 units to stage 6. We can call these numbers $f_6(0), f_6(1), f_6(2), f_6(3), ..., f_6(25)$, where $f_6(x_6)$ is the return from allocating x_6 to stage 6.

Suppose 14 units are to be allocated to the first six stages. Then the amount allocated to stage 6 can take on any value between 0 and 14. Suppose 0 units are devoted to stage 6 itself, leaving the entire 14 units to be distributed among the preceding five stages. We know how to distribute 14 units to the preceding stages from a quick look at Table 12.1. Thus, the solution for zero allocated to stage 6 is immediately known. Suppose 1 unit of the 14 units is to be allocated to stage 6, with the remaining 13 units to be divided among the five preceding stages. The payoff from this division is also immediately known. It is the return from allocating one unit to stage 6, namely $f_6(1)$, plus the optimal return from allocating 13 units to stages 1 to 5, which we already know, or a total return of $f_6(1) + 483{,}500$, and the allocation to the first five stages is (2, 4, 2, 4, 1).

Suppose 2 units of the 14 are to be devoted to stage 6, leaving 12 to be allocated to the preceding five stages. The return from this allocation is $f_6(2) + 479{,}500$. If 3 units of the 14 are devoted to stage 6, then 11 remain for stages 1 to 5 and the return is $f_6(3) + 475{,}000$. If 4 units of the 14 go to stage 6, the return is $f_6(4) + 470{,}000$. And so on. Basically we are performing a search over all possible ways to allocate 14 units to the first six stages. From this search, we will select the optimum (highest-value) distribution, which we will say is (2, 3, 1, 4, 1, 3). That is, we are claiming that the optimal way to distribute 14 units to stages 1 to 6 involves the optimal way to distribute 11 units to stages 1 to 5 plus 3 units to stage 6. We have found the optimal way to distribute 14 units by the following search over x_6, the amount allocated to the sixth stage:

$$\max_{x_6=0,1,2,\ldots,14} [f_6(x_6) + F_{1\ldots5}(14 - x_6)]$$

where $F_{1\ldots5}(14 - x_6)$ is the optimal return from distributing $14 - x_6$ units to stages 1 to 5.

Now suppose we want to distribute 15 units to stages 1 to 6. We can find the optimal way to distribute 15 units by a similar search, this time over $x_6 = 0, 1, 2, \ldots, 15$. That is, we seek the

$$\max_{x_6=0,1,2,\ldots,15} [f_6(x_6) + F_{1\ldots5}(15 - x_6)]$$

The optimal way to distribute 16 units is to find the

$$\max_{x_6=0,1,2,\ldots,16} [f_6(x_6) + F_{1\ldots5}(16 - x_6)]$$

and so on.

By conducting these successive searches over x_6 for all possible allocations 0, 1, 2, ..., 25 to the first six stages, we can create a new table like Table 12.1 but with allocations to the first six as opposed to first five stages. The table would have the same type of information as Table 12.1.

In general, if we represent the amount allocated to the first k stages as A, we can summarize the search process at stage k as finding for all $A = 0, 1, \ldots, 25$

$$F_{1\ldots k}(A) = \max_{x_k=0,1,\ldots,A} [f_k(x_k) + F_{1\ldots k-1}(A - x_k)] \tag{12.2}$$

The value of A corresponds to the values in the leftmost column of Table 12.1. Equation 12.2 is known as the dynamic programming *recursive equation*, and A is often referred to as the *state* of the system, since it represents the amount of the resource allocated at any point in the calculations.

Looking at this formula allows us the opportunity to explain how the principle of optimality is implemented. Recall the statement that "An optimal policy contains only optimal subpolicies." If A units are to be distributed to the first k stages, the optimal allocation (optimal policy) consists of the highest payoff way to distribute x_k units to the kth stage and $(A - x_k)$ units to the $k - 1$ preceding stages. The search considers only those allocations of $(A - x_k)$ to stages 1, 2, ..., $k - 1$ that are optimal. That is, the search considers only optimal subpolicies.

It is worth noting that the objective functions are not considered to have a predefined shape. They may be convex, concave, or a mixture. Some may even be linear, but if all were linear, you could determine an optimal allocation with little effort. All that would be necessary is to allocate all units to the stage with the greatest slope of the objective function.

We present next a complete and self-contained example of staged optimization. Once again, we are siting recycling centers, this time a total of 14 of them in five counties. Table 12.2 lists the payoff or return for siting 0, 1, 2, 3, ..., and so on, recycling centers in each of the five counties. There is zero payoff from locating zero centers. A glance at the return function suggests that payoffs are initially higher for the first few centers in county 2 and also in county 4. It will be interesting to see whether the optimal allocations coincides with or reinforces this observation.

Table 12.3 is an enumeration of all the possible ways to allocate optimally 0, 1, 2, 3, up to 14 centers to the first two counties. For instance, if 11 centers are to be allocated to the first two stages, the optimal division is 6 to the first and 5 to the second. If only 5 are to be allocated, the optimal allocation is 2 to the first and 3 to the second stage. Each optimal pair is established by a search over x_2 which seeks the

$$\max_{x_2} \left[f_2(x_2) + f_1(A - x_2) \right]$$

which is developed for all values of $A = 0, 1, 2, \ldots, 14$. The last column of Table 12.3 becomes the first column of Table 12.4.

Table 12.4 considers the optimal allocation of 0, 1, 2, ..., 14 units to the first three stages or counties. Its first column consists of the return or payoff as well as the actual allocation from allocating 0, 1, 2, ..., 14 centers to the first *two* counties.

TABLE 12.2 PAYOFF OR RETURN FUNCTIONS FROM SITING RECYCLING CENTERS BY COUNTY (PEOPLE, IN THOUSANDS, COVERED BY BEST ALLOCATION OF UNITS)

Number of Recycling Centers Allocated	County 1	County 2	County 3	County 4	County 5
14	910	1012	890	977	790
13	895	998	861	965	772
12	876	979	826	945	752
11	856	957	787	915	730
10	831	933	746	877	704
9	802	906	699	836	674
8	762	871	646	789	639
7	712	830	588	737	598
6	652	776	526	676	552
5	567	702	457	604	498
4	475	612	383	514	417
3	375	509	301	413	330
2	270	370	210	302	238
1	150	200	110	160	130
0	0	0	0	0	0

TABLE 12.3 OPTIMAL ALLOCATION OF CENTERS TO THE FIRST TWO STAGES

Units A to Be Allocated to Stage 1 and 2	Optimal Way to Allocate A Units Between First Two Stages	$\underset{x_2}{\text{Max}}[f_2(x_2) + f_1(A - x_2)]^*$ People (in thousands) Covered by the Best Allocation of A Units to Stages 1 and 2 (Found by Search)
14	(7, 7)	1542
13	(7, 6)	1488
12	(6, 6)	1428
11	(6, 5)	1354
10	(5, 5)	1269
9	(5, 4)	1179
8	(4, 4)	1087
7	(3, 4)	987
6	(3, 3)	884
5	(2, 3)	779
4	(1, 3)	659
3	(1, 2)	520
2	(0, 2)	370
1	(0, 1)	200
0	(0, 0)	0

*Becomes $F_{1...2}(A)$ at the next iteration.

The third column is the payoff or return for allocating centers to county 3 and is the same as column 3 of Table 12.2. The fourth and fifth columns are the optimal allocations of A units to the first three counties as well as the return from that optimal allocation. The optimal allocation of A units is established by enumerating all the ways A can be allocated between the first three stages where only optimal allocations of $A - x_3$ to the first two counties are considered. That is, we find

$$F_{1\ldots3}(A) = \underset{x_3}{\text{Max}}\,[f_3(x_3) + F_{1\ldots2}(A - x_3)]$$

for all values of $A = 0, 1, 2, \ldots, 14$.

Table 12.5 repeats the process of enumeration, this time for x_4 to stage 4 and $A - x_4$ to stages 1, 2, and 3, where $(A - x_4)$ is always allocated optimally (i.e., only optimal allocations are considered) to these first three stages. Table 12.6 goes through parallel calculations to establish the optimal allocation of $A = 0, 1, 2, \ldots,$ 14 to the five counties. The interested reader can follow the allocations/searches to verify that the procedure has been correctly implemented or to find any error in the process. The optimal return for siting 14 centers in the five counties is 1,849,000 people located within 15 minutes driving time of a recycling center. The optimal distribution of centers is (3 ,4, 1, 4, 2).

TABLE 12.4 OPTIMAL ALLOCATION OF CENTERS TO THE FIRST THREE STAGES

$F_{1...2}(A)$ The Value of the Objective When A is Optimally Allocated Between Stages 1 and 2	Units A to Be Allocated To Stages 1, 2, 3	$f_3(x_3)$ (if $x_3 = A$)	Optimal Allocation of A Units Among Stages 1, 2, 3	$F_{1...3}(A) = \underset{x_3}{\text{Max}}[f_3(x_3) + F_{1...2}(A - x_3)]$ People (in Thousands) Covered by the Best Allocation of A Units to Stages 1, 2, and 3 (Found by Search)
Value and Allocation				
1542 (7, 7)	14	890	(6, 5, 3)	1655
1488 (7, 6)	13	861	(5, 5, 3)	1570
1428 (6, 6)	12	826	(5, 4, 3)	1480
1354 (6, 5)	11	787	(5, 4, 2)	1389
1269 (5, 5)	10	746	(4, 4, 2)	1297
1179 (5, 4)	9	699	(3, 4, 2)	1197
	9		(4, 4, 1)	1197
1087 (4, 4)	8	646	(3, 4, 1)	1097
987 (3, 4)	7	588	(3, 3, 1)	994
884 (3, 3)	6	526	(2, 3, 1)	889
779 (2, 3)	5	457	(2, 3, 0)	779
659 (1, 3)	4	383	(1, 3, 0)	659
520 (1, 2)	3	301	(1, 2, 0)	520
370 (0, 2)	2	210	(0, 2, 0)	370
200 (0, 1)	1	110	(0, 1, 0)	200
0 (0, 0)	0	0	(0, 0, 0)	0

An alert reader will ask, as you probably have already asked, "Suppose the stages were presented in a different order, what happens?" The principle of optimality is such that the answer in terms of quantity of resource allocated to each stage will not change; it is not dependent on the order of presentation of the stages.

The reader interested in mastering dynamic programming is referred to the well-regarded text *The Art and Theory of Dynamic Programming* by Dreyfus and Law (Academic Press, 1977).

12.B.3 Computational Efficiency

Dynamic programming does suffer one flaw which has become known by the colorful name "the curse of dimensionality." The curse is merely the flaw that limits its application. You noted that the objective function was optimized subject to a single constraint. Dynamic programming with more than one constraint can be very difficult to implement. There are tricks to implement the technique subject

TABLE 12.5 OPTIMAL ALLOCATION OF CENTERS TO THE FIRST FOUR STAGES

$F_{1...3}(A)$ The Value of the Objective When A is Optimally Divided Between Stages 1 to 3	Units A to Be Allocated to Stages 1, 2, 3, 4	$f_4(x_4)$ (if $x_4 = A$)	Optimal Allocation of A Units to Stages 1, 2, 3, 4	$F_{1...4}(A) = \underset{x_4}{\text{Max}}[f_4(x_4) + F_{1...3}(A - x_4)]$ People (in Thousands) Covered by the Best Allocation of A Units to Stages 1, 2, 3, 4 (Found by Search)
Value and Allocation				
1655 (6, 5, 3)	14	977	(4, 4, 2, 4)	1811
1570 (5, 5, 3)	13	965	(3, 4, 2, 4)	1711
	13		(4, 4, 1, 4)	1711
1480 (5, 4, 3)	12	945	(3, 4, 1, 4)	1611
1389 (5, 4, 2)	11	915	(3, 4, 1, 3)	1510
1297 (4, 4, 2)	10	877	(3, 3, 1, 3)	1407
1197 (3, 4, 2)	9	836	(2, 3, 1, 3)	1302
1197 (4, 4, 1)	9			
1097 (3, 4, 1)	8	789	(2, 3, 0, 3)	1192
994 (3, 3, 1)	7	737	(2, 3, 0, 2)	1081
889 (2, 3, 1)	6	676	(1, 3, 0, 2)	961
779 (2, 3, 0)	5	604	(1, 2, 0, 2)	822
659 (1, 3, 0)	4	514	(1, 2, 0, 1)	680
520 (1, 2, 0)	3	413	(0, 2, 0, 1)	530
370 (0, 2, 0)	2	302	(0, 2, 0, 0)	370
200 (0, 1, 0)	1	160	(0, 1, 0, 0)	200
0 (0, 0, 0)	0	0	(0, 0, 0, 0)	0

to more than one constraint, but they are beyond the scope of this text. The text by Dreyfus and Law suggests one such method for handling more than one constraint.

Finally, we should compare dynamic programming to enumeration. Suppose we have n stages and b units of resource to allocate among the n stages. The number of ways these b units can be allocated (enumerated) is

$$E = \frac{(n + b - 1)!}{b!(n - 1)!} \tag{12.3}$$

It can be shown that dynamic programming will require a comparison of D allocations where

$$D = (1 + b)\left[n\left(1 + \frac{b}{2}\right) - b\right] \tag{12.4}$$

If $n = 10$ and $b = 20$, then $E = 10{,}015{,}005$ ways to enumerate solutions, and $D = 1890$ solutions to be compared, a clear advantage for dynamic programming.

TABLE 12.6 OPTIMAL ALLOCATION OF CENTERS TO THE FIRST FIVE STAGES

$F_{1...4}(A)$ The Value of the Objective When A is Optimally Divided Among Stages 1, 2, ..., 4	Units A to Be Allocated to Stages 1 to 5	$f_5(x_5)$ (if $x_5 = A$)	Optimal Allocation of A Units to Stages 1 to 5	$F_{1...5}(A) = \underset{x_5}{\text{Min}}[f_5(x_5) + F_{1...4}(A - x_5)]$ People (in Thousands) Covered by the Best Allocation of A Units to all Five Stages (Found by Search)
Value and Allocation				
1811 (4, 4, 2, 4)	14	790	(3, 4, 1, 4, 2)	1849
1711 (3, 4, 2, 4)	13	772	(3, 4, 1, 3, 2)	1748
1711 (4, 4, 1, 4)	13			
1611 (3, 4, 1, 4)	12	752	(3, 3, 1, 3, 2)	1645
1510 (3, 4, 1, 3)	11	730	(2, 3, 1, 3, 2)	1540
1407 (3, 3, 1, 3)	10	704	(2, 3, 1, 3, 1)	1432
1302 (2, 3, 1, 3)	9	674	(2, 3, 0, 3, 1)	1322
1192 (2, 3, 0, 3)	8	639	(2, 3, 0, 2, 1)	1211
1081 (2, 3, 0, 2)	7	598	(1, 3, 0, 2, 1)	1091
961 (1, 3, 0, 2)	6	552	(1, 3, 0, 2, 0)	961
822 (1, 2, 0, 2)	5	498	(1, 2, 0, 2, 0)	822
680 (1, 2, 0, 1)	4	417	(1, 2, 0, 1, 0)	680
530 (0, 2, 0, 1)	3	330	(0, 2, 0, 1, 0)	530
370 (0, 2, 0, 0)	2	238	(0, 2, 0, 0, 0)	370
200 (0, 1, 1, 0)	1	130	(0, 1, 0, 0, 0)	200
0 (0, 0, 0, 0)	0	0	(0, 0, 0, 0, 0)	0

12.C NONLINEAR PROGRAMMING: BACKGROUND

12.C.1 Maxima, Minima, and Saddle Points

Assume that the gain or loss from alternative engineering designs can be represented by the function of one variable shown in Figure 12.1. The engineer wants to find the highest or lowest point on the bounded function. Such a point is called the *global maximum* or *global minimum*:

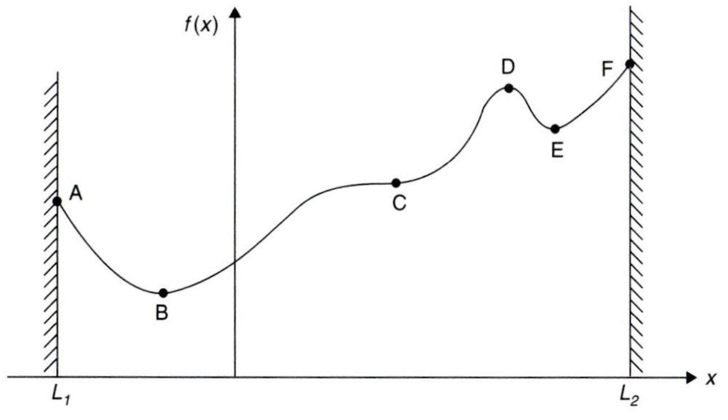

Figure 12.1 Constrained function of one variable.

A function $f(x)$ defined over a closed region has a global maximum at the point x_0 if $f(x) \leq f(x_0)$ for all x allowable.

By changing the inequality to $\geq$, the definition of global minimum is obtained. Verify that points F and B in Figure 12.1 represent the global maximum and global minimum, respectively. Note also that the global maximum occurs at the upper bound constraint on x.

Except for point C, which will be discussed later, all the labeled points are also of interest in the "local" sense; that is, within a small region they represent the best (or worst) that can be done:

A function $f(x)$ is defined at all points within a δ-neighborhood of x_0. The function has a local maximum at x_0 if there exists an ε, $0 < \varepsilon < \delta$, such that $f(x_0) \geq f(x)$ for all x in the ε-neighborhood of x_0.

Reversing the sign to $\leq$ defines a local minimum. Verify that points A, D, and F represent local maxima, while points B and E represent local minima (points F and B are simultaneously local and global optima).

Point C is of little interest to the engineer since it represents neither a local nor global optimum point. At point C, the slope is zero, and the function changes form (from concave to convex, as defined later). Such a point is called a *horizontal inflection point* in one dimension and a *saddle point* in two or more dimensions:

A function has a saddle point at x_0 if all partial derivatives are equal to zero at the point, but x_0 does not satisfy the definition of a local maximum or minimum.

Figure 12.2 explains how the point got its name.

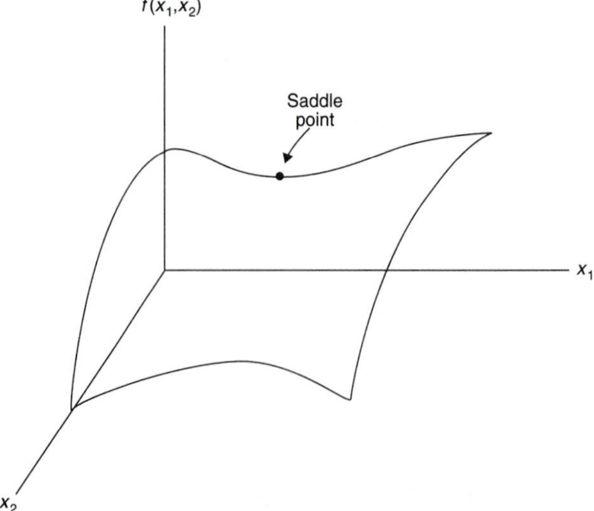

Figure 12.2 Illustration of a saddle point.

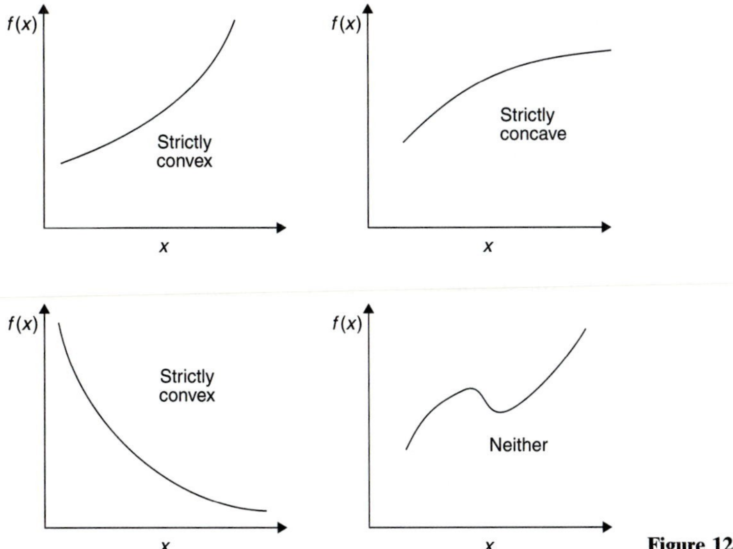

Figure 12.3 Form of a function.

12.C.2 Form of a Function

It is sometimes useful to know the form of a function when attempting to find an optimal point. Define

> A strictly convex (strictly concave) function is one for which a straight line drawn between any two points on the function will everywhere overestimate (underestimate) the value of the function.

Some examples are shown in Figure 12.3.

It can also be shown that the sum of strictly convex (strictly concave) functions is strictly convex (strictly concave) and that adding or subtracting a linear function (or a constant) from a strictly convex (strictly concave) function does not change the form of the function.

Example 12-1.

Determine the form of the function $f(x)$ for $x > 0$.

$$f(x) = \frac{1}{x} + x^3 - \sqrt{x} + 3x + 4.$$

Solution: Since the first three terms are strictly convex, and adding a linear function (or a constant) will not change the form of the function, the function $f(x)$ is strictly convex for the range specified.

These background observations will be used in the discussions that follow.

12.D UNCONSTRAINED OPTIMIZATION

12.D.1 Theory

The unconstrained optimization problem can be expressed as:

$$\text{Optimize} \quad Z = f(x_1, x_2, \ldots, x_n). \tag{12.5}$$

From calculus we know that for local optima of the unconstrained function to exist, a *necessary condition* is that the gradient vector $\nabla f = 0$; that is,

$$\nabla f = \left(\frac{\partial f}{\partial x_1}, \frac{\partial f}{\partial x_2}, \cdots, \frac{\partial f}{\partial x_n}\right) = 0 \tag{12.6}$$

which says that *each* partial derivative must equal zero *simultaneously*. Equation 12.6 defines a *critical point*, or *stationary point*, of the function:

> A function $f(x_1, x_2, \ldots, x_n)$ has a critical point, or stationary point, at x_0 if $\nabla f = 0$ at x_0.

A critical point, or stationary point, can be a local minimum, local maximum, or saddle point. Since we are not interested in the latter type of point, it is useful to have the following *sufficiency condition* to distinguish between the type of points:

> For a function of only one variable, if $f^{(1)}(x_0) = 0, f^{(2)}(x_0) = 0, \ldots, f^{(n)}(x_0) = 0$, but $f^{(n+1)}(x_0) \neq 0$; then $f(x)$ has a relative *maximum* at x_0 if n is odd and $f^{(n+1)}(x_0) < 0$; $f(x)$ has a relative *minimum* at x_0 if n is odd and $f^{(n+1)}(x_0) > 0$. If n is even, $f(x)$ has a horizontal inflection point (saddle point) at x_0.

For a function of two variables, where $\nabla f = 0$ and

$$H_0 = \begin{bmatrix} \dfrac{\partial^2 f}{\partial x_1^2} & \dfrac{\partial^2 f}{\partial x_1 \partial x_2} \\ \dfrac{\partial^2 f}{\partial x_2 \partial x_1} & \dfrac{\partial^2 f}{\partial x_2^2} \end{bmatrix} = \begin{bmatrix} A & B \\ C & D \end{bmatrix}$$

where H_0 = hessian matrix of order 2 evaluated at x_0;

If $AD - BC > 0$ and $A + D < 0$, then relative maximum at (x_1^0, x_2^0);

If $AD - BC > 0$ and $A + D > 0$, then relative minimum at (x_1^0, x_2^0);

If $AD - BC < 0$, then saddle point; and

If $AD - BC = 0$, then inconclusive.

Development of these conditions date from the mid-1700s and are attributed in part to the mathematicians Maclaurin (1692–1746) and Lagrange (1736–1813). Further development is available in the classic work by Hancock (1960).

Example 12-2.

Determine the stationary point and type of point found for

$$f(x_1, x_2) = x_1^2 - x_2^2 + 4$$

Solution:

$$\left.\begin{aligned} \frac{\partial f}{\partial x_1} &= \quad 2x_1 = 0 \\ \frac{\partial f}{\partial x_2} &= -2x_2 = 0 \end{aligned}\right\} \Rightarrow \begin{aligned} x_1^0 &= 0 \\ x_2^0 &= 0 \end{aligned}$$

$$H_o = \begin{bmatrix} 2 & 0 \\ 0 & -2 \end{bmatrix} = \begin{bmatrix} A & B \\ C & D \end{bmatrix}$$

$AD - BC = -4 < 0$, therefore *saddle point.*

Similar sufficiency conditions hold for functions of three or more variables, but are complicated and omitted here (see Gottfried and Weisman, 1973).

Given the complexity of the sufficiency conditions, the following observation is sometimes useful:

> A strictly convex (strictly concave) function will possess at most only one stationary point (it may have none), and that point, if it exists, is the global minimum (global maximum) of the function.

The reader may want to carry out a few simple sketches to prove the observation.

12.D.2 Applications

A common characteristic of unconstrained optimization problems in engineering is the presence of strong trade-offs between problem components. This is demonstrated in the following two applications.

Optimal Capacity Expansion. As population increases, wastewater treatment plants and interceptor sewers must be built to handle additional waste flow. Ideally, the capacity expansions stay just ahead of waste flow growth, as shown in Figure 12.4. The question arises as to what the optimal expansion size and timing should be.

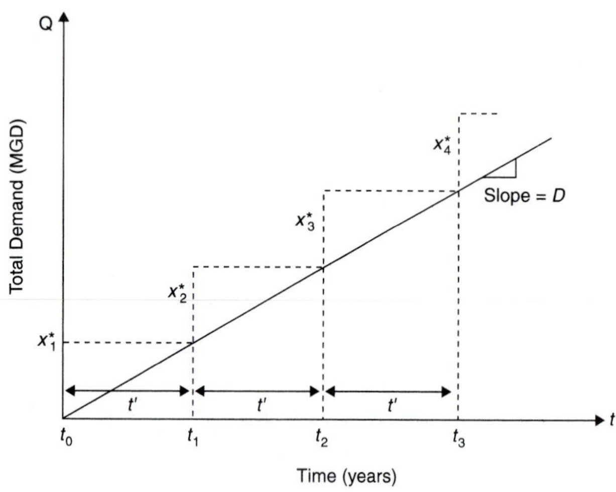

Figure 12.4 Optimal capacity expansion problem.

Considering the facility cost itself, there is a strong argument for building large, and therefore infrequently. Both wastewater treatment plants and interceptor sewers demonstrate *economies of scale* defined by

$$c(x) = ax^b$$

where

$$0 < b < 1$$

and

$c(x)$ = present worth cost of capital and operation and maintenance ($ million);

x = hydraulic design capacity (million gallons per day, MGD);

a = constant; and

b = constant, the economy of scale factor.

A typical cost function is shown in Figure 12.5. Due to the shape of the function (strictly concave), there is an incentive to build one large facility now rather than a sequence of smaller ones.

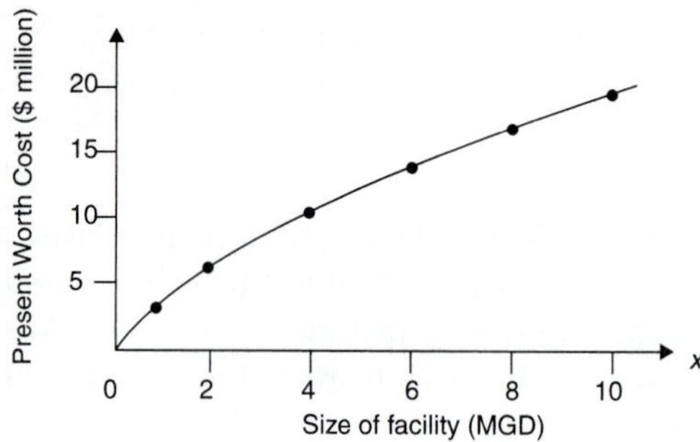

Figure 12.5 Cost function showing economics of scale (a = 4.0, b = 0.7).

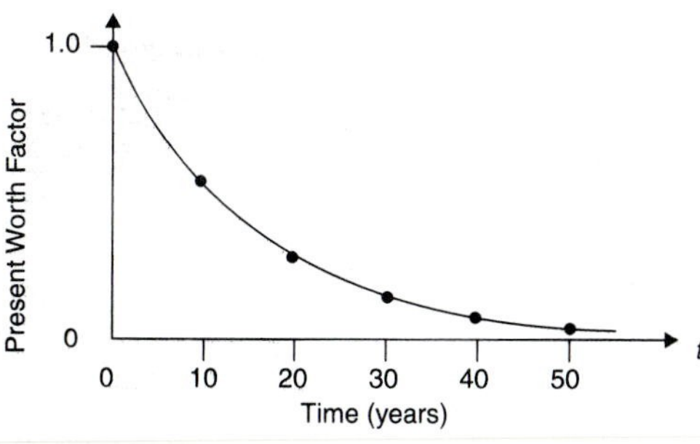

Figure 12.6 Present worth factor, e^{-rt} (r = 6 percent).

Counteracting this approach is the fact that if expenditures are put off into the future, their present worth cost is decreased. This is illustrated in Figure 12.6, where the present worth factor for continuous discounting is plotted.

The trade-off is between economies of scale (build large now) and discounting (build small now). The problem can be modeled as an unconstrained optimization problem. The solution steps will follow the early work by Manne (1961). Define

D = rate of linear demand increase (MGD per year);

t = time, in years; and

r = rate of discount selected by Congress for use in the analysis of water resource projects, expressed as a decimal.

The key to solving this problem is the observation that at any point of capacity expansion, t_0, t_1, t_2, and so on, the future situation is identical with respect to making the expansion decision. Therefore, the size of the expansion at each of these times should be identical; that is, $x_1^* = x_2^* = x_3^*$, and so on. This, of course, means that

$$(t_1 - t_0) = (t_2 - t_1) = (t_3 - t_2) = t'$$

and

$$x_i^* = Dt'.$$

Write the present worth cost (PWC) of all expansions over n periods of length t' years each:

$$PWC = ax^b e^{-rt_0} + ax^b e^{-rt'} + ax^b e^{-r2t'} + ax^b e^{-r3t'} + \cdots + ax^b e^{-rnt'}. \quad (12.7)$$

If $t_0 = 0$,

$$\begin{aligned} PWC &= ax^b (1 + e^{-rt'} + e^{-2rt'} + e^{-3rt'} + \cdots + e^{-nrt'}) \\ &= ax^b (1 + e^{-rt'} + (e^{-rt'})^2 + (e^{-rt'})^3 + \cdots + (e^{-rt'})^n]. \end{aligned}$$

The term $e^{-rt'}$ can be bounded as

$$0 < \frac{1}{e^{rt'}} < 1 \text{ for } r > 0, t' > 0.$$

Therefore, as $n \to \infty$, the geometric series converges to

$$\frac{1}{1 - e^{-rt'}}$$

and

$$PWC = \frac{ax^b}{1 - e^{-rt'}}. \tag{12.8}$$

Substitute $x = Dt'$ and take natural logs:

$$\ln (PWC) = \ln a + b \ln (Dt') - \ln (1 - e^{-rt'}).$$

Take the partial with respect to t':

$$\frac{\partial}{\partial t'} [\ln(\text{PWC})] = \frac{b}{t'_0} - \frac{re^{-rt'_0}}{1 - e^{-rt'_0}} = 0.$$

This can be solved for t'_0 for fixed values of the economy of scale factor, b, and discount rate, r. Note that the optimal time between expansions does *not* depend on the rate of demand growth, D, a remarkable result. In contrast, the *size* of the expansion *does* depend on D, since $x_0 = Dt'_0$. Evaluation of the second partial at t'_0, though laborious, confirms that a relative minimum has been found. Alternatively, one can directly plot the total cost function (equation 12.8) as total cost versus t' and see that the relative minimum is also the global minimum.

Table 12.7 summarizes optimal expansion times (design periods) for wastewater treatment plants giving secondary treatment (b typically ranging from 0.6 to 0.8). Table 12.8 does the same for interceptor sewers (b typically ranging from 0.3 to 0.5).

Traditional design periods for treatment plants and interceptor sewers have been 20 and 50 years, respectively. While these figures may have been essentially correct for the low interest rates (from 2 to 4%) prevailing in the 1950s and

TABLE 12.7 OPTIMAL EXPANSION TIME FOR TREATMENT PLANTS (YEARS; USEPA, 1975)

Discount Rate (%)	Economy of Scale Factor		
	0.6	0.7	0.8
5	19	13	9
7	13	10	6
10	9	7	4
12	8	6	4

TABLE 12.8 OPTIMAL EXPANSION TIME FOR INTERCEPTOR SEWERS (YEARS; USEPA, 1975)

Discount Rate (%)	Economy of Scale Factor		
	0.3	0.4	0.5
5	40	31	24
7	28	22	17
10	20	16	12
12	18	14	10

1960s, they should be closely questioned for currently higher rates. The federal discount rate peaked at 8.875% in the late 1980s and was 7.625% in fiscal 1996. The USEPA considered the above analysis in modifying its now historical Construction Grants Program (USEPA, 1975). Private investment decisions, for which interest rates are usually much higher than the federal discount rate, would follow a similar analysis.

Optimal Reservoir Size for Irrigation. The total benefit (TB) derived from irrigation water in a region has been estimated as $TB = 100q - 0.0005\,q^2$, where TB is expressed as dollars per year ($/yr) and q represents dependable water supply, expressed in acre-feet per year (AF/yr).

Estimates of the annual total cost (TC) of a reservoir at different sizes results in the expression $TC = 44.42\,q^{0.90} + 0.098\,q^{1.45}$, where TC has units of dollars per year.

The agency responsible for reservoir construction wishes to maximize net benefit (NB), defined as total benefit minus total cost. Algebraically,

$$\text{Maximize} \quad NB(q) = TB(q) - TC(q). \tag{12.9}$$

The necessary conditions for a stationary point are

$$\frac{\partial NB}{\partial q} = \frac{\partial TB}{\partial q} - \frac{\partial TC}{\partial q} = 0$$

or

$$MB(q) = MC(q) \tag{12.10}$$

where MB and MC are marginal benefit and marginal cost, respectively.

Taking the partial derivatives,

$$MB(q) = \quad 100 - 0.001q$$

$$MC(q) = \quad 39.98\,q^{-0.10} + 0.142q^{0.45}.$$

Equating these and solving for q_0 is difficult. A graphical approach is both easier and more instructive. Figure 12.7 shows the point of intersection of the two curves at q_0 equal to about 66,000 AF/yr. Since the area under the MB curve far exceeds

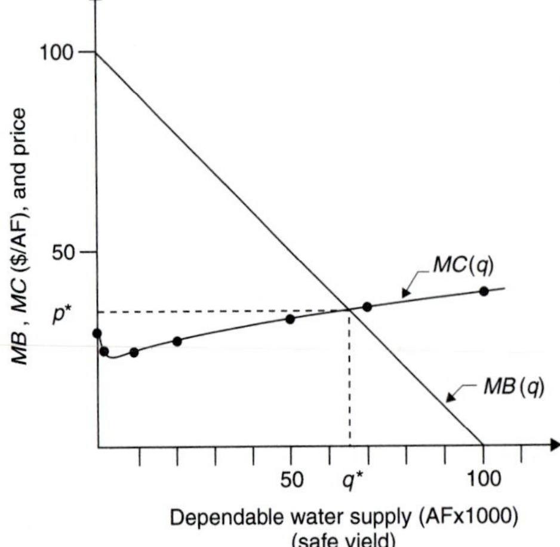

Figure 12.7 Optimal size of an irrigation reservoir.

the area under the MC curve up to the point of intersection, we know that net benefits have been maximized. The point q_0 is a global maximum, q^*, by observation.

The reader will recognize the MB curve to be the *demand curve* for irrigation water. For an amount of water, q^*, actually to be used, the price of irrigation water must be set equal to p^*, about $34.00/AF. At this price, net benefits are maximized.

The trade-off in this problem is complex; it is between benefits that increase at a declining rate as more irrigation water is made available (declining marginal benefit as less suitable land is brought under irrigation) and a reservoir cost that increases at an increasing rate in terms of q (declining marginal water yield as reservoir capacity is increased).

12.E CALCULUS WITH SUBSTITUTION

12.E.1 Theory

The calculus with substitution technique is useful for small nonlinear optimization problems having equality constraints. The problem is to

$$\text{Optimize} \quad Z = f(x_1, x_2, \ldots, x_n)$$

Subject to:

$$\begin{aligned} g_1(x_1, x_2, \ldots, x_n) &= b_1 \\ g_2(x_1, x_2, \ldots, x_n) &= b_2 \\ &\vdots \\ g_m(x_1, x_2, \ldots, x_n) &= b_m \end{aligned}$$

where $m < n$.

This problem has been referred to in the literature as the "classical optimization problem." In theory, we could use the constraint equations to solve for m variables in terms of the remaining $(n - m)$ variables and substitute these expressions into the objective function, producing an unconstrained optimization problem in $(n - m)$ variables. Calculus could then be applied as in the previous section.

In practice, however, it is usually difficult or impossible to solve the constraint equations for m variables, especially when the number of constraints becomes more than a few. The method of Lagrange multipliers, described in the next section, then becomes a useful alternative approach. First, let us look at some cases where the calculus with substitution method is the best approach, giving rapid insight into an engineering problem.

12.E.2 Applications

Best Hydraulic Section. For open channel flow, the Manning equation is commonly used:

$$Q = \frac{C_m}{n} A R^{2/3} S^{1/2} \tag{12.11}$$

where

Q = discharge (ft^3/sec);

C_m = an empirical constant,
= 1.49 $ft^{1/3}/sec$;

n = Manning roughness factor, a constant that depends on channel material;

A = flow cross sectional area (ft^2);

R = hydraulic radius (ft);
= A/P, where P is wetted channel perimeter (ft); and

S = slope of the energy grade line, also equal to channel slope for steady, uniform flow.

Assume that Q, C_m, n, and S are known. Further, assume that the channel cross section is to be rectangular, as shown in Figure 12.8. The channel bottom width, b, and height, h, are to be determined so as to minimize the cost of the channel lining. Assume cost of the lining is proportional to total wetted perimeter $(b + 2h)$. The problem is to

$$\text{Minimize} \quad P = b + 2h$$

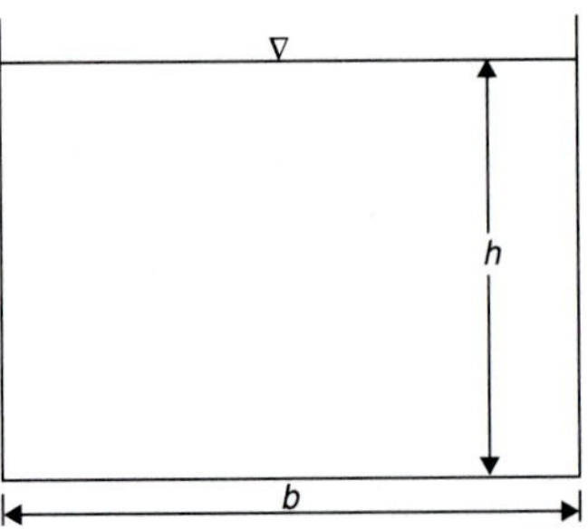

Figure 12.8 Rectangular channel cross section.

Subject to:

$$\frac{1.49}{n}(bh)\frac{(bh)^{2/3}}{(b + 2h)^{2/3}}S^{1/2} = Q \tag{12.12}$$

which is in the classical optimization format. The reader can verify, however, that there is no way to solve for one variable in terms of the other in the constraint equation. The procedure cannot continue directly, but it will be shown that by transforming the constraint, an important observation can be made that will then permit a solution.

The constraint can be written:

$$\frac{(bh)^{5/2}}{(b + 2h)} = \left(\frac{nQ}{1.49\, S^{1/2}}\right)^{3/2} = k_1.$$

or

$$(bh)^{5/2} = k_1(b + 2h)$$

$$bh = [k_1(b + 2h)]^{2/5}.$$

Since $(b + 2h)$ is to be minimized, the term on the right will take on its minimum possible value in the final solution. This means that the expression on the left will also assume its minimum possible value in the final solution. However, the expression (bh) is just A, so it is concluded that *both wetted perimeter and area will be minimized in the final optimal solution.* This is of great practical significance, since excavation costs are closely related to the cross-sectional area of flow. Both lining and excavation costs will tend to be minimized simultaneously.

By observing the last form of the constraint, it can also be said that for any prechosen value for the wetted perimeter term $(b + 2h)_0$ (note that b and h are not prechosen) and related right-hand-side constant, $[k_1(b + 2h)_0]^{2/5}$, representing cross-sectional area, the optimization problem will find the best combination of b and h so that

$$(b^* + 2h^*) = (b + 2h)_0$$

and

$$(b^*h^*) = A^* = [k_1(b + 2h)_0]^{2/5} = A_0, \text{ a minimum.}$$

In other words, for a preselected cross-sectional area of flow, the optimization problem will choose b and h to minimize the wetted perimeter. This last observation allows the problem to be recast in a simpler but equivalent form:

$$\text{Minimize} \quad P = b + 2h$$

Subject to:

$$bh = k \tag{12.13}$$

which says that for any required cross-sectional area, we seek the best values for b and h to minimize the wetted perimeter. Calculus with substitution can be used on this simpler problem. Solving for h in terms of b:

$$h = kb^{-1}$$

and substituting,

$$\text{Minimize} \quad P = b + 2kb^{-1}$$

$$\frac{\partial P}{\partial b} = 1 - 2kb^{-2} = 0$$

$$b^2 = 2k$$

$$b_0 = \sqrt{2k}$$

and

$$h_0 = \frac{k}{\sqrt{2k}} = \frac{1}{2}\sqrt{2k}.$$

Therefore,

$$b_0 = 2h_0.$$

Since the transformed objective function is strictly convex, the stationary point found is the global minimum. The best rectangular channel to minimize lining cost has a bottom width twice the length of each vertical side.

A similar approach can be taken to determine the best geometry for other cross-sectional shapes when the wetted perimeter is to be minimized.

Optimal Tank Design. It is desired to construct a vertical cylindrical steel tank as shown in Figure 12.9. The tank is open at the top, and it is known that the bottom must be twice as thick as the sides (thickness has been determined). Since a large number of tanks of different volume are to be constructed, the de-

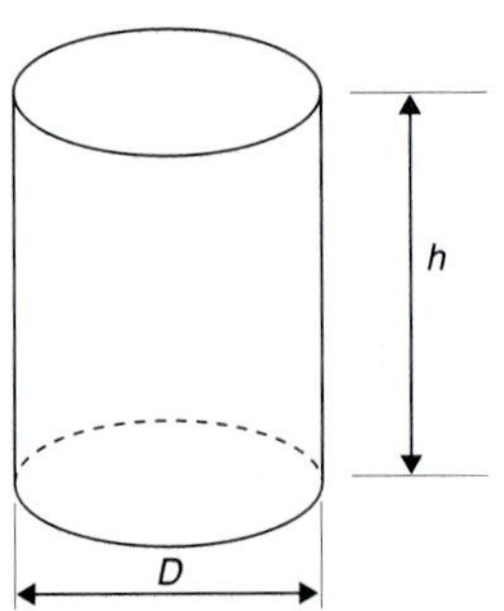

Figure 12.9 Vertical cylindrical steel tank.

signer wants to determine whether there is an optimal *shape* (diameter-to-height ratio) that all the tanks should have. Assume that the weight of material used represents total cost of the tank. The problem is to

$$\text{Minimize} \quad M = t(\pi Dh) + 2t\left(\frac{\pi D^2}{4}\right) \tag{12.14}$$

$$\text{Subject to:} \quad \frac{\pi D^2}{4} h = k$$

where M is the total amount of material used (cubic feet), t is the thickness of the sides (feet), and k is the required volume of a tank (cubic feet). Solving for h in terms of D,

$$h = \frac{4k}{\pi D^2}$$

and substituting,

$$\text{Minimize} \quad M = \frac{4tk}{D} + \frac{1}{2} t \, \pi D^2$$

which is strictly convex in terms of D. If the stationary point exists, it will be the global minimum. Take the partial derivative and set equal to zero:

$$\frac{\partial M}{\partial D} = -\frac{4tk}{D^2} + t\,\pi D = 0$$

$$D^3 = \frac{4k}{\pi}$$

$$D^* = \left(\frac{4k}{\pi}\right)^{1/3}$$

Substituting for h produces

$$h^* = \left(\frac{4k}{\pi}\right)^{1/3}$$

so that

$$\frac{D^*}{h^*} = 1.0$$

The most economical tank design under the given conditions is to have the height of the tank equal to diameter.

12.F LAGRANGE MULTIPLIERS

12.F.1 Theory

Another method to handle nonlinear mathematical programming problems is the technique known as Lagrange multipliers. Our simplified treatment will not build a theoretical foundation for the technique, but will focus on the way it is made operational.

In its most general form, the technique appends all constraints to the original objective function, creating a new problem statement with an expanded objective function and no constraints. Although we illustrate the methodology with a problem in three variables and two constraints, the reader should be able to extend the formulation to situations in which the number of variables and the number of constraints differ from these illustrative values.

Suppose the objective is

$$\text{Minimize or maximize} \quad Z = f(x_1, x_2, x_3)$$

and the constraints, which are two in number, are

$$g_1(x_1, x_2, x_3) = b_1,$$

$$g_2(x_1, x_2, x_3) = b_2.$$

We let λ_1 and λ_2 be the *Lagrange multipliers* associated with the first and second constraints, respectively, and form the new and unconstrained objective as

$$L = f(x_1, x_2, x_3) + \lambda_1 [b_1 - g_1(x_1, x_2, x_3)] + \lambda_2[b_2 - g_2(x_1, x_2, x_3)]$$

where L is referred to as the *Lagrangian function*. Any number of constraints can be folded into the objective function in this way, although the problem can become unwieldy to solve, especially if the constraints are nonlinear. The problem statement is now reduced to an objective function only, and the methods of calculus may be applied.

Five partial derivatives need to be taken to apply the calculus. These are the partial derivatives with respect to each of the variables, x_1, x_2, x_3, and with respect to each of the multipliers, λ_1 and λ_2. The derivatives, which will be set equal to zero, take the following general form:

$$\frac{\partial L}{\partial x_1} = \frac{\partial f(x_1, x_2, x_3)}{\partial x_1} + \lambda_1 \frac{\partial[b_1 - g_1(x_1, x_2, x_3)}{\partial x_1} + \lambda_2 \frac{\partial[b_2 - g_2(x_1, x_2, x_3)]}{\partial x_1}$$

$$\frac{\partial L}{\partial x_2} = \frac{\partial f(x_1, x_2, x_3)}{\partial x_2} + \lambda_1 \frac{\partial[b_1 - g_1(x_1, x_2, x_3)}{\partial x_2} + \lambda_2 \frac{\partial[b_2 - g_2(x_1, x_2, x_3)]}{\partial x_2}$$

$$\frac{\partial L}{\partial x_3} = \frac{\partial f(x_1, x_2, x_3)}{\partial x_3} + \lambda_1 \frac{\partial[b_1 - g_1(x_1, x_2, x_3)}{\partial x_3} + \lambda_2 \frac{\partial[b_2 - g_2(x_1, x_2, x_3)]}{\partial x_3}$$

$$\frac{\partial L}{\partial \lambda_1} = b_1 - g_1(x_1, x_2, x_3)$$

$$\frac{\partial L}{\partial \lambda_2} = b_2 - g_2(x_1, x_2, x_3).$$

When the five resulting equations are set equal to zero, it is often possible to solve for the five unknowns $(x_1, x_2, x_3, \lambda_1, \lambda_2)$. In fact, more than one solution may occur. When multiple solutions obtain, one may reflect a minimum, another a maximum, and still others indicate a local minimum or local maximum to a strangely shaped surface.

Without proof, we also state that the Lagrange multiplier λ_i can be interpreted to represent the change in the objective function at optimality caused by a small change in the right-hand-side constant of constraint i; that is,

$$\lambda_i = \frac{\Delta f}{\Delta b_i}. \tag{12.15}$$

The Lagrange multiplier thus has the same meaning as the dual variable in linear programming.

The reader interested in the "why" of Lagrange multipliers can have their intuition heightened by reading appropriate sections of any one of most advanced calculus texts or *Introduction to Mathematical Programming* by Winston (Duxbury Press, 1994).

12.F.2 Applications

Sheet Metal Forming. Suppose we have a square of sheet metal with side S which is to be cut and folded into a topless box of maximum volume. Figure 12.10 shows the square with the cuts as dashed lines, the folds as solid lines, and

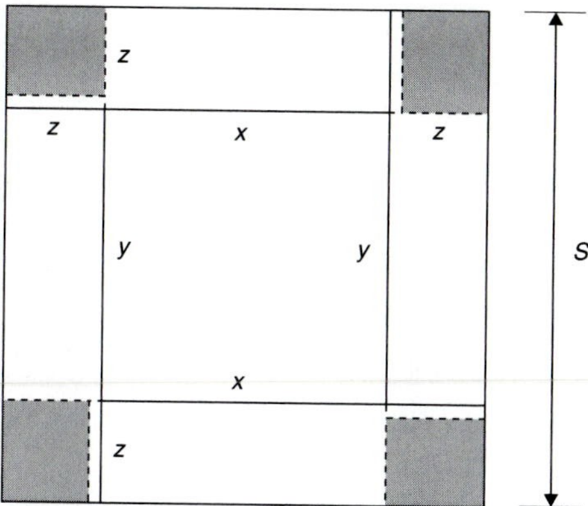

Figure 12.10 A square of sheet metal.

waste area shaded. Some small amount next to the shaded waste area is retained for soldering and connection to adjacent box sides; this "overlap area" is the small area between the dotted and solid lines. Determine the dimensions of cutting, x, y, and z.

Those quick at geometry will note that since the problem setup insists that the sides meet evenly at the top when folded up (box height equal to z), x must equal y. However, we leave the dimensions as x, y, and z for illustrative purposes.

The problem may be stated as

$$\text{Maximize} \quad xyz$$

$$\text{Subject to:} \quad x + 2z = S$$

$$y + 2z = S.$$

The Lagrangian function is

$$L = xyz + \lambda_1(S - x - 2z) + \lambda_2(S - y - 2z).$$

The partial derivatives are taken and are set equal to zero

$$\frac{\partial L}{\partial x} = yz - \lambda_1 = 0$$

$$\frac{\partial L}{\partial y} = xz - \lambda_2 = 0$$

$$\frac{\partial L}{\partial z} = xy - 2\lambda_1 - 2\lambda_2 = 0$$

$$\frac{\partial L}{\partial \lambda_1} = S - x - 2z = 0$$

$$\frac{\partial L}{\partial \lambda_2} = S - y - 2z = 0.$$

Adding the first two equations gives

$$z(x + y) = (\lambda_1 + \lambda_2).$$

The third equation can also be solved for $(\lambda_1 + \lambda_2)$; that is,

$$(\lambda_1 + \lambda_2) = \frac{1}{2}xy$$

so that

$$z(x + y) = \frac{1}{2}xy.$$

The variable x is a function of z from the fourth equation, and y is a function of z from the fifth.

$$x = (S - 2z), \qquad y = (S - 2z)$$

indicating that x and y are equal. Hence,

$$z(x + y) = \frac{1}{2}xy$$

becomes

$$2xz = \frac{1}{2}x^2$$

or

$$z = \frac{1}{4}x.$$

Since $x + 2z = S$ and $z = \frac{1}{4}x$, we have

$$x + 2\left(\frac{1}{4}\right)x = S$$

or

$$x = \frac{2}{3}S$$

giving a solution for length of the folds and depth of the cuts of

$$x = y = \frac{2}{3}S$$

and

$$z = \frac{1}{6}S.$$

That is, the box has sides of $\frac{1}{6}S$ in height and base dimensions of $\frac{2}{3}S$ by $\frac{2}{3}S$, with maximal volume of $\frac{2}{27}S^3$.

Optimal Reinforced Concrete Beam Design. Figure 12.11 shows a reinforced concrete beam of rectangular cross section with tension reinforcement only. The beam is to be designed to resist a design moment, M_u, at least cost. Total cost is the sum of the cost of steel reinforcing bars, concrete, wooden formwork, and any architectural building adjustments that might be necessitated by excessive beam width or depth. For simplicity, only the material cost of steel and concrete will be considered. Assume A615 Grade 60 steel at \$500/ton (\$0.85/in.^{2} per foot of beam) and 4000 psi normal weight concrete at \$25/yd^3 (\$0.93/ft^2 per foot of beam). Costs are based on local conditions. Take M_u to be 300 kip-feet, and assume that 2.5 in. of concrete, measured from the centroid of steel reinforcement, is needed for proper cover.

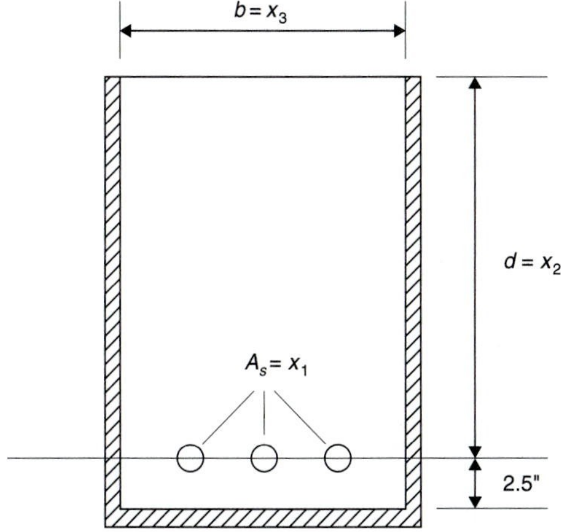

Figure 12.11 Reinforced concrete beam.

Define

x_1 = cross-sectional area of steel reinforcement (usually designated A_s) (square inches);

x_2 = effective depth of beam (depth to centroid of reinforcing steel, usually designated d) (inches); and

x_3 = beam width (usually designated b) (inches).

For a one-foot length of beam, cost can be expressed as

$$\text{Minimize} \quad C = 0.85x_1 + 0.93\left(\frac{x_2 + 2.5}{12}\right)\left(\frac{x_3}{12}\right).$$

Combining terms,

$$\text{Minimize} \quad C = 0.85\, x_1 + 0.00646\, x_2x_3 + 0.01615\, x_3.$$

Four constraints are generally needed. The first two are derived from the American Concrete Institute Code (ACI, 1992). A certain minimum amount of steel reinforcement is required by the code to ensure that the beam is stronger when acting as a reinforced beam than when acting as a plain concrete section (utilizing the strength of concrete in tension). This avoids the possible sudden failure of a plain concrete section. For the type of steel chosen, the limit is

$$\frac{x_1}{x_2x_3} \geq 0.00333.$$

On the other hand, too much steel reinforcement cannot be used or the failure mode of the beam will be through sudden compressive failure of the concrete at the top, rather than a slow excessive deflection caused by failure of the steel. For the type of steel and concrete chosen, the code limit is

$$\frac{x_1}{x_2x_3} \leq 0.0214$$

The third constraint ensures that the resisting moment of the beam is equal to the imposed design moment. The general equation is

$$M_R = \phi A_s f_y \left(d - \frac{A_s f_y}{1.7 f'_c b}\right)$$

where

M_R = resisting moment of the beam (kip-inches);

ϕ = a strength reduction factor which reduces the theoretical strength of the beam to account for minor adverse field conditions, equal to 0.90 for bending;

f_y = yield strength of steel reinforcement (psi); and

f'_c = compressive strength of concrete (psi).

Substituting x_1, x_2, and x_3, converting to kip-feet and simplifying, results in

$$4.5\, x_1 x_2 - 39.706\, \frac{x_1^2}{x_3} = 300$$

An equal sign is used since it is intuitive that a stronger beam than necessary will not result.

This model can be solved using the Lagrange multiplier technique (as explained shortly). The result, however, is a beam that is too tall and thin to adequately place the rebars with the required concrete cover and separation between bars. A fourth constraint is therefore necessary:

$$x_3 \geq 12.$$

The choice of 12 inches is an initial guess that can be refined as optimization results are obtained. This finalizes the model, but the question arises as to how it can be solved with Lagrange multipliers when inequality constraints are present. A simple trick will do. We summarize the constraint set with the trick included:

$$\frac{x_1}{x_2 x_3} - x_4^2 = 0.00333$$

$$\frac{x_1}{x_2 x_3} + x_5^2 = 0.0214$$

$$4.5\, x_1 x_2 - 39.706\, \frac{x_1^2}{x_3} = 300$$

$$x_3 - x_6^2 = 12.$$

By including squared slack and surplus variables, the goal of subtracting or adding a positive quantity is guaranteed in each inequality constraint, without further constraints on the slack or surplus variables themselves. The Lagrangian function can now be written

$$L = 0.85\, x_1 + 0.00646\, x_2 x_3 + 0.01615\, x_3$$

$$+ \lambda_1 \left(0.00333 - \frac{x_1}{x_2 x_3} + x_4^2\right)$$

$$+ \lambda_2 \left(0.0214 - \frac{x_1}{x_2 x_3} - x_5^2\right)$$

$$+ \lambda_3 \left(300 - 4.5\, x_1 x_2 + 39.706 \frac{x_1^2}{x_3} \right)$$

$$+ \lambda_4 \, (12 - x_3 + x_6^2)$$

and partials taken

$$\frac{\partial L}{\partial x_1} = 0.85 - \frac{\lambda_1}{x_2 x_3} - \frac{\lambda_2}{x_2 x_3} - 4.5\, \lambda_3 x_2 + 79.412 \frac{\lambda_3 x_1}{x_3} = 0$$

$$\frac{\partial L}{\partial x_2} = 0.00646\, x_3 + \frac{\lambda_1 x_1}{x_2^2 x_3} + \frac{\lambda_2 x_1}{x_2^2 x_3} - 4.5\, \lambda_3 x_1 = 0$$

$$\frac{\partial L}{\partial x_3} = 0.00646\, x_2 + 0.01615 + \frac{\lambda_1 x_1}{x_2 x_3^2} + \frac{\lambda_2 x_1}{x_2 x_3^2} - 39.706 \frac{\lambda_3 x_1^2}{x_3^2} - \lambda_4 = 0$$

$$\frac{\partial L}{\partial x_4} = 2\lambda_1 x_4 = 0$$

$$\frac{\partial L}{\partial x_5} = -2\lambda_2 x_5 = 0$$

$$\frac{\partial L}{\partial x_6} = 2\lambda_4 x_6 = 0$$

$$\frac{\partial L}{\partial \lambda_1} = 0.00333 - \frac{x_1}{x_2 x_3} + x_4^2 = 0$$

$$\frac{\partial L}{\partial \lambda_2} = 0.0214 - \frac{x_1}{x_2 x_3} - x_5^2 = 0$$

$$\frac{\partial L}{\partial \lambda_3} = 300 - 4.5 x_1 x_2 + 39.706 \frac{x_1^2}{x_3} = 0$$

$$\frac{\partial L}{\partial \lambda_4} = 12 - x_3 + x_6^2 = 0.$$

The ten equations in ten unknowns can be solved using any convenient numerical method that solves simultaneous nonlinear equations. The result is (Shan, 1975)

$$x_1^* = 2.38 \text{ in}^2 \qquad \lambda_1 = 0.0$$

$$x_2^* = 29.80 \text{ in.} \qquad \lambda_2 = 0.0$$

$$x_3^* = 12.0 \text{ in.} \qquad \lambda_3 = \$0.00724 \text{ per foot per k-ft}$$

$$x_4^* = 0.05767 \qquad \lambda_4 = \$0.19735 \text{ per foot per inch}$$

$$x_5^* = 0.12143 \qquad C^* = \$4.53 \text{ per foot.}$$

$$x_6^* = 0.0$$

The reader may want to verify that the solution satisfies the necessary conditions. Note that the value of λ_i for noncritical constraints is zero. The λ_i for critical constraints can be used to predict the change in optimal cost caused by a small change in the value of the right-hand-side constants. For example, if the imposed design moment were decreased by 100 kip-feet, the predicted cost of an optimally redesigned beam would be

$$\begin{aligned} C^*_{\text{new}} &= C^* + \lambda_3 \, \Delta b_3 \\ &= 4.53 + (0.00724)(-100) \\ &= 4.53 - 0.72 \\ &= \$3.83/\text{ft}. \end{aligned}$$

The actual optimal cost is \$3.71/ft, indicating that even for a rather large change in the right-hand side, the dual variable is a useful predictor in this case.

Results of such optimization models, as well as years of design experience, are incorporated in standard design procedures for reinforced concrete design (Spiegel and Limbrunner, 1992). However, since local economic conditions vary, it is always useful to periodically derive optimal designs as a check. Some optimization results are available as an aid in deriving optimal designs (Friel, 1974).

12.G GRADIENT SEARCH

The methodology of Lagrange multipliers is so versatile that we will use it immediately to derive our next procedure to handle nonlinear objective functions. We begin with an unconstrained objective function in two variables, $f(x, y)$, which may have unusual shape. It may, in fact, have multiple minima or maxima—a hill here, valleys there. Although we illustrate gradient search for a problem with a function of two variables, the method can easily be generalized to any number of independent variables. The unconstrained objective we are considering is such that the well-known analytical techniques of taking partial derivatives and setting them equal to zero, followed by solution for the variables, do not work. The technique of setting partial derivatives to zero does not work on our unconstrained objective because, let us say, there is no way to separate the variables in the partial derivatives and hence to solve for them.

Faced with such an unconstrained objective, we need to devise a strategy which can assist us in finding its optimum. Suppose we desire to

$$\text{Maximize} \quad Z = f(x, y)$$

where x and y are coordinates in the horizontal plane and Z may be thought of as the elevation above the horizontal plane. We can take partial derivatives and set them equal to zero, but we are unable to find values of x and y, unable to solve the equations.

We now derive a formula for searching the xy space to find the local maxima, that is, the points at which the partial derivatives are, in fact, zero, or very very close to zero. Some functions may have only one maximum; some may have many.

Suppose we are standing at some point, our current point c, with coordinates x^c, y^c, in xy space. As mountain climbers, we look around and declare our strategy "Fastest way to the top!" In mathematical terms, this means "follow the gradient," the direction with the greatest rate of increase given our step size; hence the name gradient search. The hill on which we are standing is presumed relatively smooth over short steps. Our plan is to take a step of length s, where s is measured in the horizontal plane, but we need to know the direction of that step, the direction that delivers the greatest increase in the objective function given the step size planned. To establish the direction which yields the maximum change in the objective given the step size, we make use of the concept of the total derivative dz, which is given by

$$dz = \frac{\partial z}{\partial x}\, dx + \frac{\partial z}{\partial y}\, dy.$$

If steps Δx and Δy, both of unknown magnitude and sign, are taken in the x and y directions, the change in the objective, Δz, is approximated by

$$\Delta z = \left(\frac{\partial z}{\partial x}\right)_c \Delta x + \left(\frac{\partial z}{\partial y}\right)_c \Delta y$$

where the partial derivatives are evaluated at c, the current position. We have previously specified that the step size is s. The question now is how far to move in each coordinate direction, given such a step size.

Our problem is to

$$\text{Maximize} \quad \Delta z = \left(\frac{\partial z}{\partial x}\right)_c \Delta x + \left(\frac{\partial z}{\partial y}\right)_c \Delta y \tag{12.16}$$

$$\text{Subject to:} \quad (\Delta x)^2 + (\Delta y)^2 = s^2.$$

The constraint, of course, is the Pythagorean theorem and indicates the required relationship between step size and movement in the two directions. Lagrange multipliers will be used to solve this problem (although it could also be solved by calculus). The Lagrangian function is established as

$$L = \left(\frac{\partial z}{\partial x}\right)_c \Delta x + \left(\frac{\partial z}{\partial y}\right)_c \Delta y + \lambda[s^2 - (\Delta x)^2 - (\Delta y)^2]$$

where λ is the Lagrange multiplier. In the following derivation we drop the notation of c for the current point but will understand that the partials have known values which are associated with the current position. We can now take the partials with respect to Δx and Δy as well as with respect to λ and set these partial derivatives equal to zero.

$$\frac{\partial L}{\partial \Delta x} = \frac{\partial z}{\partial x} - 2\lambda\Delta x = 0$$

$$\frac{\partial L}{\partial \Delta y} = \frac{\partial z}{\partial y} - 2\lambda\Delta y = 0$$

$$\frac{\partial L}{\partial \lambda} = s^2 - (\Delta x)^2 - (\Delta y)^2 = 0.$$

Solving for λ in the first two equations provides

$$\lambda = \frac{1}{2\Delta x}\frac{\partial z}{\partial x}$$

and

$$\lambda = \frac{1}{2\Delta y}\frac{\partial z}{\partial y}.$$

When these two representations of λ are equated, we arrive at the relation

$$\Delta x = \Delta y \frac{\partial z/\partial x}{\partial z/\partial y}$$

substituting for Δx in the step size equation, $(\Delta x)^2 + (\Delta y)^2 = s^2$, yields a relation which can be solved for Δy as

$$\Delta y = \frac{s(\partial z/\partial y)}{[(\partial z/\partial x)^2 + (\partial z/\partial y)^2]^{1/2}}. \tag{12.17}$$

Of course, the equation for Δx is a parallel one; that is,

$$\Delta x = \frac{s(\partial z/\partial x)}{[(\partial z/\partial x)^2 + (\partial z/\partial y)^2]^{1/2}}. \tag{12.18}$$

The values of Δx and Δy are those that maximize Δz, the change in z, given the step size s.

When the step of size s is taken, the new coordinates are

$$x = x^c + \Delta x$$

$$y = y^c + \Delta y.$$

In general, the relation for the kth step is

$$x^k = x^{k-1} + \Delta x$$

$$y^k = y^{k-1} + \Delta y \tag{12.19}$$

where Δx and Δy are evaluated by the equations above.

These relations can be generalized for functions of n variables $x_1, x_2, ..., x_n$ in which case the steps in each of the n directions are

$$\Delta x_j = \frac{s(\partial z/\partial x_j)}{\left[\sum_{j=1}^{n} (\partial z/\partial x_j)^2\right]^{1/2}} \qquad (j = 1, 2, \ldots, n). \tag{12.20}$$

The choice of a step size is important, but there are no hard and fast rules. At the first step, a value of s small in relation to the space being searched, perhaps on the order one-one hundredth of the length of the space, might be chosen. The size of the step might be decreased on every subsequent move by a factor so that the move does not consistently carry the search past the point of interest. One rule out of many possible rules might be to make the step size $(1/k)s$, where k is the number of the move. That is, the size on the fourth move is one-fourth of the size of the step on the first move. Experimentation on the rate of step size reduction may well be necessary. Too fast a rate of reduction will result in little movement from the starting point. Too slow a rate may result in the search bouncing around the space between points with high rates of change, but with no progress toward the optimal solution.

What does each step accomplish? It carries the search to a new point, and at each new point the partial derivatives are evaluated. When the partials are sufficiently close to zero, the search terminates. Partials equal to zero, of course, are the indicator of a flat spot, or hill top, of the objective function. At such a point the value of the objective is determined and recorded. The analyst defines what sufficiently close to zero means.

The search may well lead to a local maximum, the top of one of many hills, and not to the top of the highest hill. The search, to be credible, must begin at multiple random starting points and follow each path to its highest point. It is hoped that the multiple starts from random positions increases the possibility of finding the true optimal solution. In any event, from among all the highest points (from among all the local optima found), the point with the highest value of the objective is reported as "the best solution found."

EXERCISES

12.1. **(a)** In Table 12.3, show the calculations necessary to compute $F_{1\ldots 2}$ (7).
(b) In Table 12.4, show the calculations necessary to compute $F_{1\ldots 3}$ (10).
(c) In Table 12.6, show the calculations necessary to compute $F_{1\ldots 5}$ (11).

12.2. If reservoirs are located close to one another on the same river, sometimes they can be treated as one large reservoir with a capacity equal to the sum of the individual reservoir capacities. For example, to provide the required safe yield for water supply, a total capacity of 40 million acre-feet (maf) might be required. Three potential sites for reservoirs exist along a short stretch of river. Cost data are provided in the table. It is desired to achieve the required total storage at least cost (further development of this problem can be found in Wathne et al., 1975).
(a) Write the mathematical programming model for the problem.
(b) Write the dynamic programming recursive equation for the problem and define terms.
(c) Construct tables and calculate the optimal cost and associated optimal reservoir size at each site to achieve a total storage of from 0 to 50 maf (in increments of 5 maf).
(d) Plot optimal cost versus total system storage.

Reservoir Site	Reservoir Size (maf)	Reservoir Total Cost ($ million)
1	5	3
	10	5
	15	11
2	5	5
	10	7
	15	16
3	5	6
	10	9
	15	12
	20	20

12.3. Consider the case of parallel streams to be used for water supply for a city. Assume that the best site for a reservoir has been found on each stream and that the cost has been evaluated in terms of reservoir capacity at each site. Further, for each site the safe yield (based on a once-in-50-year recurrence interval shortage) has been found as a function of reservoir capacity. Cost can then be expressed in terms of safe yield for each potential reservoir site as shown in the table. The city wants to meet its water supply need at least cost (further development of this problem can be found in Wathne et al., 1975).

Define:

x_k = safe yield draft from reservoir k (on stream k);

$c_k(x_k)$ = cost as a function of safe yield draft from reservoir k;

D = total water supply draft required by the city; and

n = number of parallel streams.

(a) Write the mathematical programming model that describes this problem.
(b) Write the dynamic programming recursive equation for the problem.
(c) Construct tables and calculate the optimal cost and associated optimal draft from each stream to achieve a total draft of from 30 to 100 million gallons per day (MGD) in increments of 10 MGD.
(d) Plot optimal cost versus total system safe yield.

Stream	Safe Yield Draft (MGD)	Reservoir Total Cost ($ million)
1	10	0
	20	22
	30	44
2	20	0
	30	6
	40	18
	50	50
3	0	0
	10	10
	20	14

12.4. A water resources agency has three projects in which it can invest. According to the level of investment in each project, a benefit is achieved as in the following table.

Project	Investment ($ million)	Total Benefit ($ million)	Net Benefit ($ million)
1	0	0	0
	5	6	1
	10	12	2
	15	22	7
2	0	0	0
	5	12	7
	10	16	6
	15	18	3
3	0	0	0
	5	8	3
	10	16	6
	15	24	9

The agency takes as its goal the maximization of net benefits, defined as total benefit minus investment cost. The agency has an authorized budget of up to $25 million to allocate among the projects.
Construct tables and perform calculations to determine the best investment policy using dynamic programming. What is the best policy and associated net benefit?

12.5. The following problem is adapted from Sathaye and Hall (1976). A long aqueduct is to be sized by sections to deliver irrigation water to n irrigation districts, as shown in the sketch below. The amount of water to be used by each district is to be determined.

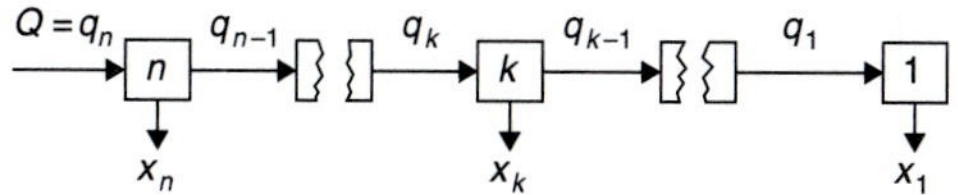

Define:

x_k = amount of water delivered annually to irrigation district k;

Q = total amount of water available;

q_k = total amount of water allocated to the first k districts;

$v_k(x_k)$ = value or net benefit (present worth) from allocating an amount of water x_k to irrigation district k; and

$c_k(q_k)$ = cost (present worth) of aqueduct section k, based on an annual total flow of q_k.

It is desired to maximize overall net benefits of the project.

(a) Write the mathematical programming model for the problem.
(b) Write the dynamic programming recursive equation that will solve the problem.
(c) If 1,100,000 acre-feet of water were available (equal to Q), x_k and q_k were broken into 100,000 AF increments, and $n = 5$ districts;
  (1) How many possible solutions are there (by enumeration)?
  (2) How many allocations will have to be compared using DP?

12.6. Are the following functions strictly convex, strictly concave, or neither?

(a) $f(x) = \ln x - x^2 + x^{0.6} + x - 10,$ for $x > 0$.

(b) $f(x) = 3x^4 - 2\ln x - 5x^{0.3} - 2x + 4,$ for $x > 0$.

(c) $f(x) = \dfrac{3}{x} - \ln x + x^3 + x^2 - 1,$ for $x > 0$.

(d) $f(x_1x_2) = x_1^2 + x_2^3 + \dfrac{1}{x_1} + x_2,$ for $x_1 > 0, x_2 > 0$.

(e) $f(x_1x_2) = x_1^2 - 2x_2^2 + 5,$ for $x_1 > 0, x_2 > 0$.

12.7. Find the stationary points(s) and type of point for

(a) $f(x) = x^3 + \dfrac{1}{x} + x,$ for $x > 0$.

(b) $f(x) = x^3$.

(c) $f(x) = x \sin x, \quad -\pi \leq x \leq \pi$.

(d) $f(x) = \dfrac{2x}{1 + x^2}$.

12.8. Find the stationary points(s) and type of point for

(a) $f(x_1, x_2) = x_1^2 - x_1x_2$.

(b) $f(x_1, x_2) = (x_1 - 2)^2 + (x_2 - 1)^2$.

(c) $f(x_1, x_2) = \ln x_1x_2 - x_1x_2$.

(d) $f(x_1, x_2) = x_1 + x_2 - (x_1 - 4)^2 - (x_2 - 3)^2$.

12.9. From past records of highway construction and maintenance costs, a state highway department determines the following cost relationships for two-lane bituminous highways within the state:

$$c_c(x) = 0.1x^{2.0}$$

$$c_m(x) = 3.0\frac{1}{x}$$

where the equations hold for $(0.5 \le x \le 10.0)$ and where

x = pavement thickness (inches);

$c_c(x)$ = annualized construction cost (millions of dollars per mile per year); and

$c_m(x)$ = annual maintenance cost (millions of dollars per mile per year).

(a) In words, describe the basic trade-off decision that the highway department must make.

(b) Determine the optimal pavement thickness, justifying the type of point found.

(c) If pavement thickness can be specified only in whole inches, what is the best pavement thickness, considering normal variance in pavement thickness caused by field construction conditions? How sensitive is the optimal cost to small changes in thickness?

12.10. Three towns have decided that they would like to jointly build an airport to serve all three communities. They also decide that the location of the airport should be such that it minimizes the sum of the population-weighted, squared distance from the airport to each of the towns.

On a map, the coordinates (x, y) (from an arbitrary origin) of the towns are (2, 6), (7, 5), and (3, 1) with respective populations of 60,000, 30,000, and 90,000 people. Find the best location (x, y) for the airport, recalling that distance from point i to point j is

$$d_{ij} = \sqrt{(x_i - x_j)^2 + (y_i - y_j)^2}\,.$$

Verify the type of point found.

12.11. As shown in the figure, three cities are to receive water from one well field which is yet to be developed. The location of the well field is to be determined. City 1 is to receive 1.0 million gallons per day, city 2 is to receive 2.0 MGD, and city 3 needs 3.0 MGD. The cost of delivery can be found from the second sketch, where cost is given in units of \$1000 per mile of pipe, depending on flow quantity.

Map coordinates (in miles) of cities 1–3 are (15, 40), (12, 10), and (50, 20), respectively.

The distance between two points i and j is given by

$$d_{ij} = \sqrt{(x_i - x_j)^2 + (y_i - y_j)^2}\,.$$

(a) Write the objective function with constants substituted where appropriate.
(b) What problems do you see in solving the simultaneous partial derivatives?
(c) Determine the solution by spreadsheet enumeration using a grid size of one mile by one mile. What are the best coordinates for the well field?

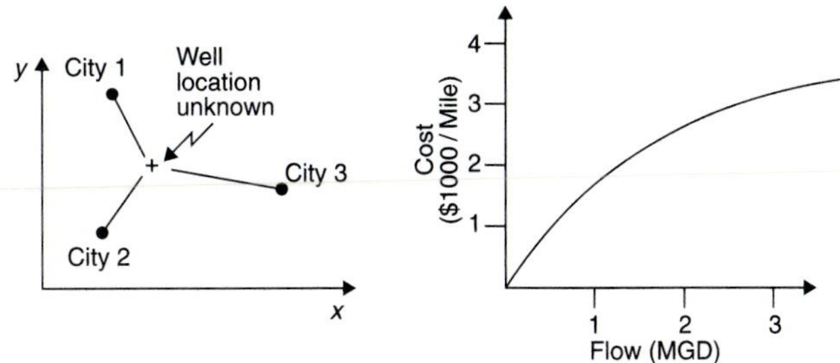

12.12. The market price of an item is \$250. The production cost, C, for a particular company is a function of its output level, q, and is given by $C = 100q^{1.2}$, where C has units of dollars and q is the number of units produced. The firm wants to maximize profit, defined as total revenue minus total cost.

(a) Determine the optimum output level to maximize profit for the firm and derive the sufficiency condition to show that the output does, in fact, represent a local maximum.
(b) Arguing from the form of the objective function, what type of stationary point has been found?

12.13. Solve the following problem using calculus with substitution. What type of critical point has been found, based on the sufficiency conditions?

$$\text{Optimize} \quad Z = \frac{1}{x_1} + \frac{1}{x_2}$$

$$\text{Subject to: } x_1 + x_2 = 1$$

12.14. For the following problem, add a squared slack variable, x_3^2, to the first constraint to produce an equation.

(a) Solve the optimization problem using calculus with substitution (ultimately a problem in one variable).
(b) For the answer found, derive the sufficiency condition and state the type of critical point.
(c) From the form of the revised objective function, is the critical point a global optimum? If so, what type?

$$\text{Optimize} \quad Z = 4x_1 x_2 + x_1$$

$$\text{Subject to: } x_1 \leq 5$$

$$x_1 + x_2 = 10$$

12.15. Using calculus with substitution, determine the optimal dimensions of a triangular channel to minimize wetted perimeter for a given cross-sectional area. Assume no freeboard.

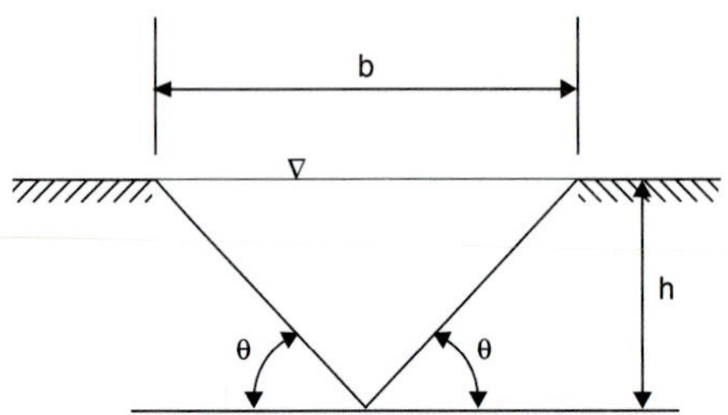

12.16. Using calculus with substitution, determine the optimal dimensions of a trapezoidal channel to minimize wetted perimeter for a given cross-sectional area.
Assume no freeboard.

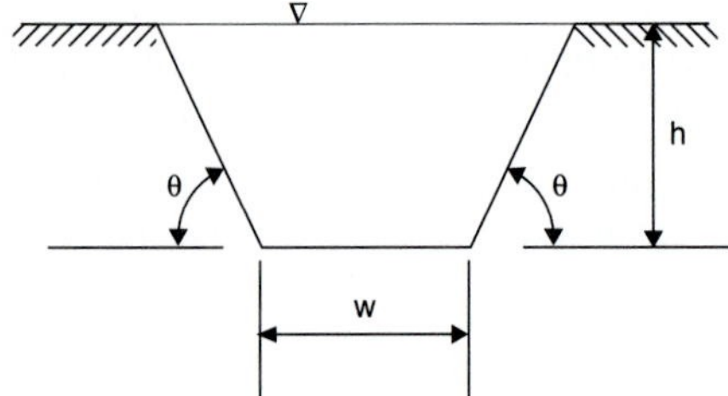

12.17. Which is the most efficient channel shape: rectangular, triangular, or trapezoidal? Use results from the text and problems 15 and 16 for the optimal wetted perimeter, along with the Manning equation. How much more efficient is the most efficient channel shape, in percentage terms?

12.18. Highway departments often store salt in sheds for use in road de-icing. Assume a conical shape as shown in the figure, and assume that the shed is filled from the top of the structure. The circular base pad costs 50 dollars per square yard. The slanted sides of the cone are made of wood and shingles, and cost 30 dollars per square yard of curved surface. Each cone must hold 300 cubic yards of salt.

(a) Write a mathematical programming model that will find the best height (h) of the cone and radius (r) of the base to minimize the total cost of the structure. Solve using calculus with substitution.

(b) If the angle of repose of dry salt is 35°, what modification to the optimal design would be necessary, if any?

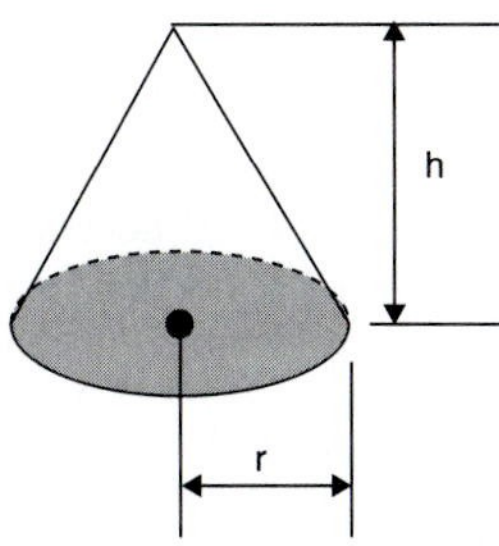

$$\text{Area of curved surface} = \pi r \sqrt{r^2 + h^2}$$

$$\text{Volume} = \frac{\pi}{3} r^2 h$$

12.19. Solve the optimization problem in problem 14 using the method of Lagrange multipliers. If one of the constraint right-hand-sides could be relaxed by one unit, which constraint would yield the greatest objective function gain? By how much would the objective function be expected to improve?

12.20. A rectangular container is open at the top and must have a volume of 10 yds^3. The side materials cost C dollars per square yard, while material for the bottom costs 2C dollars per square yard.

(a) Solve for the optimal dimensions using Lagrange multipliers.

(b) What are the units for λ_1?

(c) If it were desired to have the container hold 13 yd^3 instead of 10 yd^3, by how much would cost be expected to increase? Use the value of λ_1 to estimate the change in cost—do not resolve the problem.

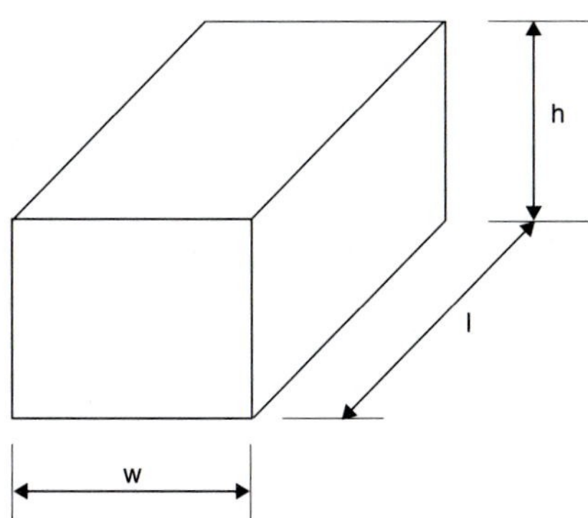

12.21. Determine the largest ellipse and smallest ellipse that will satisfy the equation of a circle, that is,

$$\text{Optimize} \quad Z = x_1^2 + 2x_2^2$$

$$\text{Subject to: } x_1^2 + x_2^2 = 1$$

(a) Solve graphically.
(b) Solve using Lagrange multipliers.

12.22. Set up the following problem in its Lagrange multiplier form. Show the Lagrangian function and derive the related necessary conditions. Verify that the following solution satisfies the necessary conditions: $x_1 = 5, x_2 = 5$. Relaxing which constraint would yield the greatest improvement in the objective function?

$$\text{Optimize} \quad Z = 4x_1 x_2 + x_1$$

$$\text{Subject to: } x_1 \leq 5$$

$$x_1 + x_2 = 10$$

$$\text{with } x_1 \text{ unrestricted, } x_2 \geq 0$$

12.23. Refer to the optimal reinforced concrete beam design example. Reformulate the mathematical programming model, including the cost of formwork, and under the following conditions:

concrete:	4,000 psi normal weight at $25/yd³
steel:	A615 Grade 60 at $500/ton;
formwork:	pine at $3.00/ft² contact area;
imposed moment:	300 kip-feet;
steel placement:	two rows at bottom (total beam height = d + 4.0 inches), and;
beam width:	no limitation (no constraint).

(a) Verify that the following solution satisfies your model.

$$x_1^* = 4.11 \text{ in}^2 \qquad C^* = \$19.37 \text{ ft.}$$

$$x_2^* = 20.01 \text{ in}$$

$$x_3^* = 9.59 \text{ in}$$

(b) Based on the model outcome, if one wanted to invest in research to reduce the cost of such beams, where would attention be most productively directed?
(c) What do you conclude about the sensitivity of the optimal beam design to inclusion of the cost of formwork in this case?

12.24. The city councils of three rural towns have agreed to jointly build a solid waste transfer station at a location between the towns that will minimize the sum of distances to the transfer station from the towns. Map coordinates, in miles, are $(x, y) = (10, 30)$, $(40, 10)$, and $(60, 50)$ for cities 1, 2, 3, respectively. Recall that

$$d_{ij} = \sqrt{(x_i - x_j)^2 + (y_i - y_j)^2}$$

Assume the location is to be determined using the method of gradient search. The search space is limited to points inside a rectangle encompassing all three cities, and a random starting point is selected as $(x, y) = (20, 20)$. Assume a step size one one-hundredth the length of the smallest side of the encompassing rectangle. Determine the optimal direction to move at the first step, and the associated improvement in the objective function.

REFERENCES

AMERICAN CONCRETE INSTITUTE, *Building Code Requirements for Reinforced Concrete* (ACI 318-89, revised 1992) *and Commentary* - ACI 318R-89, (revised 1992). Detroit: American Concrete Institute, 1992), various pagination.

DREYFUS, STUART and AVERILL LAW, *The Art and Theory of Dynamic Programming*. New York: Academic Press, 1977, 284 pp.

FRIEL, LEROY L., "Optimum Singly Reinforced Concrete Sections," *Journal of the American Concrete Institute*, Vol. 71, no. 11, November 1974, pp. 556–558.

GOTTFRIED, BYRON S. and JOEL WEISMAN, *Introduction to Optimization Theory*. Englewood Cliffs, NJ; Prentice Hall, 1973, 571 pp.

HANCOCK, HARRIS, *Theory of Maxima and Minima*. New York: Dover, 1960, 193 pp.

MANNE, ALAN S., "Capacity Expansion and Probabilistic Growth," *Econometrica*, Vol. 29, no. 4, October 1961, pp. 632–649.

SATHAYE, JAYANT and WARREN A. HALL, "Optimization of Design Capacity of an Aqueduct," *Journal of the Irrigation and Drainage Division*, Proceedings of the American Society of Civil Engineers, Vol. 102, no. IR3, September 1976, pp. 295–305.

SHAN, JOHN K.-C., "Optimal Flexural Design of Rectangular Reinforced Concrete Beams," unpublished M.S. thesis, The Ohio State University, Columbus, OH, 1975, 58 pp.

SPIEGEL, LEONARD and GEORGE F. LIMBRUNNER, *Reinforced Concrete Design*. Englewood Cliffs, NJ: Prentice Hall, 1992, 471 pp.

U.S. ENVIRONMENTAL PROTECTION AGENCY, *Economic Report: Alternative Methods of Financing Wastewater Treatment*, Report of the Administrator of the US EPA to the Congress of the United States, Report No. EPA-230/3-76-002, National Technical Information Service No. PB 251-305, October, 1975, various pagination.

WATHNE, MAGNE, REVELLE, CHARLES S., and JON LIEBMAN, "Optimal Capacities of Water Supply Reservoirs in Series and Parallel," *Water Resources Bulletin*, Vol. 11, no. 3, June 1975, pp. 536–545.

WINSTON, WAYNE, *Introduction to Mathematical Programming: Applications and Algorithms*. Belmont, CA: Duxbury Press, 1994, 818 pp.

CHAPTER 13

Engineering Economics I: Interest and Equivalence

13.A INTRODUCTION

13.A.1 Engineering Economics

Engineering economics is the application of a standard set of equivalence equations in determining the relative economic value of a small set of alternative capital investments that differ in their time streams of costs and benefits.

This chapter will develop the necessary equivalence equations and provide practice in their application. Chapter 14 will use these equations in the context of the choice between alternative capital investments common to civil and environmental engineering. Finally, Chapter 15 will include the real-world complications of depreciation, taxes, and inflation.

Together, the three chapters should provide good preparation for those portions of the Fundamentals of Engineering (FE) and Principles and Practice of Engineering (PE) exams that deal with engineering economics. The topics should also be of considerable value in personal financial planning: a special section on this topic is included at the end of Chapter 15, and all three chapters include homework problems representing situations that may be encountered in personal finance.

Most important, the tools developed are intended for use by civil and environmental engineers in those public and private projects that require engineering economic analysis to be carried out, either for the entire project to determine financial feasibility or for certain components of the project to ensure the most economical design. Please note, however, that the final choice in any project decision will include factors other than economic; such considerations as environmental and social impacts, regional economic development, and political concerns are not addressed here.

13.A.2 Cash Flow Table and Diagram

The time stream of costs and benefits that results from a financial transaction can be displayed in either a *cash flow table or cash flow diagram* format. For example, suppose a person decides to borrow \$10,000 and has agreed to pay back this amount by paying \$1000 during years 1, 2, and 3, and \$11,000 during year 4. From the viewpoint of the borrower, a cash flow table can be set up as in Table 13.1. The sign convention is that receipts are positive and disbursements are negative. Additionally, an *end-of-period convention* applies; if any receipts or disbursements occur within a time period, they do not appear until the end of that time period. Thus, the \$10,000, is received at the end of year 0 (the present), while payments are made at the end of years 1, 2, 3, and 4.

Alternatively, the same time stream can be represented in a cash flow diagram as in Figure 13.1.

Note that no attempt is made to plot the arrows to scale, although usually some attempt is made to represent relative magnitudes. As in the cash flow table, the end-of-period convention applies, and time period zero represents the present. Also note that the sign of the cash flow depends on the viewpoint taken; in this case, that of the borrower. The cash flow from the viewpoint of the lender would have the opposite sign.

TABLE 13.1 CASH FLOW TABLE

End of Year	Receipts/ Disbursements
0	\$10,000
1	−1,000
2	−1,000
3	−1,000
4	−11,000

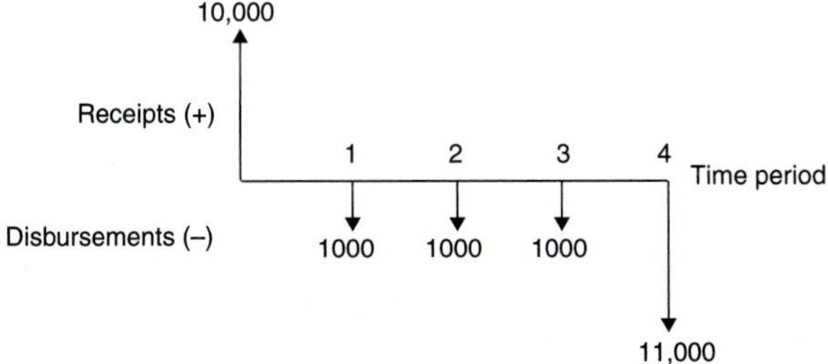

Figure 13.1 Cash flow diagram.

13.A.3 Simple Interest and Equivalence

Simple annual interest is interest computed only on the principal outstanding for a year. The *principal* is the amount given by the lender to the borrower. The cash flow table and cash flow diagram in the previous section represent a person who has borrowed $10,000 principal now and has agreed to pay simple annual interest of 10% ($1000) on the outstanding principal for four years. At the end of the fourth year, in addition to the interest, the principal is also to be repaid.

From the viewpoint of both lender and borrower, the cash flow is "optimal". The borrower is willing to pay the stated simple annual interest to obtain use of the $10,000 for four years, and the lender is willing to give up use of the same amount for four years in exchange for the annual interest payments. In other words, the current arrangement is *equivalent* to (or possibly better than) the best arrangement that either party could obtain elsewhere. To be satisfied with the cash flow, the borrower must not be able to borrow the money from any other lender at less simple annual interest; conversely, the lender cannot find any borrower who is willing to pay more simple annual interest.

If the interest rate were to be changed, equivalence would break down. At a higher interest rate, the borrower would not be willing to borrow, while at a lower interest rate, the lender would not be willing to lend.

To summarize, for a financial investment to be worthwhile, the interest earned must be at least equivalent to that which could be obtained elsewhere.

13.B COMPOUND INTEREST: SINGLE PAYMENT

13.B.1 Present Worth and Future Worth

Compound interest is interest computed on both the principal and interest outstanding for a given time period. Define

i = interest rate for a given compound interest time period, expressed as a decimal;

n = number of time periods;

P = present amount of money; and

F = future amount of money at the end of period n.

If one were to deposit P dollars in a bank today, and earn compound interest for n time periods, how much would be available to receive at the end of period n? The cash flow diagram is shown in Figure 13.2.

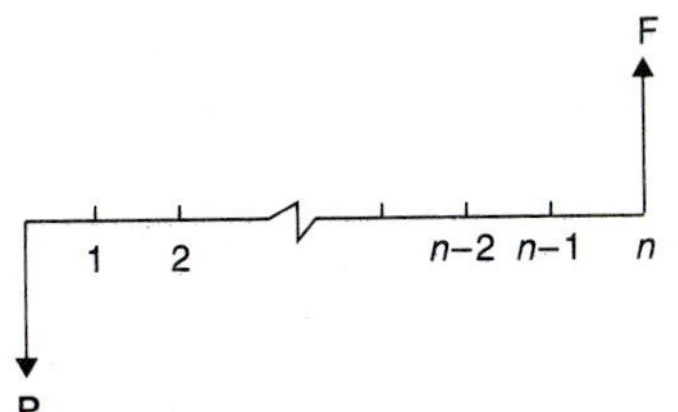

Figure 13.2 Cash flow diagram for bank deposit.

The calculation can be made as follows:

End of Period	Amount Available	
0	P	$= P$
1	$P(i + i)$	$= P(1 + i)$
2	$P(1 + i)(1 + i)$	$= P(1 + i)^2$
3	$P(1 + i)^2(1 + i)$	$= P(1 + i)^3$
⋮	⋮	⋮
n	$P(1 + i)^{n-1}(1 + i)$	$= P(1 + i)^n$

or

$$F = P(1 + i)^n \tag{13.1}$$

which is the *single-payment compound amount formula*, and the term

$$(1 + i)^n = \textit{single-payment compound amount factor}$$
$$= (F/P, i, n)$$

or

$$F = P(F/P, i, n) \tag{13.2}$$

which can be read as "the future amount, F, equals the present amount, P, times the factor for F given P compounded at interest rate i for n periods." The notation chosen enables the symbolic cancellation of "P" on the right-hand side of the functional equation.

Equation (13.1) can be solved for P in terms of F:

$$P = F(1 + i)^{-n} \tag{13.3}$$

which is the *single-payment present worth formula*, and the term

$$(1 + i)^{-n} = \textit{single-payment present worth factor}$$
$$= (P/F, i, n).$$

or

$$P = F(P/F, i, n). \tag{13.4}$$

For public sector projects, the single-payment present worth factor is often referred to as the single-payment present *value* factor, also known as the *discount factor.*

13.B.2 Equivalence

The amounts P and F are *equivalent* in the sense that if an individual could earn compound interest, i, over n periods, the present amount, P, would become F at the end of period n. If, however, the individual could earn a higher interest rate on the amount P, the offer of F in the future would no longer be equivalent to P, since the individual could earn more than the amount F at the higher rate. Whether equivalence exists between given amounts P and F, therefore, depends on the interest rate under which the comparison is being made.

The factors $(F/P, i, n)$ and $(P/F, i, n)$ are provided in the appendix for common values of i and n.

Example 13-1.

Compute the present amount, P, that would be equivalent to the two receipts shown in the cash flow diagram. The compound interest rate for a period is 5%. Demonstrate equivalence for your answer.

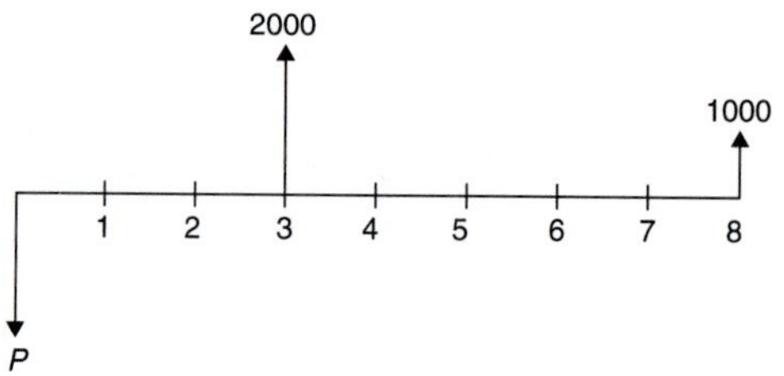

Solution: Use the tables in the appendix or perform a direct computation:

$$P = 2000(P/F, 5\%, 3) + 1000(P/F, 5\%, 8)$$

or

$$\begin{aligned} P &= 2000(1 + 0.05)^{-3} + 1000(1 + 0.05)^{-8} \\ &= 2000(0.863838) + 1000(0.676839) \\ &= 1727.68 + 676.84 \\ &= \$2404.52. \end{aligned}$$

Equivalence can be shown by going in the reverse direction. Assume a present amount, \$2404.52, and show that it will produce the two future receipts. The amount grows to a future amount, F_3, in three years,

$$\begin{aligned} F_3 &= \$2404.52(1 + 0.05)^3 \\ &= \$2783.53 \end{aligned}$$

at which time \$2000 is received, leaving \$783.53 to grow at 5% for 5 years to an amount F_8,

$$F_8 = \$783.53(1 + 0.05)^5$$
$$= \$1000.00$$

which is the desired amount. Note that at any other interest rate, the future amounts would not be produced exactly, and equivalency would not hold. Also note that the maintenance of at least six decimal places is recommended in most cases to ensure sufficient accuracy, especially when n is large. Finally observe that the *principle of superposition* was used to find P. Cash flows can be broken into subsets and treated individually using equivalence relationships.

13.B.3 Rule of 72

From tables for the single-payment compound amount factor, an interesting observation can be made: *the rule of 72; a present sum doubles in value when the product of the interest rate (in percent) and number of compounding periods is about 72.* This is shown in Table 13.2.

The rule of 72 is useful when making mental estimates of growth rates of money, population, inflation and so on. For example, if the 1980 rate of inflation approximating 12% had continued, consumer prices would have doubled in only 6 years. At the 1993 inflation rate of 3%, it would have taken 24 years.

The reverse is also true—present value is only one-half that of a future amount obeying the rule of 72. At 7% interest, a project benefit received 10 years hence is worth only one-half that amount today. Since the rule is repetitive, it can also be said that a benefit received 20 years in the future is worth only one-fourth that amount today, again assuming an interest rate of 7%

13.B.4 Nominal and Effective Interest Rate

Banks will often quote an interest rate of, say, "4% compounded quarterly." The interpretation is that for each quarter of a year 1% interest is earned. Since compounding is carried out each quarter, the actual annual *effective interest rate* earned

TABLE 13.2 COMPOUND AMOUNT FACTORS DEMONSTRATING THE RULE OF 72

i	n	$i \times n$	$(F/P, i, n)^*$	i	n	$i \times n$	$(F/P, i, n)^*$
1	72	72	2.05	7	10	70	1.97
2	36	72	2.04	8	9	72	2.00
3	24	72	2.03	9	8	72	1.99
4	18	72	2.03	10	7	70	1.95
5	14	70	1.98	11	7	77	2.08
6	12	72	2.01	12	6	72	1.97

*Rounded to two decimal spaces.

is slightly more than the *nominal interest rate* of 4%. The nominal interest rate for a time period ignores the effect of any subperiod compounding. The effective interest rate for a time period includes the effect of any subperiod compounding. Unless specified otherwise, a quoted nominal rate is assumed to apply to a time period of one year.

Define:

r = nominal interest rate per time period;

m = number of compounding subperiods per time period;

i_y = effective annual interest rate; and

i_s = effective subperiod interest rate, equal to r/m.

To derive the annual effective rate of interest due to subperiod compounding, imagine one dollar compounded over m subperiods of a year. The total amount, F, at the end of one year would be

$$F = \$1\left(1 + \frac{r}{m}\right)^m = \left(1 + \frac{r}{m}\right)^m.$$

Deducting the \$1 initial amount produces the effective annual interest rate as a decimal,

$$i_y = \left(1 + \frac{r}{m}\right)^m - 1 \tag{13.5}$$

which can also be expressed as

$$i_y = (1 + i_s)^m - 1 \tag{13.6}$$

Example 13-2.

Given 7% interest compounded monthly, find the effective annual interest rate.

Solution. $r = 0.07$, $m = 12$, $i_s = r/m = 0.07/12 = 0.005833$

$$i_y = (1 + i_s)^m - 1 = (1 + 0.005833)^{12} - 1$$

$$i_y = 0.072286 \text{ or } 7.23\%.$$

13.B.5 Continuous Compounding

If n is the number of periods, and a present sum is compounded over m subperiods of a period at a nominal period interest rate, r, the future amount after n periods becomes

$$F = P\left(1 + \frac{r}{m}\right)^{mn}. \tag{13.7}$$

If the number of compounding subperiods, m, is allowed to become very large, the solution will represent *continuous compounding*. Mathematically,

$$F = P \lim_{m \to \infty} \left(1 + \frac{r}{m}\right)^{mn}. \tag{13.8}$$

Let $x = r/m$. Then $mn = (1/x)\,(rn)$, and equation 13.8 becomes

$$F = P \lim_{x \to 0} [(1 + x)^{1/x}]^{rn} \tag{13.9}$$

From calculus,

$$\lim_{x \to 0} (1 + x)^{1/x} = e \tag{13.10}$$

Equation 13.9 can then be written as

$$\text{compound amount: } F = P\,e^{rn} = P(F/P, r, n). \tag{13.11}$$

Conversely,

$$\text{present worth: } P = \text{F}\,e^{-rn} = F(P/F, r, n). \tag{13.12}$$

To determine the effective interest rate for a one-year period, consider one dollar compounded continuously for one year:

$$F = \$1\,e^{r(1)} = e^{r}.$$

Subtracting the initial $1, the effective interest is

$$i_y = e^{r} - 1. \tag{13.13}$$

Example 13-3.

A bank offers a nominal interest rate of 7%, compounded continuously. What is the effective rate?

Solution: Since a time period is not mentioned, it is assumed that the nominal rate is for one year. The effective rate for one year is

$$i_y = e^{r} - 1 = e^{0.07} - 1 = 1.07251 - 1 = 0.07251 \text{ or } 7.25\%$$

which compares with the effective rate of 7.23% (found in Example 13-2) for monthly compounding.

Example 13-4.

At an annual nominal rate of 2, 4, 6, 8, and 10%, compounded continuously, how long does it take for a present amount to double in value? Does the rule of 72 apply for continuous compounding at these rates?

Solution: $F = P\,e^{rn}$, where $F = 2$, $P = 1$. Taking the natural log:

$$\ln 2 = \ln 1 + rn \ln e$$

or

$$0.6931 = rn$$

and

$$100rn = 69.31$$

From which it can be observed that the rule of 72 becomes the rule of 69 for continuous discounting. However, since most phenomena are described in terms of their annual effective rate of compounding, the rule of 72 is more useful and could be applied even for continuous compounding with little error.

Finally, note that r and n may apply to time periods other than one year. This is illustrated in Example 13-5.

Example 13-5.

A bank quotes a quarterly interest rate of 2%, compounded continuously. What is the (a) effective quarterly interest rate? (b) nominal annual interest rate? and (c) effective annual interest rate?

Solution:

(a) Note that the quoted rate is the nominal quarterly interest rate. The effective quarterly rate is given by

$$i_q = e^r - 1 = e^{0.02} - 1 = 1.0202 - 1 = 0.0202 = 2.02\%$$

(b) The nominal annual interest rate is the rate ignoring any compounding during subperiods. It is simply $4 \times 2\%$, or 8%.

(c) The effective annual interest rate can be found in one of three ways:

(1) Observing that the quarterly effective return will be compounded four times a year:

$$i_y = (1 + i_s)^m - 1 = (1 + 0.0202)^4 - 1 = 0.0833 = 8.33\%.$$

(2) Using the formula for future worth with $n = 4$ time periods and a nominal rate of 2% per quarter,

$$F = P\, e^{rn} = \$1\, e^{(0.02)(4)} = 1.0833$$

subtract the original \$1 to obtain the effective interest rate:

$$i_y = 1.0833 - 1 = 0.0833 = 8.33\%.$$

(3) Using the nominal annual interest rate of 8% with continuous compounding,

$$F = P\, e^{rn}$$

with

$$P = \$1,\ r = 0.08,\text{ and } n = 1$$

$$F = 1\, e^{(0.08)(1)} = 1.0833$$

$$i_y = 1.0833 - 1 = 0.0833 = 8.33\%.$$

Note that the terms r and n must refer to the same time period when used in an equation.

13.C STANDARD CASH FLOW SERIES

The single-payment compound amount or present worth formulas could be applied to each end-of-period payment in an arbitrary cash flow to resolve the series into its single-payment equivalent. However, it is inconvenient to do so. Certain cash flow series occur so frequently that standard formulas have been derived to express the equivalence relationships. These are summarized in the following sections. Derivations are available in many standard engineering economics texts, such as those listed in the end of this chapter. Results are also provided in a tabular format in the appendix.

13.C.1 Uniform Series

The standard cash flow diagram for a uniform series is shown in Table 13.3. Note the absence of a payment at the end of period zero, but the presence of a payment

TABLE 13.3 UNIFORM SERIES EQUIVALENCE FORMULAS

Item	Diagram	Equation	Symbolic Form
Present worth formula/factor	0; A; n; P=?	$P = A\left[\frac{(1+i)^n - 1}{i(1+i)^n}\right]$	$P = A(P/A, i, n)$
Capital recovery formula/factor	0; A=?; n; P	$A = P\left[\frac{i(1+i)^n}{(1+i)^n - 1}\right]$	$A = P(A/P, i, n)$
Compound amount formula/factor	0; A; n; F=?	$F = A\left[\frac{(1+i)^n - 1}{i}\right]$	$F = A(F/A, i, n)$
Sinking fund formula/factor	0; A=?; n; F	$A = F\left[\frac{i}{(1+i)^n - 1}\right]$	$A = F(A/F, i, n)$

at the end of period n. The *uniform series present worth factor* (*pwf*), $(P/A, i, n)$, finds the present worth of a uniform series of n receipts or payments of amount A each period, while the *uniform series capital recovery factor* (*crf*), $(A/P, i, n)$, permits calculation of the uniform series amount, A, that will recover an initial capital investment, P, in n periods.

The *uniform series compound amount factor* (*caf*), $(F/A, i, n)$, is used to find the future worth of a uniform series, A, while the *uniform series sinking fund factor* (*sff*), $(A/F, i, n)$, permits calculation of the uniform series amount, A, that must be deposited to finance a future capital investment, F, in period n.

Uniform series are common in civil and environmental engineering. Operation and maintenance costs, estimated benefit streams from projects, and annualized capital costs are common examples.

13.C.2 Arithmetic Gradient Series

Table 13.4 shows the standard cash flow diagram for an arithmetic gradient series. Note that there is no receipt at the end of period 1 and that the present worth term is therefore two periods removed from the first gradient amount, G.

TABLE 13.4 ARITHMETIC AND GEOMETRIC GRADIENT SERIES EQUIVALENCE FORMULAS

Item	Diagram	Equation and Symbolic Form
Arithmetic Gradient Series Present worth formula/factor	$(n-1)G$, $(n-2)G$, $3G$, $2G$, G; 0, 1, 2, 3, 4, n; $P=?$; Gradient $= G$	$$P = G\left[\frac{(1+i)^n - in - 1}{i^2(1+i)^n}\right]$$ $$P = G(P/G, i, n)$$
Geometric Gradient Series Present worth formula/factor	A_n, A_{n-1}, A_1, A_2, A_3; 0, 1, 2, 3, n; $P=?$; Growth rate $= g$	$(i \neq g)$: $$P = A_1\left[\frac{1-(1+g)^n(1+i)^{-n}}{i-g}\right]$$ $$P = A_1(P/A_1, g, i, n)$$ $(i = g)$: $$P = A_1[n(1+i)^{-1}]$$ $$P = A_1(P/A_1, g, i, n)$$

Repair costs are sometimes assumed to increase linearly with time, as are estimated benefits from projects.

Example 13-6.

Convert the following series to its equivalent future worth, F, when i is 12%.

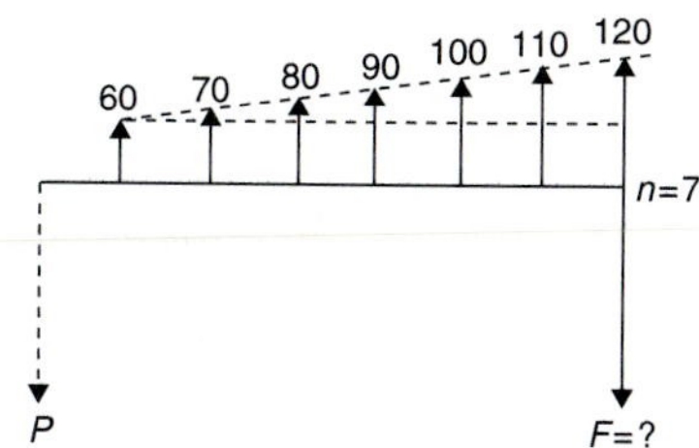

Solution: Break the series up into a uniform series and an arithmetic gradient series where $A = 60$ and $G = 10$. Then

$$F = [A(P/A, 12\%, 7) + G(P/G, 12\%, 7)](F/P, 12\%, 7)$$
$$= [60(4.563757) + 10(11.644267)](2.210681)$$
$$= \$862.76.$$

Note that the format of the arithmetic gradient series conveniently allows a uniform series to be broken out, as in this example

13.C.3 Geometric Gradient Series

Table 13.4 illustrates the format of the geometric gradient series. At the end of year 1, an amount, A_1, is received. The series then grows at a compound (geometric) rate, g, per period, given by

$$A_n = A_1(1 + g)^{n-1}. \tag{13.14}$$

Since there are three parameters (g, i, n) in the geometric gradient series present worth formula, a tabular format is inconvenient, and the equation must be used directly.

The geometric gradient series is useful in cases of "double compounding," for example, when benefits from a project grow at a compound rate and are themselves subject to compound interest (e.g., hydroelectric generation benefits may grow at a compound rate equal to that of a region's economy, while the benefit itself is compounded financially).

13.C.4 Uniform Series: Continuous Compounding

In the formulas presented in Table 13.3 for uniform series, if the interest rate, i, is replaced by the effective interest rate for a time period under continuous compounding,

TABLE 13.5 UNIFORM SERIES EQUIVALENCE FORMULAS UNDER CONTINUOUS COMPOUNDING

Item	Equation	Symbolic Form
Present worth formula/factor	$P = A\left[\dfrac{e^{rn}-1}{e^{rn}(e^r-1)}\right]$	$P = A(P/A, r, n)$
Capital recovery formula/factor	$A = P\left[\dfrac{e^{rn}(e^r-1)}{e^{rn}-1}\right]$	$A = P(A/P, r, n)$
Compound amount formula/factor	$F = A\left[\dfrac{e^{rn}-1}{e^r-1}\right]$	$F = A(F/A, r, n)$
Sinking fund formula/factor	$A = F\left[\dfrac{e^r-1}{e^{rn}-1}\right]$	$A = F(A/F, r, n)$

$$i = e^r - 1$$

the equations in Table 13.5 are produced. Note that r is the nominal interest rate for the period to which n refers.

Example 13-7.

John has set up an automatic deposit plan between his place of employment and his bank. Every month, \$300 is deposited. The bank pays 8% interest, compounded monthly. A competitor bank pays the same interest rate, but offers continuous compounding. Over three years, what would be the monetary gain of using the bank offering continuous compounding?

Solution: Under *monthly* compounding,

$$i_m = \frac{r}{m} = \frac{0.08}{12} = 0.006667$$

$$F_m = A(F/A, i, n) = \text{A}\left[\frac{(1+i)^n - 1}{i}\right]$$

$$= 300\left[\frac{(1+0.006667)^{36}-1}{0.006667}\right]$$

$$= \$12{,}160.68.$$

Under *continuous* compounding, A is a monthly deposit; therefore, r is for a month and equals 0.006667, as above:

$$F_c = A(F/A, r, n) = A\left[\frac{e^{rn}-1}{e^r-1}\right]$$

$$= 300\left[\frac{e^{(0.006667)(36)} - 1}{e^{0.006667} - 1}\right]$$

$$= \$12{,}165.60.$$

Difference: $4.92 in favor of continuous compounding over three years; negligible considering other factors.

13.C.5 Capitalized Cost

In certain cases, it is necessary to assume that a service or facility will be provided forever. Major pipelines, highways, dams, and water and wastewater treatment plants often fall into this category. Periodically, they will be replaced or renovated to maintain a desired level of service. In such situations, the analysis period (or planning horizon) is infinity.

What is the present worth of cost for such facilities? The amount of money needed to be set aside now to provide for perpetual service is the *capitalized cost* of a project. The situation is illustrated in Figure 13.3. If an amount P were set aside, in one year an amount of interest Pi would be available to spend. In the next year, a similar amount of interest, Pi, would be available, and so on. The amount, P, would never change and, in fact, could not be drawn upon or it would not last to infinity.

If A must be spent each year to maintain perpetual service, the initial principal must be made large enough such that

$$A = Pi \tag{13.15}$$

$$= P(A/P, i, \infty) \tag{13.16}$$

or, rearranging,

$$P = \frac{A}{i} \tag{13.17}$$

$$= A(P/A, i, \infty) \tag{13.18}$$

where $(A/P, i, \infty)$ is the uniform series capital recovery factor for an infinite series, equal to the interest rate, i, and $(P/A, i, \infty)$ is the uniform series present worth factor for an infinite series, equal to $1/i$.

Two complications occur in determining the capitalized cost of construction projects. One is that an initial cost of construction (at $t = 0$) is usually involved, and the second is that periodic renovation or replacement does not always occur annually, but after a number of years. The initial cost can be handled immediate-

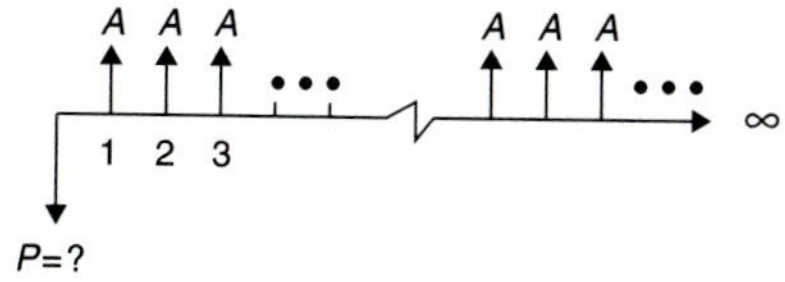

Figure 13.3 Equivalence relationship for capitalized cost.

ly by merely adding it to the necessary capitalized cost. The periodic replacement cost must be resolved into its equivalent annual series before it can be capitalized.

Example 13-8.

A dam initially costs \$30 million and must be renovated every 50 years at a cost of \$10 million. Interest is 8%. Find the capitalized cost sufficient to construct and maintain the project in perpetuity.

Solution: Draw the cash flow diagram:

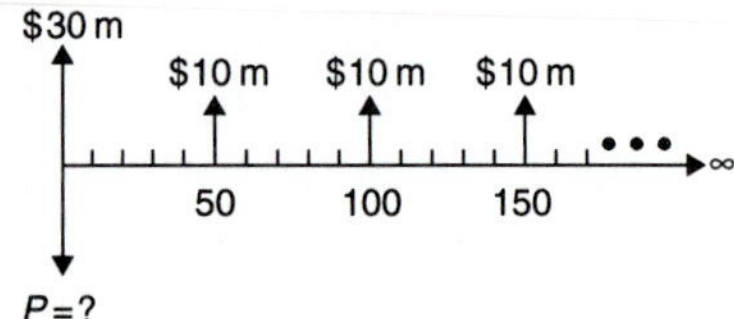

Spread each \$10 million renovation cost to the left over the preceding 50 years to produce an annual *infinite* series:

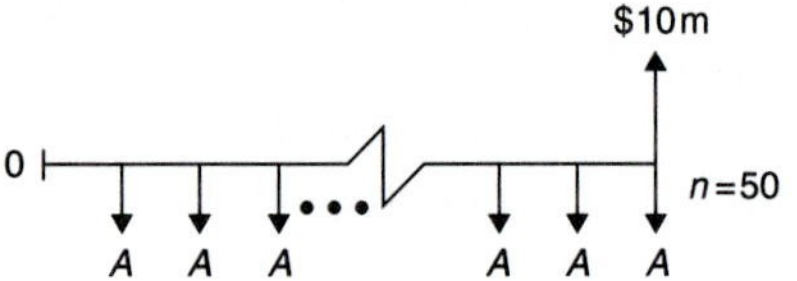

Solve for A:

$$A = F(A/F, i, n) = \$10{,}000{,}000\ (A/F, 8\%, 50)$$
$$= 10{,}000{,}000\ (0.001743)$$
$$= \$17{,}430$$

then

$$P = \$30{,}000{,}000 + \$17{,}430(1/0.08)$$
$$= \$30{,}217{,}875$$

After paying \$30 million for initial construction, the remaining \$217,875 set aside would pay for renovation at 50-year intervals in perpetuity. Check:

$$\$217{,}875(1.08)^{50} = \$10{,}218{,}689$$

which is approximately equal to \$10,217,875, the predicted amount. (Note that even with interest tables carried to six decimal places, considerable round-off error can be present when compounding occurs over long planning horizons, here 50 years.)

An alternative solution approach recognizes the following relationship,

$$P(1.08)^{50} = 10{,}000{,}000 + P$$

$$46.901612P = 10{,}000{,}000 + P$$
$$P = \$217{,}857$$

which is a more accurate estimate of the capitalized cost of renovation alone.

CHAPTER SUMMARY

This chapter has developed the concepts of simple and compound interest and equivalence. Standard cash flow series were presented along with the corresponding equivalence formulas. In Chapter 14, these equivalence relationships will we applied to determine which project, out of a small set of alternative projects, is the most economical.

EXERCISES

13.1. A bank has agreed to lend $100,000 for five years at simple annual interest of 8%. Show the cash flow table and cash flow diagram from the viewpoint of the bank, assuming that no early principal payments are made.

13.2. A company has arranged a four-year loan for $30,000 with simple annual interest of 6%. In addition, it decides to pay back $10,000, $5000, and $3000 of the principal at the end of years 1, 2, and 3, respectively. Fill in the following table from the viewpoint of the company:

End of Year	Amount Received	Principal Paid	Principal Outstanding	Interest Paid	Final Cash Flow
0					
1					
2					
3					
4					

13.3. A person promises to pay simple annual interest, i, on a loan where a constant fraction, f, of the outstanding principal is paid back each year until the final year when both the interest and remaining principal are paid. For the cash flow table shown, find the values of i and f. Solve mathematically.

End of Year	Receipt/ Disbursement
0	$10,000
1	−5,800
2	−2,900
3	−1,450
4	−725
5	−675

13.4. An engineer can purchase a corporate bond for $10,000 that pays $200 in interest, semiannually, for five years (ten payments). At the end of five years, in addition to the last $200 interest payment, the original $10,000 investment is repaid. Draw the cash flow diagram for the engineer. Is this investment equivalent to a simple annual interest of 5% that can be earned in an alternative investment?

13.5. Find the unknown quantity:

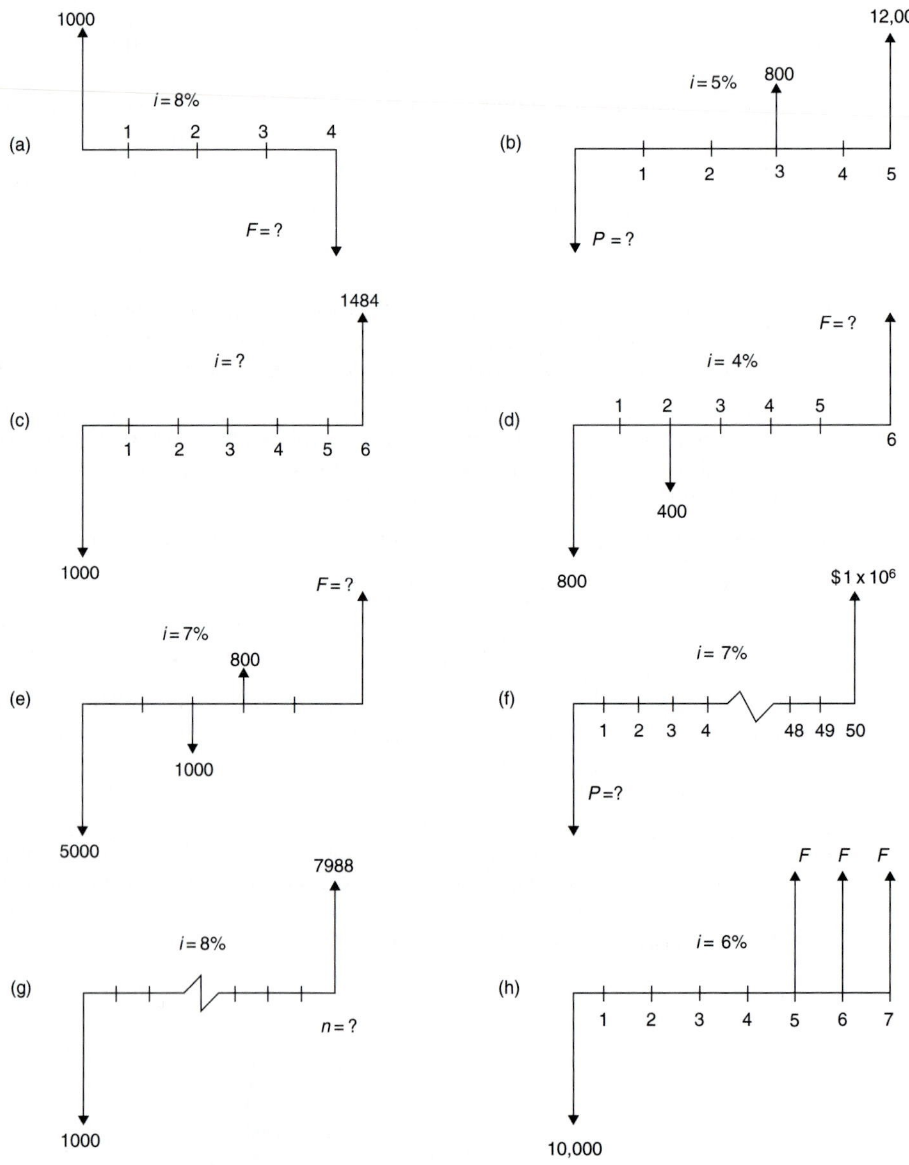

13.6. A recent graduate invests $4000 in a stock mutual fund. If a conservative estimate of after-tax compound interest is 8%, how many years would it take to buy a $16,000 car using the fund? Solve by the rule of 72.

13.7. A water resources engineer is trying to decide on the number of years over which to estimate future benefits and costs for a project. The engineer decides not to count any future benefits or costs that have a present worth amount less than 3% of their future worth amount. If the compound interest rate is 8%, use the rule of 72 to determine the number of years over which to estimate future benefits and costs.

13.8. For the period 1984–1994, operational speed and disk storage capacity of computers doubled about every two years. Using the rule of 72, what compound annual rate of increase does this represent? Check your answer analytically. What do you observe?

13.9. Compare the effective annual interest rate produced under the following frequencies of subperiod compounding: (a) yearly, (b) semiannually, (c) quarterly, (d) monthly, (e) weekly, and (f) daily. Use interest rates of 3, 8, and 30%. What do you observe?

13.10. An engineer in a developing country observes that his project bank account has grown from $1,600,000 to $1,769,279 (local currency units) in 20 days with no deposits or withdrawals being made. He knows that the account earns interest compounded daily.

(a) What is the daily compound rate of interest, in percent, earned on the account?
(b) What is the effective annual interest, in percent?
(c) What is the nominal annual interest, in percent?

13.11. Six years ago Juanita put $2000 in a special bank account. Today, her account shows $3425. She thinks that this is a good deal and would like to tell her friends about it tonight. The bank is already closed, however, and the only thing she is sure of is that the account pays interest compounded monthly. Compute the nominal annual interest and the effective annual interest for this account.

13.12. Whoops! In Exercise 13.11, just as Juanita was about to leave the house for her meeting, her husband, Enrique, learns of her plan and says he is "sure" that the account pays interest compounded continuously. Recompute both the nominal annual interest and effective annual interest, assuming continuous compounding.

13.13. At the end of five years an account paying 6% interest, compounded continuously, shows a balance of $10,798.87. If only an initial deposit was made, what was the amount? What is the effective annual interest?

13.14. Find the unknown quantity in each of the following.

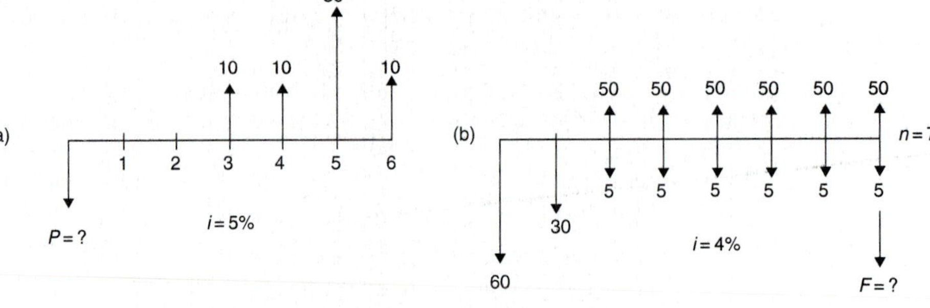

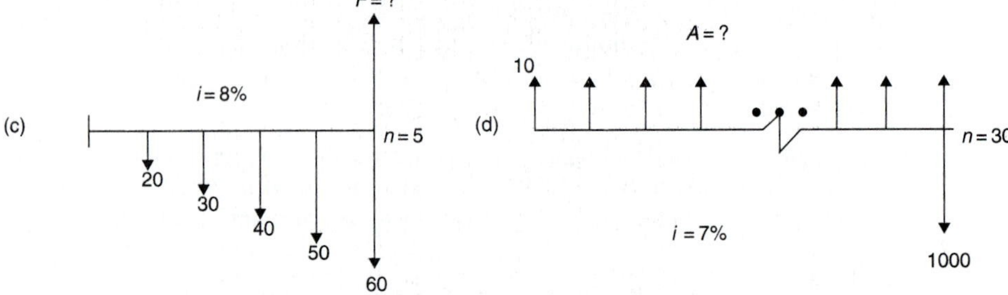

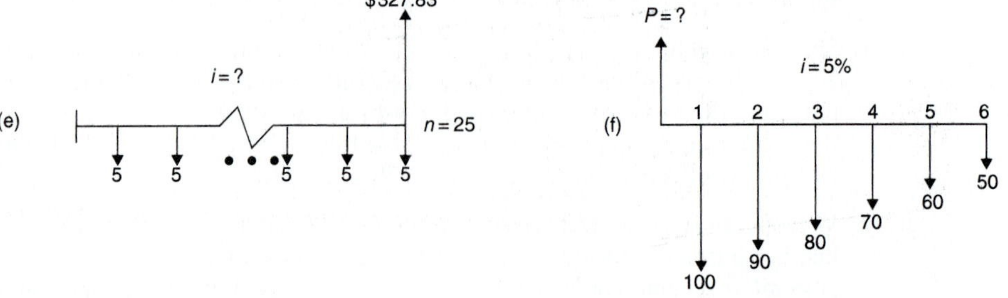

13.15. An engineer buys a piece of equipment for \$1000 and spends \$90 per year on maintenance. During the fourth year, a major overhaul cost of \$300 is expected. Benefits from use of the equipment follow an arithmetic gradient series with $G = \$50$. At the end of six years the engineer sells the equipment for \$400. Draw the cash flow diagram from the engineer's viewpoint.

13.16. For the following diagram, assuming equivalence, determine the interest rate using the compound interest tables with interpolation.

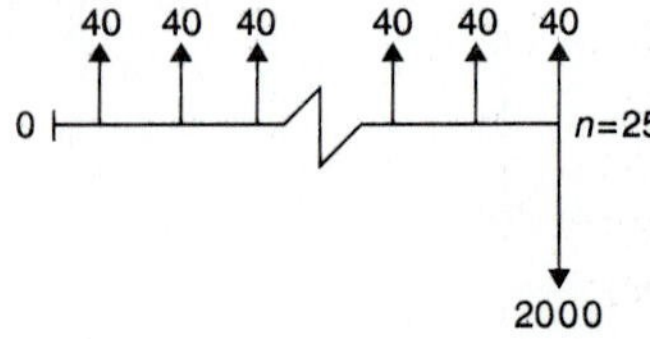

13.17. A hydroelectric dam is projected to produce annual benefits that grow in concert with the regional economic growth rate of 2%. The first benefit amount, $600,000, occurs at the end of year 5 (after the dam is constructed and the reservoir fills). If the interest rate is 8% and the benefits are assumed to continue growing through year 50 (50-year planning period), what is the present value of benefits (at $t = 0$)?

13.18. A newly employed engineer decides to place 5% of her salary every year into a retirement account. Her first year salary is $34,000, and it is anticipated that her real salary (after removing the effect of inflation) will increase 3% per year. What will be in the account at the end of year 30 if annual interest earned on the account is

(a) 5%?

(b) 3%?

13.19. Rework Example 13-7 assuming that John makes automatic deposits of $150 semimonthly (two times per month) instead of $300 per month.

13.20. Linda has just opened a savings account at a bank that pays 6% interest, compounded continuously. She deposits $30 to open the account and then deposits $30 semimonthly (two times per month) for two years. At the end of the two-year period, how much is in her account?

13.21. A university has a "sabbatical plan" for its faculty whereby six years are on campus, and the next year is off campus. Salary in the sabbatical year is not fully covered, however. A professor estimates she will have to spend a total of $15,000 of her own money in equal monthly withdrawals of $1250 in the sabbatical year. She will continue to use her current bank during the off year. The bank pays 5% interest, compounded continuously.

If she makes uniform monthly deposits to her bank over the six-year period, what amount must she deposit each month to cover her sabbatical year expenditure? Draw the cash flow diagram from the professor's viewpoint.

13.22. In an irrigation district in the West, an irrigator can purchase an extra share (allotment) which entitles the holder to 0.7 acre-feet (AF) of water per year in perpetuity. If the share costs $1600, what is the cost of one acre-foot of water to the irrigator? Assume an interest rate of 8%.

13.23. A city wants to set aside enough money to build, operate, and renovate a sewage treatment plant in perpetuity. An engineering company estimates that the plant will cost an immediate $20 million to build and will require $5 million every 20 years to replace major equipment and $10 million every 50 years to pay for major structural renovation. It is estimated that operation and maintenance costs will be $1.5 million every year. What amount will the city need to set aside? Interest earned on the annuity is 7%.

13.24. A public utility commission requires that a trust fund be established by a private company wishing to build and maintain a water treatment plant for a small city. The company must deposit enough money to build the plant and then to operate and renovate the facility in perpetuity. The plant will cost $10,000,000 to build, has an annual operating expense of $600,000, and must be renovated every 20 years at a cost of $1,000,000. The trust fund earns 6% interest. What amount must the company put in the trust fund?

REFERENCES

AU, TUNG and THOMAS P. AU, *Engineering Economics for Capital Investment Analysis*. Boston: Allyn and Bacon, 1983, 506 pp.

BLANK, LELAND T. and ANTHONY J. TARQUIN, *Engineering Economy*. New York: McGraw-Hill, 1989, 531 pp.

BUSSEY, LYNN E. and TED. G. ESCHENBACH, *The Economic Analysis of Industrial Projects*. Englewood Cliffs, NJ: Prentice Hall, 1992, 450 pp.

DEGARMO, E. PAUL, SULLIVAN, WILLIAM G., and JAMES A. BONTADELLI, *Engineering Economy*. New York: Macmillan, 1993, 726 pp.

ESCHENBACH, TED, *Cases in Engineering Economy*. New York: John Wiley and Sons, 1989, 208 pp.

FLEISCHER, G. A., *Engineering Economy: Capital Allocation Theory*. Boston: PWS, Wadsworth, 1984, 521 pp.

GRANT, EUGENE L., IRESON, W. GRANT, and RICHARD S. LEAVENWORTH, *Principles of Engineering Economy*. New York: John Wiley and Sons, 1990, 591 pp.

NEWNAN, DONALD G., *Engineering Economic Analysis*. San Jose, CA.: Engineering Press, 1991, 578 pp.

RIGGS, JAMES L., *Engineering Economics*. New York: McGraw-Hill, 1982, 789 pp.

SMITH, GERALD W., *Engineering Economy: Analysis of Capital Expenditures*. Ames: Iowa State University Press, 1987, 594 pp.

STEINER, HENRY MALCOLM, *Engineering Economic Principles*. New York: McGraw-Hill, 1992, 559 pp.

TAYLOR, GEORGE A., *Managerial and Engineering Economy: Economic Decision-Making*. Monterey, CA.: Brooks/Cole Engineering Division, Wadsworth, 1980, 538 pp.

THUESEN, G. J. and W. J. FABRYCKY, *Engineering Economy*. Englewood Cliffs, NJ: Prentice Hall, 1993, 717 pp.

WHITE, JOHN A., AGEE, MARVIN H., and KENNETH E. CASE, *Principles of Engineering Economic Analysis*. New York: John Wiley and Sons, 1984, 546 pp.

CHAPTER

14 Engineering Economics II: Choice Between Alternatives

14.A INTRODUCTION

In this chapter, four economically exact methods will be presented to determine the best alternative out of a small set of mutually exclusive alternatives. The analysis methods are (1) present worth, (2) annual cash flow, (3) incremental benefit-cost ratio, and (4) incremental rate of return. A fifth approach, payback period analysis, gives only an approximate solution to the choice problem and is often incorrect in an economic sense. It is presented, however, because it is commonly referred to in civil and environmental engineering practice and in business applications.

Breakeven analysis is a specialized technique often used in engineering practice to test the sensitivity of a solution to input parameters, either financial or physical.

14.B BASIC CONCEPTS

14.B.1 Maximization of Net Benefits

Single Project. For a single project such as a reservoir, one of the central issues is the size or *scale*, Q, at which to build. For example, the scale might represent reservoir storage in thousand acre-feet. The fundamental criterion by which to determine proper project scale is the *maximization of net benefits*:

$$\text{Maximize } NB(Q) = TB(Q) - TC(Q) \tag{14.1}$$

where

$NB(Q)$ = net benefit as a function of scale;

$TB(Q)$ = total benefit as a function of scale; and

$TC(Q)$ = total cost as a function of scale.

To maximize the above function, take the derivative and set equal to zero:

$$\frac{dNB(Q)}{dQ} = \frac{dTB(Q)}{dQ} - \frac{dTC(Q)}{dQ} = 0$$

giving

$$MB(Q) = MC(Q) \tag{14.2}$$

or

$$\frac{MB(Q)}{MC(Q)} = 1.0 \tag{14.3}$$

where

$MB(Q)$ = marginal benefit as a function of scale, and

$MC(Q)$ = marginal cost as a function of scale.

So long as marginal benefit exceeds marginal cost, project scale should be increased. When the two are exactly equal, stop. To proceed any further would produce marginal costs that exceed marginal benefits, and net benefits would be reduced. The situation is illustrated in Figure 14.1, where the optimal scale of project is Q^*. Note, however, that the same necessary condition (equation 14.2) applies as well to the *minimization* of net benefits (also shown in Figure 14.1). Analytically, the second derivative would be necessary to determine the type of point found.

It will be useful for later sections of this chapter to illustrate the economics of a single project in an alternative form (Figure 14.2). The axes of the graph are total cost (x axis) and total benefit (y axis). The line shown at a 45° slope represents points where total cost and total benefit are equal, and has a slope of 1.0. Project economic data from Figure 14.1 are transposed onto Figure 14.2. Total cost can now be interpreted (indirectly) as the scale of the project (x axis). The corresponding total benefit is also plotted (y axis). If total benefit falls below the 45° line, the project is not economically justified at that scale. Conversely, if total benefit falls above the line, the project is worthwhile. The vertical distance from

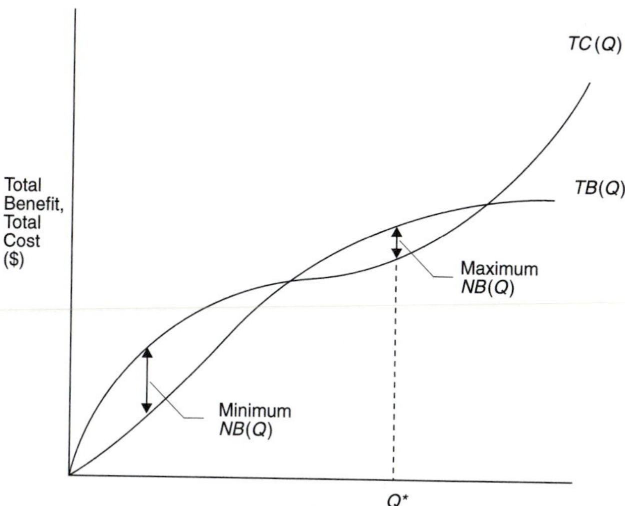

Figure 14.1 Maximization of net benefits for a single project.

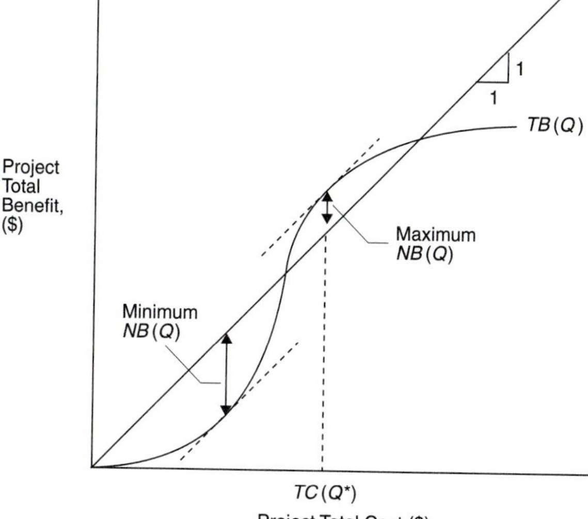

Figure 14.2 Cost-benefit plot for a single project.

the total benefit line to the 45° line represents net benefit (positive or negative). The maximum positive distance is desired, where net benefits are maximized.

The largest positive distance can be found in Figure 14.2 by moving the 45° line parallel to itself and upward until it is tangent to the curved line. The slope of the 45° line is 1.0, therefore, at the point of tangency, the necessary condition (equation 14.3) is satisfied and net benefits are maximized. Point Q^* again results. It should be noted that the necessary condition is also satisfied at the point of *minimum* net benefits, as shown in Figure 14.2.

TABLE 14.1 COST AND BENEFIT DATA FOR SIX PROJECTS

	Project					
	E	C	A	D	F	B
Benefit ($ million)	3.0	11.0	12.0	17.0	19.0	19.0
Cost ($ million)	5.0	7.0	10.0	12.0	16.0	19.0
Net benefit ($ million)	−2.0	4.0	2.0	5.0	3.0	0.0

Multiple Projects. Often, the decision problem is to choose the best project out of a set of mutually exclusive projects. Again, the economic criterion is maximization of net benefits: choose that project out of the set of projects that will maximize net benefits. Note that there is no cost constraint imposed on the problem.

Multiple projects can be analyzed using the theory developed for a single project in the last section. Cost and benefit data are provided in Table 14.1 for six hypothetical projects. Projects are arranged in order of increasing cost. Figure 14.3 is a plot of the data in the format developed in the last section: total project cost represents project scale. The optimal project can be found graphically by moving the 45° line upward and parallel to itself. Project *D* is the last project that the line passes through and is therefore optimal. Figure 14.3 will be useful in later sections of this chapter.

Although Figure 14.3 clearly identifies differences between projects, the solution can, of course, be found directly in Table 14.1. Simple subtraction of total cost from total benefit yields net benefit, which is highest for project *D*.

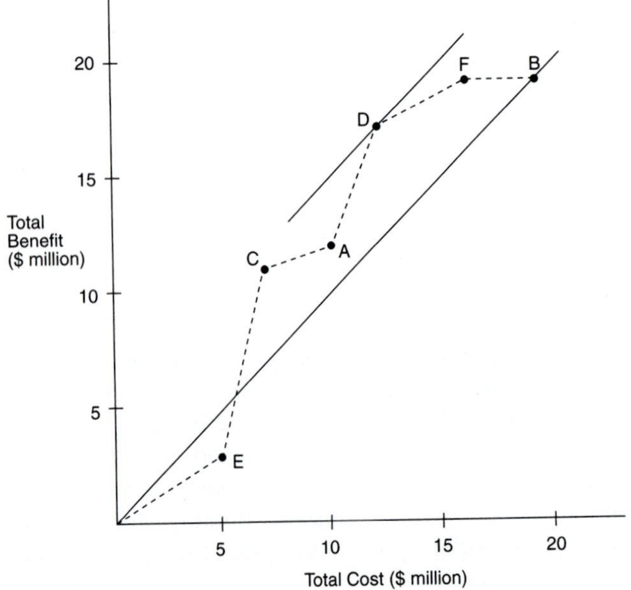

Figure 14.3 Cost-benefit plot for multiple projects.

14.B.2 Sunk Costs

Sunk costs are costs already incurred or committed to, about which nothing can be done. As such, they should have no bearing on present or future decisions; any new action should be based on current alternatives and their outcomes. Although this is correct from an economic viewpoint, it is often emotionally or politically difficult to ignore past investments of time, money, and effort.

14.B.3 Opportunity Cost

The *opportunity cost* of a resource used on a project is the value of the resource when used in the most likely alternative endeavor. In a perfectly competitive economy, the opportunity cost of a resource is equal to its market price.

14.C PRESENT WORTH ANALYSIS

14.C.1 Criterion

Present worth analysis is the most commonly applied technique in civil and environmental engineering economics. Choice of the best project from a set of projects is made using the maximization of net benefits criterion:

$$\underset{j}{\text{Maximize}} \{PWNB_j = PWB_j - PWC_j\} \tag{14.4}$$

where

$PWNB$ = present worth of project net benefit;

PWB = present worth of project benefit;

PWC = present worth of project cost; and

j = project index ($j = 1, 2, \ldots, N$).

If all projects deliver the same benefit, the criterion reduces to finding the alternative with minimum PWC. Conversely, if costs are fixed, the choice reduces to finding the alternative delivering greatest PWB.

In civil and environmental engineering, the above criterion is often expressed in terms of present *value*, which has the same meaning as present worth:

$$\underset{j}{\text{Maximize}} \{PVNB_j = PVB_j - PVC_j\} \tag{14.5}$$

14.C.2 Analysis Period

In present worth analysis, care must be taken to compare alternative projects in a "fair" manner. In particular, the lifetimes (*analysis period* or *planning horizon*)

(a) **Alternative 1 (6-year life)**

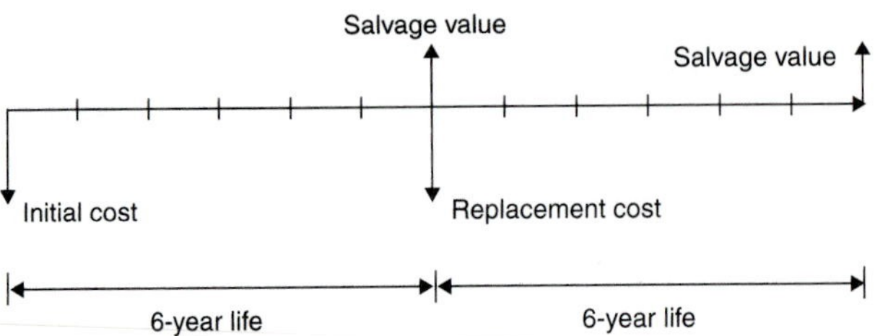

(b) **Alternative 2 (4-year life)**

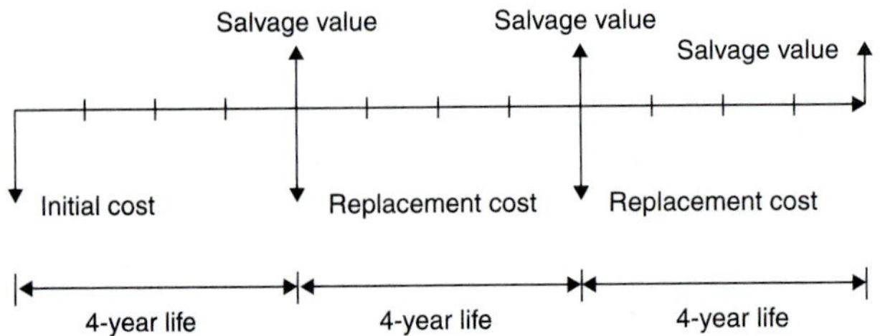

Figure 14.4 Analysis period equal to the least common multiple of the lifetimes of the alternatives.

of the alternatives must be the same. There are four ways to achieve a common analysis period:

1. Use the least common multiple of the useful lifetimes of the alternatives. At the end of each alternative's lifetime, a salvage value is received and a replacement cost is incurred, as illustrated in Figure 14.4.
2. For a fixed analysis period not equal to a common multiple, use the market value of the alternatives at the end of the analysis period, as shown in Figure 14.5.
3. Fix the analysis period at a large number of years and ignore costs or benefits (including salvage and replacement) beyond this time period. For example, water resource projects often have a 50-year planning horizon.
4. Use an infinite analysis period that includes benefits and costs occurring forever.

Example 14-1.

An engineering company is considering the purchase of one of two computer systems. System 1 is based on small, decentralized personal computers and has an initial cost of $100,000. These will be replaced in five years at the same cost, $100,000. Salvage value is $10,000 at the end of years 5 and 10. It is estimated that the benefits for the first five years will be $30,000 per year, and for the second five years, $60,000 per year.

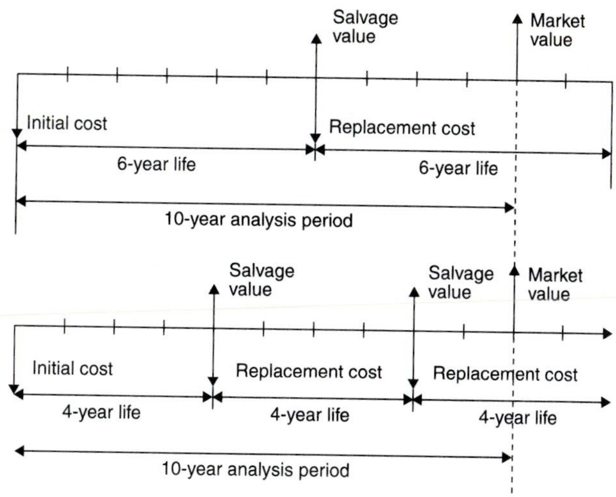

Figure 14.5 Analysis period not equal to least common multiple of the lifetimes of the alternatives.

System 2 is based on larger, more powerful work stations and has an initial cost of \$500,000. Replacement will occur at the end of year 10, with a salvage value of \$5000. Benefits are estimated to be \$90,000 per year over the ten-year period.

The firm uses a 10% rate of interest. Which system should the company purchase, if any?

Solution: Since neither costs nor benefits are fixed, choose the system that has the highest present worth of net benefits, *PWNB* (if positive). By convention, treat salvage value as a reduction in cost. Draw the cash flow diagrams for each system as a common multiple of ten years.

System 1 (small):

S=10K S=10K
30K 30K 30K 30K 30K 60K 60K 60K 60K 60K
0 n=10
100K 100K

$$PWC_1 = 100{,}000 + 100{,}000\,(P/F, 10\%, 5) - 10{,}000\,(P/F, 10\%, 5) - 10{,}000\,(P/F, 10\%, 10)$$

$$= 100{,}000 + 100{,}000\,(0.620921) - 10{,}000\,(0.620921) - 10{,}000\,(0.385543)$$

$$= \$152{,}027.$$

$$PWB_1 = 30{,}000\ (P/A, 10\%, 5) + 60{,}000\ (P/A, 10\%, 5)\ (P/F, 10\%, 5)$$
$$= 30{,}000\ (3.790787) + 60{,}000\ (3.790787)\ (0.620921)$$
$$= \$254{,}950.$$
$$PWNB_1 = PWB - PWC$$
$$= 254{,}950 - 152{,}027$$
$$= \$102{,}923.$$

System 2 (large):

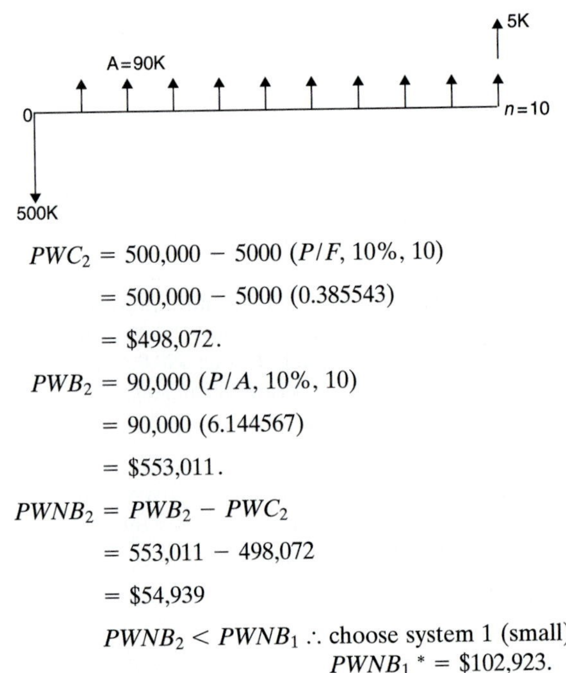

$$PWC_2 = 500{,}000 - 5000\ (P/F, 10\%, 10)$$
$$= 500{,}000 - 5000\ (0.385543)$$
$$= \$498{,}072.$$
$$PWB_2 = 90{,}000\ (P/A, 10\%, 10)$$
$$= 90{,}000\ (6.144567)$$
$$= \$553{,}011.$$
$$PWNB_2 = PWB_2 - PWC_2$$
$$= 553{,}011 - 498{,}072$$
$$= \$54{,}939$$
$$PWNB_2 < PWNB_1 \therefore \text{choose system 1 (small)}$$
$$PWNB_1{}^* = \$102{,}923.$$

Example 14-2.

A city water department has decided that two alternatives are available to meet the need for an additional 40 million gallons of water per day. One is to build a dam and reservoir, the other is to undertake a strong program of water conservation/education. The dam and related facilities can be constructed over three years at a cost of $20 million in year 1, $10 million in year 2, and $5 million in year 3. Thereafter, maintenance is $0.5 million per year.

The conservation program will cost $5 million the first year, $4 million the second year, and $3 million the third year. Thereafter, $2 million per year will have to be spent to maintain low water use.

The planning horizon for both alternatives is 50 years; any costs beyond this length of time are ignored, and an interest rate of 7% is to be used. Which alternative should be undertaken?

Solution: Note that total benefits of supplying 40 MGD are the same for both alternatives. The alternative with the minimum cost should then be selected. Draw the cash flow diagrams and find the present worth of cost for the two alternatives.

Dam

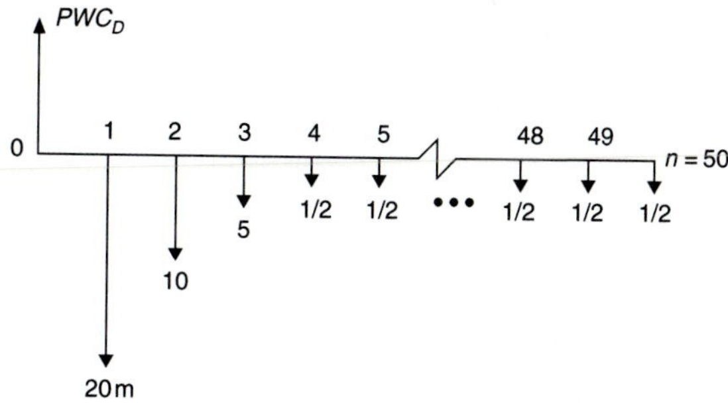

$$PWC_D = \$20\text{m}(P/F, 7\%, 1) + 10(P/F, 7\%, 2) + 5(P/F, 7\%, 3)$$
$$+ [0.5(P/A, 7\%, 47)](P/F, 7\%, 3)$$
$$= 20\ (0.934579) + 10\ (0.873439) + 5\ (0.816298)$$
$$+ 0.5(13.691608)(0.816298)$$
$$= 18.692 + 8.734 + 4.081 + 5.588$$
$$= \$37{,}095{,}000.$$

Conservation

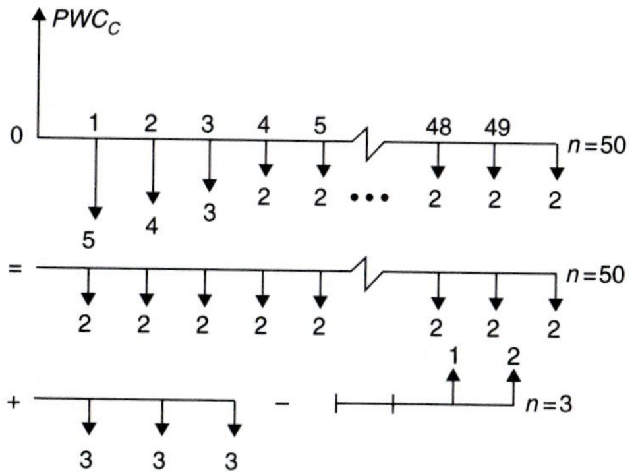

$$PWC_C = 2(P/A, 7\%, 50) + 3(P/A, 7\%\ 3) - 1(P/G, 7\%, 3)$$
$$= 2(13.800746) + 3(2.624316) - 1(2.506035)$$
$$= \$32{,}968{,}000,\ PWC_C < PWC_D \therefore \text{ choose conservation.}$$

14.D ANNUAL CASH FLOW ANALYSIS

14.D.1 Criterion

Maximization of net benefit remains the economic criterion in annual cash flow analysis. Instead of computing present worth, however, the *annual worth* (or annualized benefit, cost, and net benefit) is calculated. Since annual worth and present worth are equivalent measures, the two methods are identical. The criterion is

$$\underset{j}{\text{Maximize}} \{EUANB_j = EUAB_j - EUAC_j\} \tag{14.6}$$

where

$EUANB$ = equivalent uniform annual net benefit;

$EUAB$ = equivalent uniform annual benefit; and

$EUAC$ = equivalent uniform annual cost.

j = project index ($j = 1, 2, \ldots, N$).

14.D.2 Analysis Period

Annual cash flow analysis is unique in its ability to handle varying lifetimes of alternatives in a convenient manner. Several situations are possible.

Analysis Period Equal to Multiple of Lifetime. In this case, annual cash flow for one lifetime can be used to choose between alternatives, as illustrated in Example 14-3.

Example 14-3.

An engineer must choose between two pumps for use in a water distribution system. Both pump A and pump B will deliver water at an adequate flow rate and pressure head. Data are

	Pump A	Pump B
Initial cost ($)	200,000	300,000
Operating cost ($/yr)	30,000	23,000
Lifetime (yrs)	12	12
Salvage value ($)	10,000	60,000

The interest rate is 7%. Find the best pump.

Solution: Lifetimes and benefits are the same for the two pumps; therefore, make a direct comparison of annual cost, $EUAC$.

$$\begin{aligned} EUAC_A &= 200{,}000\,(A/P, 7\%, 12) + 30{,}000 - 10{,}000\,(A/F, 7\%, 12) \\ &= 200{,}000\,(0.125902) + 30{,}000 - 10{,}000\,(0.055902) \end{aligned}$$

$$= \$54{,}621.$$

$$EUAC_B = 300{,}000\,(A/P, 7\%, 12) + 23{,}000 - 60{,}000\,(A/F, 7\%, 12)$$
$$= 300{,}000\,(0.125902) + 23{,}000 - 60{,}000\,(0.055902)$$
$$= \$57{,}416.$$

The EUAC of pump *A* is lower: **choose pump *A*.**

If the analysis period were any integer multiple of the equal lifetimes of the alternatives, the annual cost (*EUAC*) would remain the same, assuming replacements had the same initial costs, operating costs, and salvage value. The annual cost for one lifetime would be repeated for each lifetime up to the end of the analysis period; therefore, only one lifetime need be analyzed.

Analysis Period Equal to Least Common Multiple of Lifetimes. If the analysis period is the least common multiple of the lifetimes of the alternatives, the annual cash flow computed for *one* lifetime can be used for comparison among the alternatives. The reason is that the annual cash flow for one lifetime of an alternative is repeated, as the alternative is repetitively replaced, up to the least common multiple of the lifetimes. This is illustrated in Example 14-4.

Example 14-4.

Assume that the lifetime of pump B in Example 14-3 is 16 years rather than 12. The analysis period is 48 years. Which pump is now the best?

Solution: The *EUAC* for pump A is the same, since it is repetitively replaced: $EUAC_A$ = \$54,621. The lifetime of pump B has changed, so its *EUAC* must be recomputed.

$$EUAC_B = 300{,}000\,(A/P, 7\%, 16) + 23{,}000 - 60{,}000\,(A/F, 7\%, 16)$$
$$= 300{,}000\,(0.105858) + 23{,}000 - 60{,}000\,(0.035858)$$
$$= 31{,}757 + 23{,}000 - 2151$$
$$= \$52{,}606.$$

This represents the *EUAC* for the entire 48-year period as pump B is repetitively replaced. Pump B has a lower *EUAC*; therefore, **choose pump B.**

Analysis Period Long But Undefined. Many facilities are intended to be kept in operation indefinitely. The analysis period is, therefore, long but undefined. Alternatives can be viewed as being replaced as often as necessary to maintain service. The annual cash flow for one lifetime is therefore an adequate measure for comparison between alternatives.

Infinite Analysis Period. Under an infinite analysis period, the assumption is that the facility will be replaced constantly with an identical alternative. The annual cash flow for a single lifetime is again adequate for comparison.

Analysis Period Not a Least Common Multiple of Lifetimes. For the pump example, if an analysis period of 24 years were insisted upon, special treatment would have to be given to pump B if it had a lifetime of 16 years (the lifetime of pump A remains at 12 years). The market value of pump B at the end of 8 years of its (second) life would have to be estimated in order to compare its *EUAC* with that of pump A.

14.D.3 Conclusion

Using the annual cash flow criterion is often easier than using the present worth criterion when comparing alternatives. Computing *EUAC* for *one* lifetime is often sufficient for comparison. Annual cash flow and present worth analysis give the same result due to equivalence.

14.E INCREMENTAL BENEFIT-COST RATIO ANALYSIS

14.E.1 Criterion

The benefit-cost ratio for an alternative is defined as

$$\text{B-C ratio} = \frac{B}{C} = \frac{PWB}{PWC} \tag{14.7}$$

or, equivalently,

$$\frac{B}{C} = \frac{EUAB}{EUAC} \tag{14.8}$$

and the ratio must be ≥ 1.0 for an economic project.

The B-C ratio is commonly reported for water resource development projects such as hydroelectric or flood control reservoirs and navigation improvements, and typically ranges from 1.0 to more than 3.0. It serves as a convenient way to summarize the economics of a project: \$2.00 are returned for every \$1.00 invested, if the B-C ratio is 2.0.

As important as the B-C ratio is, maximization of the ratio is *not* the proper economic criterion. As always, maximization of net benefits is the economic goal. However, the B-C ratio can be used to achieve this goal if an *incremental analysis* is conducted.

14.E.2 Procedure

Two cases are possible:

Case A. If all alternatives have either the same benefit or cost, choose that alternative with the largest B-C ratio, provided that the ratio is greater than or equal to unity. If none, choose the do-nothing alternative. Net benefit will be maximized.

Case B. If neither benefit nor cost is the same for all alternatives, conduct incremental benefit-cost analysis:

1. Compute *PWB* and *PWC* (or *EUAB* and *EUAC*) for each alternative.
2. Discard any alternative having B/C < 1.0.
3. Order the remaining alternatives from lowest cost to highest cost, and number the projects 1, 2, ..., *P*.
4. For projects 1 and 2, compute

$$\frac{\Delta B}{\Delta C} = \frac{PWB_2 - PWB_1}{PWC_2 - PWC_1}$$

or

$$\frac{\Delta B}{\Delta C} = \frac{EUAB_2 - EUAB_1}{EUAC_2 - EUAC_1}$$

If $\Delta B/\Delta C \geq 1.0$, select project 2 as best so far.

If $\Delta B/\Delta C < 1.0$, select project 1 as best so far.

5. Compute $\Delta B/\Delta C$ between the best project so far and the next most costly project not yet tested.

   If $\Delta B/\Delta C \geq 1.0$, select the more costly project as best so far.

   If $\Delta B/\Delta C < 1.0$, keep the lesser cost project as best so far.

6. Repeat step 5 until all projects have been tested. The surviving project is the best.

Example 14-5.

Use the project data in Table 14.1 and find the best alternative using incremental B-C ratio analysis.

Solution: The data are reproduced below. Projects are already ordered from lowest cost to highest cost, and the B-C ratios are shown. Alternative E is discarded.

	Project					
	E	C	A	D	F	B
Benefit ($ million)	3.0	11.0	12.0	17.0	19.0	19.0
Cost ($ million)	5.0	7.0	10.0	12.0	16.0	19.0
B-C ratio	0.60 (discard)	1.57	1.20	1.42	1.19	1.00

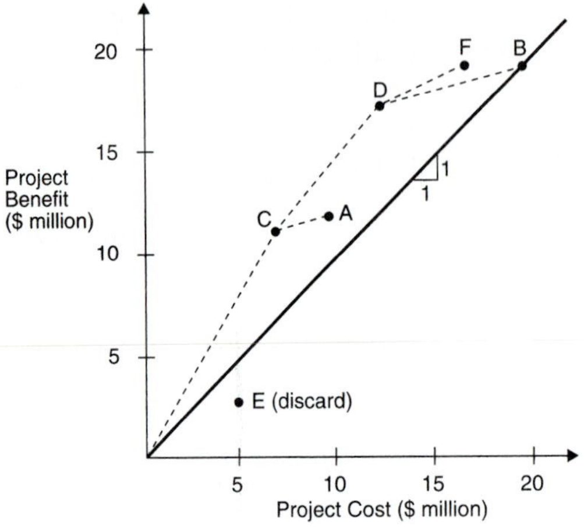

Figure 14.6 Incremental B-C ratio analysis for Example 14-5.

$$\text{Compute } \frac{\Delta B}{\Delta C} \text{ for } C \text{ to } A: \quad \frac{12.0 - 11.0}{10.0 - 7.0} = \frac{1.0}{3.0} < 1.0 \therefore \text{ keep } C;$$

$$\text{Compute } \frac{\Delta B}{\Delta C} \text{ for } C \text{ to } D: \quad \frac{17.0 - 11.0}{12.0 - 7.0} = \frac{6.0}{5.0} > 1.0 \therefore \text{ select } D;$$

$$\text{Compute } \frac{\Delta B}{\Delta C} \text{ for } D \text{ to } F: \quad \frac{19.0 - 17.0}{16.0 - 12.0} = \frac{2.0}{4.0} < 1.0 \therefore \text{ keep } D;$$

$$\text{Compute } \frac{\Delta B}{\Delta C} \text{ for } D \text{ to } B: \quad \frac{19.0 - 17.0}{19.0 - 12.0} = \frac{2.0}{7.0} < 1.0 \therefore \text{ keep } D.$$

No other projects ∴ project D is best, $B/C = 1.42$, $PWNB$ = \$5.0 million. (Note that the project with the highest *B-C* ratio, project *C*, was not selected.)

The procedure applied in Example 14-5 is shown graphically in Figure 14.6. Dashed lines represent incremental comparisons made, and *the slope of each dashed line represents the incremental B-C ratio tested.*

$$\text{slope} = \frac{\Delta B}{\Delta C} = \text{incremental B-C ratio}$$

It is therefore possible to see why certain moves were not made. If the slope of the line between the best project so far and the next project tested is less than 1.0, the move should not be made, since net benefits would be reduced. If the slope is greater than (or equal to) 1.0, the move should be made to the next project tested, since net benefits would increase (or stay the same).

14.E.3 Conclusion

Incremental B-C ratio analysis is the straightforward implementation of marginal analysis as expressed by equation 14.3:

$$\frac{MB(Q)}{MC(Q)} = 1.0.$$

The procedure therefore guarantees maximization of net benefits. Note that benefits and costs must be properly calculated using either present worth or annual cash flow analysis.

14.F INCREMENTAL RATE OF RETURN ANALYSIS

14.F.1 Internal Rate of Return

Incremental rate of return analysis is based upon the concept of the *internal rate of return (IRR)*, i^*, for a capital investment. The *IRR* is that interest rate which equates benefits to costs for the given cash flow, that is, the interest rate for which

$$PWB = PWC \tag{14.9}$$

or

$$EUAB = EUAC. \tag{14.10}$$

At the internal rate of return, net benefits are zero.

Example 14-6.

Find the internal rate of return for the following cash flow:

End of Year	Receipt/ Disbursement
0	\$−1000
1	257
2	257
3	257
4	257
5	257

Solution: Set $PWC = PWB$:

$$1000 = 257\,(P/A, i^*, 5)$$

$$3.891050 = (P/A, i^*, 5)$$

From the appendix, $i^* = 9.0\%$.

Since the internal rate of return establishes equivalence between benefits and costs, it can also be interpreted as the rate of compound interest earned on an investment. As such, it can be compared to the rate of return that can be earned in alternative investments to determine whether the investment is worthwhile.

14.F.2 Minimum Attractive Rate of Return

The *minimum attractive rate of return (MARR)* is the highest compound rate of interest that can be obtained in alternative investments. It is therefore the minimum rate of return that the investment in question must earn to be attractive. Many businesses establish a *MARR* that the *IRR* of proposed investments must meet or exceed to be considered further.

If the *IRR* for an investment exceeds the *MARR*, the net benefits would be positive if the *MARR* were used in the calculation. Conversely, if *IRR* is less than *MARR*, the net benefit would be negative if the *MARR* were used. The reason for this is important: in essentially all civil and environmental engineering projects (and most others), costs are incurred early in time, while benefits are received late in time. At a rate of interest higher than the *IRR*, benefits distant in time are "discounted" (reduced) more than up-front costs when brought back to present value, and net benefit becomes negative. This is illustrated in Example 14-7 and Figure 14.7.

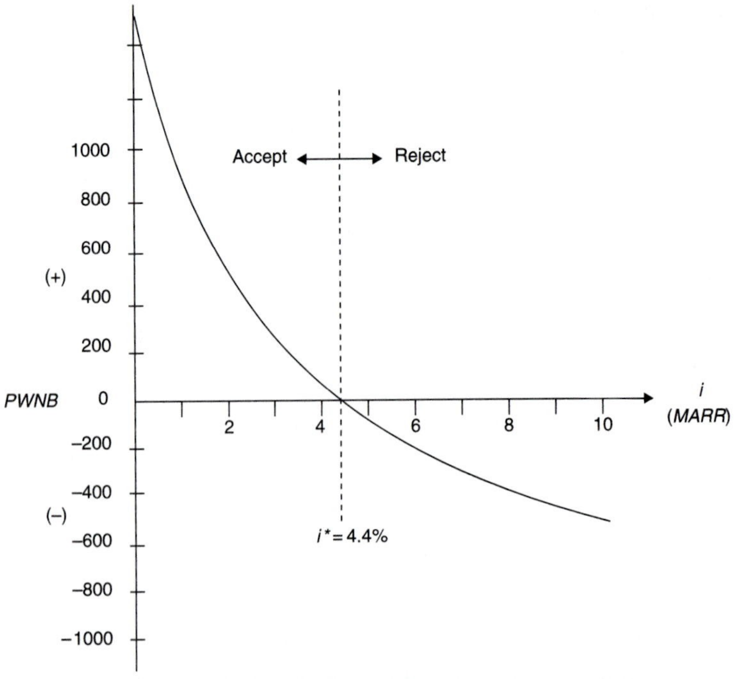

Figure 14.7 *PWNB* versus interest rate for Example 14-7.

Example 14-7.

For the cash flow shown, calculate and plot *PWNB* for different interest rates and find the *IRR*.

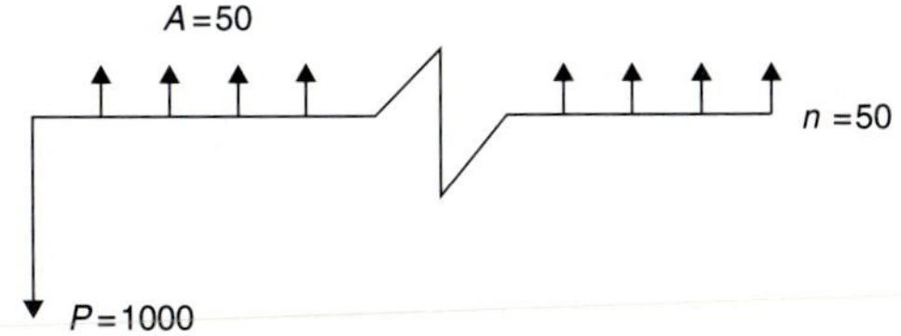

Solution: Calculate

$$PWNB = PWB - PWC$$
$$= A(P/A, i, 50) - 1000$$

for $i = 0, 1, 2, \ldots, 10\%$ and plot in Figure 14.7.

By graphical interpolation, $i^* = 4.4\%$.

In Figure 14.7, the effect of using a high interest rate to discount benefits and costs is clearly shown. If a *MARR* of 10% were established for the project in Example 14-7, the *IRR* of 4.4% would clearly indicate economic infeasibility. If, however, the *MARR* were set at 3%, the *IRR* of 4.4% would imply positive net benefits at the 3% rate.

The decision rule can be summarized for a single project:

If $IRR \geq MARR$, accept.

If $IRR < MARR$, reject.

14.F.3 Incremental Rate of Return Analysis

To choose the best project out of a set of mutually exclusive projects using rate of return, an incremental analysis must be performed. This ensures that the economic criterion of maximization of net benefits is achieved. The procedure is as follows:

1. For each of the mutually exclusive alternatives, compute the internal rate of return *IRR*, also abbreviated as *ROR*. Discard any for which $ROR < MARR$.
2. Order the remaining alternatives from lowest cost to highest cost and begin the procedure with the two lowest-cost alternatives on the list.
3. Compute the difference in cash flows between the two projects as

   {higher-cost project} − {lower-cost project} = {incremental cash flow}.
4. Compute the incremental rate of return, ΔROR, for the incremental cash flow. Follow the decision rule:

If $\Delta ROR \geq MARR$, accept the higher-cost project as best so far.

If $\Delta ROR < MARR$, accept the lower-cost project as best so far.

5. Compare the best project so far with the next highest-cost project not yet tested, by returning to step 3. If the list is exhausted, stop. The last surviving alternative is the best.

The procedure is illustrated in Example 14-8.

Example 14-8.

Use the data in the table below to determine the best project to undertake. Solve by the incremental rate of return method. Interest rate (*MARR*) is 8%.

End of Year	Project ($million)					
	E	C	A	D	F	B
0	−5.0	−7.0	−10.0	−12.0	−16.0	−19.0
1	0.447	1.639	1.788	2.533	2.833	2.833
2	↓	↓	↓	↓	↓	↓
⋮						
10	0.447	1.639	1.788	2.533	2.833	2.833
i^*	(−)	19.5%	12.3%	16.6%	12.0%	8.0%

Solution: Projects are already listed in ascending order of cost. The last row shows the calculated *IRR* for each project. Project E is discarded since $i^* < MARR$. In the next step, compare increments sequentially.

	A-C	D-C	F-D	B-D
Incremental C	3.0	5.0	4.0	7.0
Incremental annual B	0.149	0.894	0.300	0.300
ΔROR	(−)	12.3%	(−)	(−)
Decision	Keep C	Accept D	Keep D	Keep D

The increment A-C produces a negative rate of return (the sum of undiscounted benefits is less than cost); therefore, alternative C is kept as best so far. The increment D-C produces $\Delta ROR = 12.3\%$, which is greater than the *MARR* (8%); D is accepted as best so far and compared with alternative F, producing a negative rate of return. D is then compared to B, again resulting in ΔROR negative; D is accepted as best so far. No further alternatives exist; stop, D is best project: choose project D.

The steps taken in Example 14-8 are shown graphically in Figure 14.8 as dashed lines. The initial calculation of internal rates of return establishes that it is worthwhile moving from the do-nothing alternative (at the origin) to project C, the lowest-cost feasible alternative. From C, project A was tested (A-C) and found not to be worthwhile; C was returned to as best so far and tested with D, and so on.

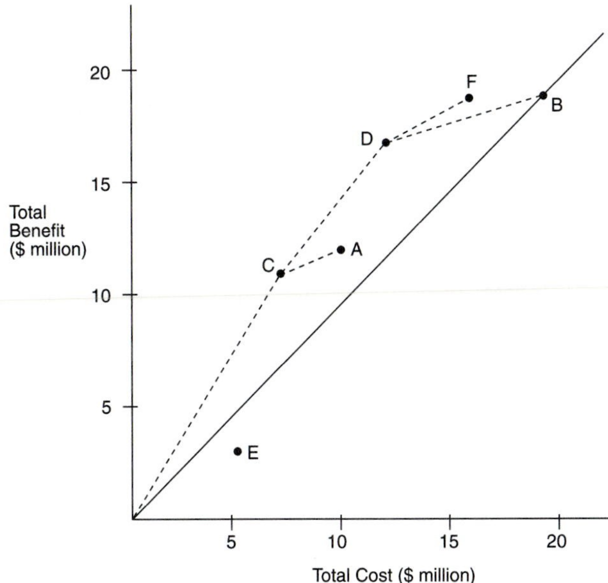

Figure 14.8 Steps in incremental rate of return analysis for Example 14-8.

The graphical representation in Figure 14.8 displays the rationale for the ΔROR procedure. As long as the slope of the incremental step is greater than (or equal to) 1.0, the ΔROR for that step will be greater than (or equal to) $MARR$, and the step will be taken; that is,

$$\frac{\Delta B}{\Delta C} \geq 1.0 \Rightarrow \Delta ROR \geq MARR. \tag{14.11}$$

Why? Recall that Figure 14.8 is drawn based on discounting annual benefits and costs at the $MARR$, 8% in this case. (Figure 14.8 is the same as Figure 14.3, and the data in Example 14-8 is the same as Table 14.1.) Any incremental steps shown in Figure 14.8 as having

$$\frac{\Delta B}{\Delta C} = 1.0 \tag{14.12}$$

automatically requires that

$$\Delta ROR = MARR \tag{14.13}$$

for that step. If any other ΔROR were used, discounted benefits would not equal discounted costs. *If an incremental slope greater than* 1.0 *is displayed in Figure* 14.8, a $\Delta ROR > MARR$ *would be required to discount the benefits so they exactly equaled cost.*

The rationale for ΔROR analysis is therefore the same as for present worth (or annual cash flow) analysis: maximization of net benefits by carrying the scale of the project to the point where marginal benefit equals marginal cost.

14.G PAYBACK PERIOD

The *payback period* is the period of time required for undiscounted benefits of an investment to equal undiscounted costs. Example 14-9 illustrates the method.

Example 14-9.

Determine the payback period for each of the following alternatives:

End of Year	Receipt/Disbursement	
	Alt. A	Alt. B
0	\$−80,000	\$−40,000
1	20,000	20,000
2	30,000	20,000
3	40,000	5,000
4	50,000	5,000

Solution: Assuming continuous receipt of benefits over time, alternative A has a payback period of 2.75 years. Alternative B has a payback period of 2.0 years, suggesting that it should be the preferred alternative. However, if interest is 8.0%, the PWNB of A is \$32,743 while that for B is \$3310, indicating that A is the preferred alternative.

Example 14-9 illustrates the fact that the payback period is only a crude estimate of whether a project is worthwhile. Payback period analysis ignores

- The time value of money,
- Any receipts or disbursements beyond the payback period.

The payback period is often quoted for business investments, but it should not serve as a substitute for more accurate methods.

14.H BREAKEVEN ANALYSIS

Breakeven analysis determines the value of a parameter necessary to make two alternatives economically equivalent. The analysis is usually expressed in the form of a *breakeven chart*, wherein the format is to plot the parameter on the x axis and the measure-of-goodness, or economic criterion, on the y axis. *Sensitivity analysis* is closely related, being the determination of the magnitude of variation in a parameter that would be necessary to change a decision. The three concepts are illustrated in Example 14-10.

Example 14-10.

Two alternatives are indicated below. At what value of annual benefit, x, would A be equivalent to B? Illustrate in a breakeven chart. If the best estimate for benefits

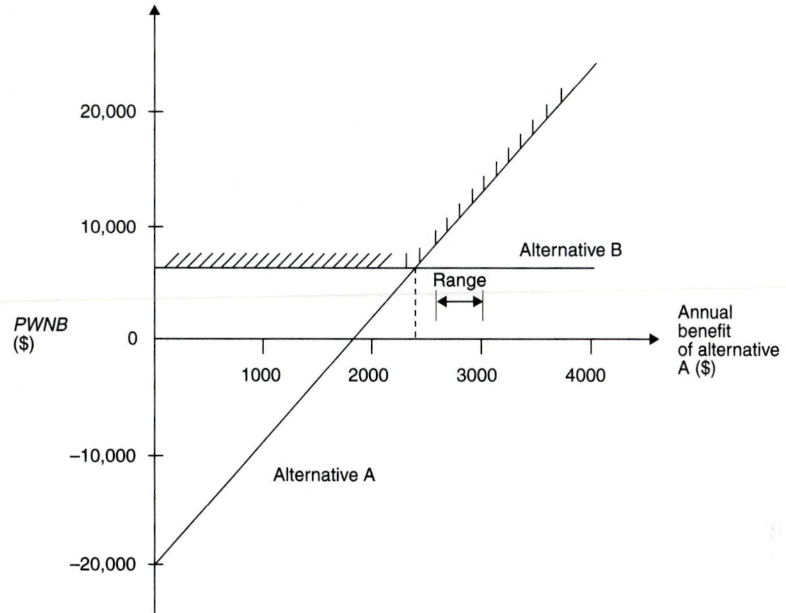

Figure 14.9 Breakeven chart for Example 14-10.

of A were in the range \$2600–3000 per year, would the choice of the best alternative be sensitive to the benefit estimate? $n = 20$ years, $i = 7\%$.

	Alternative	
	A	B
Initial cost	\$20,000	\$15,000
Annual benefit	x	2,000

Solution: Using *PWNB* as the criterion,

$$x\ (P/A, 7\%, 20) - 20{,}000 = 2000\ (P/A, 7\%, 20) - 15{,}000.$$

Solving,

$$x = \$2472.$$

The breakeven chart is shown in Figure 14.9. Since the range in parameter estimate is entirely above the breakeven point, the solution is not sensitive to the benefit estimate: choose alternative A.

CHAPTER SUMMARY

The current chapter details approaches to choosing the best alternative to implement from a set of mutually exclusive alternatives. Four methods—(1) present worth analysis, (2) annual cash flow analysis, (3) incremental benefit-cost ratio analysis, and (4) incremental rate of return analysis—maximize net benefits and produce the same answer for a given problem. Payback period analysis is an approximate method that does not always choose the best alternative. Breakeven and sensitivity analysis are useful when problem parameters are not known with certainty.

EXERCISES

14.1. Assume the following equations for the total benefit and total cost of a reservoir project as a function of scale, Q, expressed as million gallons water delivered per day (MGD):

$$TB = 1{,}000{,}000\,Q^{0.7}$$

$$TC = 5000\,Q^{2.0}$$

where TB and TC are expressed in millions of dollars, present worth. Determine the scale of the project graphically and analytically.

14.2. An engineering student has just finished the freshman year and has received an offer of \$20,000 per year in a full-time job, with prospects of salary increasing 3% per year until retirement after 33 years. If employment is taken, the student will likely not finish his engineering degree. Tuition and other costs are \$10,000 next year, increasing at 7% per year. A starting salary of \$36,000 could be expected upon graduation from the four-year program. Salary increases in the engineering job are estimated at 4% per year until retirement after 30 years.

On the basis of economics alone, should the student take the job now or finish college? Analyze as two mutually exclusive alternatives and solve with present worth analysis. Interest rate is 7%.

14.3. Rework Exercise 14.2 using only one option, that of attending college. Treat the lost salary of a job as an opportunity cost of college.

14.4. A flood control project has a construction cost during the first year of \$10 million, during the second year \$6 million, and during the third year \$2 million. It is completed at the end of the third year and thereafter incurs an annual operating cost of \$200,000 per year. Benefits from the project also begin during the fourth year and are valued at \$1.5 million in that year, growing at a 2% compound rate of increase out to the planning horizon (analysis period) of 50 years. The interest rate is taken to be 6%. Carefully draw the cash flow diagram. What is the present worth of cost? What is the present worth of benefit? Is this a viable project economically?

14.5. In addition to annual maintenance cost, the water supply dam in Example 14-2 must be renovated every 50 years at a cost of \$10 million. What is the effect on *PWC* of the dam and water conservation alternatives caused by ignoring all costs beyond the 50-year planning horizon? Does the inclusion of perpetual costs change the decision?

14.6. The activated sludge unit of a wastewater treatment plant requires an aerator. Two options exist and have the characteristics shown in the table. The city owner of the treatment plant sets an interest rate of 8%. Determine which aerator to purchase.

	Aerator A	Aerator B
Initial cost	400,000	200,000
Annual operating cost	60,000	75,000
Salvage value	60,000	20,000
Lifetime (years)	10	5

(a) Form the cash flow table for a present worth analysis, showing the correct sign for all entries.
(b) Solve by present worth analysis.
(c) Solve by annual cash flow analysis.
(d) Solve by incremental benefit-cost ratio analysis.
(e) Solve by incremental rate of return analysis. Do not find the exact value for the incremental rate of return, just check at the *MARR*.

14.7. A wastewater treatment plant must be built to improve water quality on a stream. Two options are available: a secondary treatment plant removing 85% of the pollution or a tertiary treatment plant removing 95% of the pollution. Improved water quality is given a dollar value as shown in the table. Assume that, by law, the planning horizon (analysis period) is 20 years. The *MARR* is 8%. Determine the best plant to build using the method listed below.

	Secondary ($ million)	Tertiary ($ million)
Initial cost	20.0	40.0
Annual operating cost	1.0	2.0
Annual benefit	3.5	7.0
Lifetime (years)	20	20

(a) Form the cash flow table for the alternatives showing the correct sign for all entries.
(b) Solve using present worth analysis.
(c) Solve using annual cash flow analysis.
(d) Solve using incremental benefit-cost ratio analysis.
(e) Solve using incremental rate of return analysis, assuming that each alternative has a satisfactory internal rate of return.

14.8. An engineering firm has identified five ways to cut costs in its main office. Only one of the options can be implemented, however, since each involves significant training time for staff engineers. Data are provided in the table. Each option has a lifetime of seven years, and the firm sets a *MARR* at 5%.

	Option				
	A	B	C	D	E
Capital cost ($ million)	2.713	0.375	1.650	0.088	0.950
Annual cost ($ million/yr)	0.093	0.270	0.132	0.147	0.228
Annual benefit ($ million/yr)	0.890	0.288	0.841	0.312	0.505

(a) Solve by present worth analysis.
(b) Solve by annual cash flow analysis.
(c) Solve by incremental benefit-cost ratio analysis.
(d) Solve by incremental rate of return analysis, using the full detailed procedure.

14.9. In comparing three alternatives by the rate-of-return method, an analyst is given the following information:

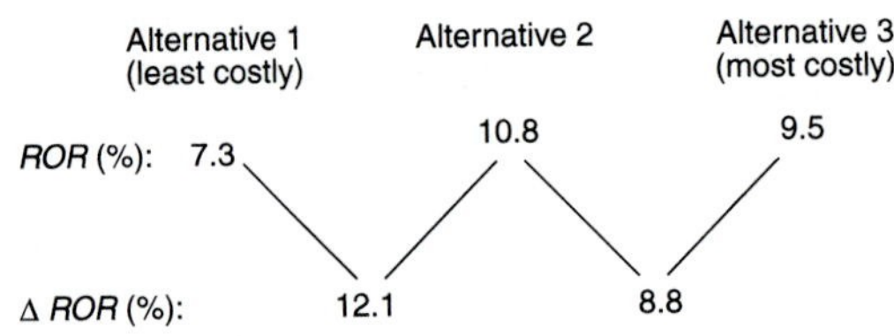

(a) If the *MARR* is 8%, which alternative should be chosen?
(b) If the *MARR* is 9%, which alternative should be chosen?

14.10. Determine the payback period for the following investments:

End of Year	A	B	C
0	$−1000	$−1000	$−1000
1	500	300	−500
2	400	400	700
3	300	500	700
4	200	600	700
5	100	700	700

How does the choice of investment using payback period compare to present worth analysis if *MARR* is 5%?

14.11. If the payback period were calculated on the basis of actual time value of money (at $n = 0$) at the prevailing *MARR*, would it be longer, shorter, or stay the same? Why?

14.12. A contractor is considering the purchase of an earthmover to avoid the cost of a rental unit. The earthmover costs $350,000 initially and will cost $7000 per year to store and maintain. When used, the operating cost is \$300 per day to the contractor. A rental unit costs \$500 per day to rent and operate. The contractor estimates that he would keep the earthmover 20 years and could then sell it at salvage for \$20,000. He requires a 12% rate of return on any investment.

Determine how many days per year he would have to operate the earthmover to make it a worthwhile purchase. Sketch the breakeven chart for this situation.

14.13. Construction cost of a sewage treatment plant shows "economies of scale," where cost increases as the capacity of the plant increases, but at a decreasing rate; that is, the cost curve is concave and can be expressed in the form

$$C = 1{,}300{,}000\,Q^{0.65}$$

where Q is the capacity of the plant in million gallons per day and C is the construction cost (in dollars).

Demand for sewage treatment for a city is projected to increase linearly from zero today to 20 MGD 20 years from now (1 MGD per year increase).

Under the assumption that either one or four plant expansions will occur (20 MGD capacity now or four expansions of 5 MGD each, with the first now), determine the breakeven interest rate at which the cost of one expansion exactly equals the cost of four expansions. Plot the breakeven chart. If the city sets the *MARR* at 6%, how many expansions (one or four) should occur?

REFERENCES

AU, TUNG and THOMAS P. AU, *Engineering Economics for Capital Investment Analysis*. Boston: Allyn and Bacon, 1983, 506 pp.

BLANK, LELAND T. and ANTHONY J. TARQUIN, *Engineering Economy*. New York: McGraw-Hill, 1989, 531 pp.

BUSSEY, LYNN E. and TED G. ESCHENBACH, *The Economic Analysis of Industrial Projects*. Englewood Cliffs, NJ: Prentice Hall, 1992, 450 pp.

DEGARMO, E. PAUL, SULLIVAN, WILLIAM G., and JAMES A. BONTADELLI, *Engineering Economy*. New York: Macmillan, 1993, 726 pp.

ESCHENBACH, TED, *Cases in Engineering Economy*. New York: John Wiley and Sons, 1989, 208 pp.

FLEISCHER, G. A., *Engineering Economy: Capital Allocation Theory*. Boston: PWS, Wadsworth, 1984, 521 pp.

GRANT, EUGENE L., IRESON, W. GRANT, and RICHARD S. LEAVENWORTH, *Principles of Engineering Economy*. New York: John Wiley and Sons, 1990, 591 pp.

NEWNAN, DONALD G., *Engineering Economic Analysis*. San Jose, CA.: Engineering Press, 1991, 578 pp.

RIGGS, JAMES L., *Engineering Economics*. New York: McGraw-Hill, 1982, 789 pp.

SMITH, GERALD W., *Engineering Economy: Analysis of Capital Expenditures*. Ames: Iowa State University Press, 1987, 594 pp.

STEINER, HENRY MALCOLM, *Engineering Economic Principles*. New York: McGraw-Hill, 1992, 559 pp.

TAYLOR, GEORGE A, *Managerial and Engineering Economy: Economic Decision-Making*. Monterey, CA.: Brooks/Cole Engineering Division, Wadsworth, 1980, 538 pp.

THUESEN, G. J. and W. J. FABRYCKY, *Engineering Economy*. Englewood Cliffs, NJ: Prentice Hall, 1993, 717 pp.

WHITE, JOHN A., AGEE, MARVIN H., and KENNETH E. CASE, *Principles of Engineering Economic Analysis*. New York: John Wiley and Sons, 1984, 546 pp.

CHAPTER

15 Engineering Economics III: Depreciation, Taxes, Inflation, and Personal Financial Planning

15.A INTRODUCTION

In practice, economic decisions are complicated by such considerations as taxes, possible tax deductions (one of which is depreciation of an asset), and inflation. These same considerations play a strong role in personal financial planning. This chapter will show how to include such factors in engineering economic calculations.

In addition, it has been the observation (and experience) of the authors that more attention should be given to personal financial planning by young engineers starting their career. Certain nontraditional topics are therefore included in this chapter in the area of personal financial planning.

15.B DEPRECIATION

The value of property acquired for use in a trade or business or for the production of income can be systematically deducted over a period of years for income tax purposes. The property must have a useful life of more than one year and must be of such a nature that it will wear out or become obsolete over time. Common examples are machinery, equipment, computers, buildings and building improvements, landscaping, and rental property. Land is not depreciable, since it is viewed as not wearing out or becoming obsolete. In addition, special tax rules apply to the depreciation of computer software and automobiles. If computers or peripheral equipment are not used entirely at a place of business, special rules also apply.

Depreciable property is classified into two broad categories: (1) *real property*—buildings and building structural components, and (2) *personal property*—all property other than real property. Note that both personal and real property may be owned and used by either an individual or a business. To be depreciable, however, the property must be used by a business or in the production of income by an individual.

Herein, *depreciation* is defined as the allocation of the cost (minus salvage value) of an asset over its useful or depreciable life for tax purposes. Thus,

$$D_{\text{total}} = P - S \tag{15.1}$$

where

D_{total} = total depreciation allowable over time;

P = initial asset value or cost; and

S = salvage value.

The *book value* of an asset in year n is its initial cost or value minus any depreciation charges taken up to that time:

$$BV_n = P - \sum_{t=0}^{n-1} D_t \tag{15.2}$$

where

BV_n = book value of an asset in year n, and

D_t = depreciation taken at the end of year t.

Note that book value in year 1, BV_1, is equal to P since no depreciation is taken at $t = 0$. Note also that book value always includes salvage value. If depreciation represented an accurate measure of how an asset wears out or goes obsolete, then book value would represent the remaining value, or worth, of the asset. In practice, depreciation charges are determined more by accounting rules and tax laws than by actual economic depreciation.

15.B.1 Straight-Line Depreciation

One approach to depreciation is to assume that the annual depreciation charge is a constant amount chosen to produce a book value of S at the end of year N:

$$SLD_t = \frac{1}{N}(P - S) \tag{15.3}$$

where

SLD_t = straight-line depreciation taken at the end of year t, and

N = useful (or depreciable) lifetime.

Example 15-1.

An asset is purchased for $10,000 and is to be depreciated over five years. Salvage value at the end of year 5 is $1000. Using straight-line depreciation, what is the annual depreciation charge and schedule of book value?

Solution: Annual depreciation is

$$SLD_t = \frac{1}{N}(P - S) = \frac{1}{5}(10{,}000 - 1000)$$

$$= \$1800$$

which gives a schedule for book value as follows:

Year	Book Value	Depreciation Amount
1	$10,000	$1800
2	8,200	1800
3	6,400	1800
4	4,600	1800
5	2,800	1800
≥ 6	1,000 (salvage value)	—

Book value is plotted in Figure 15.1. Note that the schedule of book value is a straight line only if depreciation is viewed as a continuous process.

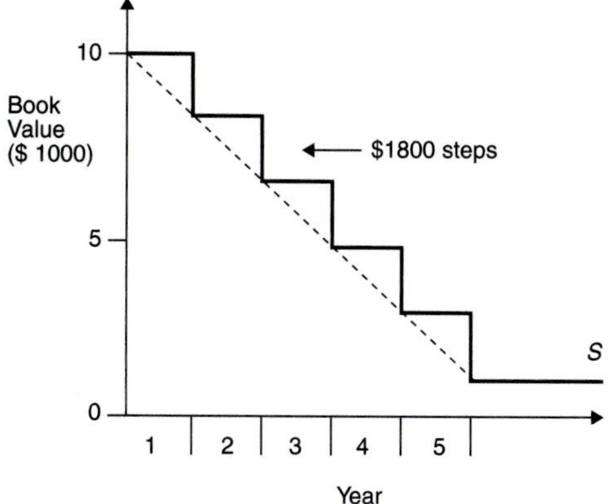

Figure 15.1 Straight-line depreciation for Example 15-1.

15.B.2 Sum-of-the-Years'-Digits Depreciation

An interesting depreciation method is the sum-of-the-years'-digits (SOYD). An advantage of this method over straight-line depreciation is in having larger depreciation amounts early in time. This permits larger tax deductions and resultant savings early in time which can be invested to earn compound interest. Overall returns are therefore higher than under straight-line depreciation for most cases.

The method uses the sum-of-the-years'-digits for the planning period of length N:

$$\text{total sum of the years' digits} = 1 + 2 + 3 + \cdots + (N - 1) + N \tag{15.4}$$

$$= \frac{N}{2}(N + 1) \tag{15.5}$$

and calculates depreciation for the end of year t as

$$SOYDD_t = \frac{\text{useful life remaining at beginning of year } t}{\text{total sum of the years' digits}}(P - S) \tag{15.6}$$

$$= \frac{N - t + 1}{(N/2)(N + 1)}(P - S) \tag{15.7}$$

where $SOYDD_t$ = sum-of-the-years'-digits depreciation at the end of year t.

Example 15-2.

Compute the $SOYD$ depreciation for the data in Example 15-1: P = \$10,000, S = \$1000, N = 5.

Solution:

$$SOYD = 1 + 2 + 3 + 4 + 5$$

$$= \frac{N}{2}(N + 1) = \frac{5}{2}(5 + 1) = 15$$

$$SOYDD_1 = \frac{5 - 1 + 1}{15}(10{,}000 - 1000) = \frac{5}{15}(9000) = \$3000$$

$$SOYDD_2 = \frac{5 - 2 + 1}{15}(10{,}000 - 1000) = \frac{4}{15}(9000) = \$2400$$

$$SOYDD_3 = \frac{5 - 3 + 1}{15}(10{,}000 - 1000) = \frac{3}{15}(9000) = \$1800$$

$$SOYDD_4 = \frac{5-4+1}{15}(10{,}000 - 1000) = \frac{2}{15}(9000) = \$1200$$

$$SOYDD_5 = \frac{5-5+1}{15}(10{,}000 - 1000) = \frac{1}{15}(9000) = \$\ 600$$

$$\text{sum} = \$9000.$$

Total depreciation is $9000, leaving only salvage value at the end of year $N = 5$. Note the higher SOYD depreciation values in years 1 and 2 as compared to straight-line depreciation ($1800 every year).

15.B.3 Declining Balance Depreciation

Declining balance (DB) depreciation is based on taking a constant fraction of the current book value as the depreciation amount every year. Double declining balance depreciation (DDBD) uses the fraction $2/N$:

$$DDBD_t = \frac{2}{N}(\text{book value}) = \frac{2}{N}(BV_t)$$

$$= \frac{2}{N}(P - \text{total depreciation taken to date}). \tag{15.8}$$

For year 1,

$$DDBD_1 = \frac{2}{N}(P - 0) = \frac{2}{N}P.$$

If $S = 0$, straight-line depreciation would give

$$SLD_1 = \frac{1}{N}(P - S) = \frac{1}{N}P$$

showing double declining balance depreciation to be twice, or 200%, that of straight-line for the first year. Note that the amount is not double for subsequent years, however.

Another depreciation scheme uses a 150% rate, or

$$(150\%)\ DBD_t = \frac{1.5}{N}(\text{book value}). \tag{15.9}$$

Example 15-3.

Compute DDB depreciation for the data of Example 15-1 and 15-2: P = \$10,000, S = \$1000, and N = 5.

Solution:

$$DDBD_t = \frac{2}{N}\text{ (book value)} = \frac{2}{N}(BV_t) = \frac{2}{5}(BV_t).$$

Year	BV_t	$DDBD_t$
1	\$10,000	\$4000
2	6000	2400
3	3600	1440
4	2160	864
5	1296	296*
≥6	1000 (salvage value)	—

*Note that the full DDBD at the end of year 5 [\$1296(0.40)] = \$518.40) cannot be taken since this would drive book value below salvage value (\$1000). Also note that DDBD in years 1 and 2 is considerably higher than SLD. DDBD is higher than SOYDD for year 1 and equal in year 2.

Example 15-3 illustrates a complication with the DDB or 150% DB method: book value is not guaranteed to be equal to salvage at the end of year N if the formula is strictly adhered to. Three cases can result:

1. For high values of salvage relative to initial cost, DDB or 150% DB are likely to give book values below salvage. To compensate, merely cut off the depreciation schedule early, taking a smaller depreciation amount than indicated for the year in question, in order to exactly reach salvage value (this was done in Example 15-3).
2. For low values of salvage relative to initial cost, DDB or 150% DB are likely to give book values above salvage. To compensate for this, tax law allows switching from a declining balance to straight-line depreciation at any time. This permits salvage value to be exactly met.
3. It is possible for DDB or 150% DB to produce book value exactly equal to salvage value in the last year, but this is rare in actual practice.

The three cases are illustrated in Figure 15.2.

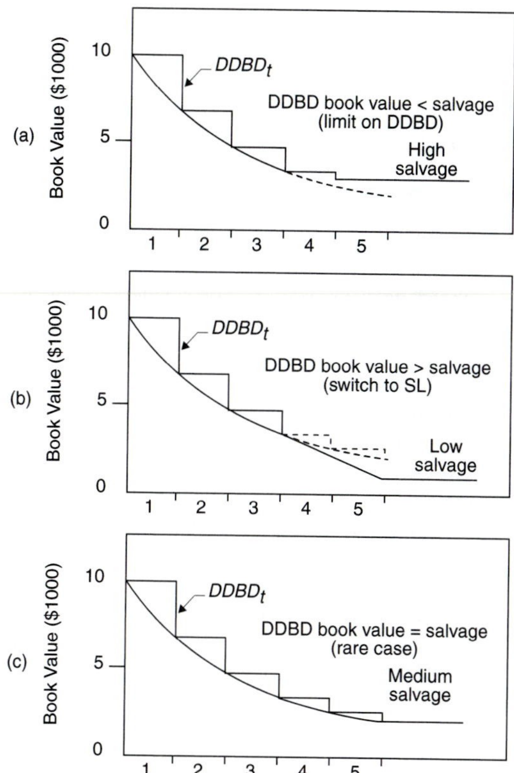

Figure 15.2 Possible relationship of book value to salvage.

15.B.4 Modified Accelerated Cost Recovery System Depreciation

The Tax Reform Act of 1986 (PL99-514) outlines the *modified accelerated cost recovery system (MACRS) depreciation* method. It applies to tangible personal or real property placed in service after 1986. From 1981 through 1986, the *accelerated cost recovery system (ACRS) depreciation* method was used. The two systems are very similar. Both provide tax advantages over previously discussed methods in that the statutory depreciation period is generally shorter than the useful life of an asset, and salvage value is assumed to be zero.

To calculate allowable depreciation, the *property class* and *recovery period* of the asset must be found by using Table 15.1. Then the prescribed percentage for depreciation in each year is read from Table 15.2 or 15.3. Table 15.2, MACRS for personal property, is based on 200% DB depreciation for the 3-, 5-, 7- and 10-year property classes, with switch to SL depreciation so as to maximize depreciation in each year. For the 15- and 20-year property classes, 150% DB depreciation is used with switch to SL. A midyear convention applies to both; that is, if the

TABLE 15.1 PROPERTY CLASS AND RECOVERY PERIOD FOR ASSETS UNDER MACRS DEPRECIATION

Class Life, L (yr)	Property Class and Recovery Period (yr)	Depreciable Asset
		Personal Property
$L \leq 4$	3-year property	Over-the-road tractor units
$4 < L < 10$	5-year property	Information systems (computers and peripheral equipment)
		Datahandling equipment except computers (typewriters, calculators, copiers, duplicating equipment)
		Airplanes and helicopters
		Automobiles,* buses, taxis
		Freight and heavy trucks (concrete ready-mix)
		Trailers and trailer-mounted containers
		Offshore oil and gas drilling equipment and vessels
		Onshore oil and gas drilling equipment
		Construction and real estate development equipment
		Equipment used in research and development
		Equipment used for qualified electrical energy generation (geothermal, solar, wind, biomass)
		Semiconductor manufacturing equipment
$10 \leq L < 16$	7-year property	Office furniture, fixtures, and equipment
		Railroad cars, locomotives, tracks
		Mining equipment
		Agricultural machinery, equipment, grain bins, fences
$16 \leq L < 20$	10-year property	Barges, tugs, vessels for water transportation
		Single-purpose agricultural structures
		Fruit or nut-bearing trees or vines
$20 \leq L < 25$	15-year property	Land improvements (sidewalks, roads, canals, waterways, drainage facilities, wharves, docks, bridges, fences, landscaping, radio and TV transmission towers)
		Telephone distribution plants
		Municipal wastewater treatment plants
$L \geq 25$	20-year property	Farm buildings (general purpose)
		Municipal sewers
		Real Property
—	27.5-yr property	Residential rental property
—	31.5-yr property	Nonresidential real property (placed in service before May 13, 1993)
—	39-yr property	Nonresidential real property (placed in service on or after May 13, 1993)
—	50-yr property	Railroad grading or tunnel bore

*Other special conditions apply to depreciation of automobiles.

TABLE 15.2 MACRS DEPRECIATION FOR PERSONAL PROPERTY

Recovery Year	Recovery period*					
	3-year	5-year	7-year	10-year	15-year	20-year
1	33.33	20.00	14.29	10.00	5.00	3.750
2	44.45	32.00	24.49	18.00	9.50	7.219
3	14.81	19.20	17.49	14.40	8.55	6.677
4	7.41	11.52	12.49	11.52	7.70	6.177
5		11.52	8.93	9.22	6.93	5.713
6		5.76	8.92	7.37	6.23	5.285
7			8.93	6.55	5.90	4.888
8			4.46	6.55	5.90	4.522
9				6.56	5.91	4.462
10				6.55	5.90	4.461
11				3.28	5.91	4.462
12					5.90	4.461
13					5.91	4.462
14					5.90	4.461
15					5.91	4.462
16					2.95	4.461
17						4.462
18						4.461
19						4.462
20						4.461
21						2.231

*For 3-, 5-, 7-, and 10-year recovery periods, the depreciation method is 200% DB with switch to SL. For 15- and 20-year recovery periods, the depreciation method is 150% DB with switch to SL.

asset is placed in service at any time during the year, the depreciation period starts at the midpoint of the year.

Table 15.3, MACRS for real property, is based on straight-line depreciation with a midmonth convention; that is, if an asset is placed in service at any time during a month, it is assumed to begin its depreciable life at the midpoint of the month (one-half month depreciation).

Example 15-4.

An engineering company purchases files, cabinets, desks, and chairs for use in the office. The $10,000 purchase is made March 10. Find the allowable depreciation amounts for current and future years.

TABLE 15.3 MACRS FOR REAL PROPERTY

	Recovery Year	Month Property Placed in Service											
		Jan	Feb	Mar	Apr	May	Jun	Jul	Aug	Sep	Oct	Nov	Dec
		27.5-Year Residential Rental Real Property*											
	1	3.485	3.182	2.879	2.576	2.273	1.970	1.667	1.364	1.061	0.758	0.455	0.152
	2–9	3.636	3.636	3.636	3.636	3.636	3.636	3.636	3.636	3.636	3.636	3.636	3.636
Even years	10–26	3.637	3.637	3.637	3.637	3.637	3.637	3.636	3.636	3.636	3.636	3.636	6.636
Odd years	11–27	3.636	3.636	3.636	3.636	3.636	3.636	3.637	3.637	3.637	3.637	3.637	3.637
	28	1.970	2.273	2.576	2.879	3.182	3.485	3.636	3.636	3.636	3.636	3.636	3.636
	29	0.000	0.000	0.000	0.000	0.000	0.000	0.152	0.455	0.758	1.061	1.364	1.667
		31.5-Year Nonresidential Real Property (placed in service before 05/13/93)*											
	1	3.042	2.778	2.513	2.249	1.984	1.720	1.455	1.190	0.926	0.661	0.397	0.132
	2–7	3.175	3.175	3.175	3.175	3.175	3.175	3.175	3.175	3.175	3.175	3.175	3.175
	8	3.175	3.174	3.175	3.174	3.175	3.174	3.175	3.175	3.175	3.175	3.175	3.175
Odd years	9–31	3.174	3.175	3.174	3.175	3.174	3.175	3.174	3.175	3.174	3.175	3.174	3.175
Even years	10–30	3.175	3.174	3.175	3.174	3.175	3.174	3.175	3.174	3.175	3.174	3.175	3.174
	32	1.720	1.984	2.249	2.513	2.778	3.042	3.175	3.174	3.175	3.174	3.175	3.174
	33	0.000	0.000	0.000	0.000	0.000	0.000	0.132	0.397	0.661	0.926	1.190	1.455
		39-Year Nonresidential Real Property (placed in service on or after 05/13/93)*											
	1	2.461	2.247	2.033	1.819	1.605	1.391	1.177	0.963	0.749	0.535	0.321	0.107
	2–39	2.564	2.564	2.564	2.564	2.564	2.564	2.564	2.564	2.564	2.564	2.564	2.564
	40	0.107	0.321	0.535	0.749	0.963	1.177	1.391	1.605	1.819	2.033	2.247	2.461

*Depreciation method is SL with midmonth convention.

Solution: Table 15.1 shows office equipment and furniture to be seven-year property. Table 15.2 provides allowable depreciation percentages:

Year		MACRS%	Depreciation Amount
1		14.29	$1429
2		24.49	2449
3		17.49	1749
4		12.49	1249
5		8.93	893
6		8.92	892
7		8.93	893
8		4.46	446
	Total:	100	$10,000

Example 15-5.

Determine the year in which to switch from 200% DB to SL depreciation for seven-year property. Assume a $100 depreciable asset.

Solution: Compute the 200% DB depreciation amount and compare to the alternative SL depreciation amount. $N = 7$ for 200% DBD, while for SLD, N is remaining depreciable life, which varies.

End of year	Book Value	200% DB Depreciation $(2/N)(BV)$	Alternative SL Depreciation $(1/N)(BV)$	Comments
1	$100	$14.29*	$7.14	$N = 7.0$ for SL; use half of full-year amount.
2	85.71	24.49*	13.19	$N = 6.5$ for SL.
3	61.22	17.49*	11.13	$N = 5.5$ for SL.
4	43.73	12.49*	9.72	$N = 4.5$ for SL.
5	31.24	8.92	8.93*	$N = 3.5$ for SL.
6	22.31	—	8.92*	Continue with SL.
7	13.39	—	8.93*	Continue with SL.
8	4.46	—	4.46*	Continue with SL.

*Maximum depreciation amount; alternate digits in SL to eliminate effect of round-off.

Note that since the asset value is $100, the depreciation amounts correspond to the depreciation factors in Table 15.2, which are expressed in percent.

Example 15-6.

On October 29, 1994, an engineering company purchased a new office site for $200,000, of which land value represents $50,000. Determine the depreciation schedule for current and future years.

Solution: The acquisition is nonresidential real property placed in service on or after May 13, 1993, therefore, it represents 39-year property. However, land is not depreciable, leaving only $150,000 to be depreciated. Table 15.3 can be used by entering the column for October:

Recovery Year	Factor (%) for 39-Year Property	Depreciation Amount
1	0.535	$ 802.50
2–39	2.564	3,846.00
40	2.033	3,049.50
	Total = 100.00	$150,000.00

Note that the effect of round-off has already been taken into account in Table 15.3, and the midmonth convention applies.

15.B.5 Other Property and Depreciation Methods

Intangible Personal Property. Intangible personal property includes patents, copyrights, franchises, videocassettes, software programs, books, and similar revenue-producing assets. Related costs may be depreciated over the useful life (as estimated by the owner) or by the *income forecast depreciation* method. Using the latter, fractional depreciation in any year is given by the ratio of income received from the asset in that year to the total income expected to be earned from the asset.

Example 15-7.

An engineer has written a software program and received copyright. Expected revenues from its sale are \$10,000, and capitalized costs of production are \$4000. If sales in 1994 amounted to \$2500, what is allowable depreciation using the income forecast method?

Solution: Multiply by the ratio of income to total expected income:

$$\frac{2500}{10{,}000}(4000) = \$1000 \text{ depreciation.}$$

Units of Production. The units of production depreciation method can be elected in place of MACRS for assets that have lifetimes easily measured in terms of production units. Depreciation in any year is given by the ratio of production in that year to total lifetime production of the asset.

Example 15-8.

For \$3000, a company purchases a photocopier machine having a 1,000,000-copy lifetime as estimated by the manufacturer. In the first year, 50,000 copies are made. What is allowable depreciation using the units of production method?

Solution: Multiply by the ratio of production in the year to total lifetime production:

$$\frac{50{,}000}{1{,}000{,}000}(3000) = \$150 \text{ depreciation.}$$

Computer Software. Under certain conditions, computer software purchased for a business after August 10, 1993 is depreciated using a straight-line method over 36 months, beginning with the month of purchase. Thus, if a package is purchased on July 7, 1994 for \$300, depreciation for 1994 is \$50 (= (6/36)(\$300)).

15.B.6 Choice of Depreciation Method

The most advantageous depreciation method to use in any given situation depends on many factors. Among these are (1) interest rate used in comparisons, (2) length of depreciation period, (3) investment opportunities resulting from tax savings over time, and (4) pattern of tax liabilities for the person or firm year by year. Since some of these factors are difficult to estimate, a person or firm may choose to use just one of the methods as standard practice unless unusual circumstances arise. As noted, it is generally advisable to obtain depreciation deductions early in time in order to maximize present worth of benefits from tax savings. Tax laws limit the degree of choice between the methods.

15.B.7 Amortization and Depletion

Amortization. Certain business expenses cannot be deducted entirely in the year of their occurrence, and yet do not represent depreciable assets (they do not wear out or become obsolete). The expense of acquiring and holding *intangible assets* such as goodwill, going-concern value, know-how, government-granted licenses or permits, leases, patents, copyrights, trade names, and similar items can be *amortized* over prescribed periods of time, usually 15 years. Business start-up expenses, at the election of the taxpayer, may be amortized over a 60-month period, beginning with the month in which business begins. The deductions are prorated equally over each month of the 60-month period.

Depletion. Assets of such natural resources as mineral ores, oil and gas wells, and timber are used up over time. The investment in these resources is allowed to be deducted from income under certain rules. Two calculation methods are used in most cases: cost depletion and percentage depletion. The method yielding the largest depletion allowance must be used in that year. For most oil and gas wells and for timber, only cost depletion is allowed.

Cost depletion is based on the number of physical units of the resource recoverable at the beginning of the year and the number of units sold during the year. The calculation is

$$\text{cost depletion} = (\text{adjusted basis})\left(\frac{\text{units sold in year}}{\text{units recoverable at beginning of year}}\right).$$

The adjusted cost basis is the updated asset cost minus any depletion allowances taken to date. Also, updated estimates of units recoverable should be used as they become available.

Example 15-9.

For \$100,000 a company purchased a tract of land having 1,000,000 board-feet of timber. The timber itself was valued at \$80,000. Depletion credits of \$20,000 have been taken to date on 250,000 board-feet cut and sold. In the current year, 125,000 board-feet were sold. What is the current year depletion allowance?

Solution: Calculate using updated cost basis and recoverable units:

TABLE 15.4 PERCENTAGE DEPLETION RATES FOR CERTAIN NATURAL RESOURCES

Natural Resource	Percentage Depletion Rate
Sulfur and uranium U.S. deposits of asbestos, bauxite, graphite, mica, ores of antimony, cadmium, cobalt, lead, lithium, manganese, mercury, molybdenum, nickel, platinum, tin, tungsten, and zinc	22
U.S. deposits of gold, silver, copper, iron ore, and oil shale Oil and gas from small independent producers U.S. geothermal deposits	15
Metal mines not qualifying for 22% or 15% rates Clays sold for purposes dependent on refractory properties	14
Coal, lignite, sodium chloride	10
Clay and shale used for manufacture of sewer pipe or brick Clay, shale, and slate used for lightweight aggregates	7.5
Gravel, peat, sand, stone (other than dimension stone or ornamental stone), clay used in manufacture of drainage and roofing tile, and if from brine wells—bromine, calcium chloride, and magnesium chloride Any mineral used as rip rap, ballast, road material, rubble, or concrete aggregate if not classified elsewhere	5
All other minerals (e.g., borax, calcium carbonates, diatomaceous earth, dolomite, granite, limestone, marble, clam shells and oyster shells, potash, dimension stone or ornamental stone.	14

Source: Internal Revenue Code (1994), Sections 611-613A.

$$\text{cost depletion} = (80{,}000 - 20{,}000)\left(\frac{125{,}000}{1{,}000{,}000 - 250{,}000}\right)$$

$$= \$10{,}000.$$

Percentage depletion is computed as a percentage of gross income and is unusual in that the depletion allowance can be claimed even if the adjusted basis in the resource is zero. The percentages allowed are prescribed by law and vary by resource. Selected resources and their applicable percentages are provided in Table 15.4. A depletion allowance cannot be greater than 50% (100% for oil and gas produced by small independent producers) of the taxable income before deducting depletion.

Example 15-10.

A gold mine has a total income of $800,000 and expenses of $400,000 for a given year. What is the allowable depletion deduction?

Solution: Table 15.4 gives 15% as the percentage depletion rate. Calculate the depletion amount without considering the 50% limitation:

$$\text{depletion amount} = (0.15)(800{,}000) = \$120{,}000.$$

Determine limitation amount:

$$\text{gross income - expenses} = \text{taxable income}$$

$$800{,}000 - 400{,}000 = 400{,}000 \text{ taxable}$$

$$(0.50)\ (\$400{,}000) = \$200{,}000 \text{ limitation.}$$

Since the limitation is greater than the indicated percentage depletion amount, the full $120,000 can be claimed. Answer = $120,000.

15.C TAXES

15.C.1 Corporate Income Tax Base and Rate

Taxable income for corporations is generally given by

$$\begin{aligned}\text{taxable income} = {} & \text{gross income} + \text{capital gains net of capital losses} \\ & - \text{current expenses} - \text{depreciation} - \text{depletion allowance.}\end{aligned} \tag{15.10}$$

Current expenses are by far the largest category of deductions. These include salary and wages, materials, routine maintenance, heat, light, telephone, and expendable supplies. Capital gains and losses are defined by

$$\text{capital gain or loss} = \text{sale price} - \text{book value.} \tag{15.11}$$

Capital gains are taxed as regular income for corporations, and capital losses can be deducted only to the extent of capital gains. Any excess of capital losses over capital gains for a year can be carried back three years and, if not entirely used up, can be carried forward up to five years.

Note that expenditures for capital assets are not deducted in the year of their occurrence but are depreciated, amortized, or depleted in the current and later years. Similarly, salvage value, to the extent that it is equal to book value, is not treated as income.

Corporate federal income tax rates for 1994 are given in Table 15.5. In addition, states and municipalities often levy corporate income taxes, generally totaling from 1 to 12%.

Example 15-11.

An engineering company has revenues of $1,000,000 for the calendar year. Expenses are $600,000, capital gains are $300,000, capital losses are $200,000, depreciation is $100,000, and depletion is $50,000. What is the firm's taxable income and tax payment due?

TABLE 15.5 CORPORATE FEDERAL INCOME TAX RATES FOR 1994

Taxable Income		Incremental	
Over	But Not Over	Tax Rate (%)	Tax on Income Falling Within a Bracket
0	50,000	15	\$0 + 15% of excess over \$0
50,000	75,000	25	\$7,500 + 25% of excess over \$50,000
75,000	100,000	34	\$13,750 + 34% of excess over \$75,000
100,000	335,000	39	\$22,250 + 39% of excess over \$100,000
335,000	10,000,000	34	\$113,900 + 34% of excess over \$335,000
10,000,000	15,000,000	35	\$3,400,000 + 35% of excess over \$10,000,000
15,000,000	18,333,333	38	\$5,150,000 + 38% of excess over \$15,000,000
18,333,333	—	35	\$6,416,667 + 35% of excess over \$18,333,333

Solution: Capital losses of \$200,000 can offset a portion of capital gains, leaving a net capital gain of \$100,000.

$$\begin{aligned} \text{taxable income} &= 1{,}000{,}000 + 100{,}000 \\ &\quad - 600{,}000 - 100{,}000 - 50{,}000 \\ &= \$350{,}000. \end{aligned}$$

$$\begin{aligned} \text{tax payment} &= 0.15(50{,}000) = \$7500 \\ &+ 0.25(25{,}000) = \$6250 \\ &+ 0.34(25{,}000) = \$8500 \\ &+ 0.39(235{,}000) = \$91{,}650 \\ &+ 0.34(15{,}000) = \$5100 \\ &\text{Total} = \$119{,}000 \end{aligned}$$

which is 34.0% of taxable income.

15.C.2 Individual Income Tax Base and Rate

Determination of federal income tax due from individuals requires the computation of

- gross, or total income,
- adjusted gross income (AGI), and
- taxable income.

Gross income is the sum of all sources of income for the year:

gross income = wages + salary + tips + interest + dividends + refund of state and local taxes + alimony received + business income + capital gains or losses + taxable pensions or annuities + unemployment compensation + social security benefits + other income.

Certain adjustments are made to gross income:

adjustments = qualifying payments to a retirement plan + one-half of any self-employment (Social Security) tax payment + self-employed health insurance premiums + alimony paid + interest penalties on early withdrawal of savings.

Adjusted gross income is then given by

$$\text{adjusted gross income} = \text{gross income} - \text{adjustments}. \tag{15.12}$$

Taxable income can then be found as

$$\begin{aligned}\text{taxable income} = \ & \text{adjusted gross income} \\ & - \text{larger of itemized deductions or standard deduction} \\ & - \text{exemptions}.\end{aligned} \tag{15.13}$$

The standard deduction for the 1994 tax year was \$3800 for single taxpayers or \$6350 for married taxpayers filing a joint return. The exemption amount was \$2450 for each person dependent on the gross income for their living. The itemized deduction amount is given by

itemized deductions = medical and dental expenses exceeding 7.5% of adjusted gross income + state and local income taxes + real estate (property) taxes + home mortgage interest and points + charitable contributions + casualty and theft losses + work-related moving expenses + sum of unreimbursed employee expenses (including any depreciable items) and other miscellaneous deductions exceeding 2% of adjusted gross income.

Note that capital losses can be directly deducted by individuals (up to a limit of \$3000). They do not have to offset capital gains as with corporations.

Federal income tax is computed on taxable income using the tax rate schedule provided in Table 15.6. As is the case for corporate taxes, the schedule is *graduated*, having progressively higher incremental tax rates with rising income.

Example 15-12.

Brian, a recent graduate, accepts a job 300 miles from his current home and incurs moving expenses of \$1000. His salary is \$32,000, interest income for the tax year is \$150, dividends from stocks are \$60, capital gain from the sale of certain stocks is \$300,

TABLE 15.6 INCREMENTAL 1994 FEDERAL INCOME TAX RATES FOR INDIVIDUALS

Taxable Income Bracket	Incremental Tax Rate (%)	Tax for Income Falling Within Bracket
		Single
\$0–22,750	15	\$0 + 15% of amount over \$0
\$22,750–55,100	28	\$3,412.50 + 28% of amount over \$22,750
\$55,100–115,000	31	\$12,470.50 + 31% of amount over \$55,100
\$115,000–250,000	36	\$31,039.50 + 36% of amount over \$115,000
Over \$250,000	39.6	\$79,639.50 + 39.6% of amount over \$250,000
		Married—Filing Jointly
\$0–38,000	15	\$0 + 15% of amount over \$0
\$38,000–91,850	28	\$5,700.00 + 28% of amount over \$38,000
\$91,850–140,000	31	\$20,778.00 + 31% of amount over \$91,850
\$140,000–250,000	36	\$35,704.50 + 36% of amount over \$140,000
Over \$250,000	39.6	\$75,304.50 + 39.6% of amount over \$250,000

and capital loss from the sale of other stocks is \$600. Interest penalty on early withdrawal of savings from a certificate of deposit is \$40. Charitable contributions are \$400, state and local income taxes are \$2200, and medical expenses do not exceed the 7.5% limit. Employee business expenses are \$500, which includes a \$100 depreciation allowance for a qualified capital asset. Brian rents an apartment and, therefore, does not qualify for property tax or home mortgage deductions. If he is single, what is Brian's (a) gross income, (b) adjusted gross income, (c) taxable income, and (d) tax owed?

Solution:

(a) Gross income: = \$32,000 wages
+ 150 interest
+ 60 dividends
− 300 net capital loss

\$31,910 gross income

(b) Adjusted gross income = \$31,910 gross income
− 40 interest penalty

\$31,870 AGI

(c) Taxable income: Check employee business expense (EBE) allowed:

$0.02\,(AGI) = 0.02\,(31{,}870) = \$637.40$

$EBE = \$500$, which does not exceed \$637.40; therefore, $EBE = 0$.

Calculate itemized deductions:

$$
\begin{array}{rl}
\$\quad 0 & \text{medical and dental} \\
2200 & \text{state and local taxes} \\
400 & \text{charitable contributions} \\
1000 & \text{moving expense} \\
0 & \text{employee business expense} \\
\hline
\$3600 &
\end{array}
$$

which is less than the standard deduction of \$3800 for single taxpayers. Take the standard deduction.

(a) Taxable income = \$31,870 AGI

$$
\begin{array}{rl}
& -\ 3800 \text{ standard deduction} \\
& -\ 2450 \text{ personal exemption (one)} \\
\hline
& \$25{,}620 \text{ taxable income}
\end{array}
$$

(b) Tax owed: Compute two ways for illustration. Using incremental tax rates:

$$0.15(22{,}750) + 0.28(25{,}620 - 22{,}750) = \$4216.10.$$

Using the formula in Table 15.6,

$$3412.50 + 0.28(25{,}620 - 22{,}750) = \$4216.10.$$

This represents an average federal tax rate of

$$\frac{\$4216.10}{\$25{,}620} = 16.5\% \text{ on taxable income}$$

or

$$\frac{\$4216.00}{\$31{,}910} = 13.2\% \text{ on gross income.}$$

Note also that, to qualify, the moving expense must be work related, and the distance between the new job location and old residence must be at least 50 miles greater than the distance between the old job location and old residence. Since the total move in this example was 300 miles, it is assumed that the test is met.

15.C.3 Combined Incremental Tax Rate

Many business and personal economic decisions are made "at the margin," that is, based on the incremental tax rate that would apply to a small change in income when combined federal, state, and local taxes are considered. As noted above, federal taxable income excludes taxes paid to state and local government. The reverse is not true, however. For an increment of income, the applicable incremental tax can be computed using the incremental tax rates for the appropriate bracket:

$$\Delta \text{ tax} = \Delta \text{ income } (1 - \Delta \text{ state and local tax rate}) \, \Delta \text{ federal tax rate}$$
$$+ \, \Delta \text{ income } (\Delta \text{ state and local tax rate})$$

$$= \Delta \text{ income } [(1 - \Delta \text{ state and local tax rate}) \, \Delta \text{ federal tax rate} + \Delta \text{ state and local tax rate}]$$

from which

$$\Delta \text{ rate} = (1 - \Delta \text{ state and local tax rate}) \, \Delta \text{ federal tax rate} + \Delta \text{ state and local tax rate}. \tag{15.14}$$

Example 15-13.

An engineer is married, files jointly, and projects taxable income in the 28% federal bracket. An opportunity to earn $1000 arises. What incremental tax rate would apply if the incremental state tax rate is 5% and local taxes are 2%?

Solution: Simple addition would produce a total tax rate of 35%. Due to federal tax credits, the actual incremental rate will be less:

$$\Delta \text{ rate} = (1 - 0.07)(0.28) + 0.07$$

$$= 33.0\%.$$

15.C.4 Before-Tax and After-Tax Rate of Return

Since taxes are levied on investment income, the after-tax rate of return will always be less than the before-tax return. Depreciation serves to moderate the impact of taxes, however.

Example 15-14.

For a corporation, calculate the before- and after-tax rate of return on the following investment, excluding and then including depreciation (straight-line, $N = 5$, $S = 0$). Incremental tax rate is 35% and the firm is profitable.

Year	Cash Flow
0	$−10,000
1	1200
2	2400
3	3600
4	4800
5	6000

Solution: Calculate rate of return before taxes:

$$10{,}000 = 1200 \, (P/A, i^*, 5) + 1200 \, (P/G, i^*, 5).$$

Note that the sum of the two factors must be 8.333333. Interpolating from the tables in the appendix, i^* (before) = 18.1%.

Form the after-tax cash flow:

Year	Cash Flow	Tax	After-Tax Cash Flow
0	$−10,000	—	$−10,000
1	1200	−420	780
2	2400	−840	1560
3	3600	−1260	2340
4	4800	−1680	3120
5	6000	−2100	3900

Calculate rate of return after taxes:

$$\$10{,}000 = 780(P/A, i^*, 5) + 780\ (P/G, i^*, 5).$$

The sum of the two factors must now be 12.820513. Interpolating from the tables in the appendix, i^*(after) = 4.4%.

Including depreciation,

Year	Cash Flow	Depreciation	Taxable Income	Tax	After-Tax Cash Flow
0	$−10,000	—	—	—	$−10,000
1	1200	$2000	$−800	$280	1480
2	2400	2000	400	−140	2260
3	3600	2000	1600	−560	3040
4	4800	2000	2800	−980	3820
5	6000	2000	4000	−1400	4600

Calculate rate of return including depreciation and taxes:

$$10{,}000 = 1480\ (P/A, i^*, 5) + 780\ (P/G, i^*, 5).$$

Using the appendix with trial values of i^* and interpolating,

$$i^*(\text{after, w/deprec}) = 13.1\%.$$

Note that since the firm is profitable, the excess depreciation amount at the end of year 1 can be used to offset income elsewhere in the firm, creating a tax savings ($280). Conclusion: The rate of return is strongly influenced by both taxes and depreciation.

15.D INFLATION

15.D.1 Historical Inflation Rates

Cost Indices. The *Consumers Price Index (CPI)* follows the cost of a standard bundle of consumer goods over time, as shown in Figure 15.3. The relative importance of components of the index in December 1993 was housing (41.4%), transportation (17.0%), food (15.8%), medical care (7.1%), energy (7.0%), apparel and upkeep (5.9%), and other (5.8%). A shortcoming of the index (and all such indices) is immediately apparent. Not everyone purchases the same bundle of goods; therefore, the index may not apply to a particular individual. Another difficulty is that quality of goods changes over time, and this is not reflected in the index. Further, buying patterns change over time, while the package of commodities in the index stays constant over long periods. Nevertheless, it is a very useful indicator of price inflation and often serves as an index to adjust salary, wages, and pension benefits. Figure 15.3 indicates that inflation is quite variable on a year-to-year percentage basis, but has a rather smooth growth pattern overall.

Other indices are more appropriate for measuring price inflation in the construction industry. The *Construction Cost Index (CCI)* is provided in Table 15.7. It measures the cost of fixed amounts of common labor, structural steel, cement,

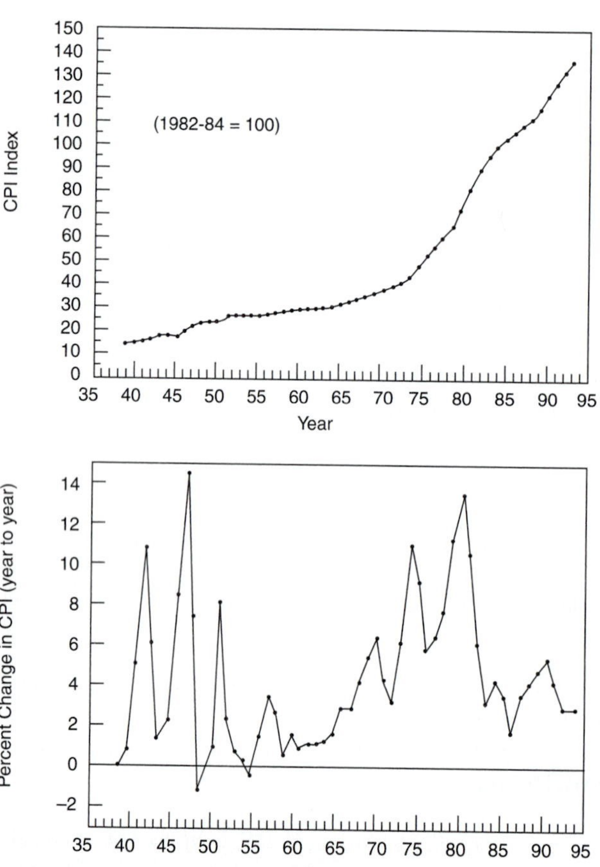

Figure 15.3 CPI index and percentage change year to year (Source: Joint Economic Committee, 1950–1994).

and lumber. Also shown in Table 15.7 is the *Building Cost Index (BCI)*, which substitutes skilled labor in place of common labor in the CCI (the materials list is the same in the BCI and CCI). The BCI is more appropriate for residential and commercial building construction. Both the CCI and BCI are available on a monthly basis, published in the *Engineering News Record.*

The U.S. Army Corps of Engineers maintains the *Civil Works Construction Cost Index (CWCCI)*, which is specific to water resource development projects. Subindices are available for specific project components such as reservoirs, dams, locks, power plants, roads, railroads and bridges, dredging, and so on. The composite index provided in Table 15.7 includes the effect of 20 project components and can be used for general project cost updates. The index is also available on a quarterly basis. Since the CWCCI is not available before 1967, the CCI can be used to update to 1967, after which the CWCCI can be used. For simplicity, the fiscal year index can be treated as a calendar year index in long-term cost updates.

Example 15-15.

A multiple-purpose reservoir project had a construction cost of $14,500,000 in 1954. What would the same project cost if built in 1993?

Solution: First use the CCI to update to 1967; then use the CWCCI.

$$C_{1967} = C_{1954}\frac{CCI_{1967}}{CCI_{1954}}$$

$$= 14{,}500{,}000\,\frac{1074}{628}$$

$$= 24{,}797{,}771.$$

$$C_{1993} = C_{1967}\frac{CWCCI_{1993}}{CWCCI_{1967}}$$

$$= 24{,}797{,}771\,\frac{428.49}{100.00}$$

$$= 100{,}626{,}000.$$

Note: Original cost is usually expressed at the midpoint of a multiyear construction period, in 1954 dollars in this case.

Other indices are also available. The U.S. Environmental Protection Agency (EPA) maintains indices for wastewater treatment plants and sanitary sewers, while the Bureau of Reclamation uses indices appropriate to water resource development projects of that agency.

Inflation Rate. Since cost indices measure the price of a standard package of goods, they reflect pure price inflation. The *inflation rate*, *f*, is the com-

TABLE 15.7 COMMON COST INDICES

Construction Cost Index (CCI) (calendar year average: 1913 = 100)

Year	CCI	Year	CCI	Year	CCI	Year	CCI	Year	CCI	Year	CCI
		1920	251	1935	196	1950	510	1965	971	1980	3237
1906	95	1921	202	1936	206	1951	543	1966	1019	1981	3535
1907	101	1922	174	1937	235	1952	569	1967	1074	1982	3825
1908	97	1923	214	1938	236	1953	600	1968	1155	1983	4066
1909	91	1924	215	1939	236	1954	628	1969	1269	1984	4146
1910	96	1925	207	1940	242	1955	660	1970	1381	1985	4195
1911	93	1926	208	1941	258	1956	692	1971	1581	1986	4295
1912	91	1927	206	1942	276	1957	724	1972	1753	1987	4406
1913	100	1928	207	1943	290	1958	759	1973	1895	1988	4519
1914	89	1929	207	1944	299	1959	797	1974	2020	1989	4615
1915	93	1930	203	1945	308	1960	824	1975	2212	1990	4732
1916	130	1931	181	1946	346	1961	847	1976	2401	1991	4835
1917	181	1932	157	1947	413	1962	872	1977	2576	1992	4985
1918	189	1933	170	1948	461	1963	901	1978	2776	1993	5210
1919	198	1934	198	1949	477	1964	936	1979	3003		

Building Cost Index (BCI) (calendar year average: 1913 = 100)

Year	BCI	Year	BCI	Year	BCI	Year	BCI	Year	BCI	Year	BCI
		1925	183	1940	203	1955	469	1970	836	1985	2428
		1926	185	1941	211	1956	491	1971	948	1986	2483
		1927	186	1942	222	1957	509	1972	1048	1987	2541
1913	100	1928	188	1943	229	1958	525	1973	1138	1988	2598
1914	92	1929	191	1944	235	1959	548	1974	1205	1989	2634
1915	95	1930	185	1945	239	1960	559	1975	1306	1990	2702
1916	131	1931	168	1946	262	1961	568	1976	1425	1991	2751
1917	167	1932	131	1947	313	1962	580	1977	1545	1992	2834
1918	159	1933	148	1948	341	1963	594	1978	1674	1993	2996
1919	159	1934	167	1949	352	1964	612	1979	1819		
1920	207	1935	166	1950	375	1965	627	1980	1941		
1921	166	1936	172	1951	401	1966	650	1981	2097		
1922	155	1937	196	1952	416	1967	676	1982	2234		
1923	186	1938	197	1953	431	1968	721	1983	2384		
1924	186	1939	197	1954	446	1969	790	1984	2417		

Civil Works Construction Cost Index (CWCCI): Composite Index
(fiscal year average: 1967 = 100; example: FY1988 = Oct. 1, 1987–Sept. 30, 1988)

Year	CWCCI	Year	CWCCI	Year	CWCCI
		1975	189.80	1985	354.31
		1976	203.43	1986	356.24
1967	100.00	1977	215.68	1987	361.43
1968	104.98	1978	234.58	1988	374.45
1969	112.09	1979	255.68	1989	388.68
1970	119.92	1980	280.71	1990	398.34
1971	132.17	1981	308.09	1991	406.78
1972	142.49	1982	329.87	1992	415.22
1973	149.16	1983	340.21	1993	428.49
1974	166.25	1984	349.63		

Source: CCI and BCI, *Engineering News Record*; CWCCI, U.S. Army Corps of Engineers EM1110-2-1304

pound rate of increase in price of a good or fixed package of goods and is usually expressed on an annual basis. In general,

$$F_a = P_a(1 + f)^n \tag{15.15}$$

where

F_a = actual future price of a good;

P_a = actual price in period zero;

f = inflation rate per period; and

n = number of periods.

Deflation is said to occur if f is negative.

Example 15-16.

What was the overall annual rate of inflation in consumer goods for the period 1970–1990?

Solution: $P_a = P_{1970} = CPI_{1970}$; $F_a = F_{1990} = CPI_{1990}$. From Figure 15.3, CPI_{1970} is about 39; CPI_{1990} is about 131.

$$131 = 39(1 + f)^{20}$$

$$f = 6.2\%.$$

15.D.2 Inflation-Adjusted (Real) Rate of Return

If future returns from an investment are expressed in future (actual, or current) dollars rather than today's (constant or real) dollars, they must be converted to the latter before the real (inflation-adjusted) rate of return can be found. The analysis is complicated by the presence of depreciation and taxes. Depreciation write-offs are in today's dollars, while taxes are based upon actual (current) taxable income.

Example 15-17.

Find the nominal and real rates of return for the investment shown, if the inflation rate is 5%. Depreciation is straight line over five years, with no salvage, and a marginal tax rate of 34% applies.

Year	Actual Cash Flow
0	$-1000
1	350
2	400
3	450
4	500
5	550

Solution: Account for depreciation and taxes:

Year	Actual Cash Flow	Deprec.	Taxable Income	Tax	Actual After-Tax Cash Flow
0	\$−1000	0	—	—	\$−1000
1	350	−200	150	−51	299
2	400	−200	200	−68	332
3	450	−200	250	−85	365
4	500	−200	300	−102	398
5	550	−200	350	−119	431

Convert actual after-tax cash flow to real dollars:

Year	Actual After-Tax Cash Flow	Multiply by	Real Cash Flow
0	\$−1000	$(1 + f)^0$	\$−1000
1	299	$(1 + f)^{-1}$	285
2	332	$(1 + f)^{-2}$	301
3	365	$(1 + f)^{-3}$	315
4	398	$(1 + f)^{-4}$	327
5	431	$(1 + f)^{-5}$	338

Find the nominal rate of return:

$$1000 = 299(1 + i^*)^{-1} + 332(1 + i^*)^{-2} + 365(1 + i^*)^{-3} + 398(1 + i^*)^{-4} + 431(1 + i^*)^{-5}.$$

By trial and error, $i^*_{\text{nominal}} = 22.36\%$.
Find the real rate of return:

$$1000 = 285(1 + i^*)^{-1} + 301(1 + i^*)^{-2} + 315(1 + i^*)^{-3} + 327(1 + i^*)^{-4} + 338(1 + i^*)^{-5}.$$

By trial and error, $i^*_{\text{real}} = 16.53\%$.

Note that real rate of return is not simply 5% less than the nominal rate.

Relationship Between Nominal and Real Rate of Return. Consider an investment, P, with a single actual payment, F_a, in period n. The inflation-adjusted payment is

$$F_0 = F_{\text{real}} = F_a(1 + f)^{-n}$$

and the real rate of return, i^*, is that rate that produces

$$P = \frac{F_a(1 + f)^{-n}}{(1 + i^*)^n}.$$

Which can be written as

$$P = \frac{F_a}{(1 + i^*)^n(1 + f)^n}$$

$$= \frac{F_a}{[(1 + i^*)(1 + f)]^n}$$

$$= \frac{F_a}{(1 + i^* + f + i^*f)^n}$$

$$= \frac{F_a}{(1 + u)^n}.$$

Where u is the nominal, or overall rate of return, given by

$$u = i^* + f + i^*f. \tag{15.16}$$

Example 15-18.

Use the overall rate of return found in Example 15-17 to calculate the real rate of return.

Solution: $u = 22.36\%$; $f = 0.05$; find i^*:

$$u = i^* + f + i^*f$$

$$i^* = \frac{u - f}{1 + f}$$

$$= \frac{0.2236 - 0.05}{1.05}$$

$$= 16.53\%$$

which was found by trial and error in Example 15-17.

Example 15-19.

A three-year Treasury note is purchased for \$10,000 and pays 6% interest (\$600) each year for three years. At the end of year 3, in addition to the interest, the \$10,000 original payment is returned. Inflation is expected to be 3, 4, and 5% in years 1, 2, and 3, respectively. Neglecting taxes, what is the nominal and real rate of return on the note?

Solution: Since inflation varies by year, create price indices to deflate future payments.

Year	Actual Cash Flow	Price Index	Real Cash Flow
0	$−10,000	$100.00	$−10,000.00
1	600	103.00	582.52
2	600	107.12	560.12
3	600	112.48	533.45
	10,000		8,890.79

Determine the nominal, overall rate of return using actual dollars:

$$10{,}000 = 600(1+u)^{-1} + 600(1+u)^{-2} + 600(1+u)^{-3} + 10{,}000(1+u)^{-3}$$

giving $u = 6.00\%$.

The real rate of return is

$$10{,}000 = 582.52(1+i^*)^{-1} + 560.12(1+i^*)^{-2} + 533.45(1+i^*)^{-3} + 8890.79(1+i^*)^{-3}.$$

An approximation can be found by using $u = 0.06$ and $f = 0.04$ (midpoint inflation) in

$$i^* = \frac{u-f}{1+f}$$

yielding 1.92%. Trial and error produces the true value for real rate of return:

$$i^* = 1.96\%.$$

15.E PERSONAL FINANCIAL PLANNING

Most of the engineering economics and mathematical modeling topics discussed in this book are directly applicable to personal financial planning. Illustration of some specific application areas follows. Also, it should be noted that there are many useful guides to personal financial planning in the popular literature (some are listed in the bibliography at the end of this chapter). The authors encourage all recent engineering graduates to become familiar with these guides and to develop personal financial plans early in their career. An example adapted from *The Wealthy Barber* (Chilton, 1991) explains why.

Example 15-20.

At age 22, one of two twins (twin A) starts investing $2000 per year in a retirement account earning 8% compounded annually tax free. He invests for only eight years and never saves a penny thereafter. The other twin (twin B) waits 10 years, then invests $2000 in a similar account every year for 33 years until age 65. The twins meet at age 65 and discuss retirement plans. What is the relative value of their accounts?

Solution:

$$\text{Twin A:} \quad F_{65} = A(P/A, 8\%, 8)(F/P, 8\%, 43)$$

$$= 2{,}000\ (5.746639)(27.36664)$$

$$= \$314{,}532.$$

$$\text{Twin B:} \quad F_{65} = A(F/A, 8\%, 33)$$

$$= 2000\ (145.95061)$$

$$= \$291{,}901.$$

Conclusion: Twin A is slightly better off, even though contributions were made for only eight years. More of twin A's final amount is composed of interest earnings than is the case for twin B, illustrating the advantage of investing early.

15.E.1 Mortgage Interest and Principal Payments

A *mortgage* is the commitment of an asset (such as a home) as security for a loan. Interest on a home mortgage is tax deductible if itemizing. The interest deduction applies to both the principal and second residence, if any. Loans used to acquire, construct, or improve a first or second home are called *home acquisition loans*, for which the interest on up to $1 million in debt is deductible. Loans secured by a home mortgage but used for any other purpose are called *home equity loans*, for which the interest on a loan amount up to $100,000 is deductible. The amount of the home equity loan subject to interest deductibility cannot exceed the fair market value of the home minus any outstanding loan principal. Further, a reduction in the interest deduction must be made if the adjusted gross income for a single taxpayer exceeds $111,800 (in 1994).

Example 15-21.

A single taxpayer with an adjusted gross income of $90,000 obtains a home equity loan to purchase certain consumer goods. The outstanding principal on his original mortgage is $40,000 and he obtains a $70,000 loan. Fair market value of the home is $100,000. What is the loan amount for which an interest deduction may be taken?

Solution: The taxpayer's AGI is less than $111,800, so no reduction in interest deduction is necessary. The loan amount subject to an interest deduction is fair market value minus any unpaid loan principal:

$$\$100{,}000 - \$40{,}000 = \$60{,}000.$$

This is less than the $100,000 allowable ceiling; therefore, interest on the entire $60,000 is tax deductible. Note, however, that interest on $10,000 of the $70,000 loan is not tax deductible.

A loan payment is composed of an interest payment and a payment that reduces the principal of the loan. These two amounts vary over time, as computed in Example 15-22.

Example 15-22.

As an investment, an engineer purchases a second residence for $130,000, of which $100,000 is the mortgage loan. A 30-year fixed-rate loan is obtained at 8% compounded monthly with monthly payments.

(a) What is the monthly mortgage payment?

(b) What is the division of the payment between interest and principal for the first nine months if a windfall payment of $50,000 is made in month 4 and another of $40,000 is made in month 7?

Solution:

(a)

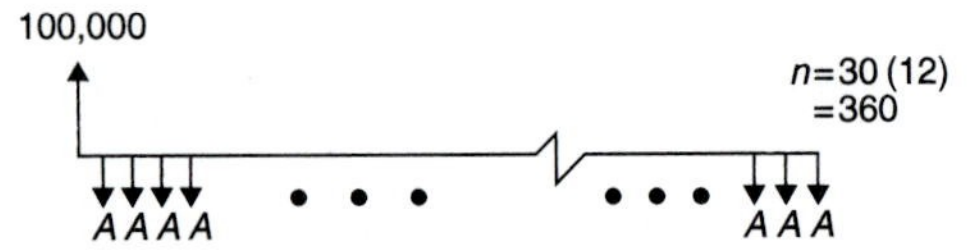

100,000

n=30 (12)
=360

A A A A • • • • • • A A A

$$i_s = \frac{r}{m} = \frac{0.08}{12} = 0.00666\overline{6};$$

$$A = P(A/P, i, n)$$

$$= 100{,}000\left(\frac{i(1+i)^n}{(1+i)^n - 1}\right)$$

$$= 100{,}000\left(\frac{0.00666\overline{6}\,(1+0.00666\overline{6})^{360}}{(1+0.00666\overline{6})^{360} - 1}\right)$$

$$= \$733.76 \text{ per month.}$$

(b) For the first month, interest is owed in the amount of

$$(0.00666\overline{6})\ (\$100{,}000) = \$666.67$$

but a payment of $733.76 is made. The difference, $67.09, goes toward reducing the principal to $99,932.91, on which interest is owed at the end of month 2 in the amount of

$$(0.00666\overline{6})\ (\$99{,}932.91) = \$666.22$$

and so on, as summarized in the following table.

End of Month	Payment	One Month Interest	Payment on Principal	Payment on Principal as Percentage of Payment	Principal Outstanding
0	—	—	—	—	100,000.00
1	733.76	666.67	67.09	9.14	99,932.91
2	733.76	666.22	67.54	9.20	99,865.37
3	733.76	665.77	67.99	9.27	99,797.38
4*	50,000.00	665.32	49,334.68	98.67	50,462.70
5	733.76	336.42	397.34	54.15	50,065.36
6	733.76	333.77	399.99	54.51	49,665.37
7*	40,000.00	331.10	39,668.90	99.17	9,996.47
8	733.76	66.64	667.12	90.92	9,329.35
9	733.76	62.20	671.56	91.52	8,657.79
⋮	⋮	⋮	⋮	⋮	⋮

*Windfall payments.

Note:

- Initially, monthly payments are almost entirely (over 90%) devoted to paying interest.
- When principal outstanding is cut in half to about $50,000, over half of the payment goes toward reducing the principal.
- When only about 10% of the principal is outstanding ($10,000), about 90% of the payment goes toward reducing principal.

For the loan amount in Example 15-22, under a normal repayment plan (no windfall payments), a total of about $7970 in interest would be paid in the first year, which is far above the standard deduction of $3800 for a single taxpayer or $6350 for a couple filing jointly. Also, property taxes are deductible and could well be $2000 per year on this house. These figures indicate the considerable tax benefits of home ownership. It should also be noted that lending institutions report the amount of mortgage interest to the borrower and to the Internal Revenue Service (IRS) each year (usually in January), for the previous tax year.

15.E.2 Home Finance

There are many rules-of-thumb used to determine how much an individual or family can afford to pay on a home mortgage, or what the purchase price of an affordable house might be. All such rules have shortcomings and should be used with care. The final decision depends on the personal outlook and habits of the prospective buyer and the financial assessment of prospective mortgage lenders. Some commonly used rules are

1. You can afford to buy a home that costs 2.5 times your annual income (Morris and Siegel, 1992).
2. Monthly "home costs" should not exceed one to one-and-a-half week's take-home pay (Bamford et al., 1992). Home costs are taken to mean mortgage, insurance, and taxes.
3. Real estate experts suggest that a middle-income family can afford to pay 35% of its gross annual income for housing—including utilities, heating, insurance, taxes, maintenance, repairs, and mortgage payments (Bamford et al., 1992).
4. Lenders often use guidelines established by the Federal National Mortgage Association that say the monthly debt obligation on a house—mortgage payment, property taxes, and insurance—should not exceed 28% of monthly gross income, and total housing and other debts should not exceed 36% of monthly gross income, all based on a 10% initial down payment.

Example 15-23.

An engineer has a gross annual income of $62,000, of which $43,800 is take-home. Her car and other consumer purchases are from savings (zero debt). Aside from mortgage payments, other annual housing costs are estimated as heating (gas), $750; electricity, $1000; water, $460; phone, $800; insurance, $340; maintenance and repairs, $600; and property tax, $2300. Estimate the maximum price of an affordable home for this individual if the mortgage interest rate is 9% compounded monthly over 30 years, assuming a 10% down payment.

Solution:

Rule 1: Simply 2.5(62,000) = $155,000.

Rule 2: Home costs could be up to $(1.5)\left(\frac{\$43{,}800}{52}\right) = \$1263/\text{month}.$

Subtracting nonmortgage home costs,

$$1263 - \frac{340 + 2300}{12} = 1263 - 220 = \$1043/\text{month for mortgage only.}$$

At 9% over 30 years, compounded monthly,

$$\begin{aligned} P &= A\ (P/A, 0.0075, 360) \\ &= 1{,}043\ (124.28187) \\ &= \$129{,}626. \end{aligned}$$

Considering the down payment,

$$(x - 0.1x) = 129{,}626$$

$$x = \$144{,}029.$$

Rule 3: Using 35% of gross annual income;

$$\frac{0.35(62{,}000)}{12} = \$1808/\text{month}.$$

Subtract nonmortgage payments

$$1808 - \frac{340 + 2300 + 750 + 1000 + 460 + 800 + 600}{12} = \$1287/\text{month mortgage}.$$

At 9% interest as before,

$$P = 1287(124.28187)$$

$$= \$159{,}951.$$

Again, considering down payment,

$$(x - 0.1x) = 159{,}951$$

$$x = \$177{,}723.$$

Rule 4: Using 28% of monthly gross income,

$$(0.28)\left(\frac{62{,}000}{12}\right) = \$1447/\text{month}.$$

Subtracting nonmortgage debt obligations,

$$1447 - \frac{2300 + 340}{12} = \$1227/\text{month mortgage}.$$

As before,

$$P = 1227(124.28187) = \$152{,}494$$

with down payment, $x = \$169{,}438$.
There are no other debts, so the 36% limit does not apply:

$$x = \$169{,}438.$$

The estimates have a range of over $30,000, but still serve to provide some guidance.

15.E.3 Retirement Accounts

General. In retirement, three sources of income may be available: employer-sponsored pensions, Social Security payments, and private savings. If these are not adequate, continued work—either on a full-time or part-time basis—is an option. Private savings are becoming an increasingly important part of retirement income. However, it is important to distinguish between *retirement* savings and *normal* savings. Normal savings are generally used up and replenished many times during a working life. Savings for cars, college, vacation trips, and even a home are rather short term when compared to saving for retirement. To be effective, retirement savings should be set aside in accounts that are difficult (carry penalties) to access. Otherwise, it is likely that such savings will be spent before they are needed in retirement. An employee must be on constant alert to find the best financial instrument for private retirement savings. Often these are associated with employer-sponsored pension plans.

A *pension* is a lump sum or series of payments made by an employer to a former employee who has met certain employment conditions, usually service for a period of years. An *annuity* is a series of payments resulting from an investment governed by a contract. Annuities can be purchased by individuals from life insurance companies or an employer may offer an annuity contract as an alternative form for payment of a pension. An *individual retirement account (IRA)* is an account established by an individual with any bank, brokerage firm, insurance company, or credit union approved by the IRS to serve as custodian. The characteristic feature of pension plans and IRAs is that they are designed for long-term savings for retirement. Generally, withdrawals before age $59\frac{1}{2}$ are subject to a 10% penalty, payable to the IRS. Interest earned in a pension or IRA account accumulates tax free until withdrawn. However, withdrawals must begin by April 1 following the year in which the account holder turns $70\frac{1}{2}$, and after that date a certain minimum annual withdrawal is required by law.

Employer Pension Plans. There are many types of employer pension plans. So-called *company qualified plans* are monitored by the federal government. The plans can be of the *defined benefit* or, more commonly, *defined contribution* type. Defined benefit plans are funded entirely by the employer using actuarial data on expected lifetimes of the employee group and the level of future benefits desired. Defined contribution plans establish a separate account for each employee, and some plans permit employee contributions. Types of defined contribution plans offered include *subsidized thrift, stock ownership, profit sharing*, and *deferred compensation*—401(*k*) for private sector; 403(b) for charitable, educational, or nonprofit organizations (Bamford et al., 1992). Deferred compensation plans offer the additional benefit of tax deferral of employee contributions (the contribution is not counted as employee income for the year). Since benefits are not defined, their final level depends on how well the investment (in stocks, bonds, mutual funds, certificates of deposit, etc.) performs.

If employees contribute to a pension plan they may withdraw their contributions at any time, including at termination of employment. However, unless the employee is *vested*, he or she cannot withdraw any contributions made by the company nor any interest earned in the account (including interest on the employee's own contributions). The Tax Reform Act of 1986 requires that, at the employer's choice, employees either be 100% vested after five years or 20% vested after three years and incrementally graduated to full vesting after seven years. If any distributions are made by the company, they should be immediately *rolled over* by the employee into another retirement account to avoid payment of taxes and early withdrawal penalty. Vesting is an important consideration when deciding whether and when to leave one employment to accept another.

A *simplified employee pension plan (SEP)*, if established by an employer, may allow the employee to contribute tax-deferred income to the pension account in addition to employer contributions. Also, if a person is self-employed, or earns self-employment income in addition to his place of employment, he or she may set up a *Keogh plan* and make tax-deferred contributions to the account, within certain limits.

Last, anyone who has *earned income* can set up an IRA. Whether contributions are fully, partially, or not tax deferred at all depends on filing status, income, and pension coverage at place of employment. A single taxpayer may contribute up to $2000 to an IRA. If married and both spouses work, each may contribute $2000. If only one spouse works and a joint return is filed, the limit for IRA contributions is $2250. If neither spouse actively participates in a pension plan at their place of employment, then contributions to an IRA are fully tax deductible. If either participates in a pension plan, deductions are gradually phased out with rising income, although nondeductible contributions, along with deductible contributions, may be made up to the maximum limit allowed. Drawbacks of an IRA are (1) the relatively small amounts that can be contributed (as opposed to employer-based plans) and (2) the fact that in most cases contributions are not fully tax deferred.

To summarize, certain key advantages accrue to pension plans: (1) interest and/or dividends accumulate tax deferred in all of the plans mentioned, (2) in some plans employee contributions may be tax deferred in the year of contribution, and (3) employers usually make deposits on behalf of employees and/or match employee contributions. The effect of the latter practice is the same as the second; deferral of income taxes on the initial deposit, since employer contributions are not reported as income to the employee. Example 15-24 demonstrates the advantages of tax-deferred contributions and accumulations.

Example 15-24.

With annual contributions of $2000, calculate the amount accumulated after 30 years in a retirement savings account at 8% interest with

(a) tax on contributions and accumulations (incremental tax rate of 28%);

(b) tax on contributions but not accumulations;

(c) tax-deferred contributions and accumulations.

What are the final amounts normalized in percentage terms?

Solution:

(a) Each year $(1 - 0.28)(2000) = \$1440$ is contributed, and the after-tax rate of return is $(1 - 0.28)(0.08) = 5.76\%$:

$$F = \$1440\ (F/A, 5.76\%, 30)$$

$$= 1440\ \frac{(1 + 0.0576)^{30} - 1}{0.0576} = \$109{,}148.$$

(b) Each year \$1440 is contributed, and the rate of return is 8%:

$$F = 1440\ (F/A, 8\%, 30)$$

$$= 1440\ \frac{(1 + 0.08)^{30} - 1}{0.08} = \$163{,}128.$$

(c) Each year \$2000 is contributed, and the rate of return is 8%:

$$F = 2000\ (F/A, 8\%, 30)$$

$$= 2000\ \frac{(1 + 0.08)^{30} - 1}{0.08} = \$226{,}566.$$

(d) Normalized amounts are

(a) 48.2%, (b) 72.0%, and (c) 100.0%.

The benefits of tax deferral are obviously very strong. Note that pension payments, after retirement, would be in the same proportion as the above percentages, assuming a uniform series of payments over a fixed number of years.

Pension Amount. Given a pension fund of initial amount, P, earning an interest rate, i, compounded annually, how long will the fund last if equal annual withdrawals of w% of P are made starting at the end of year 1?

Obviously, w must be larger than i, or the fund would last forever. The calculation can be made as follows:

End of Year	Amount Left in Pension Fund, Assuming $i = 5\%$, $w = 10\%$	Calculation
0	$P = \$1$	
1	0.9500	$P(1 + i) - wP$
2	0.8975	$[P(1 + i) - wP](1 + i) - wP$ $= P(1 + i)^2 - wP(1 + i) - wP$ $= P(1 + i)^2 - wP[(1 + i) + 1]$
3	0.8424	$[P(1 + i)^2 - wP[(1 + i) + 1]](1 + i) - wP$ $= P(1 + i)^3 - wP[(1 + i)^2 + (1 + i)] - wP$ $= P(1 + i)^3 - wP[(1 + i)^2 + (1 + i) + 1]$
⋮		
n	—	$P(1 + i)^n - wP[(1 + i)^{n-1} + (1 + i)^{n-2} + \cdots + (1 + i) + 1]$
⋮		
13	0.1144	$P(1 + i)^{13} - wP[(1 + i)^{12} + (1 + i)^{11} + \cdots + (1 + i) + 1]$
14*	0.0201	$P(1 + i)^{14} - wP[(1 + i)^{13} + (1 + i)^{12} + \cdots + (1 + i) + 1]$

*Last year in which 10% of P can be withdrawn.

TABLE 15.8 NUMBER OF FULL YEARS A PENSION FUND WILL LAST

Annual Withdrawal, w (%) of P	Annual Effective Interest, i, on Fund (%)							
	3	4	5	6	7	8	9	10
15	7	7	8	8	9	9	10	11
14	8	8	9	9	10	11	11	13
13	8	9	9	10	11	12	13	15
12	9	10	11	11	12	14	16	18
11	10	11	12	13	14	16	19	25
10	12	13	14	15	17	20	26	∞
9	13	14	16	18	22	28	∞	∞
8	15	17	20	23	30	∞	∞	∞
7	18	21	25	33	∞	∞	∞	∞
6	23	28	36	∞	∞	∞	∞	∞

As shown, the general equation for the amount remaining in the pension fund after a withdrawal at the end of period n is

$$P_n = P(1 + i)^n - wP[(1 + i)^{n-1} + (1 + i)^{n-2} + \cdots + (1 + i)+1]. \qquad (15.17)$$

The sum within brackets can be found by adding one to the sum of the first $(n - 1)$ future worth terms in the appendix. The pension fund will last slightly more than 14 years if $i = 5$ and $w = 10\%$. For convenience, Table 15.8 is developed using equation 15.17.

Example 15-25.

Charlotte and Dick are both 60 years old and married. Their joint life expectancy is 26 years, and they have \$50,000 in an IRA earning 6%. If they want to deplete the account in 26 years, how much must they withdraw each year?

Solution:

From Table 15.8, with $i = 6\%$, they can withdraw between 7 and 8% of the initial amount every year. By interpolation,

$$8\% - \left(\frac{26 - 23}{33 - 23}\right)(1\%) = 7.7\%$$

or

$$0.077(\$50{,}000) = \$3850 \text{ yearly}.$$

15.E.4 Risk-Return Trade-off

Table 15.9 shows the trade-off of investment risk and return for some common investment options. Risk of loss of investment principal generally decreases downward in the table, with progressively safer and safer investments. On the other hand, percentage return on an investment also decreases downward.

TABLE 15.9 TRADE-OFF OF INVESTMENT RISK AND RETURN

Type of Investment	Investment Option
Speculative (high risk, high return)	• State lotteries • Futures trading • Options trading • Collectibles • Precious metals • Junk bonds
Growth	• Foreign stocks, bonds • Small-capitalization stocks • Home purchase • Commercial real estate • Medium-capitalization stocks • Blue-chip stocks • Equity mutual funds
Income	• Zero coupon bonds • Municipal bonds • Corporate bonds • Agency bonds • Bond mutual funds • U.S. Treasury bonds • U.S. Treasury notes
Stability (low risk, low return)	• Cash value life insurance • Fixed annuities • U.S. savings bonds • U.S. Treasury bills • Insured bank savings accounts • Cash

Investments giving low returns with high reliability causes another type of risk: that earnings will not keep up with inflation. For example, insured bank passport savings accounts in 1994 commonly provided a return about equal to inflation (near 3%). If an investment does not grow considerably faster than inflation, it will not achieve such long-range goals as providing adequate retirement income.

Portfolio Allocation. Every investor must decide on the level of risk that can be tolerated in his or her portfolio of investments. Generally the younger the investor, the greater risk that can be tolerated. This is primarily due to the dilution of risk over time. For example, over a one-year period, a single stock may experience considerable growth or decline due to such short-term factors as the weather, poor management, economic recession or recovery, military conflict, and so on. But generally, over the long run, if the company is basically sound, one would expect the price of the stock to appreciate at a level about equal to the general stock market. Therefore, if an investor is young, more of the portfolio can be

TABLE 15.10 HISTORICAL STOCK PRICES (S&P 500 STOCK INDEX, AVERAGE OF DAILY PRICES)

Year	Price Index	Year	Price Index	Year	Price Index
1959	57.38				
1960	55.85	1975	86.16	1990	334.59
1961	66.27	1976	102.01	1991	376.18
1962	62.38	1977	98.20	1992	415.74
1963	69.87	1978	96.02	1993	451.41
1964	81.37	1979	103.01		
1965	88.17	1980	118.78		
1966	85.26	1981	128.05		
1967	91.93	1982	119.71		
1968	98.70	1983	160.41		
1969	97.84	1984	160.46		
1970	83.22	1985	186.84		
1971	98.29	1986	236.34		
1972	109.20	1987	286.83		
1973	107.43	1988	265.79		
1974	82.85	1989	322.84		

Source: U.S. Council of Economic Advisers.

devoted to growth-type investments. The closer the investor gets to calling upon (liquidating) the investment portfolio—such as at retirement—the less risk that can be tolerated. In retirement itself, maximum emphasis must be placed upon stability and income, while still leaving a minor portion of the portfolio dedicated to growth (especially in the early years of retirement).

15.E.5 Stocks

Table 15.10 lists historical stock prices as measured by the Standard & Poor's 500 Stock Index.

Example 15-26.

On July 1, 1991 a person retires. Thirty years previously, this person purchased $10,000 in an S&P 500 Stock Index fund. Neglecting management charges and dividend returns, (a) how much should be in the retirement account? and (b) what compound annual interest does this represent?

Solution:

(a) Use the ratio of indices for 1961 and 1991 (both represent stock price at the midpoint of the year):

$$10{,}000\left(\frac{376.18}{66.27}\right) = \$56{,}765.$$

(b)

$$F = P(F/P, i, 30)$$

$$376.18 = 66.27\ (F/P, i, 30)$$

$$5.6765 = (F/P, i, 30).$$

By interpolation in the appendix,

$$i = 5.95\%.$$

Note: Total return, with reinvestment of dividends, would be considerably more than this. For all common stocks on the New York Stock Exchange, the total annual return for the 1961–1991 period was 12.44% (Engel and Hecht, 1994).

15.E.6 Bonds

There are basically three types of bonds: U.S. government, municipal, and corporate. A bond is nothing more than a certificate that promises to pay the owner (or bearer) a certain rate of interest (usually in equal semiannual payments) over a period of time and then a final lump-sum amount on the bond's maturity date. The lump-sum payment is for the *face*, or *par, value* of the bond, usually some multiple of $1000. For new bond issues, the face value is usually the purchase price (an exception is the zero coupon bond, as explained later).

U.S. government bonds are by far the most risk free in terms of default (loss of initial capital investment). The full faith and credit of the U.S. Treasury stands behind each certificate. These bonds also have the advantage of being exempt from state and local income tax (although federal income tax must still be paid) and of being generally *noncallable*; that is, they cannot be paid off before their maturity date (or very close to it). The bearer, therefore, does not have to be concerned with finding another investment sometime in the near future.

There are several varieties of U.S. bonds: *Treasury bills* have maturities of 3, 6, or 12 months; *Treasury notes* have maturities of 2 to 10 years; and *Treasury bonds* have maturities from 10 to 30 years. Government agencies also issue bonds from time to time. All these instruments are traded in financial markets. Two other government bonds are exclusively bought from, and sold back to, the U.S. government at fixed prices: Series EE and HH savings bonds.

Table 15.11 lists historical annual interest rates associated with government and other types of bonds, as well as mortgages and loans. Higher interest is usually associated with a higher risk of default and with longer maturities.

Also evident is that interest rates have fluctuated significantly over time. This is another source of risk, especially when purchasing intermediate to long-term notes and bonds (say, 5 to 30 years). If, for some reason, a bond holder has to sell a bond before it matures, and prevailing interest rates have risen since the bond was purchased, the bond will be *discounted*, that is, sold for less than its face value.

TABLE 15.11 HISTORICAL INTEREST RATES

Year	3 Month U.S. Treasury Bills	3–5 Year U.S. Treasury Notes/Bonds	10 Year U.S. Treasury Bonds	High-Grade Municipal Bonds	AAA Corporate Bonds	Bank Prime Rate	New-Home Mortgage Rate
1959	3.41	4.33	4.08	3.95	4.38	—	5.77
1960	2.93	3.99	4.02	3.73	4.41	—	6.16
1961	2.38	3.60	3.90	3.46	4.35	—	5.78
1962	2.78	3.57	3.95	3.18	4.33	—	5.60
1963	3.16	3.72	4.00	3.23	4.26	—	5.47
1964	3.55	4.06	4.15	3.22	4.40	—	5.45
1965	3.95	4.22	4.21	3.27	4.49	—	5.46
1966	4.88	5.16	4.65	3.82	5.13	—	6.29
1967	4.32	5.07	4.85	3.98	5.51	—	6.55
1968	5.34	5.59	5.26	4.51	6.18	—	7.13
1969	6.68	6.85	6.12	5.81	7.03	—	8.19
1970	6.46	7.37	6.58	6.51	8.04	—	9.05
1971	4.35	5.77	5.74	5.70	7.39	—	7.78
1972	4.07	5.85	5.63	5.27	7.21	—	7.53
1973	7.04	6.92	6.30	5.18	7.44	8.03	7.95
1974	7.89	7.82	7.56	6.09	8.57	10.81	8.92
1975	5.84	7.49	7.99	6.89	8.83	7.86	9.01
1976	4.99	6.77	7.61	6.49	8.43	6.84	8.99
1977	5.27	6.69	7.42	5.56	8.02	6.83	9.01
1978	7.22	8.29	8.41	5.90	8.73	9.06	9.54
1979	10.04	9.71	9.44	6.39	9.63	12.67	10.77
1980	11.51	11.55	11.46	8.51	11.94	15.27	12.65
1981	14.03	14.44	13.91	11.23	14.17	18.87	14.70
1982	10.69	12.92	13.00	11.57	13.79	14.86	15.14
1983	8.63	10.45	11.10	9.47	12.04	10.79	12.57
1984	9.58	11.89	12.44	10.15	12.71	12.04	12.38
1985	7.48	9.64	10.62	8.18	11.37	9.93	11.55
1986	5.98	7.06	7.68	7.38	9.02	8.33	10.17
1987	5.82	7.68	8.39	7.73	9.38	8.21	9.31
1988	6.69	8.26	8.85	7.76	9.71	9.32	9.19
1989	8.12	8.55	8.49	7.24	9.26	10.87	10.13
1990	7.51	8.26	8.55	7.25	9.32	10.01	10.05
1991	5.42	6.82	7.86	6.89	8.77	8.46	9.32
1992	3.45	5.30	7.01	6.41	8.14	6.25	8.24
1993	3.02	4.44	5.87	5.63	7.22	6.00	7.20

Source: U.S. Council of Economic Advisers.

Example 15-27.

A 20-year, 6% government bond with face value \$10,000 has 9 years left to maturity. It is sold on the open market when comparable bonds and other investments are bringing 12%.

(a) What is a fair estimate of the discounted price?

(b) For the 11 years that the bearer held the bond, what is the rate of return, given the estimate of sale price in (a)?

Solution:

(a) A new purchaser will be willing to pay a price that will make the rate of return over the remaining 9 years equal to 12%. For simplicity, assume the semiannual interest payments are received at the end of a year and can be reinvested at 12%.

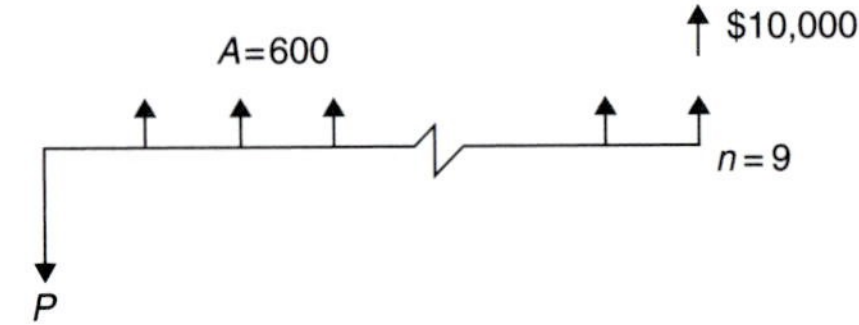

$$
\begin{aligned}
P &= \$600\,(P/A, 12\%, 9) + \$10{,}000\,(P/F, 12\%, 9)\\
&= 600\,(5.328250) + 10{,}000\,(0.360610)\\
&= \$6803.
\end{aligned}
$$

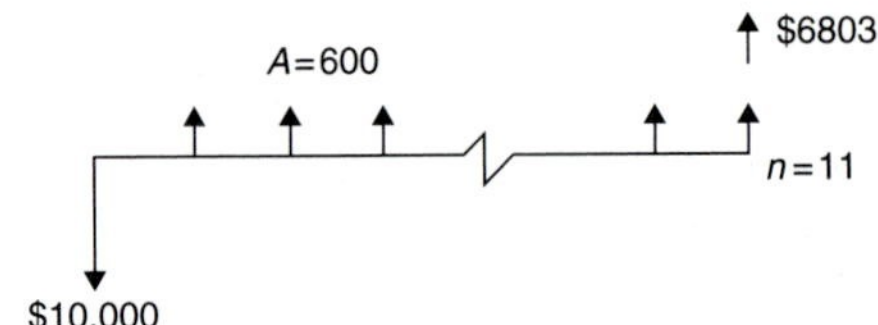

(b) $$\$10{,}000 = 600(P/A, i^*, 11) + 6803\,(P/F, i^*, 11).$$

By trial and error and interpolation, $i^* = 3.6\%$.

Conclusion: Significant interest rate loss will occur if a bond holder is forced to sell at a time when interest rates have risen significantly. Conversely, selling a high-yield bond during a period of low rates will result in a gain.

Municipal bonds, or *municipals*, are bonds issued by state agencies, cities, counties, school districts, water or sanitation districts, and other units of local government. The bonds can be used for the construction of roads, sewers, water systems, government housing, airports, schools, and so on. Interest on municipals is generally exempt from federal income tax, and is sometimes exempt from state and local income tax, especially in the state of issuance. This feature makes municipals particularly attractive to individuals in higher income tax brackets. For example, a "triple-exempt" bond offering a 6% return to a person in a combined incremental tax bracket of 36% would be equivalent to 9.38% taxable: $(1 - 0.36)(9.38) = 6.0$.

General obligation municipal bonds are backed by the taxing power of the issuing local government. *Revenue bonds* are backed only by the income of the project in question; through water and/or sewer charges, for example. As such, revenue bonds carry a risk premium (higher interest rate yield) as opposed to general obligation bonds. Most municipal bonds are rated by investment firms such as Moody's or Standard & Poor's, and such ratings should be seriously considered by potential investors.

The price of municipals is often quoted in terms of yield, maturity, and coupon interest rate. The investment firm in charge of issuing the municipal must then compute the actual sale price, as in Example 15-28.

Example 15-28.

A municipal bond has a 5.4% coupon rate, seven-year maturity, and a yield to maturity of 5.8%. It is sold in units of $1000 face value. What is the sale price?

Solution: Annual interest payments are $54 and face value is received at maturity:

$$P = \$54(P/A, 5.8\%, 7) + 1000(P/F, 5.8\%, 7)$$

$$= 54\left(\frac{(1.058)^7 - 1}{0.058(1.058)^7}\right) + 1000(1.058)^{-7}$$

$$= \$977.51.$$

Some U.S. Treasury bonds and municipal bonds are issued as *zero coupon* bonds. This means that no semiannual interest payments are paid. Rather, the purchase price is heavily discounted from face value to provide the desired yield to maturity. The resulting small initial investment is attractive to individual investors, especially at intermediate maturities for college expenses. The Series EE savings bond is a zero coupon bond.

Example 15-29.

A zero coupon 15-year bond has a 10% yield to maturity. If the face value is $10,000, what is the purchase price?

Solution: Draw the cash flow diagram:

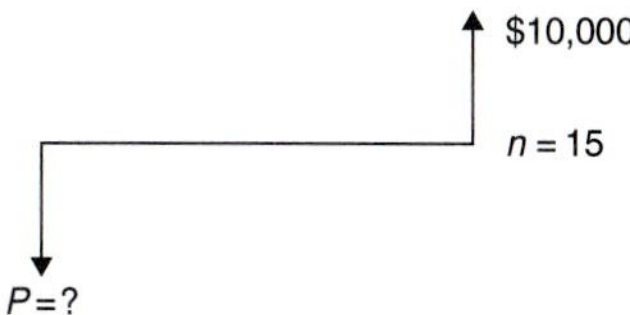

$$P = \$10{,}000\,(P/F, 10\%, 15)$$
$$= 10{,}000(0.239392)$$
$$= \$2393.92.$$

Corporate bonds are also rated by investment service companies and the ratings should be of great value to individual investors. Bonds with the lowest ratings (in the C range) have been dubbed "junk bonds" and offer very high interest rate yields. As seen in Table 15-11, even the highest-rated corporate bonds generally have yields above U.S. Treasury bonds or high-grade municipals. However, interest is not generally tax deductible on corporate bonds.

CHAPTER SUMMARY

This chapter places engineering economics in the context of corporate and personal financial planning. Depreciation, taxes, and inflation significantly affect the standard engineering economics calculations discussed in Chapters 13 and 14. In addition, the IRS code must be considered in any corporate or personal financial plan. Since this code is complex and undergoes continuous modification, the topics and examples in this chapter should be considered only as a starting point for more detailed investigation, planning, and analysis.

EXERCISES

15.1. An asset with initial cost of \$20,000 is to be depreciated over eight years using straight-line depreciation. Salvage value is estimated at \$2000. What is the annual depreciation charge and schedule of book value?

15.2. Compute the SOYD depreciation and schedule of book value for Exercise 15.1.

15.3. Compute the DDB depreciation and schedule of book value for Exercise 15.1.

15.4. Compute the DDB depreciation and schedule of book value if the initial cost of an asset is \$100,000, depreciation period is six years, and salvage is zero.

15.5. A barge company on the Ohio River purchases new barges and tugs for \$1,000,000 on August 28. Find the allowable depreciation amounts for the current and future years.

15.6. Derive the MACRS depreciation schedule for 15-year personal property (150% DB method), showing where to switch to SL depreciation. Assume a \$100 depreciable asset.

15.7. On July 7, 1990, an engineering company purchased a site for a branch office. Total cost was \$500,000, of which \$100,000 was for land. Determine the depreciation period and amounts allowed.

15.8. On October 8, 1994, an engineer purchases a two-apartment housing unit for rental income and as an investment. Total cost was \$300,000, of which \$75,000 was for land. Determine the depreciation period and amounts allowed.

15.9. An engineering company has published a software program and related instruction manual for wastewater treatment plant design. It expects to sell 1000 copies at $500 each before extensive revisions will be necessary. If development and publishing costs are $300,000 and 200 copies are sold the first year, what is the allowable depreciation?

15.10. A photocopier rated at 1,000,000 copies is purchased by an engineering company for $10,000. In the third year of use, the digital counter shows that 150,000 copies were made. What is allowable depreciation for the third year, assuming that the units of production depreciation method has been elected?

15.11. For $10,000,000, a major energy company has purchased oil extraction rights in a certain land tract. Originally, an estimate of 1,000,000 barrels total production was made, but in September 1995, this was raised by 200,000 barrels. By the beginning of 1995, 400,000 barrels had been extracted and $3,500,000 cost depletion had been taken. In 1995 an additional 100,000 barrels were removed. What is the allowable cost depletion for 1995?

15.12. For $800,000, a manufacturer purchased rights to extract clay and shale for making sewer pipe. In a certain year, gross income is $3,000,000 and taxable income is $300,000. What is the allowable depletion amount using percentage depletion?

15.13. For a certain year, an engineering company has a total income of $5,000,000, expenses of $4,000,000, capital gains of $100,000, capital losses of $200,000, and depreciation and depletion allowances of $50,000 and $30,000, respectively. What is the firm's taxable income and tax payment?

15.14. Susan is a self-employed environmental consultant and receives $40,000 in business income. Interest on several savings accounts totals $600, and stock dividends are $300. Sale of stock results in a $1000 capital gain, and refund of state and local taxes were $100. This year she contributed $4000 to her company's retirement plan, which is tax deferred. One-half of her Social Security tax payment is $3800, and she made payments of $2400 for self-employment health insurance premiums. Susan is single and makes mortgage interest payments of $7500, pays state and local taxes of $2500 and property taxes of $1500. Miscellaneous deductions total $900, and charitable contributions are $800. Medical and dental expenses do not exceed 7.5% of AGI. Determine Susan's (a) gross income, (b) adjusted gross income, (c) taxable income, and (d) tax owed.

15.15. Compute the combined incremental tax rate if a person is in the 31% federal tax bracket and state and local taxes total 8%.

15.16. A profitable corporation is in the 39% incremental tax bracket. Calculate the before- and after-tax rate of return on the following investment, excluding and then including depreciation (straight line, $N = 4$, $S = 0$):

Year	Cash Flow
0	$-20,000
1	3,000
2	6,000
3	9,000
4	12,000

15.17. The CPI was 72.6 in 1979 and 144.5 in 1993, an approximate doubling. Use the rule of 72 to estimate the annual compound rate of inflation for this period. Check by finding the exact value using the single-payment compound amount formula.

15.18. If the inflation rate is 6%, find the nominal and real rates of inflation for the investment shown. Depreciation is straight line over five years, with salvage equal to $200 at the end of year 5. A marginal tax rate of 39% applies. After the nominal rate of return is found, use the equational relationship between nominal and real rate of return to find the latter.

Year	Actual Cash Flow
0	$−1000
1	300
2	350
3	400
4	450
5	500
	200 (salvage value)

15.19. A five-year U.S. Treasury note is purchased for $10,000 and pays 7% interest ($700) each year for five years. At the end of year 5, in addition to the interest, the $10,000 original payment is returned. Inflation is expected to be 3% per year for the five year period. Neglecting taxes, what is the nominal and real rate of return on the note?

15.20. An engineer is single with income of $80,000 and holds a first mortgage on his home with $50,000 principal outstanding. Fair market value of the home is $140,000. A home equity loan of $100,000 is taken out to start a business. What is the loan amount for which an interest deduction can be taken?

15.21. A young engineer purchases a "starter home" on which a $50,000 mortgage is obtained at 9% interest compounded monthly. It is a 30-year, fixed-rate mortgage.

(a) What is the monthly mortgage payment?

(b) What is the division of the payment between interest and principal for the first four months if a windfall payment of $20,000 is made in month 3?

15.22. An individual takes out a car loan for $10,000 repayable in annual installments at 10% compound interest, with an option to pay off the loan early with no penalty charge.

(a) What is the payment required to pay off the loan at the end of year 3 after two annual installments have been made? Assume the payment must be equivalent to the last three normal annual payments at 10% interest.

(b) Develop a table showing the total payment, interest payment, principal payment, and principal outstanding for each year under the five-year payment plan and a similar table for the three-year payment plan.

(c) If the individual earns 5% interest in a savings account, and starts with $20,000 in the account, what will be in the account at the end of year 5 under the two payment plans? What is the savings for paying off the loan early, expressed in end of year 5 dollars?

(d) What would be the savings over the two payment plans at the end of year 5 if cash had been paid for the car initially?

15.23. A young engineer earns $32,000 gross income, of which $24,000 is take-home. A car payment of $210 per month is made and is expected to continue at this rate for at least five years. For average-sized two-bedroom homes in this geographic area, annual costs are estimated as follows: heating, $600; electricity, $800; water, $400; phone, $500; insurance, $280; maintenance and repairs, $300; and property tax $1800. Estimate the maximum affordable house for this individual if the mortgage interest rate is 8% compounded monthly over 30 years, assuming a 10% down payment.

15.24. At age 60, an individual who is now 25 years old wants to have accumulated a $200,000 retirement account. If the person expects to remain in the 28% incremental tax bracket, and the available pension funds are assumed to earn 9% interest, what amount would have to be contributed each year from this individual's annual income under the following pension plans:

(a) tax on contributions and accumulations;

(b) tax on contributions but not accumulations;

(c) tax-deferred contributions and accumulations.

What are the final amounts, normalized in percentage terms?

15.25. At age 60, male life expectancy is 18.2 years. At the same age, female life expectancy is 21.7 years. If married, the joint life expectancy of two 60-year-olds is 26.0 years. If a pension fund earns 7% and is to be depleted at the end of the life expectancy, what is the percentage reduction in annual pension amount for taking a "joint survivor" pension payout, as opposed to a "single life" payout if:

(a) the pension is held by the male;

(b) the pension is held by the female.

15.26. An investor plans to supplement retirement income with stock dividends and does not want the total dividend amount to vary greatly year to year. Initially, assume that there are only two stocks to choose from and that each costs $100 per share. The variance of dividend payments from stock 1 is $25 per share, while the variance for stock 2 is only $9. Dividend payments for the two stocks are uncorrelated. A total of $10,000 is to be invested. To minimize variance of total stock dividends,

(a) How much of each stock should be purchased? (Hint: Develop a small mathematical programming model and solve by trial and error.)

(b) Write a generalized (multistock) mathematical programming model. What technique would you use to solve this larger model?

15.27. For the decade of the 1980s, it has often been quipped that it was impossible to lose money in the stock market. What was the annual rate of price appreciation for the S&P 500 Stock Index from mid-1979 to mid-1989?

15.28. Use the stock index data in Table 15.10 to illustrate the observation that risk declines with longer-held investments. Risk in this case can be measured by the spread (variance) in annual rate of return in stock price appreciation. Form 1-, 5-, 10-, 15-, 20-, and 25-year pairs of stock prices, and plot annual effective rates of price appreciation versus time, with time as the x axis.

15.29. A $10,000, 30-year U.S. Treasury bond was purchased in 1980 at 11%, and placed on the market in 1993 when comparable instruments were bringing only 6%.

(a) What is a fair estimate of sale price?

(b) For the 13-year period that the bond was held, what was the rate of return to the bondholder, given the sale price found in (a)?

15.30. A municipal bond has a 7.1% coupon rate, 12-year maturity, and a yield to maturity of 7.5%. Par value is $5000. What is the sale price?

15.31. A zero coupon 10-year bond has a 6% yield to maturity. If the face value is $5000, what is the purchase price?

15.32. A zero coupon $10,000 bond has five years left to maturity and must be placed on the market by a bondholder when prevailing yields to maturity for 5-year bonds of similar quality are 5%. What is a fair estimate of the sale price of the bond?

REFERENCES

APOSTOLOU, NICHOLAS G. and D. LARRY CRUMBLEY, *Keys to Understanding the Financial News*. Hauppauge, NY: Barron's Educational Series, 1994, 140 pp.

BAMFORD, JANET et al., *The Consumer Reports Money Book: How to Get It, Save It, and Spend It Wisely*. Yonkers, NY: Consumer Reports Books, 1992, 584 pp.

BOROSON, WARREN, *Keys to Investing in Mutual Funds*. Hauppauge, NY: Barron's Educational Series, 1992, 158 pp.

CHILTON, DAVID, *The Wealthy Barber: Everyone's Common-Sense Guide to Becoming Financially Independent*. Rocklin, CA: Prima, 1991, 200 pp.

ENGEL, LOUIS and HENRY HECHT, *How to Buy Stocks*. Boston: Little, Brown, 1994, 398 pp.

H&R BLOCK TAX SERVICES, *H&R Block 1995 Income Tax Guide*. New York: Fireside, 1994, 512 pp.

J.K. LASSER INSTITUTE, *Your Income Tax 1995*. New York: Macmillan, 1994, 687 pp.

LYNCH, PETER, *Beating the Street*. New York: Fireside, 1994, 332 pp.

MALKIEL, BURTON G., *A Random Walk Down Wall Street*. New York: W.W. Norton, 1990, 440 pp.

MORRIS, KENNETH M. and ALAN M. SIEGEL, *The Wall Street Journal Guide to Understanding Personal Finance*. New York: Lightbulb Press, 1992, 173 pp.

MORRIS, KENNETH M. and ALAN M. SIEGEL, *The Wall Street Journal Guide to Understanding Money and Investing*. New York: Lightbulb Press, 1993, 155 pp.

SEASE, DOUGLAS and JOHN PRESTBO, *Barron's Guide to Making Investment Decisions*. Englewood Cliffs, NJ: Prentice Hall, 1994, 333 pp.

TOBIAS, ANDREW, *Still the Only Investment Guide You'll Ever Need*. New York: Bantam Books, 1987, 173 pp.

TOBIAS, ANDREW, *The Only Other Investment Guide You'll Ever Need*. Toronto: Bantam Books, 1989, 271 pp.

U.S. COUNCIL OF ECONOMIC ADVISERS, *Economic Indicators*. Washington, DC: U.S. Government Printing Office, various dates.

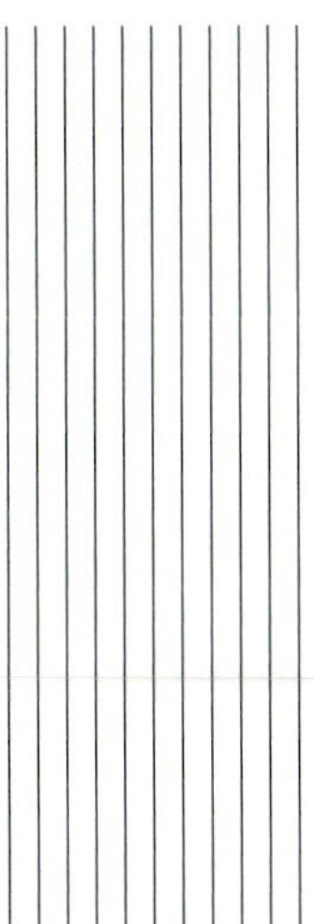

Appendix
Compound Interest Tables

0.25%	Single Payment		Uniform Series				Arithmetic Gradient Series	0.25%
	Compound Amount Factor (CAF)	Present Worth Factor (PWF)	Present Worth Factor (PWF)	Capital Recovery Factor (CRF)	Compound Amount Factor (CAF)	Sinking Fund Factor (SFF)	Present Worth Factor (PWF)	
n	*F/P*	*P/F*	*P/A*	*A/P*	*F/A*	*A/F*	*P/G*	*n*
1	1.002 500	0.997 506	0.997 506	1.002 500	1.000 000	1.000 000	0.000 000	1
2	1.005 006	0.995 019	1.992 525	0.501 876	2.002 500	0.499 376	0.995 019	2
3	1.007 519	0.992 537	2.985 062	0.335 001	3.007 506	0.332 501	2.980 093	3
4	1.010 038	0.990 062	3.975 124	0.251 564	4.015 025	0.249 064	5.950 280	4
5	1.012 563	0.987 593	4.962 718	0.201 502	5.025 063	0.199 002	9.900 653	5
6	1.015 094	0.985 130	5.947 848	0.168 128	6.037 625	0.165 628	14.826 305	6
7	1.017 632	0.982 674	6.930 522	0.144 289	7.052 719	0.141 789	20.722 347	7
8	1.020 176	0.980 223	7.910 745	0.126 410	8.070 351	0.123 910	27.583 909	8
9	1.022 726	0.977 779	8.888 524	0.112 505	9.090 527	0.110 005	35.406 138	9
10	1.025 283	0.975 340	9.863 864	0.101 380	10.113 253	0.098 880	44.184 201	10
11	1.027 846	0.972 908	10.836 772	0.092 278	11.138 536	0.089 778	53.913 282	11
12	1.030 416	0.970 482	11.807 254	0.084 694	12.166 383	0.082 194	64.588 583	12
13	1.032 992	0.968 062	12.775 316	0.078 276	13.196 799	0.075 776	76.205 323	13
14	1.035 574	0.965 648	13.740 963	0.072 775	14.229 791	0.070 275	88.758 742	14
15	1.038 163	0.963 239	14.704 203	0.068 008	15.265 365	0.065 508	102.244 095	15
16	1.040 759	0.960 837	15.665 040	0.063 836	16.303 529	0.061 336	116.656 656	16
17	1.043 361	0.958 441	16.623 481	0.060 156	17.344 287	0.057 656	131.991 716	17
18	1.045 969	0.956 051	17.579 533	0.056 884	18.387 648	0.054 384	148.244 586	18
19	1.048 584	0.953 667	18.533 200	0.053 957	19.433 617	0.051 457	165.410 592	19
20	1.051 206	0.951 289	19.484 488	0.051 323	20.482 201	0.048 823	183.485 079	20
21	1.053 834	0.948 916	20.433 405	0.048 939	21.533 407	0.046 439	202.463 409	21
22	1.056 468	0.946 550	21.379 955	0.046 773	22.587 240	0.044 273	222.340 961	22
23	1.059 109	0.944 190	22.324 145	0.044 795	23.643 708	0.042 295	243.113 133	23
24	1.061 757	0.941 835	23.265 980	0.042 981	24.702 818	0.040 481	264.775 340	24
25	1.064 411	0.939 486	24.205 466	0.041 313	25.764 575	0.038 813	287.323 012	25
26	1.067 072	0.937 143	25.142 609	0.039 773	26.828 986	0.037 273	310.751 599	26
27	1.069 740	0.934 806	26.077 416	0.038 347	27.896 059	0.035 847	335.056 566	27
28	1.072 414	0.932 475	27.009 891	0.037 023	28.965 799	0.034 523	360.233 399	28
29	1.075 096	0.930 150	27.940 041	0.035 791	30.038 213	0.033 291	386.277 596	29
30	1.077 783	0.927 830	28.867 871	0.034 641	31.113 309	0.032 141	413.184 675	30
31	1.080 478	0.925 517	29.793 388	0.033 564	32.191 092	0.031 064	440.950 171	31
32	1.083 179	0.923 209	30.716 596	0.032 556	33.271 570	0.030 056	469.569 635	32
33	1.085 887	0.920 906	31.637 503	0.031 608	34.354 749	0.029 108	499.038 635	33
34	1.088 602	0.918 610	32.556 112	0.030 716	35.440 636	0.028 216	529.352 756	34
35	1.091 323	0.916 319	33.472 431	0.029 875	36.529 237	0.027 375	560.507 599	35

(cont.)

(cont.)

0.25%	Single Payment		Uniform Series				Arithmetic Gradient Series	0.25%
	Compound Amount Factor (CAF)	Present Worth Factor (PWF)	Present Worth Factor (PWF)	Capital Recovery Factor (CRF)	Compound Amount Factor (CAF)	Sinking Fund Factor (SFF)	Present Worth Factor (PWF)	
n	*F/P*	*P/F*	*P/A*	*A/P*	*F/A*	*A/F*	*P/G*	*n*
36	1.094 051	0.914 034	34.386 465	0.029 081	37.620 560	0.026 581	592.498 783	36
37	1.096 787	0.911 754	35.298 220	0.028 330	38.714 612	0.025 830	625.321 943	37
38	1.099 528	0.909 481	36.207 700	0.027 618	39.811 398	0.025 118	658.972 731	38
39	1.102 277	0.907 213	37.114 913	0.026 943	40.910 927	0.024 443	693.446 814	39
40	1.105 033	0.904 950	38.019 863	0.026 302	42.013 204	0.023 802	728.739 878	40
41	1.107 796	0.902 694	38.922 557	0.025 692	43.118 237	0.023 192	764.847 622	41
42	1.110 565	0.900 443	39.822 999	0.025 111	44.226 033	0.022 611	801.765 765	42
43	1.113 341	0.898 197	40.721 196	0.024 557	45.336 598	0.022 057	839.490 039	43
44	1.116 125	0.895 957	41.617 154	0.024 029	46.449 939	0.021 529	878.016 195	44
45	1.118 915	0.893 723	42.510 876	0.023 523	47.566 064	0.021 023	917.339 999	45
46	1.121 712	0.891 494	43.402 370	0.023 040	48.684 979	0.020 540	957.457 232	46
47	1.124 517	0.889 271	44.291 641	0.022 578	49.806 692	0.020 078	998.363 693	47
48	1.127 328	0.887 053	45.178 695	0.022 134	50.931 208	0.019 634	1 040.055 196	48
49	1.130 146	0.884 841	46.063 536	0.021 709	52.058 536	0.019 209	1 082.527 572	49
50	1.132 972	0.882 635	46.946 170	0.021 301	53.188 683	0.018 801	1 125.776 666	50
60	1.161 617	0.860 869	55.652 358	0.017 969	64.646 713	0.015 469	1 600.084 536	60
70	1.190 986	0.839 640	64.143 853	0.015 590	76.394 437	0.013 090	2 147.611 092	70
72	1.196 948	0.835 458	65.816 858	0.015 194	78.779 387	0.012 694	2 265.556 855	72
80	1.221 098	0.818 935	72.425 952	0.013 807	88.439 181	0.011 307	2 764.456 812	80
84	1.233 355	0.810 797	75.681 321	0.013 213	93.341 920	0.010 713	3 029.759 228	84
90	1.251 971	0.798 740	80.503 816	0.012 422	100.788 454	0.009 922	3 446.869 973	90
96	1.270 868	0.786 863	85.254 603	0.011 730	108.347 387	0.009 230	3 886.283 161	96
100	1.283 625	0.779 044	88.382 483	0.011 314	113.449 955	0.008 814	4 191.241 729	100
108	1.309 523	0.763 637	94.545 300	0.010 577	123.809 259	0.008 077	4 829.012 470	108
120	1.349 354	0.741 096	103.561 753	0.009 656	139.741 419	0.007 156	5 852.111 603	120
240	1.820 755	0.549 223	180.310 914	0.005 546	328.301 998	0.003 046	19 398.985 224	240
360	2.456 842	0.407 027	237.189 382	0.004 216	582.736 885	0.001 716	36 263.929 943	360
480	3.315 149	0.301 646	279.341 764	0.003 580	926.059 501	0.001 080	53 820.752 482	480
∞	∞	0	400.000 000	0.002 500	∞	0	160 000.000 000	∞

0.50%	Single Payment		Uniform Series				Arithmetic Gradient Series	0.50%
	Compound Amount Factor (CAF)	Present Worth Factor (PWF)	Present Worth Factor (PWF)	Capital Recovery Factor (CRF)	Compound Amount Factor (CAF)	Sinking Fund Factor (SFF)	Present Worth Factor (PWF)	
n	*F/P*	*P/F*	*P/A*	*A/P*	*F/A*	*A/F*	*P/G*	*n*
1	1.005 000	0.995 025	0.995 025	1.005 000	1.000 000	1.000 000	0.000 000	1
2	1.010 025	0.990 075	1.985 099	0.503 753	2.005 000	0.498 753	0.990 075	2
3	1.015 075	0.985 149	2.970 248	0.336 672	3.015 025	0.331 672	2.960 372	3
4	1.020 151	0.980 248	3.950 496	0.253 133	4.030 100	0.248 133	5.901 115	4
5	1.025 251	0.975 371	4.925 866	0.203 010	5.050 251	0.198 010	9.802 597	5
6	1.030 378	0.970 518	5.896 384	0.169 595	6.075 502	0.164 595	14.655 188	6
7	1.035 529	0.965 690	6.862 074	0.145 729	7.105 879	0.140 729	20.449 325	7
8	1.040 707	0.960 885	7.822 959	0.127 829	8.141 409	0.122 829	27.175 522	8
9	1.045 911	0.956 105	8.779 064	0.113 907	9.182 116	0.108 907	34.824 359	9
10	1.051 140	0.951 348	9.730 412	0.102 771	10.228 026	0.097 771	43.386 491	10
11	1.056 396	0.946 615	10.677 027	0.093 659	11.279 167	0.088 659	52.852 639	11
12	1.061 678	0.941 905	11.618 932	0.086 066	12.335 562	0.081 066	63.213 598	12
13	1.066 986	0.937 219	12.556 151	0.079 642	13.397 240	0.074 642	74.460 229	13
14	1.072 321	0.932 556	13.488 708	0.074 136	14.464 226	0.069 136	86.583 463	14
15	1.077 683	0.927 917	14.416 625	0.069 364	15.536 548	0.064 364	99.574 299	15
16	1.083 071	0.923 300	15.339 925	0.065 189	16.614 230	0.060 189	113.423 805	16
17	1.088 487	0.918 707	16.258 632	0.061 506	17.697 301	0.056 506	128.123 114	17
18	1.093 929	0.914 136	17.172 768	0.058 232	18.785 788	0.053 232	143.663 429	18
19	1.099 399	0.909 588	18.082 356	0.055 303	19.879 717	0.050 303	160.036 017	19
20	1.104 896	0.905 063	18.987 419	0.052 666	20.979 115	0.047 666	177.232 212	20
21	1.110 420	0.900 560	19.887 979	0.050 282	22.084 011	0.045 282	195.243 414	21
22	1.115 972	0.896 080	20.784 059	0.048 114	23.194 431	0.043 114	214.061 088	22
23	1.121 552	0.891 622	21.675 681	0.046 135	24.310 403	0.041 135	233.676 763	23
24	1.127 160	0.887 186	22.562 866	0.044 321	25.431 955	0.039 321	254.082 034	24
25	1.132 796	0.882 772	23.445 638	0.042 652	26.559 115	0.037 652	275.268 557	25
26	1.138 460	0.878 380	24.324 018	0.041 112	27.691 911	0.036 112	297.228 055	26
27	1.144 152	0.874 010	25.198 028	0.039 686	28.830 370	0.034 686	319.952 311	27
28	1.149 873	0.869 662	26.067 689	0.038 362	29.974 522	0.033 362	343.433 173	28
29	1.155 622	0.865 335	26.933 024	0.037 129	31.124 395	0.032 129	367.662 550	29
30	1.161 400	0.861 030	27.794 054	0.035 979	32.280 017	0.030 979	392.632 412	30
31	1.167 207	0.856 746	28.650 800	0.034 903	33.441 417	0.029 903	418.334 792	31
32	1.173 043	0.852 484	29.503 284	0.033 895	34.608 624	0.028 895	444.761 783	32
33	1.178 908	0.848 242	30.351 526	0.032 947	35.781 667	0.027 947	471.905 539	33
34	1.184 803	0.844 022	31.195 548	0.032 056	36.960 575	0.027 056	499.758 273	34
35	1.190 727	0.839 823	32.035 371	0.031 215	38.145 378	0.026 215	528.312 260	35

(cont.)

(cont.)

0.50%	Single Payment		Uniform Series				Arithmetic Gradient Series	0.50%
	Compound Amount Factor (CAF)	Present Worth Factor (PWF)	Present Worth Factor (PWF)	Capital Recovery Factor (CRF)	Compound Amount Factor (CAF)	Sinking Fund Factor (SFF)	Present Worth Factor (PWF)	
n	*F/P*	*P/F*	*P/A*	*A/P*	*F/A*	*A/F*	*P/G*	*n*
36	1.196 681	0.835 645	32.871 016	0.030 422	39.336 105	0.025 422	557.559 832	36
37	1.202 664	0.831 487	33.702 504	0.029 671	40.532 785	0.024 671	587.493 382	37
38	1.208 677	0.827 351	34.529 854	0.028 960	41.735 449	0.023 960	618.105 359	38
39	1.214 721	0.823 235	35.353 089	0.028 286	42.944 127	0.023 286	649.388 272	39
40	1.220 794	0.819 139	36.172 228	0.027 646	44.158 847	0.022 646	681.334 687	40
41	1.226 898	0.815 064	36.987 291	0.027 036	45.379 642	0.022 036	713.937 229	41
42	1.233 033	0.811 009	37.798 300	0.026 456	46.606 540	0.021 456	747.188 578	42
43	1.239 198	0.806 974	38.605 274	0.025 903	47.839 572	0.020 903	781.081 470	43
44	1.245 394	0.802 959	39.408 232	0.025 375	49.078 770	0.020 375	815.608 700	44
45	1.251 621	0.798 964	40.207 196	0.024 871	50.324 164	0.019 871	850.763 117	45
46	1.257 879	0.794 989	41.002 185	0.024 389	51.575 785	0.019 389	886.537 625	46
47	1.264 168	0.791 034	41.793 219	0.023 927	52.833 664	0.018 927	922.925 185	47
48	1.270 489	0.787 098	42.580 318	0.023 485	54.097 832	0.018 485	959.918 810	48
49	1.276 842	0.783 182	43.363 500	0.023 061	55.368 321	0.018 061	997.511 570	49
50	1.283 226	0.779 286	44.142 786	0.022 654	56.645 163	0.017 654	1 035.696 587	50
60	1.348 850	0.741 372	51.725 561	0.019 333	69.770 031	0.014 333	1 448.645 795	60
70	1.417 831	0.705 303	58.939 418	0.016 967	83.566 105	0.011 967	1 913.642 742	70
72	1.432 044	0.698 302	60.339 514	0.016 573	86.408 856	0.011 573	2 012.347 790	72
80	1.490 339	0.670 988	65.802 305	0.015 197	98.067 714	0.010 197	2 424.645 508	80
84	1.520 370	0.657 735	68.453 042	0.014 609	104.073 927	0.009 609	2 640.664 052	84
90	1.566 555	0.638 344	72.331 300	0.013 825	113.310 936	0.008 825	2 976.076 879	90
96	1.614 143	0.619 524	76.095 218	0.013 141	122.828 542	0.008 141	3 324.184 603	96
100	1.646 668	0.607 287	78.542 645	0.012 732	129.333 698	0.007 732	3 562.793 430	100
108	1.713 699	0.583 533	83.293 424	0.012 006	142.739 900	0.007 006	4 054.374 734	108
120	1.819 397	0.549 633	90.073 453	0.011 102	163.879 347	0.006 102	4 823.505 065	120
240	3.310 204	0.302 096	139.580 772	0.007 164	462.040 895	0.002 164	13 415.539 540	240
360	6.022 575	0.166 042	166.791 614	0.005 996	1 004.515 042	0.000 996	21 403.304 060	360
480	10.957 454	0.091 262	181.747 584	0.005 502	1 991.490 734	0.000 502	27 588.357 289	480
∞	∞	0	200.000 000	0.005 000	∞	0	40 000.000 000	∞

0.75%	Single Payment		Uniform Series				Arithmetic Gradient Series	0.75%
	Compound Amount Factor (CAF)	Present Worth Factor (PWF)	Present Worth Factor (PWF)	Capital Recovery Factor (CRF)	Compound Amount Factor (CAF)	Sinking Fund Factor (SFF)	Present Worth Factor (PWF)	
n	*F/P*	*P/F*	*P/A*	*A/P*	*F/A*	*A/F*	*P/G*	*n*
1	1.007 500	0.992 556	0.992 556	1.007 500	1.000 000	1.000 000	0.000 000	1
2	1.015 056	0.985 167	1.977 723	0.505 632	2.007 500	0.498 132	0.985 167	2
3	1.022 669	0.977 833	2.955 556	0.338 346	3.022 556	0.330 846	2.940 834	3
4	1.030 339	0.970 554	3.926 110	0.254 705	4.045 225	0.247 205	5.852 496	4
5	1.038 067	0.963 329	4.889 440	0.204 522	5.075 565	0.197 022	9.705 813	5
6	1.045 852	0.956 158	5.845 598	0.171 069	6.113 631	0.163 569	14.486 603	6
7	1.053 696	0.949 040	6.794 638	0.147 175	7.159 484	0.139 675	20.180 844	7
8	1.061 599	0.941 975	7.736 613	0.129 256	8.213 180	0.121 756	26.774 672	8
9	1.069 561	0.934 963	8.671 576	0.115 319	9.274 779	0.107 819	34.254 378	9
10	1.077 583	0.928 003	9.599 580	0.104 171	10.344 339	0.096 671	42.606 406	10
11	1.085 664	0.921 095	10.520 675	0.095 051	11.421 922	0.087 551	51.817 355	11
12	1.093 807	0.914 238	11.434 913	0.087 451	12.507 586	0.079 951	61.873 975	12
13	1.102 010	0.907 432	12.342 345	0.081 022	13.601 393	0.073 522	72.763 164	13
14	1.110 276	0.900 677	13.243 022	0.075 511	14.703 404	0.068 011	84.471 969	14
15	1.118 603	0.893 973	14.136 995	0.070 736	15.813 679	0.063 236	96.987 585	15
16	1.126 992	0.887 318	15.024 313	0.066 559	16.932 282	0.059 059	110.297 350	16
17	1.135 445	0.880 712	15.905 025	0.062 873	18.059 274	0.055 373	124.388 747	17
18	1.143 960	0.874 156	16.779 181	0.059 598	19.194 718	0.052 098	139.249 401	18
19	1.152 540	0.867 649	17.646 830	0.056 667	20.338 679	0.049 167	154.867 079	19
20	1.161 184	0.861 190	18.508 020	0.054 031	21.491 219	0.046 531	171.229 686	20
21	1.169 893	0.854 779	19.362 799	0.051 645	22.652 403	0.044 145	188.325 267	21
22	1.178 667	0.848 416	20.211 215	0.049 477	23.822 296	0.041 977	206.142 000	22
23	1.187 507	0.842 100	21.053 315	0.047 498	25.000 963	0.039 998	224.668 203	23
24	1.196 414	0.835 831	21.889 146	0.045 685	26.188 471	0.038 185	243.892 326	24
25	1.205 387	0.829 609	22.718 755	0.044 016	27.384 884	0.036 516	263.802 950	25
26	1.214 427	0.823 434	23.542 189	0.042 477	28.590 271	0.034 977	284.388 789	26
27	1.223 535	0.817 304	24.359 493	0.041 052	29.804 698	0.033 552	305.638 688	27
28	1.232 712	0.811 220	25.170 713	0.039 729	31.028 233	0.032 229	327.541 619	28
29	1.241 957	0.805 181	25.975 893	0.038 497	32.260 945	0.030 997	350.086 681	29
30	1.251 272	0.799 187	26.775 080	0.037 348	33.502 902	0.029 848	373.263 101	30
31	1.260 656	0.793 238	27.568 318	0.036 274	34.754 174	0.028 774	397.060 230	31
32	1.270 111	0.787 333	28.355 650	0.035 266	36.014 830	0.027 766	421.467 541	32
33	1.279 637	0.781 472	29.137 122	0.034 320	37.284 941	0.026 820	446.474 632	33
34	1.289 234	0.775 654	29.912 776	0.033 431	38.564 578	0.025 931	472.071 220	34
35	1.298 904	0.769 880	30.682 656	0.032 592	39.853 813	0.025 092	498.247 142	35

(cont.)

(cont.)

0.75%	Single Payment		Uniform Series				Arithmetic Gradient Series	0.75%
	Compound Amount Factor (CAF)	Present Worth Factor (PWF)	Present Worth Factor (PWF)	Capital Recovery Factor (CRF)	Compound Amount Factor (CAF)	Sinking Fund Factor (SFF)	Present Worth Factor (PWF)	
n	*F/P*	*P/F*	*P/A*	*A/P*	*F/A*	*A/F*	*P/G*	*n*
36	1.308 645	0.764 149	31.446 805	0.031 800	41.152 716	0.024 300	524.992 356	36
37	1.318 460	0.758 461	32.205 266	0.031 051	42.461 361	0.023 551	552.296 934	37
38	1.328 349	0.752 814	32.958 080	0.030 342	43.779 822	0.022 842	580.151 067	38
39	1.338 311	0.747 210	33.705 290	0.029 669	45.108 170	0.022 169	608.545 059	39
40	1.348 349	0.741 648	34.446 938	0.029 030	46.446 482	0.021 530	637.469 330	40
41	1.358 461	0.736 127	35.183 065	0.028 423	47.794 830	0.020 923	666.914 410	41
42	1.368 650	0.730 647	35.913 713	0.027 845	49.153 291	0.020 345	696.870 943	42
43	1.378 915	0.725 208	36.638 921	0.027 293	50.521 941	0.019 793	727.329 683	43
44	1.389 256	0.719 810	37.358 730	0.026 768	51.900 856	0.019 268	758.281 493	44
45	1.399 676	0.714 451	38.073 181	0.026 265	53.290 112	0.018 765	789.717 343	45
46	1.410 173	0.709 133	38.782 314	0.025 785	54.689 788	0.018 285	821.628 312	46
47	1.420 750	0.703 854	39.486 168	0.025 325	56.099 961	0.017 825	854.005 584	47
48	1.431 405	0.698 614	40.184 782	0.024 885	57.520 711	0.017 385	886.840 449	48
49	1.442 141	0.693 414	40.878 195	0.024 463	58.952 116	0.016 963	920.124 298	49
50	1.452 957	0.688 252	41.566 447	0.024 058	60.394 257	0.016 558	953.848 629	50
60	1.565 681	0.638 700	48.173 374	0.020 758	75.424 137	0.013 258	1 313.518 881	60
70	1.687 151	0.592 715	54.304 622	0.018 415	91.620 073	0.010 915	1 708.606 494	70
72	1.712 553	0.583 924	55.476 849	0.018 026	95.007 028	0.010 526	1 791.246 287	72
80	1.818 044	0.550 042	59.994 440	0.016 668	109.072 531	0.009 168	2 132.147 226	80
84	1.873 202	0.533 845	62.153 965	0.016 089	116.426 928	0.008 589	2 308.128 298	84
90	1.959 092	0.510 440	65.274 609	0.015 320	127.878 995	0.007 820	2 577.996 051	90
96	2.048 921	0.488 062	68.258 439	0.014 650	139.856 164	0.007 150	2 853.935 242	96
100	2.111 084	0.473 690	70.174 623	0.014 250	148.144 512	0.006 750	3 040.745 301	100
108	2.241 124	0.446 205	73.839 382	0.013 543	165.483 223	0.006 043	3 419.904 093	108
120	2.451 357	0.407 937	78.941 693	0.012 668	193.514 277	0.005 168	3 998.562 143	120
240	6.009 152	0.166 413	111.144 954	0.008 997	667.886 870	0.001 497	9 494.116 170	240
360	14.730 576	0.067 886	124.281 866	0.008 046	1 830.743 483	0.000 546	13 312.387 067	360
480	36.109 902	0.027 693	129.640 902	0.007 714	4 681.320 273	0.000 214	15 513.086 567	480
∞	∞	0	133.333 333	0.007 500	∞	0	17 777.777 778	∞

1%	Single Payment		Uniform Series				Arithmetic Gradient Series	1%
	Compound Amount Factor (CAF)	Present Worth Factor (PWF)	Present Worth Factor (PWF)	Capital Recovery Factor (CRF)	Compound Amount Factor (CAF)	Sinking Fund Factor (SFF)	Present Worth Factor (PWF)	
n	*F/P*	*P/F*	*P/A*	*A/P*	*F/A*	*A/F*	*P/G*	*n*
1	1.010 000	0.990 099	0.990 099	1.010 000	1.000 000	1.000 000	0.000 000	1
2	1.020 100	0.980 296	1.970 395	0.507 512	2.010 000	0.497 512	0.980 296	2
3	1.030 301	0.970 590	2.940 985	0.340 022	3.030 100	0.330 022	2.921 476	3
4	1.040 604	0.960 980	3.901 966	0.256 281	4.060 401	0.246 281	5.804 417	4
5	1.051 010	0.951 466	4.853 431	0.206 040	5.101 005	0.196 040	9.610 280	5
6	1.061 520	0.942 045	5.795 476	0.172 548	6.152 015	0.162 548	14.320 506	6
7	1.072 135	0.932 718	6.728 195	0.148 628	7.213 535	0.138 628	19.916 815	7
8	1.082 857	0.923 483	7.651 678	0.130 690	8.285 671	0.120 690	26.381 197	8
9	1.093 685	0.914 340	8.566 018	0.116 740	9.368 527	0.106 740	33.695 916	9
10	1.104 622	0.905 287	9.471 305	0.105 582	10.462 213	0.095 582	41.843 498	10
11	1.115 668	0.896 324	10.367 628	0.096 454	11.566 835	0.086 454	50.806 736	11
12	1.126 825	0.887 449	11.255 077	0.088 849	12.682 503	0.078 849	60.568 677	12
13	1.138 093	0.878 663	12.133 740	0.082 415	13.809 328	0.072 415	71.112 628	13
14	1.149 474	0.869 963	13.003 703	0.076 901	14.947 421	0.066 901	82.422 147	14
15	1.160 969	0.861 349	13.865 053	0.072 124	16.096 896	0.062 124	94.481 039	15
16	1.172 579	0.852 821	14.717 874	0.067 945	17.257 864	0.057 945	107.273 358	16
17	1.184 304	0.844 377	15.562 251	0.064 258	18.430 443	0.054 258	120.783 398	17
18	1.196 147	0.836 017	16.398 269	0.060 982	19.614 748	0.050 982	134.995 693	18
19	1.208 109	0.827 740	17.226 008	0.058 052	20.810 895	0.048 052	149.895 011	19
20	1.220 190	0.819 544	18.045 553	0.055 415	22.019 004	0.045 415	165.466 356	20
21	1.232 392	0.811 430	18.856 983	0.053 031	23.239 194	0.043 031	181.694 959	21
22	1.244 716	0.803 396	19.660 379	0.050 864	24.471 586	0.040 864	198.566 280	22
23	1.257 163	0.795 442	20.455 821	0.048 886	25.716 302	0.038 886	216.065 999	23
24	1.269 735	0.787 566	21.243 387	0.047 073	26.973 465	0.037 073	234.180 020	24
25	1.282 432	0.779 768	22.023 156	0.045 407	28.243 200	0.035 407	252.894 463	25
26	1.295 256	0.772 048	22.795 204	0.043 869	29.525 631	0.033 869	272.195 662	26
27	1.308 209	0.764 404	23.559 608	0.042 446	30.820 888	0.032 446	292.070 164	27
28	1.321 291	0.756 836	24.316 443	0.041 124	32.129 097	0.031 124	312.504 724	28
29	1.334 504	0.749 342	25.065 785	0.039 895	33.450 388	0.029 895	333.486 304	29
30	1.347 849	0.741 923	25.807 708	0.038 748	34.784 892	0.028 748	355.002 069	30
31	1.361 327	0.734 577	26.542 285	0.037 676	36.132 740	0.027 676	377.039 383	31
32	1.374 941	0.727 304	27.269 589	0.036 671	37.494 068	0.026 671	399.585 810	32
33	1.388 690	0.720 103	27.989 693	0.035 727	38.869 009	0.025 727	422.629 109	33
34	1.402 577	0.712 973	28.702 666	0.034 840	40.257 699	0.024 840	446.157 229	34
35	1.416 603	0.705 914	29.408 580	0.034 004	41.660 276	0.024 004	470.158 312	35

(cont.)

(cont.)

1%	Single Payment		Uniform Series				Arithmetic Gradient Series	1%
	Compound Amount Factor (CAF)	Present Worth Factor (PWF)	Present Worth Factor (PWF)	Capital Recovery Factor (CRF)	Compound Amount Factor (CAF)	Sinking Fund Factor (SFF)	Present Worth Factor (PWF)	
n	*F/P*	*P/F*	*P/A*	*A/P*	*F/A*	*A/F*	*P/G*	*n*
36	1.430 769	0.698 925	30.107 505	0.033 214	43.076 878	0.023 214	494.620 685	36
37	1.445 076	0.692 005	30.799 510	0.032 468	44.507 647	0.022 468	519.532 861	37
38	1.459 527	0.685 153	31.484 663	0.031 761	45.952 724	0.021 761	544.883 536	38
39	1.474 123	0.678 370	32.163 033	0.031 092	47.412 251	0.021 092	570.661 584	39
40	1.488 864	0.671 653	32.834 686	0.030 456	48.886 373	0.020 456	596.856 056	40
41	1.503 752	0.665 003	33.499 689	0.029 851	50.375 237	0.019 851	623.456 180	41
42	1.518 790	0.658 419	34.158 108	0.029 276	51.878 989	0.019 276	650.451 356	42
43	1.533 978	0.651 900	34.810 008	0.028 727	53.397 779	0.018 727	677.831 153	43
44	1.549 318	0.645 445	35.455 454	0.028 204	54.931 757	0.018 204	705.585 308	44
45	1.564 811	0.639 055	36.094 508	0.027 705	56.481 075	0.017 705	733.703 724	45
46	1.580 459	0.632 728	36.727 236	0.027 228	58.045 885	0.017 228	762.176 468	46
47	1.596 263	0.626 463	37.353 699	0.026 771	59.626 344	0.016 771	790.993 766	47
48	1.612 226	0.620 260	37.973 959	0.026 334	61.222 608	0.016 334	820.146 005	48
49	1.628 348	0.614 119	38.588 079	0.025 915	62.834 834	0.015 915	849.623 727	49
50	1.644 632	0.608 039	39.196 118	0.025 513	64.463 182	0.015 513	879.417 630	50
60	1.816 697	0.550 450	44.955 038	0.022 244	81.669 670	0.012 244	1 192.806 145	60
70	2.006 763	0.498 315	50.168 514	0.019 933	100.676 337	0.009 933	1 528.647 439	70
72	2.047 099	0.488 496	51.150 391	0.019 550	104.709 931	0.009 550	1 597.867 334	72
80	2.216 715	0.451 118	54.888 206	0.018 219	121.671 522	0.008 219	1 879.877 099	80
84	2.306 723	0.433 515	56.648 453	0.017 653	130.672 274	0.007 653	2 023.315 308	84
90	2.448 633	0.408 391	59.160 881	0.016 903	144.863 267	0.006 903	2 240.567 482	90
96	2.599 273	0.384 723	61.527 703	0.016 253	159.927 293	0.006 253	2 459.429 786	96
100	2.704 814	0.369 711	63.028 879	0.015 866	170.481 383	0.005 866	2 605.775 753	100
108	2.928 926	0.341 422	65.857 790	0.015 184	192.892 579	0.005 184	2 898.420 284	108
120	3.300 387	0.302 995	69.700 522	0.014 347	230.038 689	0.004 347	3 334.114 847	120
240	10.892 554	0.091 806	90.819 416	0.011 011	989.255 365	0.001 011	6 878.601 558	240
360	35.949 641	0.027 817	97.218 331	0.010 286	3 494.964 133	0.000 286	8 720.432 296	360
480	118.647 725	0.008 428	99.157 169	0.010 085	11 764.772 510	0.000 085	9 511.157 926	480
∞	∞	0	100.000 000	0.010 000	∞	0	10 000.000 000	∞

1.25%	Single Payment		Uniform Series				Arithmetic Gradient Series	1.25%
	Compound Amount Factor (CAF)	Present Worth Factor (PWF)	Present Worth Factor (PWF)	Capital Recovery Factor (CRF)	Compound Amount Factor (CAF)	Sinking Fund Factor (SFF)	Present Worth Factor (PWF)	
n	*F/P*	*P/F*	*P/A*	*A/P*	*F/A*	*A/F*	*P/G*	*n*
1	1.012 500	0.987 654	0.987 654	1.012 500	1.000 000	1.000 000	0.000 000	1
2	1.025 156	0.975 461	1.963 115	0.509 394	2.012 500	0.496 894	0.975 461	2
3	1.037 971	0.963 418	2.926 534	0.341 701	3.037 656	0.329 201	2.902 298	3
4	1.050 945	0.951 524	3.878 058	0.257 861	4.075 627	0.245 361	5.756 871	4
5	1.064 082	0.939 777	4.817 835	0.207 562	5.126 572	0.195 062	9.515 979	5
6	1.077 383	0.928 175	5.746 010	0.174 034	6.190 654	0.161 534	14.156 853	6
7	1.090 850	0.916 716	6.662 726	0.150 089	7.268 038	0.137 589	19.657 149	7
8	1.104 486	0.905 398	7.568 124	0.132 133	8.358 888	0.119 633	25.994 938	8
9	1.118 292	0.894 221	8.462 345	0.118 171	9.463 374	0.105 671	33.148 703	9
10	1.132 271	0.883 181	9.345 526	0.107 003	10.581 666	0.094 503	41.097 332	10
11	1.146 424	0.872 277	10.217 803	0.097 868	11.713 937	0.085 368	49.820 106	11
12	1.160 755	0.861 509	11.079 312	0.090 258	12.860 361	0.077 758	59.296 701	12
13	1.175 264	0.850 873	11.930 185	0.083 821	14.021 116	0.071 321	69.507 173	13
14	1.189 955	0.840 368	12.770 553	0.078 305	15.196 380	0.065 805	80.431 958	14
15	1.204 829	0.829 993	13.600 546	0.073 526	16.386 335	0.061 026	92.051 863	15
16	1.219 890	0.819 746	14.420 292	0.069 347	17.591 164	0.056 847	104.348 058	16
17	1.235 138	0.809 626	15.229 918	0.065 660	18.811 053	0.053 160	117.302 074	17
18	1.250 577	0.799 631	16.029 549	0.062 385	20.046 192	0.049 885	130.895 795	18
19	1.266 210	0.789 759	16.819 308	0.059 455	21.296 769	0.046 955	145.111 451	19
20	1.282 037	0.780 009	17.599 316	0.056 820	22.562 979	0.044 320	159.931 613	20
21	1.298 063	0.770 379	18.369 695	0.054 437	23.845 016	0.041 937	175.339 190	21
22	1.314 288	0.760 868	19.130 563	0.052 272	25.143 078	0.039 772	191.317 417	22
23	1.330 717	0.751 475	19.882 037	0.050 297	26.457 367	0.037 797	207.849 857	23
24	1.347 351	0.742 197	20.624 235	0.048 487	27.788 084	0.035 987	224.920 389	24
25	1.364 193	0.733 034	21.357 269	0.046 822	29.135 435	0.034 322	242.513 209	25
26	1.381 245	0.723 984	22.081 253	0.045 287	30.499 628	0.032 787	260.612 817	26
27	1.398 511	0.715 046	22.796 299	0.043 867	31.880 873	0.031 367	279.204 020	27
28	1.415 992	0.706 219	23.502 518	0.042 549	33.279 384	0.030 049	298.271 920	28
29	1.433 692	0.697 500	24.200 018	0.041 322	34.695 377	0.028 822	317.801 914	29
30	1.451 613	0.688 889	24.888 906	0.040 179	36.129 069	0.027 679	337.779 685	30
31	1.469 759	0.680 384	25.569 290	0.039 109	37.580 682	0.026 609	358.191 202	31
32	1.488 131	0.671 984	26.241 274	0.038 108	39.050 441	0.025 608	379.022 708	32
33	1.506 732	0.663 688	26.904 962	0.037 168	40.538 571	0.024 668	400.260 723	33
34	1.525 566	0.655 494	27.560 456	0.036 284	42.045 303	0.023 784	421.892 035	34
35	1.544 636	0.647 402	28.207 858	0.035 451	43.570 870	0.022 951	443.903 695	35

(cont.)

(cont.)

1.25%	Single Payment		Uniform Series				Arithmetic Gradient Series	1.25%
	Compound Amount Factor (CAF)	Present Worth Factor (PWF)	Present Worth Factor (PWF)	Capital Recovery Factor (CRF)	Compound Amount Factor (CAF)	Sinking Fund Factor (SFF)	Present Worth Factor (PWF)	
n	*F/P*	*P/F*	*P/A*	*A/P*	*F/A*	*A/F*	*P/G*	*n*
36	1.563 944	0.639 409	28.847 267	0.034 665	45.115 505	0.022 165	466.283 015	36
37	1.583 493	0.631 515	29.478 783	0.033 923	46.679 449	0.021 423	489.017 563	37
38	1.603 287	0.623 719	30.102 501	0.033 220	48.262 942	0.020 720	512.095 156	38
39	1.623 328	0.616 019	30.718 520	0.032 554	49.866 229	0.020 054	535.503 860	39
40	1.643 619	0.608 413	31.326 933	0.031 921	51.489 557	0.019 421	559.231 980	40
41	1.664 165	0.600 902	31.927 835	0.031 321	53.133 177	0.018 821	583.268 062	41
42	1.684 967	0.593 484	32.521 319	0.030 749	54.797 341	0.018 249	607.600 886	42
43	1.706 029	0.586 157	33.107 475	0.030 205	56.482 308	0.017 705	632.219 462	43
44	1.727 354	0.578 920	33.686 395	0.029 686	58.188 337	0.017 186	657.113 024	44
45	1.748 946	0.571 773	34.258 168	0.029 190	59.915 691	0.016 690	682.271 032	45
46	1.770 808	0.564 714	34.822 882	0.028 717	61.664 637	0.016 217	707.683 160	46
47	1.792 943	0.557 742	35.380 624	0.028 264	63.435 445	0.015 764	733.339 301	47
48	1.815 355	0.550 856	35.931 481	0.027 831	65.228 388	0.015 331	759.229 556	48
49	1.838 047	0.544 056	36.475 537	0.027 416	67.043 743	0.014 916	785.344 234	49
50	1.861 022	0.537 339	37.012 876	0.027 018	68.881 790	0.014 518	811.673 848	50
60	2.107 181	0.474 568	42.034 592	0.023 790	88.574 508	0.011 290	1 084.842 851	60
70	2.385 900	0.419 129	46.469 676	0.021 519	110.871 998	0.009 019	1 370.451 343	70
72	2.445 920	0.408 844	47.292 474	0.021 145	115.673 621	0.008 645	1 428.456 095	72
80	2.701 485	0.370 167	50.386 657	0.019 847	136.118 795	0.007 347	1 661.865 129	80
84	2.839 113	0.352 223	51.822 185	0.019 297	147.129 040	0.006 797	1 778.838 393	84
90	3.058 813	0.326 924	53.846 060	0.018 571	164.705 008	0.006 071	1 953.830 260	90
96	3.295 513	0.303 443	55.724 570	0.017 945	183.641 059	0.005 445	2 127.524 375	96
100	3.463 404	0.288 733	56.901 339	0.017 574	197.072 342	0.005 074	2 242.241 085	100
108	3.825 282	0.261 419	59.086 509	0.016 924	226.022 551	0.004 424	2 468.263 607	108
120	4.440 213	0.225 214	61.982 847	0.016 133	275.217 058	0.003 633	2 796.569 449	120
240	19.715 494	0.050 722	75.942 278	0.013 168	1 497.239 481	0.000 668	5 101.528 827	240
360	87.540 995	0.011 423	79.086 142	0.012 644	6 923.279 611	0.000 144	5 997.902 675	360
480	388.700 685	0.002 573	79.794 186	0.012 532	31 016.054 774	0.000 032	6 284.744 222	480
∞	∞	0	80.000 000	0.012 500	∞	0	6 400.000 000	∞

1.50%	Single Payment		Uniform Series				Arithmetic Gradient Series	1.50%
	Compound Amount Factor (CAF)	Present Worth Factor (PWF)	Present Worth Factor (PWF)	Capital Recovery Factor (CRF)	Compound Amount Factor (CAF)	Sinking Fund Factor (SFF)	Present Worth Factor (PWF)	
n	*F/P*	*P/F*	*P/A*	*A/P*	*F/A*	*A/F*	*P/G*	*n*
1	1.015 000	0.985 222	0.985 222	1.015 000	1.000 000	1.000 000	0.000 000	1
2	1.030 225	0.970 662	1.955 883	0.511 278	2.015 000	0.496 278	0.970 662	2
3	1.045 678	0.956 317	2.912 200	0.343 383	3.045 225	0.328 383	2.883 296	3
4	1.061 364	0.942 184	3.854 385	0.259 445	4.090 903	0.244 445	5.709 848	4
5	1.077 284	0.928 260	4.782 645	0.209 089	5.152 267	0.194 089	9.422 890	5
6	1.093 443	0.914 542	5.697 187	0.175 525	6.229 551	0.160 525	13.995 601	6
7	1.109 845	0.901 027	6.598 214	0.151 556	7.322 994	0.136 556	19.401 761	7
8	1.126 493	0.887 711	7.485 925	0.133 584	8.432 839	0.118 584	25.615 739	8
9	1.143 390	0.874 592	8.360 517	0.119 610	9.559 332	0.104 610	32.612 477	9
10	1.160 541	0.861 667	9.222 185	0.108 434	10.702 722	0.093 434	40.367 482	10
11	1.177 949	0.848 933	10.071 118	0.099 294	11.863 262	0.084 294	48.856 815	11
12	1.195 618	0.836 387	10.907 505	0.091 680	13.041 211	0.076 680	58.057 076	12
13	1.213 552	0.824 027	11.731 532	0.085 240	14.236 830	0.070 240	67.945 400	13
14	1.231 756	0.811 849	12.543 382	0.079 723	15.450 382	0.064 723	78.499 441	14
15	1.250 232	0.799 852	13.343 233	0.074 944	16.682 138	0.059 944	89.697 362	15
16	1.268 986	0.788 031	14.131 264	0.070 765	17.932 370	0.055 765	101.517 828	16
17	1.288 020	0.776 385	14.907 649	0.067 080	19.201 355	0.052 080	113.939 992	17
18	1.307 341	0.764 912	15.672 561	0.063 806	20.489 376	0.048 806	126.943 489	18
19	1.326 951	0.753 607	16.426 168	0.060 878	21.796 716	0.045 878	140.508 423	19
20	1.346 855	0.742 470	17.168 639	0.058 246	23.123 667	0.043 246	154.615 361	20
21	1.367 058	0.731 498	17.900 137	0.055 865	24.470 522	0.040 865	169.245 320	21
22	1.387 564	0.720 688	18.620 824	0.053 703	25.837 580	0.038 703	184.379 761	22
23	1.408 377	0.710 037	19.330 861	0.051 731	27.225 144	0.036 731	200.000 576	23
24	1.429 503	0.699 544	20.030 405	0.049 924	28.633 521	0.034 924	216.090 087	24
25	1.450 945	0.689 206	20.719 611	0.048 263	30.063 024	0.033 263	232.631 027	25
26	1.472 710	0.679 021	21.398 632	0.046 732	31.513 969	0.031 732	249.606 540	26
27	1.494 800	0.668 986	22.067 617	0.045 315	32.986 678	0.030 315	267.000 169	27
28	1.517 222	0.659 099	22.726 717	0.044 001	34.481 479	0.029 001	284.795 849	28
29	1.539 981	0.649 359	23.376 076	0.042 779	35.998 701	0.027 779	302.977 897	29
30	1.563 080	0.639 762	24.015 838	0.041 639	37.538 681	0.026 639	321.531 007	30
31	1.586 526	0.630 308	24.646 146	0.040 574	39.101 762	0.025 574	340.440 242	31
32	1.610 324	0.620 993	25.267 139	0.039 577	40.688 288	0.024 577	359.691 022	32
33	1.634 479	0.611 816	25.878 954	0.038 641	42.298 612	0.023 641	379.269 124	33
34	1.658 996	0.602 774	26.481 728	0.037 762	43.933 092	0.022 762	399.160 668	34
35	1.683 881	0.593 866	27.075 595	0.036 934	45.592 088	0.021 934	419.352 115	35

(cont.)

(cont.)

1.50%	Single Payment		Uniform Series				Arithmetic Gradient Series	1.50%
	Compound Amount Factor (CAF)	Present Worth Factor (PWF)	Present Worth Factor (PWF)	Capital Recovery Factor (CRF)	Compound Amount Factor (CAF)	Sinking Fund Factor (SFF)	Present Worth Factor (PWF)	
n	*F/P*	*P/F*	*P/A*	*A/P*	*F/A*	*A/F*	*P/G*	*n*
36	1.709 140	0.585 090	27.660 684	0.036 152	47.275 969	0.021 152	439.830 256	36
37	1.734 777	0.576 443	28.237 127	0.035 414	48.985 109	0.020 414	460.582 207	37
38	1.760 798	0.567 924	28.805 052	0.034 716	50.719 885	0.019 716	481.595 403	38
39	1.787 210	0.559 531	29.364 583	0.034 055	52.480 684	0.019 055	502.857 591	39
40	1.814 018	0.551 262	29.915 845	0.033 427	54.267 894	0.018 427	524.356 822	40
41	1.841 229	0.543 116	30.458 961	0.032 831	56.081 912	0.017 831	546.081 445	41
42	1.868 847	0.535 089	30.994 050	0.032 264	57.923 141	0.017 264	568.020 105	42
43	1.896 880	0.527 182	31.521 232	0.031 725	59.791 988	0.016 725	590.161 729	43
44	1.925 333	0.519 391	32.040 622	0.031 210	61.688 868	0.016 210	612.495 527	44
45	1.954 213	0.511 715	32.552 337	0.030 720	63.614 201	0.015 720	635.010 985	45
46	1.983 526	0.504 153	33.056 490	0.030 251	65.568 414	0.015 251	657.697 854	46
47	2.013 279	0.496 702	33.553 192	0.029 803	67.551 940	0.014 803	680.546 152	47
48	2.043 478	0.489 362	34.042 554	0.029 375	69.565 219	0.014 375	703.546 151	48
49	2.074 130	0.482 130	34.524 683	0.028 965	71.608 698	0.013 965	726.688 379	49
50	2.105 242	0.475 005	34.999 688	0.028 572	73.682 828	0.013 572	749.963 609	50
60	2.443 220	0.409 296	39.380 269	0.025 393	96.214 652	0.010 393	988.167 392	60
70	2.835 456	0.352 677	43.154 872	0.023 172	122.363 753	0.008 172	1 231.165 816	70
72	2.921 158	0.342 330	43.844 667	0.022 808	128.077 197	0.007 808	1 279.793 793	72
80	3.290 663	0.303 890	46.407 323	0.021 548	152.710 852	0.006 548	1 473.074 112	80
84	3.492 590	0.286 321	47.578 633	0.021 018	166.172 636	0.006 018	1 568.514 041	84
90	3.818 949	0.261 852	49.209 855	0.020 321	187.929 900	0.005 321	1 709.543 875	90
96	4.175 804	0.239 475	50.701 675	0.019 723	211.720 235	0.004 723	1 847.472 533	96
100	4.432 046	0.225 629	51.624 704	0.019 371	228.803 043	0.004 371	1 937.450 611	100
108	4.992 667	0.200 294	53.313 749	0.018 757	266.177 771	0.003 757	2 112.134 789	108
120	5.969 323	0.167 523	55.498 454	0.018 019	331.288 191	0.003 019	2 359.711 434	120
240	35.632 816	0.028 064	64.795 732	0.015 433	2 308.854 370	0.000 433	3 870.691 174	240
360	212.703 781	0.004 701	66.353 242	0.015 071	14 113.585 393	0.000 071	4 310.716 476	360
480	1 269.697 544	0.000 788	66.614 161	0.015 012	84 579.836 287	0.000 012	4 415.741 196	480
∞	∞	0	66.666 667	0.015 000	∞	0	4 444.444 444	∞

2%	Single Payment		Uniform Series				Arithmetic Gradient Series	2%
	Compound Amount Factor (CAF)	Present Worth Factor (PWF)	Present Worth Factor (PWF)	Capital Recovery Factor (CRF)	Compound Amount Factor (CAF)	Sinking Fund Factor (SFF)	Present Worth Factor (PWF)	
n	*F/P*	*P/F*	*P/A*	*A/P*	*F/A*	*A/F*	*P/G*	*n*
1	1.020 000	0.980 392	0.980 392	1.020 000	1.000 000	1.000 000	0.000 000	1
2	1.040 400	0.961 169	1.941 561	0.515 050	2.020 000	0.495 050	0.961 169	2
3	1.061 208	0.942 322	2.883 883	0.346 755	3.060 400	0.326 755	2.845 813	3
4	1.082 432	0.923 845	3.807 729	0.262 624	4.121 608	0.242 624	5.617 350	4
5	1.104 081	0.905 731	4.713 460	0.212 158	5.204 040	0.192 158	9.240 273	5
6	1.126 162	0.887 971	5.601 431	0.178 526	6.308 121	0.158 526	13.680 130	6
7	1.148 686	0.870 560	6.471 991	0.154 512	7.434 283	0.134 512	18.903 491	7
8	1.171 659	0.853 490	7.325 481	0.136 510	8.582 969	0.116 510	24.877 924	8
9	1.195 093	0.836 755	8.162 237	0.122 515	9.754 628	0.102 515	31.571 966	9
10	1.218 994	0.820 348	8.982 585	0.111 327	10.949 721	0.091 327	38.955 100	10
11	1.243 374	0.804 263	9.786 848	0.102 178	12.168 715	0.082 178	46.997 731	11
12	1.268 242	0.788 493	10.575 341	0.094 560	13.412 090	0.074 560	55.671 156	12
13	1.293 607	0.773 033	11.348 374	0.088 118	14.680 332	0.068 118	64.947 546	13
14	1.319 479	0.757 875	12.106 249	0.082 602	15.973 938	0.062 602	74.799 921	14
15	1.345 868	0.743 015	12.849 264	0.077 825	17.293 417	0.057 825	85.202 128	15
16	1.372 786	0.728 446	13.577 709	0.073 650	18.639 285	0.053 650	96.128 815	16
17	1.400 241	0.714 163	14.291 872	0.069 970	20.012 071	0.049 970	107.555 416	17
18	1.428 246	0.700 159	14.992 031	0.066 702	21.412 312	0.046 702	119.458 125	18
19	1.456 811	0.686 431	15.678 462	0.063 782	22.840 559	0.043 782	131.813 879	19
20	1.485 947	0.672 971	16.351 433	0.061 157	24.297 370	0.041 157	144.600 334	20
21	1.515 666	0.659 776	17.011 209	0.058 785	25.783 317	0.038 785	157.795 850	21
22	1.545 980	0.646 839	17.658 048	0.056 631	27.298 984	0.036 631	171.379 470	22
23	1.576 899	0.634 156	18.292 204	0.054 668	28.844 963	0.034 668	185.330 900	23
24	1.608 437	0.621 721	18.913 926	0.052 871	30.421 862	0.032 871	199.630 495	24
25	1.640 606	0.609 531	19.523 456	0.051 220	32.030 300	0.031 220	214.259 236	25
26	1.673 418	0.597 579	20.121 036	0.049 699	33.670 906	0.029 699	229.198 718	26
27	1.706 886	0.585 862	20.706 898	0.048 293	35.344 324	0.028 293	244.431 131	27
28	1.741 024	0.574 375	21.281 272	0.046 990	37.051 210	0.026 990	259.939 244	28
29	1.775 845	0.563 112	21.844 385	0.045 778	38.792 235	0.025 778	275.706 388	29
30	1.811 362	0.552 071	22.396 456	0.044 650	40.568 079	0.024 650	291.716 444	30
31	1.847 589	0.541 246	22.937 702	0.043 596	42.379 441	0.023 596	307.953 823	31
32	1.884 541	0.530 633	23.468 335	0.042 611	44.227 030	0.022 611	324.403 456	32
33	1.922 231	0.520 229	23.988 564	0.041 687	46.111 570	0.021 687	341.050 775	33
34	1.960 676	0.510 028	24.498 592	0.040 819	48.033 802	0.020 819	357.881 704	34
35	1.999 890	0.500 028	24.998 619	0.040 002	49.994 478	0.020 002	374.882 643	35

(cont.)

(cont.)

2%	Single Payment		Uniform Series				Arithmetic Gradient Series	2%
	Compound Amount Factor (CAF)	Present Worth Factor (PWF)	Present Worth Factor (PWF)	Capital Recovery Factor (CRF)	Compound Amount Factor (CAF)	Sinking Fund Factor (SFF)	Present Worth Factor (PWF)	
n	*F/P*	*P/F*	*P/A*	*A/P*	*F/A*	*A/F*	*P/G*	*n*
36	2.039 887	0.490 223	25.488 842	0.039 233	51.994 367	0.019 233	392.040 453	36
37	2.080 685	0.480 611	25.969 453	0.038 507	54.034 255	0.018 507	409.342 447	37
38	2.122 299	0.471 187	26.440 641	0.037 821	56.114 940	0.017 821	426.776 373	38
39	2.164 745	0.461 948	26.902 589	0.037 171	58.237 238	0.017 171	444.330 405	39
40	2.208 040	0.452 890	27.355 479	0.036 556	60.401 983	0.016 556	461.993 132	40
41	2.252 200	0.444 010	27.799 489	0.035 972	62.610 023	0.015 972	479.753 540	41
42	2.297 244	0.435 304	28.234 794	0.035 417	64.862 223	0.015 417	497.601 009	42
43	2.343 189	0.426 769	28.661 562	0.034 890	67.159 468	0.014 890	515.525 297	43
44	2.390 053	0.418 401	29.079 963	0.034 388	69.502 657	0.014 388	533.516 529	44
45	2.437 854	0.410 197	29.490 160	0.033 910	71.892 710	0.013 910	551.565 188	45
46	2.486 611	0.402 154	29.892 314	0.033 453	74.330 564	0.013 453	569.662 106	46
47	2.536 344	0.394 268	30.286 582	0.033 018	76.817 176	0.013 018	587.798 450	47
48	2.587 070	0.386 538	30.673 120	0.032 602	79.353 519	0.012 602	605.965 718	48
49	2.638 812	0.378 958	31.052 078	0.032 204	81.940 590	0.012 204	624.155 723	49
50	2.691 588	0.371 528	31.423 606	0.031 823	84.579 401	0.011 823	642.360 589	50
60	3.281 031	0.304 782	34.760 887	0.028 768	114.051 539	0.008 768	823.697 534	60
70	3.999 558	0.250 028	37.498 619	0.026 668	149.977 911	0.006 668	999.834 315	70
72	4.161 140	0.240 319	37.984 063	0.026 327	158.057 019	0.006 327	1 034.055 703	72
80	4.875 439	0.205 110	39.744 514	0.025 161	193.771 958	0.005 161	1 166.786 767	80
84	5.277 332	0.189 490	40.525 516	0.024 676	213.866 607	0.004 676	1 230.419 116	84
90	5.943 133	0.168 261	41.586 929	0.024 046	247.156 656	0.004 046	1 322.170 082	90
96	6.692 933	0.149 411	42.529 434	0.023 513	284.646 659	0.003 513	1 409.297 343	96
100	7.244 646	0.138 033	43.098 352	0.023 203	312.232 306	0.003 203	1 464.752 746	100
108	8.488 258	0.117 810	44.109 510	0.022 671	374.412 879	0.002 671	1 569.302 514	108
120	10.765 163	0.092 892	45.355 389	0.022 048	488.258 152	0.002 048	1 710.416 045	120
240	115.888 735	0.008 629	49.568 552	0.020 174	5 744.436 758	0.000 174	2 374.879 987	240
360	1 247.561 128	0.000 802	49.959 922	0.020 016	62 328.056 387	0.000 016	2 483.567 939	360
480	13 430.198 935	0.000 074	49.996 277	0.020 001	671 459.946 767	0.000 001	2 498.026 835	480
∞	∞	0	50.000 000	0.020 000	∞	0	2 500.000 000	∞

3%	Single Payment		Uniform Series				Arithmetic Gradient Series	3%
	Compound Amount Factor (CAF)	Present Worth Factor (PWF)	Present Worth Factor (PWF)	Capital Recovery Factor (CRF)	Compound Amount Factor (CAF)	Sinking Fund Factor (SFF)	Present Worth Factor (PWF)	
n	*F/P*	*P/F*	*P/A*	*A/P*	*F/A*	*A/F*	*P/G*	*n*
1	1.030 000	0.970 874	0.970 874	1.030 000	1.000 000	1.000 000	0.000 000	1
2	1.060 900	0.942 596	1.913 470	0.522 611	2.030 000	0.492 611	0.942 596	2
3	1.092 727	0.915 142	2.828 611	0.353 530	3.090 900	0.323 530	2.772 879	3
4	1.125 509	0.888 487	3.717 098	0.269 027	4.183 627	0.239 027	5.438 340	4
5	1.159 274	0.862 609	4.579 707	0.218 355	5.309 136	0.188 355	8.888 776	5
6	1.194 052	0.837 484	5.417 191	0.184 598	6.468 410	0.154 598	13.076 197	6
7	1.229 874	0.813 092	6.230 283	0.160 506	7.662 462	0.130 506	17.954 746	7
8	1.266 770	0.789 409	7.019 692	0.142 456	8.892 336	0.112 456	23.480 611	8
9	1.304 773	0.766 417	7.786 109	0.128 434	10.159 106	0.098 434	29.611 944	9
10	1.343 916	0.744 094	8.530 203	0.117 231	11.463 879	0.087 231	36.308 790	10
11	1.384 234	0.722 421	9.252 624	0.108 077	12.807 796	0.078 077	43.533 002	11
12	1.425 761	0.701 380	9.954 004	0.100 462	14.192 030	0.070 462	51.248 181	12
13	1.468 534	0.680 951	10.634 955	0.094 030	15.617 790	0.064 030	59.419 597	13
14	1.512 590	0.661 118	11.296 073	0.088 526	17.086 324	0.058 526	68.014 129	14
15	1.557 967	0.641 862	11.937 935	0.083 767	18.598 914	0.053 767	77.000 196	15
16	1.604 706	0.623 167	12.561 102	0.079 611	20.156 881	0.049 611	86.347 700	16
17	1.652 848	0.605 016	13.166 118	0.075 953	21.761 588	0.045 953	96.027 963	17
18	1.702 433	0.587 395	13.753 513	0.072 709	23.414 435	0.042 709	106.013 671	18
19	1.753 506	0.570 286	14.323 799	0.069 814	25.116 868	0.039 814	116.278 820	19
20	1.806 111	0.553 676	14.877 475	0.067 216	26.870 374	0.037 216	126.798 659	20
21	1.860 295	0.537 549	15.415 024	0.064 872	28.676 486	0.034 872	137.549 645	21
22	1.916 103	0.521 893	15.936 917	0.062 747	30.536 780	0.032 747	148.509 387	22
23	1.973 587	0.506 692	16.443 608	0.060 814	32.452 884	0.030 814	159.656 606	23
24	2.032 794	0.491 934	16.935 542	0.059 047	34.426 470	0.029 047	170.971 082	24
25	2.093 778	0.477 606	17.413 148	0.057 428	36.459 264	0.027 428	182.433 615	25
26	2.156 591	0.463 695	17.876 842	0.055 938	38.553 042	0.025 938	194.025 984	26
27	2.221 289	0.450 189	18.327 031	0.054 564	40.709 634	0.024 564	205.730 899	27
28	2.287 928	0.437 077	18.764 108	0.053 293	42.930 923	0.023 293	217.531 971	28
29	2.356 566	0.424 346	19.188 455	0.052 115	45.218 850	0.022 115	229.413 669	29
30	2.427 262	0.411 987	19.600 441	0.051 019	47.575 416	0.021 019	241.361 285	30
31	2.500 080	0.399 987	20.000 428	0.049 999	50.002 678	0.019 999	253.360 900	31
32	2.575 083	0.388 337	20.388 766	0.049 047	52.502 759	0.019 047	265.399 348	32
33	2.652 335	0.377 026	20.765 792	0.048 156	55.077 841	0.018 156	277.464 188	33
34	2.731 905	0.366 045	21.131 837	0.047 322	57.730 177	0.017 322	289.543 669	34
35	2.813 862	0.355 383	21.487 220	0.046 539	60.462 082	0.016 539	301.626 705	35

(cont.)

(cont.)

3%	Single Payment		Uniform Series				Arithmetic Gradient Series	3%
	Compound Amount Factor (CAF)	Present Worth Factor (PWF)	Present Worth Factor (PWF)	Capital Recovery Factor (CRF)	Compound Amount Factor (CAF)	Sinking Fund Factor (SFF)	Present Worth Factor (PWF)	
n	*F/P*	*P/F*	*P/A*	*A/P*	*F/A*	*A/F*	*P/G*	*n*
36	2.898 278	0.345 032	21.832 252	0.045 804	63.275 944	0.015 804	313.702 840	36
37	2.985 227	0.334 983	22.167 235	0.045 112	66.174 223	0.015 112	325.762 226	37
38	3.074 783	0.325 226	22.492 462	0.044 459	69.159 449	0.014 459	337.795 593	38
39	3.167 027	0.315 754	22.808 215	0.043 844	72.234 233	0.013 844	349.794 228	39
40	3.262 038	0.306 557	23.114 772	0.043 262	75.401 260	0.013 262	361.749 945	40
41	3.359 899	0.297 628	23.412 400	0.042 712	78.663 298	0.012 712	373.655 065	41
42	3.460 696	0.288 959	23.701 359	0.042 192	82.023 196	0.012 192	385.502 393	42
43	3.564 517	0.280 543	23.981 902	0.041 698	85.483 892	0.011 698	397.285 196	43
44	3.671 452	0.272 372	24.254 274	0.041 230	89.048 409	0.011 230	408.997 183	44
45	3.781 596	0.264 439	24.518 713	0.040 785	92.719 861	0.010 785	420.632 482	45
46	3.895 044	0.256 737	24.775 449	0.040 363	96.501 457	0.010 363	432.185 626	46
47	4.011 895	0.249 259	25.024 708	0.039 961	100.396 501	0.009 961	443.651 529	47
48	4.132 252	0.241 999	25.266 707	0.039 578	104.408 396	0.009 578	455.025 473	48
49	4.256 219	0.234 950	25.501 657	0.039 213	108.540 648	0.009 213	466.303 087	49
50	4.383 906	0.228 107	25.729 764	0.038 865	112.796 867	0.008 865	477.480 334	50
60	5.891 603	0.169 733	27.675 564	0.036 133	163.053 437	0.006 133	583.052 609	60
70	7.917 822	0.126 297	29.123 421	0.034 337	230.594 064	0.004 337	676.086 873	70
72	8.400 017	0.119 047	29.365 088	0.034 054	246.667 242	0.004 054	693.122 552	72
80	10.640 891	0.093 977	30.200 763	0.033 112	321.363 019	0.003 112	756.086 524	80
84	11.976 416	0.083 497	30.550 086	0.032 733	365.880 536	0.002 733	784.543 373	84
90	14.300 467	0.069 928	31.002 407	0.032 256	443.348 904	0.002 256	823.630 214	90
96	17.075 506	0.058 563	31.381 219	0.031 866	535.850 186	0.001 866	858.637 702	96
100	19.218 632	0.052 033	31.598 905	0.031 647	607.287 733	0.001 647	879.854 045	100
108	24.345 588	0.041 075	31.964 160	0.031 285	778.186 267	0.001 285	917.601 260	108
120	34.710 987	0.028 809	32.373 023	0.030 890	1 123.699 571	0.000 890	963.863 466	120
240	1 204.852 628	0.000 830	33.305 667	0.030 025	40 128.420 931	0.000 025	1 103.549 098	240
∞	∞	0	33.333 333	0.030 000	∞	0	1 111.111 111	∞

4%	Single Payment		Uniform Series				Arithmetic Gradient Series	4%
	Compound Amount Factor (CAF)	Present Worth Factor (PWF)	Present Worth Factor (PWF)	Capital Recovery Factor (CRF)	Compound Amount Factor (CAF)	Sinking Fund Factor (SFF)	Present Worth Factor (PWF)	
n	*F/P*	*P/F*	*P/A*	*A/P*	*F/A*	*A/F*	*P/G*	*n*
1	1.040 000	0.961 538	0.961 538	1.040 000	1.000 000	1.000 000	0.000 000	1
2	1.081 600	0.924 556	1.886 095	0.530 196	2.040 000	0.490 196	0.924 556	2
3	1.124 864	0.888 996	2.775 091	0.360 349	3.121 600	0.320 349	2.702 549	3
4	1.169 859	0.854 804	3.629 895	0.275 490	4.246 464	0.235 490	5.266 962	4
5	1.216 653	0.821 927	4.451 822	0.224 627	5.416 323	0.184 627	8.554 670	5
6	1.265 319	0.790 315	5.242 137	0.190 762	6.632 975	0.150 762	12.506 243	6
7	1.315 932	0.759 918	6.002 055	0.166 610	7.898 294	0.126 610	17.065 749	7
8	1.368 569	0.730 690	6.732 745	0.148 528	9.214 226	0.108 528	22.180 581	8
9	1.423 312	0.702 587	7.435 332	0.134 493	10.582 795	0.094 493	27.801 275	9
10	1.480 244	0.675 564	8.110 896	0.123 291	12.006 107	0.083 291	33.881 352	10
11	1.539 454	0.649 581	8.760 477	0.114 149	13.486 351	0.074 149	40.377 162	11
12	1.601 032	0.624 597	9.385 074	0.106 552	15.025 805	0.066 552	47.247 729	12
13	1.665 074	0.600 574	9.985 648	0.100 144	16.626 838	0.060 144	54.454 618	13
14	1.731 676	0.577 475	10.563 123	0.094 669	18.291 911	0.054 669	61.961 794	14
15	1.800 944	0.555 265	11.118 387	0.089 941	20.023 588	0.049 941	69.735 497	15
16	1.872 981	0.533 908	11.652 296	0.085 820	21.824 531	0.045 820	77.744 120	16
17	1.947 900	0.513 373	12.165 669	0.082 199	23.697 512	0.042 199	85.958 092	17
18	2.025 817	0.493 628	12.659 297	0.078 993	25.645 413	0.038 993	94.349 770	18
19	2.106 849	0.474 642	13.133 939	0.076 139	27.671 229	0.036 139	102.893 334	19
20	2.191 123	0.456 387	13.590 326	0.073 582	29.778 079	0.033 582	111.564 686	20
21	2.278 768	0.438 834	14.029 160	0.071 280	31.969 202	0.031 280	120.341 358	21
22	2.369 919	0.421 955	14.451 115	0.069 199	34.247 970	0.029 199	129.202 421	22
23	2.464 716	0.405 726	14.856 842	0.067 309	36.617 889	0.027 309	138.128 400	23
24	2.563 304	0.390 121	15.246 963	0.065 587	39.082 604	0.025 587	147.101 194	24
25	2.665 836	0.375 117	15.622 080	0.064 012	41.645 908	0.024 012	156.103 997	25
26	2.772 470	0.360 689	15.982 769	0.062 567	44.311 745	0.022 567	165.121 228	26
27	2.883 369	0.346 817	16.329 586	0.061 239	47.084 214	0.021 239	174.138 459	27
28	2.998 703	0.333 477	16.663 063	0.060 013	49.967 583	0.020 013	183.142 351	28
29	3.118 651	0.320 651	16.983 715	0.058 880	52.966 286	0.018 880	192.120 590	29
30	3.243 398	0.308 319	17.292 033	0.057 830	56.084 938	0.017 830	201.061 832	30
31	3.373 133	0.296 460	17.588 494	0.056 855	59.328 335	0.016 855	209.955 639	31
32	3.508 059	0.285 058	17.873 551	0.055 949	62.701 469	0.015 949	218.792 435	32
33	3.648 381	0.274 094	18.147 646	0.055 104	66.209 527	0.015 104	227.563 449	33
34	3.794 316	0.263 552	18.411 198	0.054 315	69.857 909	0.014 315	236.260 668	34
35	3.946 089	0.253 415	18.664 613	0.053 577	73.652 225	0.013 577	244.876 794	35

(cont.)

(cont.)

4%	Single Payment		Uniform Series				Arithmetic Gradient Series	4%
	Compound Amount Factor (CAF)	Present Worth Factor (PWF)	Present Worth Factor (PWF)	Capital Recovery Factor (CRF)	Compound Amount Factor (CAF)	Sinking Fund Factor (SFF)	Present Worth Factor (PWF)	
n	*F/P*	*P/F*	*P/A*	*A/P*	*F/A*	*A/F*	*P/G*	*n*
36	4.103 933	0.243 669	18.908 282	0.052 887	77.598 314	0.012 887	253.405 199	36
37	4.268 090	0.234 297	19.142 579	0.052 240	81.702 246	0.012 240	261.839 886	37
38	4.438 813	0.225 285	19.367 864	0.051 632	85.970 336	0.011 632	270.175 447	38
39	4.616 366	0.216 621	19.584 485	0.051 061	90.409 150	0.011 061	278.407 030	39
40	4.801 021	0.208 289	19.792 774	0.050 523	95.025 516	0.010 523	286.530 302	40
41	4.993 061	0.200 278	19.993 052	0.050 017	99.826 536	0.010 017	294.541 420	41
42	5.192 784	0.192 575	20.185 627	0.049 540	104.819 598	0.009 540	302.436 992	42
43	5.400 495	0.185 168	20.370 795	0.049 090	110.012 382	0.009 090	310.214 056	43
44	5.616 515	0.178 046	20.548 841	0.048 665	115.412 877	0.008 665	317.870 049	44
45	5.841 176	0.171 198	20.720 040	0.048 262	121.029 392	0.008 262	325.402 779	45
46	6.074 823	0.164 614	20.884 654	0.047 882	126.870 568	0.007 882	332.810 403	46
47	6.317 816	0.158 283	21.042 936	0.047 522	132.945 390	0.007 522	340.091 400	47
48	6.570 528	0.152 195	21.195 131	0.047 181	139.263 206	0.007 181	347.244 554	48
49	6.833 349	0.146 341	21.341 472	0.046 857	145.833 734	0.006 857	354.268 928	49
50	7.106 683	0.140 713	21.482 185	0.046 550	152.667 084	0.006 550	361.163 846	50
60	10.519 627	0.095 060	22.623 490	0.044 202	237.990 685	0.004 202	422.996 648	60
70	15.571 618	0.064 219	23.394 515	0.042 745	364.290 459	0.002 745	472.478 923	70
72	16.842 262	0.059 374	23.515 639	0.042 525	396.056 560	0.002 525	481.016 968	72
80	23.049 799	0.043 384	23.915 392	0.041 814	551.244 977	0.001 814	511.116 144	80
84	26.965 005	0.037 085	24.072 872	0.041 541	649.125 119	0.001 541	523.943 092	84
90	34.119 333	0.029 309	24.267 278	0.041 208	827.983 334	0.001 208	540.736 923	90
96	43.171 841	0.023 163	24.420 919	0.040 949	1 054.296 034	0.000 949	554.931 180	96
100	50.504 948	0.019 800	24.504 999	0.040 808	1 237.623 705	0.000 808	563.124 875	100
108	69.119 509	0.014 468	24.638 308	0.040 587	1 702.987 724	0.000 587	576.894 913	108
120	110.662 561	0.009 036	24.774 088	0.040 365	2 741.564 020	0.000 365	592.242 761	120
240	12 246.202 364	0.000 082	24.997 959	0.040 003	306 130.059 094	0.000 003	624.459 016	240
∞	∞	0	25.000 000	0.040 000	∞	0	625.000 000	∞

5%	Single Payment		Uniform Series				Arithmetic Gradient Series	5%
	Compound Amount Factor (CAF)	Present Worth Factor (PWF)	Present Worth Factor (PWF)	Capital Recovery Factor (CRF)	Compound Amount Factor (CAF)	Sinking Fund Factor (SFF)	Present Worth Factor (PWF)	
n	*F/P*	*P/F*	*P/A*	*A/P*	*F/A*	*A/F*	*P/G*	*n*
1	1.050 000	0.952 381	0.952 381	1.050 000	1.000 000	1.000 000	0.000 000	1
2	1.102 500	0.907 029	1.859 410	0.537 805	2.050 000	0.487 805	0.907 029	2
3	1.157 625	0.863 838	2.723 248	0.367 209	3.152 500	0.317 209	2.634 705	3
4	1.215 506	0.822 702	3.545 951	0.282 012	4.310 125	0.232 012	5.102 812	4
5	1.276 282	0.783 526	4.329 477	0.230 975	5.525 631	0.180 975	8.236 917	5
6	1.340 096	0.746 215	5.075 692	0.197 017	6.801 913	0.147 017	11.967 994	6
7	1.407 100	0.710 681	5.786 373	0.172 820	8.142 008	0.122 820	16.232 082	7
8	1.477 455	0.676 839	6.463 213	0.154 722	9.549 109	0.104 722	20.969 957	8
9	1.551 328	0.644 609	7.107 822	0.140 690	11.026 564	0.090 690	26.126 829	9
10	1.628 895	0.613 913	7.721 735	0.129 505	12.577 893	0.079 505	31.652 048	10
11	1.710 339	0.584 679	8.306 414	0.120 389	14.206 787	0.070 389	37.498 841	11
12	1.795 856	0.556 837	8.863 252	0.112 825	15.917 127	0.062 825	43.624 052	12
13	1.885 649	0.530 321	9.393 573	0.106 456	17.712 983	0.056 456	49.987 909	13
14	1.979 932	0.505 068	9.898 641	0.101 024	19.598 632	0.051 024	56.553 792	14
15	2.078 928	0.481 017	10.379 658	0.096 342	21.578 564	0.046 342	63.288 031	15
16	2.182 875	0.458 112	10.837 770	0.092 270	23.657 492	0.042 270	70.159 704	16
17	2.292 018	0.436 297	11.274 066	0.088 699	25.840 366	0.038 699	77.140 451	17
18	2.406 619	0.415 521	11.689 587	0.085 546	28.132 385	0.035 546	84.204 302	18
19	2.526 950	0.395 734	12.085 321	0.082 745	30.539 004	0.032 745	91.327 514	19
20	2.653 298	0.376 889	12.462 210	0.080 243	33.065 954	0.030 243	98.488 414	20
21	2.785 963	0.358 942	12.821 153	0.077 996	35.719 252	0.027 996	105.667 261	21
22	2.925 261	0.341 850	13.163 003	0.075 971	38.505 214	0.025 971	112.846 108	22
23	3.071 524	0.325 571	13.488 574	0.074 137	41.430 475	0.024 137	120.008 677	23
24	3.225 100	0.310 068	13.798 642	0.072 471	44.501 999	0.022 471	127.140 239	24
25	3.386 355	0.295 303	14.093 945	0.070 952	47.727 099	0.020 952	134.227 505	25
26	3.555 673	0.281 241	14.375 185	0.069 564	51.113 454	0.019 564	141.258 524	26
27	3.733 456	0.267 848	14.643 034	0.068 292	54.669 126	0.018 292	148.222 580	27
28	3.920 129	0.255 094	14.898 127	0.067 123	58.402 583	0.017 123	155.110 108	28
29	4.116 136	0.242 946	15.141 074	0.066 046	62.322 712	0.016 046	161.912 605	29
30	4.321 942	0.231 377	15.372 451	0.065 051	66.438 848	0.015 051	168.622 551	30
31	4.538 039	0.220 359	15.592 811	0.064 132	70.760 790	0.014 132	175.233 336	31
32	4.764 941	0.209 866	15.802 677	0.063 280	75.298 829	0.013 280	181.739 187	32
33	5.003 189	0.199 873	16.002 549	0.062 490	80.063 771	0.012 490	188.135 108	33
34	5.253 348	0.190 355	16.192 904	0.061 755	85.066 959	0.011 755	194.416 816	34
35	5.516 015	0.181 290	16.374 194	0.061 072	90.320 307	0.011 072	200.580 686	35

(cont.)

(cont.)

5%	Single Payment		Uniform Series				Arithmetic Gradient Series	5%
	Compound Amount Factor (CAF)	Present Worth Factor (PWF)	Present Worth Factor (PWF)	Capital Recovery Factor (CRF)	Compound Amount Factor (CAF)	Sinking Fund Factor (SFF)	Present Worth Factor (PWF)	
n	*F/P*	*P/F*	*P/A*	*A/P*	*F/A*	*A/F*	*P/G*	*n*
36	5.791 816	0.172 657	16.546 852	0.060 434	95.836 323	0.010 434	206.623 696	36
37	6.081 407	0.164 436	16.711 287	0.059 840	101.628 139	0.009 840	212.543 378	37
38	6.385 477	0.156 605	16.867 893	0.059 284	107.709 546	0.009 284	218.337 777	38
39	6.704 751	0.149 148	17.017 041	0.058 765	114.095 023	0.008 765	224.005 400	39
40	7.039 989	0.142 046	17.159 086	0.058 278	120.799 774	0.008 278	229.545 181	40
41	7.391 988	0.135 282	17.294 368	0.057 822	127.839 763	0.007 822	234.956 445	41
42	7.761 588	0.128 840	17.423 208	0.057 395	135.231 751	0.007 395	240.238 870	42
43	8.149 667	0.122 704	17.545 912	0.056 993	142.993 339	0.006 993	245.392 455	43
44	8.557 150	0.116 861	17.662 773	0.056 616	151.143 006	0.006 616	250.417 492	44
45	8.985 008	0.111 297	17.774 070	0.056 262	159.700 156	0.006 262	255.314 538	45
46	9.434 258	0.105 997	17.880 066	0.055 928	168.685 164	0.005 928	260.084 389	46
47	9.905 971	0.100 949	17.981 016	0.055 614	178.119 422	0.005 614	264.728 053	47
48	10.401 270	0.096 142	18.077 158	0.055 318	188.025 393	0.005 318	269.246 732	48
49	10.921 333	0.091 564	18.168 722	0.055 040	198.426 663	0.005 040	273.641 800	49
50	11.467 400	0.087 204	18.255 925	0.054 777	209.347 996	0.004 777	277.914 782	50
60	18.679 186	0.053 536	18.929 290	0.052 828	353.583 718	0.002 828	314.343 162	60
70	30.426 426	0.032 866	19.342 677	0.051 699	588.528 511	0.001 699	340.840 898	70
72	33.545 134	0.029 811	19.403 788	0.051 536	650.902 683	0.001 536	345.148 528	72
80	49.561 441	0.020 177	19.596 460	0.051 030	971.228 821	0.001 030	359.646 048	80
84	60.242 241	0.016 600	19.668 007	0.050 844	1 184.844 828	0.000 844	365.472 732	84
90	80.730 365	0.012 387	19.752 262	0.050 627	1 594.607 301	0.000 627	372.748 792	90
96	108.186 410	0.009 243	19.815 134	0.050 466	2 143.728 205	0.000 466	378.555 532	96
100	131.501 258	0.007 604	19.847 910	0.050 383	2 610.025 157	0.000 383	381.749 224	100
108	194.287 249	0.005 147	19.897 060	0.050 259	3 865.744 985	0.000 259	386.823 634	108
120	348.911 986	0.002 866	19.942 679	0.050 144	6 958.239 713	0.000 144	391.975 054	120
∞	∞	0	20.000 000	0.050 000	∞	0	400	∞

6%	Single Payment		Uniform Series				Arithmetic Gradient Series	6%
	Compound Amount Factor (CAF)	Present Worth Factor (PWF)	Present Worth Factor (PWF)	Capital Recovery Factor (CRF)	Compound Amount Factor (CAF)	Sinking Fund Factor (SFF)	Present Worth Factor (PWF)	
n	*F/P*	*P/F*	*P/A*	*A/P*	*F/A*	*A/F*	*P/G*	*n*
1	1.060 000	0.943 396	0.943 396	1.060 000	1.000 000	1.000 000	0.000 000	1
2	1.123 600	0.889 996	1.833 393	0.545 437	2.060 000	0.485 437	0.889 996	2
3	1.191 016	0.839 619	2.673 012	0.374 110	3.183 600	0.314 110	2.569 235	3
4	1.262 477	0.792 094	3.465 106	0.288 591	4.374 616	0.228 591	4.945 516	4
5	1.338 226	0.747 258	4.212 364	0.237 396	5.637 093	0.177 396	7.934 549	5
6	1.418 519	0.704 961	4.917 324	0.203 363	6.975 319	0.143 363	11.459 351	6
7	1.503 630	0.665 057	5.582 381	0.179 135	8.393 838	0.119 135	15.449 694	7
8	1.593 848	0.627 412	6.209 794	0.161 036	9.897 468	0.101 036	19.841 581	8
9	1.689 479	0.591 898	6.801 692	0.147 022	11.491 316	0.087 022	24.576 768	9
10	1.790 848	0.558 395	7.360 087	0.135 868	13.180 795	0.075 868	29.602 321	10
11	1.898 299	0.526 788	7.886 875	0.126 793	14.971 643	0.066 793	34.870 197	11
12	2.012 196	0.496 969	8.383 844	0.119 277	16.869 941	0.059 277	40.336 860	12
13	2.132 928	0.468 839	8.852 683	0.112 960	18.882 138	0.052 960	45.962 928	13
14	2.260 904	0.442 301	9.294 984	0.107 585	21.015 066	0.047 585	51.712 840	14
15	2.396 558	0.417 265	9.712 249	0.102 963	23.275 970	0.042 963	57.554 551	15
16	2.540 352	0.393 646	10.105 895	0.098 952	25.672 528	0.038 952	63.459 246	16
17	2.692 773	0.371 364	10.477 260	0.095 445	28.212 880	0.035 445	69.401 076	17
18	2.854 339	0.350 344	10.827 603	0.092 357	30.905 653	0.032 357	75.356 921	18
19	3.025 600	0.330 513	11.158 116	0.089 621	33.759 992	0.029 621	81.306 155	19
20	3.207 135	0.311 805	11.469 921	0.087 185	36.785 591	0.027 185	87.230 445	20
21	3.399 564	0.294 155	11.764 077	0.085 005	39.992 727	0.025 005	93.113 553	21
22	3.603 537	0.277 505	12.041 582	0.083 046	43.392 290	0.023 046	98.941 160	22
23	3.819 750	0.261 797	12.303 379	0.081 278	46.995 828	0.021 278	104.700 700	23
24	4.048 935	0.246 979	12.550 358	0.079 679	50.815 577	0.019 679	110.381 206	24
25	4.291 871	0.232 999	12.783 356	0.078 227	54.864 512	0.018 227	115.973 173	25
26	4.549 383	0.219 810	13.003 166	0.076 904	59.156 383	0.016 904	121.468 424	26
27	4.822 346	0.207 368	13.210 534	0.075 697	63.705 766	0.015 697	126.859 991	27
28	5.111 687	0.195 630	13.406 164	0.074 593	68.528 112	0.014 593	132.142 005	28
29	5.418 388	0.184 557	13.590 721	0.073 580	73.639 798	0.013 580	137.309 593	29
30	5.743 491	0.174 110	13.764 831	0.072 649	79.058 186	0.012 649	142.358 787	30
31	6.088 101	0.164 255	13.929 086	0.071 792	84.801 677	0.011 792	147.286 432	31
32	6.453 387	0.154 957	14.084 043	0.071 002	90.889 778	0.011 002	152.090 112	32
33	6.840 590	0.146 186	14.230 230	0.070 273	97.343 165	0.010 273	156.768 071	33
34	7.251 025	0.137 912	14.368 141	0.069 598	104.183 755	0.009 598	161.319 151	34
35	7.686 087	0.130 105	14.498 246	0.068 974	111.434 780	0.008 974	165.742 729	35

(cont.)

(cont.)

6%	Single Payment		Uniform Series				Arithmetic Gradient Series	6%
	Compound Amount Factor (CAF)	Present Worth Factor (PWF)	Present Worth Factor (PWF)	Capital Recovery Factor (CRF)	Compound Amount Factor (CAF)	Sinking Fund Factor (SFF)	Present Worth Factor (PWF)	
n	*F/P*	*P/F*	*P/A*	*A/P*	*F/A*	*A/F*	*P/G*	*n*
36	8.147 252	0.122 741	14.620 987	0.068 395	119.120 867	0.008 395	170.038 656	36
37	8.636 087	0.115 793	14.736 780	0.067 857	127.268 119	0.007 857	174.207 210	37
38	9.154 252	0.109 239	14.846 019	0.067 358	135.904 206	0.007 358	178.249 048	38
39	9.703 507	0.103 056	14.949 075	0.066 894	145.058 458	0.006 894	182.165 157	39
40	10.285 718	0.097 222	15.046 297	0.066 462	154.761 966	0.006 462	185.956 823	40
41	10.902 861	0.091 719	15.138 016	0.066 059	165.047 684	0.006 059	189.625 585	41
42	11.557 033	0.086 527	15.224 543	0.065 683	175.950 545	0.005 683	193.173 208	42
43	12.250 455	0.081 630	15.306 173	0.065 333	187.507 577	0.005 333	196.601 652	43
44	12.985 482	0.077 009	15.383 182	0.065 006	199.758 032	0.005 006	199.913 043	44
45	13.764 611	0.072 650	15.455 832	0.064 700	212.743 514	0.004 700	203.109 646	45
46	14.590 487	0.068 538	15.524 370	0.064 415	226.508 125	0.004 415	206.193 847	46
47	15.465 917	0.064 658	15.589 028	0.064 148	241.098 612	0.004 148	209.168 129	47
48	16.393 872	0.060 998	15.650 027	0.063 898	256.564 529	0.003 898	212.035 054	48
49	17.377 504	0.057 546	15.707 572	0.063 664	272.958 401	0.003 664	214.797 246	49
50	18.420 154	0.054 288	15.761 861	0.063 444	290.335 905	0.003 444	217.457 376	50
60	32.987 691	0.030 314	16.161 428	0.061 876	533.128 181	0.001 876	239.042 791	60
70	59.075 930	0.016 927	16.384 544	0.061 033	967.932 170	0.001 033	253.327 135	70
72	66.377 715	0.015 065	16.415 578	0.060 918	1 089.628 586	0.000 918	255.514 616	72
80	105.795 993	0.009 452	16.509 131	0.060 573	1 746.599 891	0.000 573	262.549 308	80
84	133.565 004	0.007 487	16.541 883	0.060 453	2 209.416 737	0.000 453	265.216 270	84
90	189.464 511	0.005 278	16.578 699	0.060 318	3 141.075 187	0.000 318	268.394 607	90
96	268.759 030	0.003 721	16.604 653	0.060 224	4 462.650 505	0.000 224	270.790 932	96
100	339.302 084	0.002 947	16.617 546	0.060 177	5 638.368 059	0.000 177	272.047 060	100
108	540.795 972	0.001 849	16.635 848	0.060 111	8 996.599 542	0.000 111	273.935 704	108
120	1 088.187 748	0.000 919	16.651 351	0.060 055	18 119.795 797	0.000 055	275.684 593	120
∞	∞	0	16.666 667	0.060 000	∞	0	277.777 778	∞

7%	Single Payment		Uniform Series				Arithmetic Gradient Series	7%
	Compound Amount Factor (CAF)	Present Worth Factor (PWF)	Present Worth Factor (PWF)	Capital Recovery Factor (CRF)	Compound Amount Factor (CAF)	Sinking Fund Factor (SFF)	Present Worth Factor (PWF)	
n	*F/P*	*P/F*	*P/A*	*A/P*	*F/A*	*A/F*	*P/G*	*n*
1	1.070 000	0.934 579	0.934 579	1.070 000	1.000 000	1.000 000	0.000 000	1
2	1.144 900	0.873 439	1.808 018	0.553 092	2.070 000	0.483 092	0.873 439	2
3	1.225 043	0.816 298	2.624 316	0.381 052	3.214 900	0.311 052	2.506 034	3
4	1.310 796	0.762 895	3.387 211	0.295 228	4.439 943	0.225 228	4.794 720	4
5	1.402 552	0.712 986	4.100 197	0.243 891	5.750 739	0.173 891	7.646 665	5
6	1.500 730	0.666 342	4.766 540	0.209 796	7.153 291	0.139 796	10.978 376	6
7	1.605 781	0.622 750	5.389 289	0.185 553	8.654 021	0.115 553	14.714 874	7
8	1.718 186	0.582 009	5.971 299	0.167 468	10.259 803	0.097 468	18.788 938	8
9	1.838 459	0.543 934	6.515 232	0.153 486	11.977 989	0.083 486	23.140 408	9
10	1.967 151	0.508 349	7.023 582	0.142 378	13.816 448	0.072 378	27.715 552	10
11	2.104 852	0.475 093	7.498 674	0.133 357	15.783 599	0.063 357	32.466 480	11
12	2.252 192	0.444 012	7.942 686	0.125 902	17.888 451	0.055 902	37.350 611	12
13	2.409 845	0.414 964	8.357 651	0.119 651	20.140 643	0.049 651	42.330 185	13
14	2.578 534	0.387 817	8.745 468	0.114 345	22.550 488	0.044 345	47.371 809	14
15	2.759 032	0.362 446	9.107 914	0.109 795	25.129 022	0.039 795	52.446 053	15
16	2.952 164	0.338 735	9.446 649	0.105 858	27.888 054	0.035 858	57.527 072	16
17	3.158 815	0.316 574	9.763 223	0.102 425	30.840 217	0.032 425	62.592 262	17
18	3.379 932	0.295 864	10.059 087	0.099 413	33.999 033	0.029 413	67.621 949	18
19	3.616 528	0.276 508	10.335 595	0.096 753	37.378 965	0.026 753	72.599 099	19
20	3.869 684	0.258 419	10.594 014	0.094 393	40.995 492	0.024 393	77.509 060	20
21	4.140 562	0.241 513	10.835 527	0.092 289	44.865 177	0.022 289	82.339 322	21
22	4.430 402	0.225 713	11.061 240	0.090 406	49.005 739	0.020 406	87.079 298	22
23	4.740 530	0.210 947	11.272 187	0.088 714	53.436 141	0.018 714	91.720 129	23
24	5.072 367	0.197 147	11.469 334	0.087 189	58.176 671	0.017 189	96.254 502	24
25	5.427 433	0.184 249	11.653 583	0.085 811	63.249 038	0.015 811	100.676 482	25
26	5.807 353	0.172 195	11.825 779	0.084 561	68.676 470	0.014 561	104.981 369	26
27	6.213 868	0.160 930	11.986 709	0.083 426	74.483 823	0.013 426	109.165 559	27
28	6.648 838	0.150 402	12.137 111	0.082 392	80.697 691	0.012 392	113.226 419	28
29	7.114 257	0.140 563	12.277 674	0.081 449	87.346 529	0.011 449	117.162 177	29
30	7.612 255	0.131 367	12.409 041	0.080 586	94.460 786	0.010 586	120.971 824	30
31	8.145 113	0.122 773	12.531 814	0.079 797	102.073 041	0.009 797	124.655 014	31
32	8.715 271	0.114 741	12.646 555	0.079 073	110.218 154	0.009 073	128.211 989	32
33	9.325 340	0.107 235	12.753 790	0.078 408	118.933 425	0.008 408	131.643 499	33
34	9.978 114	0.100 219	12.854 009	0.077 797	128.258 765	0.007 797	134.950 738	34
35	10.676 581	0.093 663	12.947 672	0.077 234	138.236 878	0.007 234	138.135 278	35

(cont.)

(cont.)

7%	Single Payment		Uniform Series				Arithmetic Gradient Series	7%
	Compound Amount Factor (CAF)	Present Worth Factor (PWF)	Present Worth Factor (PWF)	Capital Recovery Factor (CRF)	Compound Amount Factor (CAF)	Sinking Fund Factor (SFF)	Present Worth Factor (PWF)	
n	*F/P*	*P/F*	*P/A*	*A/P*	*F/A*	*A/F*	*P/G*	*n*
36	11.423 942	0.087 535	13.035 208	0.076 715	148.913 460	0.006 715	141.199 019	36
37	12.223 618	0.081 809	13.117 017	0.076 237	160.337 402	0.006 237	144.144 137	37
38	13.079 271	0.076 457	13.193 473	0.075 795	172.561 020	0.005 795	146.973 041	38
39	13.994 820	0.071 455	13.264 928	0.075 387	185.640 292	0.005 387	149.688 331	39
40	14.974 458	0.066 780	13.331 709	0.075 009	199.635 112	0.005 009	152.292 766	40
41	16.022 670	0.062 412	13.394 120	0.074 660	214.609 570	0.004 660	154.789 229	41
42	17.144 257	0.058 329	13.452 449	0.074 336	230.632 240	0.004 336	157.180 700	42
43	18.344 355	0.054 513	13.506 962	0.074 036	247.776 496	0.004 036	159.470 233	43
44	19.628 460	0.050 946	13.557 908	0.073 758	266.120 851	0.003 758	161.660 929	44
45	21.002 452	0.047 613	13.605 522	0.073 500	285.749 311	0.003 500	163.755 923	45
46	22.472 623	0.044 499	13.650 020	0.073 260	306.751 763	0.003 260	165.758 359	46
47	24.045 707	0.041 587	13.691 608	0.073 037	329.224 386	0.003 037	167.671 383	47
48	25.728 907	0.038 867	13.730 474	0.072 831	353.270 093	0.002 831	169.498 122	48
49	27.529 930	0.036 324	13.766 799	0.072 639	378.999 000	0.002 639	171.241 679	49
50	29.457 025	0.033 948	13.800 746	0.072 460	406.528 929	0.002 460	172.905 119	50
60	57.946 427	0.017 257	14.039 181	0.071 229	813.520 383	0.001 229	185.767 743	60
70	113.989 392	0.008 773	14.160 389	0.070 620	1 614.134 174	0.000 620	193.518 530	70
72	130.506 455	0.007 662	14.176 251	0.070 541	1 850.092 216	0.000 541	194.636 483	72
80	224.234 388	0.004 460	14.222 005	0.070 314	3 189.062 680	0.000 314	198.074 799	80
84	293.925 541	0.003 402	14.237 111	0.070 239	4 184.650 579	0.000 239	199.304 635	84
90	441.102 980	0.002 267	14.253 328	0.070 159	6 287.185 427	0.000 159	200.704 199	90
96	661.976 630	0.001 511	14.264 134	0.070 106	9 442.523 288	0.000 106	201.701 624	96
100	867.716 326	0.001 152	14.269 251	0.070 081	12 381.661 794	0.000 081	202.200 081	100
108	1 490.898 199	0.000 671	14.276 132	0.070 047	21 284.259 980	0.000 047	202.909 897	108
120	3 357.788 383	0.000 298	14.281 460	0.070 021	47 954.119 756	0.000 021	203.510 314	120
∞	∞	0	14.285 714	0.070 000	∞	0	204.081 633	∞

8%	Single Payment		Uniform Series				Arithmetic Gradient Series	8%
	Compound Amount Factor (CAF)	Present Worth Factor (PWF)	Present Worth Factor (PWF)	Capital Recovery Factor (CRF)	Compound Amount Factor (CAF)	Sinking Fund Factor (SFF)	Present Worth Factor (PWF)	
n	*F/P*	*P/F*	*P/A*	*A/P*	*F/A*	*A/F*	*P/G*	*n*
1	1.080 000	0.925 926	0.925 926	1.080 000	1.000 000	1.000 000	0.000 000	1
2	1.166 400	0.857 339	1.783 265	0.560 769	2.080 000	0.480 769	0.857 339	2
3	1.259 712	0.793 832	2.577 097	0.388 034	3.246 400	0.308 034	2.445 003	3
4	1.360 489	0.735 030	3.312 127	0.301 921	4.506 112	0.221 921	4.650 093	4
5	1.469 328	0.680 583	3.992 710	0.250 456	5.866 601	0.170 456	7.372 426	5
6	1.586 874	0.630 170	4.622 880	0.216 315	7.335 929	0.136 315	10.523 274	6
7	1.713 824	0.583 490	5.206 370	0.192 072	8.922 803	0.112 072	14.024 216	7
8	1.850 930	0.540 269	5.746 639	0.174 015	10.636 628	0.094 015	17.806 098	8
9	1.999 005	0.500 249	6.246 888	0.160 080	12.487 558	0.080 080	21.808 090	9
10	2.158 925	0.463 193	6.710 081	0.149 029	14.486 562	0.069 029	25.976 831	10
11	2.331 639	0.428 883	7.138 964	0.140 076	16.645 487	0.060 076	30.265 660	11
12	2.518 170	0.397 114	7.536 078	0.132 695	18.977 126	0.052 695	34.633 911	12
13	2.719 624	0.367 698	7.903 776	0.126 522	21.495 297	0.046 522	39.046 287	13
14	2.937 194	0.340 461	8.244 237	0.121 297	24.214 920	0.041 297	43.472 280	14
15	3.172 169	0.315 242	8.559 479	0.116 830	27.152 114	0.036 830	47.885 664	15
16	3.425 943	0.291 890	8.851 369	0.112 977	30.324 283	0.032 977	52.264 021	16
17	3.700 018	0.270 269	9.121 638	0.109 629	33.750 226	0.029 629	56.588 324	17
18	3.996 019	0.250 249	9.371 887	0.106 702	37.450 244	0.026 702	60.842 558	18
19	4.315 701	0.231 712	9.603 599	0.104 128	41.446 263	0.024 128	65.013 375	19
20	4.660 957	0.214 548	9.818 147	0.101 852	45.761 964	0.021 852	69.089 791	20
21	5.033 834	0.198 656	10.016 803	0.099 832	50.422 921	0.019 832	73.062 906	21
22	5.436 540	0.183 941	10.200 744	0.098 032	55.456 755	0.018 032	76.925 656	22
23	5.871 464	0.170 315	10.371 059	0.096 422	60.893 296	0.016 422	80.672 593	23
24	6.341 181	0.157 699	10.528 758	0.094 978	66.764 759	0.014 978	84.299 677	24
25	6.848 475	0.146 018	10.674 776	0.093 679	73.105 940	0.013 679	87.804 107	25
26	7.396 353	0.135 202	10.809 978	0.092 507	79.954 415	0.012 507	91.184 151	26
27	7.988 061	0.125 187	10.935 165	0.091 448	87.350 768	0.011 448	94.439 008	27
28	8.627 106	0.115 914	11.051 078	0.090 489	95.338 830	0.010 489	97.568 679	28
29	9.317 275	0.107 328	11.158 406	0.089 619	103.965 936	0.009 619	100.573 849	29
30	10.062 657	0.099 377	11.257 783	0.088 827	113.283 211	0.008 827	103.455 792	30
31	10.867 669	0.092 016	11.349 799	0.088 107	123.345 868	0.008 107	106.216 274	31
32	11.737 083	0.085 200	11.434 999	0.087 451	134.213 537	0.007 451	108.857 475	32
33	12.676 050	0.078 889	11.513 888	0.086 852	145.950 620	0.006 852	111.381 921	33
34	13.690 134	0.073 045	11.586 934	0.086 304	158.626 670	0.006 304	113.792 416	34
35	14.785 344	0.067 635	11.654 568	0.085 803	172.316 804	0.005 803	116.091 990	35

(cont.)

(cont.)

8%	Single Payment		Uniform Series				Arithmetic Gradient Series	8%
	Compound Amount Factor (CAF)	Present Worth Factor (PWF)	Present Worth Factor (PWF)	Capital Recovery Factor (CRF)	Compound Amount Factor (CAF)	Sinking Fund Factor (SFF)	Present Worth Factor (PWF)	
n	*F/P*	*P/F*	*P/A*	*A/P*	*F/A*	*A/F*	*P/G*	*n*
36	15.968 172	0.062 625	11.717 193	0.085 345	187.102 148	0.005 345	118.283 850	36
37	17.245 626	0.057 986	11.775 179	0.084 924	203.070 320	0.004 924	120.371 336	37
38	18.625 276	0.053 690	11.828 869	0.084 539	220.315 945	0.004 539	122.357 884	38
39	20.115 298	0.049 713	11.878 582	0.084 185	238.941 221	0.004 185	124.246 994	39
40	21.724 521	0.046 031	11.924 613	0.083 860	259.056 519	0.003 860	126.042 200	40
41	23.462 483	0.042 621	11.967 235	0.083 561	280.781 040	0.003 561	127.747 049	41
42	25.339 482	0.039 464	12.006 699	0.083 287	304.243 523	0.003 287	129.365 078	42
43	27.366 640	0.036 541	12.043 240	0.083 034	329.583 005	0.003 034	130.899 793	43
44	29.555 972	0.033 834	12.077 074	0.082 802	356.949 646	0.002 802	132.354 660	44
45	31.920 449	0.031 328	12.108 402	0.082 587	386.505 617	0.002 587	133.733 086	45
46	34.474 085	0.029 007	12.137 409	0.082 390	418.426 067	0.002 390	135.038 415	46
47	37.232 012	0.026 859	12.164 267	0.082 208	452.900 152	0.002 208	136.273 911	47
48	40.210 573	0.024 869	12.189 136	0.082 040	490.132 164	0.002 040	137.442 758	48
49	43.427 419	0.023 027	12.212 163	0.081 886	530.342 737	0.001 886	138.548 050	49
50	46.901 613	0.021 321	12.233 485	0.081 743	573.770 156	0.001 743	139.592 790	50
60	101.257 064	0.009 876	12.376 552	0.080 798	1 253.213 296	0.000 798	147.300 007	60
70	218.606 406	0.004 574	12.442 820	0.080 368	2 720.080 074	0.000 368	151.532 618	70
72	254.982 512	0.003 922	12.450 977	0.080 315	3 174.781 398	0.000 315	152.107 559	72
80	471.954 834	0.002 119	12.473 514	0.080 170	5 886.935 428	0.000 170	153.800 083	80
84	642.089 342	0.001 557	12.480 532	0.080 125	8 013.616 770	0.000 125	154.371 367	84
90	1 018.915 089	0.000 981	12.487 732	0.080 079	12 723.938 616	0.000 079	154.992 535	90
96	1 616.890 192	0.000 618	12.492 269	0.080 050	20 198.627 405	0.000 050	155.411 198	96
100	2 199.761 256	0.000 455	12.494 318	0.080 036	27 484.515 704	0.000 036	155.610 726	100
108	4 071.604 565	0.000 246	12.496 930	0.080 020	50 882.557 060	0.000 020	155.880 060	108
120	10 252.992 943	0.000 098	12.498 781	0.080 008	128 149.911 781	0.000 008	156.088 462	120
∞	∞	0	12.500 000	0.080 000	∞	0	156.250 000	∞

9%	Single Payment		Uniform Series				Arithmetic Gradient Series	9%
	Compound Amount Factor (CAF)	Present Worth Factor (PWF)	Present Worth Factor (PWF)	Capital Recovery Factor (CRF)	Compound Amount Factor (CAF)	Sinking Fund Factor (SFF)	Present Worth Factor (PWF)	
n	*F/P*	*P/F*	*P/A*	*A/P*	*F/A*	*A/F*	*P/G*	*n*
1	1.090 000	0.917 431	0.917 431	1.090 000	1.000 000	1.000 000	0.000 000	1
2	1.188 100	0.841 680	1.759 111	0.568 469	2.090 000	0.478 469	0.841 680	2
3	1.295 029	0.772 183	2.531 295	0.395 055	3.278 100	0.305 055	2.386 047	3
4	1.411 582	0.708 425	3.239 720	0.308 669	4.573 129	0.218 669	4.511 323	4
5	1.538 624	0.649 931	3.889 651	0.257 092	5.984 711	0.167 092	7.111 048	5
6	1.677 100	0.596 267	4.485 919	0.222 920	7.523 335	0.132 920	10.092 385	6
7	1.828 039	0.547 034	5.032 953	0.198 691	9.200 435	0.108 691	13.374 590	7
8	1.992 563	0.501 866	5.534 819	0.180 674	11.028 474	0.090 674	16.887 654	8
9	2.171 893	0.460 428	5.995 247	0.166 799	13.021 036	0.076 799	20.571 076	9
10	2.367 364	0.422 411	6.417 658	0.155 820	15.192 930	0.065 820	24.372 774	10
11	2.580 426	0.387 533	6.805 191	0.146 947	17.560 293	0.056 947	28.248 102	11
12	2.812 665	0.355 535	7.160 725	0.139 651	20.140 720	0.049 651	32.158 984	12
13	3.065 805	0.326 179	7.486 904	0.133 567	22.953 385	0.043 567	36.073 128	13
14	3.341 727	0.299 246	7.786 150	0.128 433	26.019 189	0.038 433	39.963 332	14
15	3.642 482	0.274 538	8.060 688	0.124 059	29.360 916	0.034 059	43.806 865	15
16	3.970 306	0.251 870	8.312 558	0.120 300	33.003 399	0.030 300	47.584 911	16
17	4.327 633	0.231 073	8.543 631	0.117 046	36.973 705	0.027 046	51.282 082	17
18	4.717 120	0.211 994	8.755 625	0.114 212	41.301 338	0.024 212	54.885 975	18
19	5.141 661	0.194 490	8.950 115	0.111 730	46.018 458	0.021 730	58.386 789	19
20	5.604 411	0.178 431	9.128 546	0.109 546	51.160 120	0.019 546	61.776 976	20
21	6.108 808	0.163 698	9.292 244	0.107 617	56.764 530	0.017 617	65.050 938	21
22	6.658 600	0.150 182	9.442 425	0.105 905	62.873 338	0.015 905	68.204 754	22
23	7.257 874	0.137 781	9.580 207	0.104 382	69.531 939	0.014 382	71.235 944	23
24	7.911 083	0.126 405	9.706 612	0.103 023	76.789 813	0.013 023	74.143 258	24
25	8.623 081	0.115 968	9.822 580	0.101 806	84.700 896	0.011 806	76.926 486	25
26	9.399 158	0.106 393	9.928 972	0.100 715	93.323 977	0.010 715	79.586 298	26
27	10.245 082	0.097 608	10.026 580	0.099 735	102.723 135	0.009 735	82.124 101	27
28	11.167 140	0.089 548	10.116 128	0.098 852	112.968 217	0.008 852	84.541 910	28
29	12.172 182	0.082 155	10.198 283	0.098 056	124.135 356	0.008 056	86.842 237	29
30	13.267 678	0.075 371	10.273 654	0.097 336	136.307 539	0.007 336	89.028 000	30
31	14.461 770	0.069 148	10.342 802	0.096 686	149.575 217	0.006 686	91.102 434	31
32	15.763 329	0.063 438	10.406 240	0.096 096	164.036 987	0.006 096	93.069 024	32
33	17.182 028	0.058 200	10.464 441	0.095 562	179.800 315	0.005 562	94.931 435	33
34	18.728 411	0.053 395	10.517 835	0.095 077	196.982 344	0.005 077	96.693 464	34
35	20.413 968	0.048 986	10.566 821	0.094 636	215.710 755	0.004 636	98.358 990	35

(cont.)

(cont.)

9%	Single Payment		Uniform Series				Arithmetic Gradient Series	9%
	Compound Amount Factor (CAF)	Present Worth Factor (PWF)	Present Worth Factor (PWF)	Capital Recovery Factor (CRF)	Compound Amount Factor (CAF)	Sinking Fund Factor (SFF)	Present Worth Factor (PWF)	
n	*F/P*	*P/F*	*P/A*	*A/P*	*F/A*	*A/F*	*P/G*	*n*
36	22.251 225	0.044 941	10.611 763	0.094 235	236.124 723	0.004 235	99.931 937	36
37	24.253 835	0.041 231	10.652 993	0.093 870	258.375 948	0.003 870	101.416 239	37
38	26.436 680	0.037 826	10.690 820	0.093 538	282.629 783	0.003 538	102.815 809	38
39	28.815 982	0.034 703	10.725 523	0.093 236	309.066 463	0.003 236	104.134 522	39
40	31.409 420	0.031 838	10.757 360	0.092 960	337.882 445	0.002 960	105.376 188	40
41	34.236 268	0.029 209	10.786 569	0.092 708	369.291 865	0.002 708	106.544 539	41
42	37.317 532	0.026 797	10.813 366	0.092 478	403.528 133	0.002 478	107.643 219	42
43	40.676 110	0.024 584	10.837 950	0.092 268	440.845 665	0.002 268	108.675 766	43
44	44.336 960	0.022 555	10.860 505	0.092 077	481.521 775	0.002 077	109.645 611	44
45	48.327 286	0.020 692	10.881 197	0.091 902	525.858 734	0.001 902	110.556 070	45
46	52.676 742	0.018 984	10.900 181	0.091 742	574.186 021	0.001 742	111.410 337	46
47	57.417 649	0.017 416	10.917 597	0.091 595	626.862 762	0.001 595	112.211 484	47
48	62.585 237	0.015 978	10.933 575	0.091 461	684.280 411	0.001 461	112.962 460	48
49	68.217 908	0.014 659	10.948 234	0.091 339	746.865 648	0.001 339	113.666 088	49
50	74.357 520	0.013 449	10.961 683	0.091 227	815.083 556	0.001 227	114.325 066	50
60	176.031 292	0.005 681	11.047 991	0.090 514	1 944.792 133	0.000 514	118.968 250	60
70	416.730 086	0.002 400	11.084 449	0.090 216	4 619.223 180	0.000 216	121.294 156	70
72	495.117 015	0.002 020	11.088 670	0.090 182	5 490.189 060	0.000 182	121.591 662	72
80	986.551 668	0.001 014	11.099 849	0.090 091	10 950.574 090	0.000 091	122.430 644	80
84	1 392.598 192	0.000 718	11.103 132	0.090 065	15 462.202 134	0.000 065	122.697 928	84
90	2 335.526 582	0.000 428	11.106 354	0.090 039	25 939.184 247	0.000 039	122.975 761	90
96	3 916.911 890	0.000 255	11.108 274	0.090 023	43 510.132 110	0.000 023	123.152 948	96
100	5 529.040 792	0.000 181	11.109 102	0.090 016	61 422.675 465	0.000 016	123.233 502	100
108	11 016.960 126	0.000 091	11.110 103	0.090 008	122 399.556 957	0.000 008	123.336 661	108
120	30 987.015 749	0.000 032	11.110 753	0.090 003	344 289.063 880	0.000 003	123.409 777	120
∞	∞	0	11.111 111	0.090 000	∞	0	123.456 790	∞

10%	Single Payment		Uniform Series				Arithmetic Gradient Series	10%
	Compound Amount Factor (CAF)	Present Worth Factor (PWF)	Present Worth Factor (PWF)	Capital Recovery Factor (CRF)	Compound Amount Factor (CAF)	Sinking Fund Factor (SFF)	Present Worth Factor (PWF)	
n	*F/P*	*P/F*	*P/A*	*A/P*	*F/A*	*A/F*	*P/G*	*n*
1	1.100 000	0.909 091	0.909 091	1.100 000	1.000 000	1.000 000	0.000 000	1
2	1.210 000	0.826 446	1.735 537	0.576 190	2.100 000	0.476 190	0.826 446	2
3	1.331 000	0.751 315	2.486 852	0.402 115	3.310 000	0.302 115	2.329 076	3
4	1.464 100	0.683 013	3.169 865	0.315 471	4.641 000	0.215 471	4.378 116	4
5	1.610 510	0.620 921	3.790 787	0.263 797	6.105 100	0.163 797	6.861 802	5
6	1.771 561	0.564 474	4.355 261	0.229 607	7.715 610	0.129 607	9.684 171	6
7	1.948 717	0.513 158	4.868 419	0.205 405	9.487 171	0.105 405	12.763 120	7
8	2.143 589	0.466 507	5.334 926	0.187 444	11.435 888	0.087 444	16.028 672	8
9	2.357 948	0.424 098	5.759 024	0.173 641	13.579 477	0.073 641	19.421 453	9
10	2.593 742	0.385 543	6.144 567	0.162 745	15.937 425	0.062 745	22.891 342	10
11	2.853 117	0.350 494	6.495 061	0.153 963	18.531 167	0.053 963	26.396 281	11
12	3.138 428	0.318 631	6.813 692	0.146 763	21.384 284	0.046 763	29.901 220	12
13	3.452 271	0.289 664	7.103 356	0.140 779	24.522 712	0.040 779	33.377 193	13
14	3.797 498	0.263 331	7.366 687	0.135 746	27.974 983	0.035 746	36.800 499	14
15	4.177 248	0.239 392	7.606 080	0.131 474	31.772 482	0.031 474	40.151 988	15
16	4.594 973	0.217 629	7.823 709	0.127 817	35.949 730	0.027 817	43.416 425	16
17	5.054 470	0.197 845	8.021 553	0.124 664	40.544 703	0.024 664	46.581 939	17
18	5.559 917	0.179 859	8.201 412	0.121 930	45.599 173	0.021 930	49.639 539	18
19	6.115 909	0.163 508	8.364 920	0.119 547	51.159 090	0.019 547	52.582 683	19
20	6.727 500	0.148 644	8.513 564	0.117 460	57.274 999	0.017 460	55.406 912	20
21	7.400 250	0.135 131	8.648 694	0.115 624	64.002 499	0.015 624	58.109 523	21
22	8.140 275	0.122 846	8.771 540	0.114 005	71.402 749	0.014 005	60.689 288	22
23	8.954 302	0.111 678	8.883 218	0.112 572	79.543 024	0.012 572	63.146 208	23
24	9.849 733	0.101 526	8.984 744	0.111 300	88.497 327	0.011 300	65.481 297	24
25	10.834 706	0.092 296	9.077 040	0.110 168	98.347 059	0.010 168	67.696 401	25
26	11.918 177	0.083 905	9.160 945	0.109 159	109.181 765	0.009 159	69.794 037	26
27	13.109 994	0.076 278	9.237 223	0.108 258	121.099 942	0.008 258	71.777 257	27
28	14.420 994	0.069 343	9.306 567	0.107 451	134.209 936	0.007 451	73.649 527	28
29	15.863 093	0.063 039	9.369 606	0.106 728	148.630 930	0.006 728	75.414 631	29
30	17.449 402	0.057 309	9.426 914	0.106 079	164.494 023	0.006 079	77.076 579	30
31	19.194 342	0.052 099	9.479 013	0.105 496	181.943 425	0.005 496	78.639 539	31
32	21.113 777	0.047 362	9.526 376	0.104 972	201.137 767	0.004 972	80.107 775	32
33	23.225 154	0.043 057	9.569 432	0.104 499	222.251 544	0.004 499	81.485 591	33
34	25.547 670	0.039 143	9.608 575	0.104 074	245.476 699	0.004 074	82.777 294	34
35	28.102 437	0.035 584	9.644 159	0.103 690	271.024 368	0.003 690	83.987 154	35

(cont.)

(cont.)

10%	Single Payment		Uniform Series				Arithmetic Gradient Series	10%
	Compound Amount Factor (CAF)	Present Worth Factor (PWF)	Present Worth Factor (PWF)	Capital Recovery Factor (CRF)	Compound Amount Factor (CAF)	Sinking Fund Factor (SFF)	Present Worth Factor (PWF)	
n	*F/P*	*P/F*	*P/A*	*A/P*	*F/A*	*A/F*	*P/G*	*n*
36	30.912 681	0.032 349	9.676 508	0.103 343	299.126 805	0.003 343	85.119 375	36
37	34.003 949	0.029 408	9.705 917	0.103 030	330.039 486	0.003 030	86.178 076	37
38	37.404 343	0.026 735	9.732 651	0.102 747	364.043 434	0.002 747	87.167 266	38
39	41.144 778	0.024 304	9.756 956	0.102 491	401.447 778	0.002 491	88.090 834	39
40	45.259 256	0.022 095	9.779 051	0.102 259	442.592 556	0.002 259	88.952 536	40
41	49.785 181	0.020 086	9.799 137	0.102 050	487.851 811	0.002 050	89.755 988	41
42	54.763 699	0.018 260	9.817 397	0.101 860	537.636 992	0.001 860	90.504 659	42
43	60.240 069	0.016 600	9.833 998	0.101 688	592.400 692	0.001 688	91.201 869	43
44	66.264 076	0.015 091	9.849 089	0.101 532	652.640 761	0.001 532	91.850 788	44
45	72.890 484	0.013 719	9.862 808	0.101 391	718.904 837	0.001 391	92.454 433	45
46	80.179 532	0.012 472	9.875 280	0.101 263	791.795 321	0.001 263	93.015 674	46
47	88.197 485	0.011 338	9.886 618	0.101 147	871.974 853	0.001 147	93.537 231	47
48	97.017 234	0.010 307	9.896 926	0.101 041	960.172 338	0.001 041	94.021 681	48
49	106.718 957	0.009 370	9.906 296	0.100 946	1 057.189 572	0.000 946	94.471 460	49
50	117.390 853	0.008 519	9.914 814	0.100 859	1 163.908 529	0.000 859	94.888 869	50
60	304.481 640	0.003 284	9.967 157	0.100 330	3 034.816 395	0.000 330	97.701 011	60
70	789.746 957	0.001 266	9.987 338	0.100 127	7 887.469 568	0.000 127	98.987 017	70
72	955.593 818	0.001 046	9.989 535	0.100 105	9 545.938 177	0.000 105	99.141 895	72
80	2 048.400 215	0.000 488	9.995 118	0.100 049	20 474.002 146	0.000 049	99.560 633	80
84	2 999.062 754	0.000 333	9.996 666	0.100 033	29 980.627 542	0.000 033	99.686 569	84
90	5 313.022 612	0.000 188	9.998 118	0.100 019	53 120.226 118	0.000 019	99.811 783	90
96	9 412.343 651	0.000 106	9.998 938	0.100 011	94 113.436 513	0.000 011	99.887 382	96
100	13 780.612 340	0.000 073	9.999 274	0.100 007	137 796.123 398	0.000 007	99.920 178	100
108	29 539.966 407	0.000 034	9.999 661	0.100 003	295 389.664 066	0.000 003	99.960 054	108
120	92 709.068 818	0.000 011	9.999 892	0.100 001	927 080.688 178	0.000 001	99.985 978	120
∞	∞	0	10.000 000	0.100 000	∞	0	100.000 000	∞

11%	Single Payment		Uniform Series				Arithmetic Gradient Series	11%
	Compound Amount Factor (CAF)	Present Worth Factor (PWF)	Present Worth Factor (PWF)	Capital Recovery Factor (CRF)	Compound Amount Factor (CAF)	Sinking Fund Factor (SFF)	Present Worth Factor (PWF)	
n	*F/P*	*P/F*	*P/A*	*A/P*	*F/A*	*A/F*	*P/G*	*n*
1	1.110 000	0.900 901	0.900 901	1.110 000	1.000 000	1.000 000	0.000 000	1
2	1.232 100	0.811 622	1.712 523	0.583 934	2.110 000	0.473 934	0.811 622	2
3	1.367 631	0.731 191	2.443 715	0.409 213	3.342 100	0.299 213	2.274 005	3
4	1.518 070	0.658 731	3.102 446	0.322 326	4.709 731	0.212 326	4.250 198	4
5	1.685 058	0.593 451	3.695 897	0.270 570	6.227 801	0.160 570	6.624 003	5
6	1.870 415	0.534 641	4.230 538	0.236 377	7.912 860	0.126 377	9.297 208	6
7	2.076 160	0.481 658	4.712 196	0.212 215	9.783 274	0.102 215	12.187 158	7
8	2.304 538	0.433 926	5.146 123	0.194 321	11.859 434	0.084 321	15.224 644	8
9	2.558 037	0.390 925	5.537 048	0.180 602	14.163 972	0.070 602	18.352 042	9
10	2.839 421	0.352 184	5.889 232	0.169 801	16.722 009	0.059 801	21.521 702	10
11	3.151 757	0.317 283	6.206 515	0.161 121	19.561 430	0.051 121	24.694 535	11
12	3.498 451	0.285 841	6.492 356	0.154 027	22.713 187	0.044 027	27.838 784	12
13	3.883 280	0.257 514	6.749 870	0.148 151	26.211 638	0.038 151	30.928 955	13
14	4.310 441	0.231 995	6.981 865	0.143 228	30.094 918	0.033 228	33.944 888	14
15	4.784 589	0.209 004	7.190 870	0.139 065	34.405 359	0.029 065	36.870 949	15
16	5.310 894	0.188 292	7.379 162	0.135 517	39.189 948	0.025 517	39.695 332	16
17	5.895 093	0.169 633	7.548 794	0.132 471	44.500 843	0.022 471	42.409 454	17
18	6.543 553	0.152 822	7.701 617	0.129 843	50.395 936	0.019 843	45.007 431	18
19	7.263 344	0.137 678	7.839 294	0.127 563	56.939 488	0.017 563	47.485 628	19
20	8.062 312	0.124 034	7.963 328	0.125 576	64.202 832	0.015 576	49.842 273	20
21	8.949 166	0.111 742	8.075 070	0.123 838	72.265 144	0.013 838	52.077 118	21
22	9.933 574	0.100 669	8.175 739	0.122 313	81.214 309	0.012 313	54.191 160	22
23	11.026 267	0.090 693	8.266 432	0.120 971	91.147 884	0.010 971	56.186 396	23
24	12.239 157	0.081 705	8.348 137	0.119 787	102.174 151	0.009 787	58.065 610	24
25	13.585 464	0.073 608	8.421 745	0.118 740	114.413 307	0.008 740	59.832 204	25
26	15.079 865	0.066 314	8.488 058	0.117 813	127.998 771	0.007 813	61.490 044	26
27	16.738 650	0.059 742	8.547 800	0.116 989	143.078 636	0.006 989	63.043 336	27
28	18.579 901	0.053 822	8.601 622	0.116 257	159.817 286	0.006 257	64.496 519	28
29	20.623 691	0.048 488	8.650 110	0.115 605	178.397 187	0.005 605	65.854 181	29
30	22.892 297	0.043 683	8.693 793	0.115 025	199.020 878	0.005 025	67.120 982	30
31	25.410 449	0.039 354	8.733 146	0.114 506	221.913 174	0.004 506	68.301 599	31
32	28.205 599	0.035 454	8.768 600	0.114 043	247.323 624	0.004 043	69.400 672	32
33	31.308 214	0.031 940	8.800 541	0.113 629	275.529 222	0.003 629	70.422 768	33
34	34.752 118	0.028 775	8.829 316	0.113 259	306.837 437	0.003 259	71.372 350	34
35	38.574 851	0.025 924	8.855 240	0.112 927	341.589 555	0.002 927	72.253 753	35

(cont.)

(cont.)

11%	Single Payment		Uniform Series				Arithmetic Gradient Series	11%
	Compound Amount Factor (CAF)	Present Worth Factor (PWF)	Present Worth Factor (PWF)	Capital Recovery Factor (CRF)	Compound Amount Factor (CAF)	Sinking Fund Factor (SFF)	Present Worth Factor (PWF)	
n	*F/P*	*P/F*	*P/A*	*A/P*	*F/A*	*A/F*	*P/G*	*n*
36	42.818 085	0.023 355	8.878 594	0.112 630	380.164 406	0.002 630	73.071 165	36
37	47.528 074	0.021 040	8.899 635	0.112 364	422.982 490	0.002 364	73.828 612	37
38	52.756 162	0.018 955	8.918 590	0.112 125	470.510 564	0.002 125	74.529 952	38
39	58.559 340	0.017 077	8.935 666	0.111 911	523.266 726	0.001 911	75.178 866	39
40	65.000 867	0.015 384	8.951 051	0.111 719	581.826 066	0.001 719	75.778 858	40
41	72.150 963	0.013 860	8.964 911	0.111 546	646.826 934	0.001 546	76.333 251	41
42	80.087 569	0.012 486	8.977 397	0.111 391	718.977 896	0.001 391	76.845 191	42
43	88.897 201	0.011 249	8.988 646	0.111 251	799.065 465	0.001 251	77.317 647	43
44	98.675 893	0.010 134	8.998 780	0.111 126	887.962 666	0.001 126	77.753 417	44
45	109.530 242	0.009 130	9.007 910	0.111 014	986.638 559	0.001 014	78.155 133	45
46	121.578 568	0.008 225	9.016 135	0.110 912	1 096.168 801	0.000 912	78.525 264	46
47	134.952 211	0.007 410	9.023 545	0.110 821	1 217.747 369	0.000 821	78.866 125	47
48	149.796 954	0.006 676	9.030 221	0.110 739	1 352.699 580	0.000 739	79.179 883	48
49	166.274 619	0.006 014	9.036 235	0.110 666	1 502.496 533	0.000 666	79.468 562	49
50	184.564 827	0.005 418	9.041 653	0.110 599	1 668.771 152	0.000 599	79.734 051	50
60	524.057 242	0.001 908	9.073 562	0.110 210	4 755.065 839	0.000 210	81.446 096	60
70	1 488.019 132	0.000 672	9.084 800	0.110 074	13 518.355 744	0.000 074	82.161 430	70
72	1 833.388 372	0.000 545	9.085 951	0.110 060	16 658.076 112	0.000 060	82.242 536	72
80	4 225.112 750	0.000 237	9.088 757	0.110 026	38 401.025 004	0.000 026	82.452 937	80
84	6 414.018 645	0.000 156	9.089 492	0.110 017	58 300.169 504	0.000 017	82.512 686	84
90	11 996.873 812	0.000 083	9.090 151	0.110 009	109 053.398 293	0.000 009	82.569 540	90
96	22 439.127 359	0.000 045	9.090 504	0.110 005	203 982.975 989	0.000 005	82.602 052	96
100	34 064.175 270	0.000 029	9.090 642	0.110 003	309 665.229 724	0.000 003	82.615 514	100
108	78 502.178 503	0.000 013	9.090 793	0.110 001	713 647.077 302	0.000 001	82.631 068	108
∞	∞	0	9.090 909	0.110 000	∞	0	82.644 628	∞

12%	Single Payment		Uniform Series				Arithmetic Gradient Series	12%
	Compound Amount Factor (CAF)	Present Worth Factor (PWF)	Present Worth Factor (PWF)	Capital Recovery Factor (CRF)	Compound Amount Factor (CAF)	Sinking Fund Factor (SFF)	Present Worth Factor (PWF)	
n	*F/P*	*P/F*	*P/A*	*A/P*	*F/A*	*A/F*	*P/G*	*n*
1	1.120 000	0.892 857	0.892 857	1.120 000	1.000 000	1.000 000	0.000 000	1
2	1.254 400	0.797 194	1.690 051	0.591 698	2.120 000	0.471 698	0.797 194	2
3	1.404 928	0.711 780	2.401 831	0.416 349	3.374 400	0.296 349	2.220 754	3
4	1.573 519	0.635 518	3.037 349	0.329 234	4.779 328	0.209 234	4.127 309	4
5	1.762 342	0.567 427	3.604 776	0.277 410	6.352 847	0.157 410	6.397 016	5
6	1.973 823	0.506 631	4.111 407	0.243 226	8.115 189	0.123 226	8.930 172	6
7	2.210 681	0.452 349	4.563 757	0.219 118	10.089 012	0.099 118	11.644 267	7
8	2.475 963	0.403 883	4.967 640	0.201 303	12.299 693	0.081 303	14.471 450	8
9	2.773 079	0.360 610	5.328 250	0.187 679	14.775 656	0.067 679	17.356 330	9
10	3.105 848	0.321 973	5.650 223	0.176 984	17.548 735	0.056 984	20.254 089	10
11	3.478 550	0.287 476	5.937 699	0.168 415	20.654 583	0.048 415	23.128 850	11
12	3.895 976	0.256 675	6.194 374	0.161 437	24.133 133	0.041 437	25.952 276	12
13	4.363 493	0.229 174	6.423 548	0.155 677	28.029 109	0.035 677	28.702 366	13
14	4.887 112	0.204 620	6.628 168	0.150 871	32.392 602	0.030 871	31.362 424	14
15	5.473 566	0.182 696	6.810 864	0.146 824	37.279 715	0.026 824	33.920 171	15
16	6.130 394	0.163 122	6.973 986	0.143 390	42.753 280	0.023 390	36.366 996	16
17	6.866 041	0.145 644	7.119 630	0.140 457	48.883 674	0.020 457	38.697 306	17
18	7.689 966	0.130 040	7.249 670	0.137 937	55.749 715	0.017 937	40.907 979	18
19	8.612 762	0.116 107	7.365 777	0.135 763	63.439 681	0.015 763	42.997 901	19
20	9.646 293	0.103 667	7.469 444	0.133 879	72.052 442	0.013 879	44.967 569	20
21	10.803 848	0.092 560	7.562 003	0.132 240	81.698 736	0.012 240	46.818 762	21
22	12.100 310	0.082 643	7.644 646	0.130 811	92.502 584	0.010 811	48.554 254	22
23	13.552 347	0.073 788	7.718 434	0.129 560	104.602 894	0.009 560	50.177 589	23
24	15.178 629	0.065 882	7.784 316	0.128 463	118.155 241	0.008 463	51.692 878	24
25	17.000 064	0.058 823	7.843 139	0.127 500	133.333 870	0.007 500	53.104 637	25
26	19.040 072	0.052 521	7.895 660	0.126 652	150.333 934	0.006 652	54.417 657	26
27	21.324 881	0.046 894	7.942 554	0.125 904	169.374 007	0.005 904	55.636 890	27
28	23.883 866	0.041 869	7.984 423	0.125 244	190.698 887	0.005 244	56.767 361	28
29	26.749 930	0.037 383	8.021 806	0.124 660	214.582 754	0.004 660	57.814 092	29
30	29.959 922	0.033 378	8.055 184	0.124 144	241.332 684	0.004 144	58.782 052	30
31	33.555 113	0.029 802	8.084 986	0.123 686	271.292 606	0.003 686	59.676 104	31
32	37.581 726	0.026 609	8.111 594	0.123 280	304.847 719	0.003 280	60.500 973	32
33	42.091 533	0.023 758	8.135 352	0.122 920	342.429 446	0.002 920	61.261 221	33
34	47.142 517	0.021 212	8.156 564	0.122 601	384.520 979	0.002 601	61.961 226	34
35	52.799 620	0.018 940	8.175 504	0.122 317	431.663 496	0.002 317	62.605 170	35

(cont.)

(cont.)

12%	Single Payment		Uniform Series				Arithmetic Gradient Series	12%
	Compound Amount Factor (CAF)	Present Worth Factor (PWF)	Present Worth Factor (PWF)	Capital Recovery Factor (CRF)	Compound Amount Factor (CAF)	Sinking Fund Factor (SFF)	Present Worth Factor (PWF)	
n	*F/P*	*P/F*	*P/A*	*A/P*	*F/A*	*A/F*	*P/G*	*n*
36	59.135 574	0.016 910	8.192 414	0.122 064	484.463 116	0.002 064	63.197 030	36
37	66.231 843	0.015 098	8.207 513	0.121 840	543.598 690	0.001 840	63.740 575	37
38	74.179 664	0.013 481	8.220 993	0.121 640	609.830 533	0.001 640	64.239 364	38
39	83.081 224	0.012 036	8.233 030	0.121 462	684.010 197	0.001 462	64.696 748	39
40	93.050 970	0.010 747	8.243 777	0.121 304	767.091 420	0.001 304	65.115 873	40
41	104.217 087	0.009 595	8.253 372	0.121 163	860.142 391	0.001 163	65.499 687	41
42	116.723 137	0.008 567	8.261 939	0.121 037	964.359 478	0.001 037	65.850 946	42
43	130.729 914	0.007 649	8.269 589	0.120 925	1 081.082 615	0.000 925	66.172 219	43
44	146.417 503	0.006 830	8.276 418	0.120 825	1 211.812 529	0.000 825	66.465 900	44
45	163.987 604	0.006 098	8.282 516	0.120 736	1 358.230 032	0.000 736	66.734 212	45
46	183.666 116	0.005 445	8.287 961	0.120 657	1 522.217 636	0.000 657	66.979 222	46
47	205.706 050	0.004 861	8.292 822	0.120 586	1 705.883 752	0.000 586	67.202 842	47
48	230.390 776	0.004 340	8.297 163	0.120 523	1 911.589 803	0.000 523	67.406 844	48
49	258.037 669	0.003 875	8.301 038	0.120 467	2 141.980 579	0.000 467	67.592 863	49
50	289.002 190	0.003 460	8.304 498	0.120 417	2 400.018 249	0.000 417	67.762 412	50
60	897.596 933	0.001 114	8.324 049	0.120 134	7 471.641 112	0.000 134	68.810 034	60
70	2 787.799 828	0.000 359	8.330 344	0.120 043	23 223.331 897	0.000 043	69.210 289	70
72	3 497.016 104	0.000 286	8.330 950	0.120 034	29 133.467 532	0.000 034	69.253 011	72
80	8 658.483 100	0.000 115	8.332 371	0.120 014	72 145.692 501	0.000 014	69.359 428	80
84	13 624.290 786	0.000 073	8.332 722	0.120 009	113 527.423 218	0.000 009	69.387 969	84
90	26 891.934 223	0.000 037	8.333 023	0.120 004	224 091.118 528	0.000 004	69.413 973	90
96	53 079.909 819	0.000 019	8.333 176	0.120 002	442 324.248 488	0.000 002	69.428 065	96
100	83 522.265 727	0.000 012	8.333 234	0.120 001	696 010.547 721	0.000 001	69.433 636	100
∞	∞	0	8.333 333	0.120 000	∞	0	69.444 444	∞

13%	Single Payment		Uniform Series				Arithmetic Gradient Series	13%
	Compound Amount Factor (CAF)	Present Worth Factor (PWF)	Present Worth Factor (PWF)	Capital Recovery Factor (CRF)	Compound Amount Factor (CAF)	Sinking Fund Factor (SFF)	Present Worth Factor (PWF)	
n	*F/P*	*P/F*	*P/A*	*A/P*	*F/A*	*A/F*	*P/G*	*n*
1	1.130 000	0.884 956	0.884 956	1.130 000	1.000 000	1.000 000	0.000 000	1
2	1.276 900	0.783 147	1.668 102	0.599 484	2.130 000	0.469 484	0.783 147	2
3	1.442 897	0.693 050	2.361 153	0.423 522	3.406 900	0.293 522	2.169 247	3
4	1.630 474	0.613 319	2.974 471	0.336 194	4.849 797	0.206 194	4.009 203	4
5	1.842 435	0.542 760	3.517 231	0.284 315	6.480 271	0.154 315	6.180 243	5
6	2.081 952	0.480 319	3.997 550	0.250 153	8.322 706	0.120 153	8.581 836	6
7	2.352 605	0.425 061	4.422 610	0.226 111	10.404 658	0.096 111	11.132 199	7
8	2.658 444	0.376 160	4.798 770	0.208 387	12.757 263	0.078 387	13.765 318	8
9	3.004 042	0.332 885	5.131 655	0.194 869	15.415 707	0.064 869	16.428 397	9
10	3.394 567	0.294 588	5.426 243	0.184 290	18.419 749	0.054 290	19.079 692	10
11	3.835 861	0.260 698	5.686 941	0.175 841	21.814 317	0.045 841	21.686 669	11
12	4.334 523	0.230 706	5.917 647	0.168 986	25.650 178	0.038 986	24.224 434	12
13	4.898 011	0.204 165	6.121 812	0.163 350	29.984 701	0.033 350	26.674 408	13
14	5.534 753	0.180 677	6.302 488	0.158 667	34.882 712	0.028 667	29.023 203	14
15	6.254 270	0.159 891	6.462 379	0.154 742	40.417 464	0.024 742	31.261 673	15
16	7.067 326	0.141 496	6.603 875	0.151 426	46.671 735	0.021 426	33.384 117	16
17	7.986 078	0.125 218	6.729 093	0.148 608	53.739 060	0.018 608	35.387 604	17
18	9.024 268	0.110 812	6.839 905	0.146 201	61.725 138	0.016 201	37.271 413	18
19	10.197 423	0.098 064	6.937 969	0.144 134	70.749 406	0.014 134	39.036 565	19
20	11.523 088	0.086 782	7.024 752	0.142 354	80.946 829	0.012 354	40.685 428	20
21	13.021 089	0.076 798	7.101 550	0.140 814	92.469 917	0.010 814	42.221 398	21
22	14.713 831	0.067 963	7.169 513	0.139 479	105.491 006	0.009 479	43.648 627	22
23	16.626 629	0.060 144	7.229 658	0.138 319	120.204 837	0.008 319	44.971 805	23
24	18.788 091	0.053 225	7.282 883	0.137 308	136.831 465	0.007 308	46.195 985	24
25	21.230 542	0.047 102	7.329 985	0.136 426	155.619 556	0.006 426	47.326 432	25
26	23.990 513	0.041 683	7.371 668	0.135 655	176.850 098	0.005 655	48.368 511	26
27	27.109 279	0.036 888	7.408 556	0.134 979	200.840 611	0.004 979	49.327 592	27
28	30.633 486	0.032 644	7.441 200	0.134 387	227.949 890	0.004 387	50.208 980	28
29	34.615 839	0.028 889	7.470 088	0.133 867	258.583 376	0.003 867	51.017 858	29
30	39.115 898	0.025 565	7.495 653	0.133 411	293.199 215	0.003 411	51.759 245	30
31	44.200 965	0.022 624	7.518 277	0.133 009	332.315 113	0.003 009	52.437 963	31
32	49.947 090	0.020 021	7.538 299	0.132 656	376.516 078	0.002 656	53.058 620	32
33	56.440 212	0.017 718	7.556 016	0.132 345	426.463 168	0.002 345	53.625 592	33
34	63.777 439	0.015 680	7.571 696	0.132 071	482.903 380	0.002 071	54.143 016	34
35	72.068 506	0.013 876	7.585 572	0.131 829	546.680 819	0.001 829	54.614 789	35

(cont.)

(cont.)

13%	Single Payment		Uniform Series				Arithmetic Gradient Series	13%
	Compound Amount Factor (CAF)	Present Worth Factor (PWF)	Present Worth Factor (PWF)	Capital Recovery Factor (CRF)	Compound Amount Factor (CAF)	Sinking Fund Factor (SFF)	Present Worth Factor (PWF)	
n	*F/P*	*P/F*	*P/A*	*A/P*	*F/A*	*A/F*	*P/G*	*n*
36	81.437 412	0.012 279	7.597 851	0.131 616	618.749 325	0.001 616	55.044 567	36
37	92.024 276	0.010 867	7.608 718	0.131 428	700.186 738	0.001 428	55.435 768	37
38	103.987 432	0.009 617	7.618 334	0.131 262	792.211 014	0.001 262	55.791 581	38
39	117.505 798	0.008 510	7.626 844	0.131 116	896.198 445	0.001 116	56.114 969	39
40	132.781 552	0.007 531	7.634 376	0.130 986	1 013.704 243	0.000 986	56.408 684	40
41	150.043 153	0.006 665	7.641 040	0.130 872	1 146.485 795	0.000 872	56.675 274	41
42	169.548 763	0.005 898	7.646 938	0.130 771	1 296.528 948	0.000 771	56.917 093	42
43	191.590 103	0.005 219	7.652 158	0.130 682	1 466.077 712	0.000 682	57.136 311	43
44	216.496 816	0.004 619	7.656 777	0.130 603	1 657.667 814	0.000 603	57.334 928	44
45	244.641 402	0.004 088	7.660 864	0.130 534	1 874.164 630	0.000 534	57.514 783	45
46	276.444 784	0.003 617	7.664 482	0.130 472	2 118.806 032	0.000 472	57.677 564	46
47	312.382 606	0.003 201	7.667 683	0.130 417	2 395.250 816	0.000 417	57.824 820	47
48	352.992 345	0.002 833	7.670 516	0.130 369	2 707.633 422	0.000 369	57.957 967	48
49	398.881 350	0.002 507	7.673 023	0.130 327	3 060.625 767	0.000 327	58.078 303	49
50	450.735 925	0.002 219	7.675 242	0.130 289	3 459.507 117	0.000 289	58.187 015	50
60	1 530.053 473	0.000 654	7.687 280	0.130 085	11 761.949 792	0.000 085	58.831 276	60
70	5 193.869 624	0.000 193	7.690 827	0.130 025	39 945.150 956	0.000 025	59.056 533	70
72	6 632.052 123	0.000 151	7.691 148	0.130 020	51 008.093 256	0.000 020	59.079 165	72
80	17 630.940 454	0.000 057	7.691 871	0.130 007	135 614.926 571	0.000 007	59.133 338	80
84	28 746.783 130	0.000 035	7.692 040	0.130 005	221 121.408 693	0.000 005	59.147 062	84
90	59 849.415 520	0.000 017	7.692 179	0.130 002	460 372.427 073	0.000 002	59.159 041	90
96	124 603.595 533	0.000 008	7.692 246	0.130 001	958 481.504 103	0.000 001	59.165 196	96
∞	∞	0	7.692 308	0.130 000	∞	0	59.171 598	∞

14%	Single Payment		Uniform Series				Arithmetic Gradient Series	14%
	Compound Amount Factor (CAF)	Present Worth Factor (PWF)	Present Worth Factor (PWF)	Capital Recovery Factor (CRF)	Compound Amount Factor (CAF)	Sinking Fund Factor (SFF)	Present Worth Factor (PWF)	
n	*F/P*	*P/F*	*P/A*	*A/P*	*F/A*	*A/F*	*P/G*	*n*
1	1.140 000	0.877 193	0.877 193	1.140 000	1.000 000	1.000 000	0.000 000	1
2	1.299 600	0.769 468	1.646 661	0.607 290	2.140 000	0.467 290	0.769 468	2
3	1.481 544	0.674 972	2.321 632	0.430 731	3.439 600	0.290 731	2.119 411	3
4	1.688 960	0.592 080	2.913 712	0.343 205	4.921 144	0.203 205	3.895 651	4
5	1.925 415	0.519 369	3.433 081	0.291 284	6.610 104	0.151 284	5.973 126	5
6	2.194 973	0.455 587	3.888 668	0.257 157	8.535 519	0.117 157	8.251 059	6
7	2.502 269	0.399 637	4.288 305	0.233 192	10.730 491	0.093 192	10.648 883	7
8	2.852 586	0.350 559	4.638 864	0.215 570	13.232 760	0.075 570	13.102 796	8
9	3.251 949	0.307 508	4.946 372	0.202 168	16.085 347	0.062 168	15.562 860	9
10	3.707 221	0.269 744	5.216 116	0.191 714	19.337 295	0.051 714	17.990 554	10
11	4.226 232	0.236 617	5.452 733	0.183 394	23.044 516	0.043 394	20.356 728	11
12	4.817 905	0.207 559	5.660 292	0.176 669	27.270 749	0.036 669	22.639 878	12
13	5.492 411	0.182 069	5.842 362	0.171 164	32.088 654	0.031 164	24.824 710	13
14	6.261 349	0.159 710	6.002 072	0.166 609	37.581 065	0.026 609	26.900 940	14
15	7.137 938	0.140 096	6.142 168	0.162 809	43.842 414	0.022 809	28.862 291	15
16	8.137 249	0.122 892	6.265 060	0.159 615	50.980 352	0.019 615	30.705 666	16
17	9.276 464	0.107 800	6.372 859	0.156 915	59.117 601	0.016 915	32.430 461	17
18	10.575 169	0.094 561	6.467 420	0.154 621	68.394 066	0.014 621	34.038 000	18
19	12.055 693	0.082 948	6.550 369	0.152 663	78.969 235	0.012 663	35.531 071	19
20	13.743 490	0.072 762	6.623 131	0.150 986	91.024 928	0.010 986	36.913 544	20
21	15.667 578	0.063 826	6.686 957	0.149 545	104.768 418	0.009 545	38.190 065	21
22	17.861 039	0.055 988	6.742 944	0.148 303	120.435 996	0.008 303	39.365 808	22
23	20.361 585	0.049 112	6.792 056	0.147 231	138.297 035	0.007 231	40.446 274	23
24	23.212 207	0.043 081	6.835 137	0.146 303	158.658 620	0.006 303	41.437 132	24
25	26.461 916	0.037 790	6.872 927	0.145 498	181.870 827	0.005 498	42.344 096	25
26	30.166 584	0.033 149	6.906 077	0.144 800	208.332 743	0.004 800	43.172 828	26
27	34.389 906	0.029 078	6.935 155	0.144 193	238.499 327	0.004 193	43.928 864	27
28	39.204 493	0.025 507	6.960 662	0.143 664	272.889 233	0.003 664	44.617 560	28
29	44.693 122	0.022 375	6.983 037	0.143 204	312.093 725	0.003 204	45.244 055	29
30	50.950 159	0.019 627	7.002 664	0.142 803	356.786 847	0.002 803	45.813 238	30
31	58.083 181	0.017 217	7.019 881	0.142 453	407.737 006	0.002 453	46.329 739	31
32	66.214 826	0.015 102	7.034 983	0.142 147	465.820 186	0.002 147	46.797 912	32
33	75.484 902	0.013 248	7.048 231	0.141 880	532.035 012	0.001 880	47.221 838	33
34	86.052 788	0.011 621	7.059 852	0.141 646	607.519 914	0.001 646	47.605 324	34
35	98.100 178	0.010 194	7.070 045	0.141 442	693.572 702	0.001 442	47.951 908	35

(cont.)

(cont.)

14%	Single Payment		Uniform Series				Arithmetic Gradient Series	14%
	Compound Amount Factor (CAF)	Present Worth Factor (PWF)	Present Worth Factor (PWF)	Capital Recovery Factor (CRF)	Compound Amount Factor (CAF)	Sinking Fund Factor (SFF)	Present Worth Factor (PWF)	
n	*F/P*	*P/F*	*P/A*	*A/P*	*F/A*	*A/F*	*P/G*	*n*
36	111.834 203	0.008 942	7.078 987	0.141 263	791.672 881	0.001 263	48.264 871	36
37	127.490 992	0.007 844	7.086 831	0.141 107	903.507 084	0.001 107	48.547 244	37
38	145.339 731	0.006 880	7.093 711	0.140 970	1 030.998 076	0.000 970	48.801 820	38
39	165.687 293	0.006 035	7.099 747	0.140 850	1 176.337 806	0.000 850	49.031 168	39
40	188.883 514	0.005 294	7.105 041	0.140 745	1 342.025 099	0.000 745	49.237 644	40
41	215.327 206	0.004 644	7.109 685	0.140 653	1 530.908 613	0.000 653	49.423 408	41
42	245.473 015	0.004 074	7.113 759	0.140 573	1 746.235 819	0.000 573	49.590 433	42
43	279.839 237	0.003 573	7.117 332	0.140 502	1 991.708 833	0.000 502	49.740 519	43
44	319.016 730	0.003 135	7.120 467	0.140 440	2 271.548 070	0.000 440	49.875 308	44
45	363.679 072	0.002 750	7.123 217	0.140 386	2 590.564 800	0.000 386	49.996 294	45
46	414.594 142	0.002 412	7.125 629	0.140 338	2 954.243 872	0.000 338	50.104 834	46
47	472.637 322	0.002 116	7.127 744	0.140 297	3 368.838 014	0.000 297	50.202 160	47
48	538.806 547	0.001 856	7.129 600	0.140 260	3 841.475 336	0.000 260	50.289 390	48
49	614.239 464	0.001 628	7.131 228	0.140 228	4 380.281 883	0.000 228	50.367 535	49
50	700.232 988	0.001 428	7.132 656	0.140 200	4 994.521 346	0.000 200	50.437 512	50
60	2 595.918 660	0.000 385	7.140 106	0.140 054	18 535.133 283	0.000 054	50.835 660	60
70	9 623.644 985	0.000 104	7.142 115	0.140 015	68 733.178 463	0.000 015	50.963 151	70
72	12 506.889 022	0.000 080	7.142 286	0.140 011	89 327.778 731	0.000 011	50.975 209	72
80	35 676.981 807	0.000 028	7.142 657	0.140 004	254 828.441 480	0.000 004	51.002 961	80
84	60 257.000 902	0.000 017	7.142 739	0.140 002	430 400.006 439	0.000 002	51.009 604	84
90	132 262.467 379	0.000 008	7.142 803	0.140 001	944 724.766 995	0.000 001	51.015 162	90
∞	∞	0	7.142 857	0.140 000	∞	0	51.020 408	∞

15%	Single Payment		Uniform Series				Arithmetic Gradient Series	15%
	Compound Amount Factor (CAF)	Present Worth Factor (PWF)	Present Worth Factor (PWF)	Capital Recovery Factor (CRF)	Compound Amount Factor (CAF)	Sinking Fund Factor (SFF)	Present Worth Factor (PWF)	
n	*F/P*	*P/F*	*P/A*	*A/P*	*F/A*	*A/F*	*P/G*	*n*
1	1.150 000	0.869 565	0.869 565	1.150 000	1.000 000	1.000 000	0.000 000	1
2	1.322 500	0.756 144	1.625 709	0.615 116	2.150 000	0.465 116	0.756 144	2
3	1.520 875	0.657 516	2.283 225	0.437 977	3.472 500	0.287 977	2.071 176	3
4	1.749 006	0.571 753	2.854 978	0.350 265	4.993 375	0.200 265	3.786 436	4
5	2.011 357	0.497 177	3.352 155	0.298 316	6.742 381	0.148 316	5.775 143	5
6	2.313 061	0.432 328	3.784 483	0.264 237	8.753 738	0.114 237	7.936 781	6
7	2.660 020	0.375 937	4.160 420	0.240 360	11.066 799	0.090 360	10.192 403	7
8	3.059 023	0.326 902	4.487 322	0.222 850	13.726 819	0.072 850	12.480 715	8
9	3.517 876	0.284 262	4.771 584	0.209 574	16.785 842	0.059 574	14.754 815	9
10	4.045 558	0.247 185	5.018 769	0.199 252	20.303 718	0.049 252	16.979 477	10
11	4.652 391	0.214 943	5.233 712	0.191 069	24.349 276	0.041 069	19.128 909	11
12	5.350 250	0.186 907	5.420 619	0.184 481	29.001 667	0.034 481	21.184 888	12
13	6.152 788	0.162 528	5.583 147	0.179 110	34.351 917	0.029 110	23.135 223	13
14	7.075 706	0.141 329	5.724 476	0.174 688	40.504 705	0.024 688	24.972 496	14
15	8.137 062	0.122 894	5.847 370	0.171 017	47.580 411	0.021 017	26.693 019	15
16	9.357 621	0.106 865	5.954 235	0.167 948	55.717 472	0.017 948	28.295 990	16
17	10.761 264	0.092 926	6.047 161	0.165 367	65.075 093	0.015 367	29.782 805	17
18	12.375 454	0.080 805	6.127 966	0.163 186	75.836 357	0.013 186	31.156 492	18
19	14.231 772	0.070 265	6.198 231	0.161 336	88.211 811	0.011 336	32.421 267	19
20	16.366 537	0.061 100	6.259 331	0.159 761	102.443 583	0.009 761	33.582 173	20
21	18.821 518	0.053 131	6.312 462	0.158 417	118.810 120	0.008 417	34.644 786	21
22	21.644 746	0.046 201	6.358 663	0.157 266	137.631 638	0.007 266	35.614 999	22
23	24.891 458	0.040 174	6.398 837	0.156 278	159.276 384	0.006 278	36.498 836	23
24	28.625 176	0.034 934	6.433 771	0.155 430	184.167 841	0.005 430	37.302 324	24
25	32.918 953	0.030 378	6.464 149	0.154 699	212.793 017	0.004 699	38.031 388	25
26	37.856 796	0.026 415	6.490 564	0.154 070	245.711 970	0.004 070	38.691 771	26
27	43.535 315	0.022 970	6.513 534	0.153 526	283.568 766	0.003 526	39.288 987	27
28	50.065 612	0.019 974	6.533 508	0.153 057	327.104 080	0.003 057	39.828 280	28
29	57.575 454	0.017 369	6.550 877	0.152 651	377.169 693	0.002 651	40.314 598	29
30	66.211 772	0.015 103	6.565 980	0.152 300	434.745 146	0.002 300	40.752 587	30
31	76.143 538	0.013 133	6.579 113	0.151 996	500.956 918	0.001 996	41.146 579	31
32	87.565 068	0.011 420	6.590 533	0.151 733	577.100 456	0.001 733	41.500 602	32
33	100.699 829	0.009 931	6.600 463	0.151 505	664.665 524	0.001 505	41.818 378	33
34	115.804 803	0.008 635	6.609 099	0.151 307	765.365 353	0.001 307	42.103 340	34
35	133.175 523	0.007 509	6.616 607	0.151 135	881.170 156	0.001 135	42.358 642	35

(cont.)

(cont.)

15%	Single Payment		Uniform Series				Arithmetic Gradient Series	15%
	Compound Amount Factor (CAF)	Present Worth Factor (PWF)	Present Worth Factor (PWF)	Capital Recovery Factor (CRF)	Compound Amount Factor (CAF)	Sinking Fund Factor (SFF)	Present Worth Factor (PWF)	
n	*F/P*	*P/F*	*P/A*	*A/P*	*F/A*	*A/F*	*P/G*	*n*
36	153.151 852	0.006 529	6.623 137	0.150 986	1 014.345 680	0.000 986	42.587 174	36
37	176.124 630	0.005 678	6.628 815	0.150 857	1 167.497 532	0.000 857	42.791 574	37
38	202.543 324	0.004 937	6.633 752	0.150 744	1 343.622 161	0.000 744	42.974 251	38
39	232.924 823	0.004 293	6.638 045	0.150 647	1 546.165 485	0.000 647	43.137 394	39
40	267.863 546	0.003 733	6.641 778	0.150 562	1 779.090 308	0.000 562	43.282 991	40
41	308.043 078	0.003 246	6.645 025	0.150 489	2 046.953 854	0.000 489	43.412 843	41
42	354.249 540	0.002 823	6.647 848	0.150 425	2 354.996 933	0.000 425	43.528 580	42
43	407.386 971	0.002 455	6.650 302	0.150 369	2 709.246 473	0.000 369	43.631 676	43
44	468.495 017	0.002 134	6.652 437	0.150 321	3 116.633 443	0.000 321	43.723 460	44
45	538.769 269	0.001 856	6.654 293	0.150 279	3 585.128 460	0.000 279	43.805 127	45
46	619.584 659	0.001 614	6.655 907	0.150 242	4 123.897 729	0.000 242	43.877 757	46
47	712.522 358	0.001 403	6.657 310	0.150 211	4 743.482 388	0.000 211	43.942 316	47
48	819.400 712	0.001 220	6.658 531	0.150 183	5 456.004 746	0.000 183	43.999 675	48
49	942.310 819	0.001 061	6.659 592	0.150 159	6 275.405 458	0.000 159	44.050 614	49
50	1 083.657 442	0.000 923	6.660 515	0.150 139	7 217.716 277	0.000 139	44.095 831	50
60	4 383.998 746	0.000 228	6.665 146	0.150 034	29 219.991 638	0.000 034	44.343 066	60
70	17 735.720 039	0.000 056	6.666 291	0.150 008	118 231.466 926	0.000 008	44.415 626	70
72	23 455.489 751	0.000 043	6.666 382	0.150 006	156 363.265 009	0.000 006	44.422 085	72
80	71 750.879 401	0.000 014	6.666 574	0.150 002	478 332.529 343	0.000 002	44.436 392	80
84	125 492.736 516	0.000 008	6.666 614	0.150 001	836 611.576 774	0.000 001	44.439 628	84
∞	∞	0	6.666 667	0.150 000	∞	0	44.444 444	∞

16%	Single Payment		Uniform Series				Arithmetic Gradient Series	16%
	Compound Amount Factor (CAF)	Present Worth Factor (PWF)	Present Worth Factor (PWF)	Capital Recovery Factor (CRF)	Compound Amount Factor (CAF)	Sinking Fund Factor (SFF)	Present Worth Factor (PWF)	
n	*F/P*	*P/F*	*P/A*	*A/P*	*F/A*	*A/F*	*P/G*	*n*
1	1.160 000	0.862 069	0.862 069	1.160 000	1.000 000	1.000 000	0.000 000	1
2	1.345 600	0.743 163	1.605 232	0.622 963	2.160 000	0.462 963	0.743 163	2
3	1.560 896	0.640 658	2.245 890	0.445 258	3.505 600	0.285 258	2.024 478	3
4	1.810 639	0.552 291	2.798 181	0.357 375	5.066 496	0.197 375	3.681 352	4
5	2.100 342	0.476 113	3.274 294	0.305 409	6.877 135	0.145 409	5.585 804	5
6	2.436 396	0.410 442	3.684 736	0.271 390	8.977 477	0.111 390	7.638 015	6
7	2.826 220	0.353 830	4.038 565	0.247 613	11.413 873	0.087 613	9.760 992	7
8	3.278 415	0.305 025	4.343 591	0.230 224	14.240 093	0.070 224	11.896 170	8
9	3.802 961	0.262 953	4.606 544	0.217 082	17.518 508	0.057 082	13.999 794	9
10	4.411 435	0.226 684	4.833 227	0.206 901	21.321 469	0.046 901	16.039 947	10
11	5.117 265	0.195 417	5.028 644	0.198 861	25.732 904	0.038 861	17.994 116	11
12	5.936 027	0.168 463	5.197 107	0.192 415	30.850 169	0.032 415	19.847 207	12
13	6.885 791	0.145 227	5.342 334	0.187 184	36.786 196	0.027 184	4.000 000	13
14	7.987 518	0.125 195	5.467 529	0.182 898	43.671 987	0.022 898	23.217 465	14
15	9.265 521	0.107 927	5.575 456	0.179 358	51.659 505	0.019 358	24.728 443	15
16	10.748 004	0.093 041	5.668 497	0.176 414	60.925 026	0.016 414	26.124 051	16
17	12.467 685	0.080 207	5.748 704	0.173 952	71.673 030	0.013 952	27.407 369	17
18	14.462 514	0.069 144	5.817 848	0.171 885	84.140 715	0.011 885	28.582 822	18
19	16.776 517	0.059 607	5.877 455	0.170 142	98.603 230	0.010 142	29.655 750	19
20	19.460 759	0.051 385	5.928 841	0.168 667	115.379 747	0.008 667	30.632 074	20
21	22.574 481	0.044 298	5.973 139	0.167 416	134.840 506	0.007 416	31.518 030	21
22	26.186 398	0.038 188	6.011 326	0.166 353	157.414 987	0.006 353	32.319 973	22
23	30.376 222	0.032 920	6.044 247	0.165 447	183.601 385	0.005 447	33.044 224	23
24	35.236 417	0.028 380	6.072 627	0.164 673	213.977 607	0.004 673	33.696 957	24
25	40.874 244	0.024 465	6.097 092	0.164 013	249.214 024	0.004 013	34.284 124	25
26	47.414 123	0.021 091	6.118 183	0.163 447	290.088 267	0.003 447	34.811 393	26
27	55.000 382	0.018 182	6.136 364	0.162 963	337.502 390	0.002 963	35.284 117	27
28	63.800 444	0.015 674	6.152 038	0.162 548	392.502 773	0.002 548	35.707 312	28
29	74.008 515	0.013 512	6.165 550	0.162 192	456.303 216	0.002 192	36.085 647	29
30	85.849 877	0.011 648	6.177 198	0.161 886	530.311 731	0.001 886	36.423 446	30
31	99.585 857	0.010 042	6.187 240	0.161 623	616.161 608	0.001 623	36.724 693	31
32	115.519 594	0.008 657	6.195 897	0.161 397	715.747 465	0.001 397	36.993 046	32
33	134.002 729	0.007 463	6.203 359	0.161 203	831.267 059	0.001 203	37.231 847	33
34	155.443 166	0.006 433	6.209 792	0.161 036	965.269 789	0.001 036	37.444 143	34
35	180.314 073	0.005 546	6.215 338	0.160 892	1 120.712 955	0.000 892	37.632 703	35

(cont.)

(cont.)

16%	Single Payment		Uniform Series				Arithmetic Gradient Series	16%
	Compound Amount Factor (CAF)	Present Worth Factor (PWF)	Present Worth Factor (PWF)	Capital Recovery Factor (CRF)	Compound Amount Factor (CAF)	Sinking Fund Factor (SFF)	Present Worth Factor (PWF)	
n	*F/P*	*P/F*	*P/A*	*A/P*	*F/A*	*A/F*	*P/G*	*n*
36	209.164 324	0.004 781	6.220 119	0.160 769	1 301.027 028	0.000 769	37.800 036	36
37	242.630 616	0.004 121	6.224 241	0.160 662	1 510.191 352	0.000 662	37.948 409	37
38	281.451 515	0.003 553	6.227 794	0.160 571	1 752.821 968	0.000 571	38.079 871	38
39	326.483 757	0.003 063	6.230 857	0.160 492	2 034.273 483	0.000 492	38.196 262	39
40	378.721 158	0.002 640	6.233 497	0.160 424	2 360.757 241	0.000 424	38.299 241	40
41	439.316 544	0.002 276	6.235 773	0.160 365	2 739.478 399	0.000 365	38.390 291	41
42	509.607 191	0.001 962	6.237 736	0.160 315	3 178.794 943	0.000 315	38.470 745	42
43	591.144 341	0.001 692	6.239 427	0.160 271	3 688.402 134	0.000 271	38.541 794	43
44	685.727 436	0.001 458	6.240 886	0.160 234	4 279.546 475	0.000 234	38.604 501	44
45	795.443 826	0.001 257	6.242 143	0.160 201	4 965.273 911	0.000 201	38.659 816	45
46	922.714 838	0.001 084	6.243 227	0.160 174	5 760.717 737	0.000 174	38.708 585	46
47	1 070.349 212	0.000 934	6.244 161	0.160 150	6 683.432 575	0.000 150	38.751 562	47
48	1 241.605 086	0.000 805	6.244 966	0.160 129	7 753.781 787	0.000 129	38.789 416	48
49	1 440.261 900	0.000 694	6.245 661	0.160 111	8 995.386 873	0.000 111	38.822 743	49
50	1 670.703 804	0.000 599	6.246 259	0.160 096	10 435.648 773	0.000 096	38.852 072	50
60	7 370.201 365	0.000 136	6.249 152	0.160 022	46 057.508 533	0.000 022	39.006 319	60
70	32 513.164 839	0.000 031	6.249 808	0.160 005	203 201.030 246	0.000 005	39.047 842	70
72	43 749.714 608	0.000 023	6.249 857	0.160 004	273 429.466 299	0.000 004	39.051 321	72
80	143 429.715 890	0.000 007	6.249 956	0.160 001	896 429.474 315	0.000 001	39.058 742	80
∞	∞	0	6.250 000	0.160 000	∞	0	39.062 500	∞

17%	Single Payment		Uniform Series				Arithmetic Gradient Series	17%
	Compound Amount Factor (CAF)	Present Worth Factor (PWF)	Present Worth Factor (PWF)	Capital Recovery Factor (CRF)	Compound Amount Factor (CAF)	Sinking Fund Factor (SFF)	Present Worth Factor (PWF)	
n	*F/P*	*P/F*	*P/A*	*A/P*	*F/A*	*A/F*	*P/G*	*n*
1	1.170 000	0.854 701	0.854 701	1.170 000	1.000 000	1.000 000	0.000 000	1
2	1.368 900	0.730 514	1.585 214	0.630 829	2.170 000	0.460 829	0.730 514	2
3	1.601 613	0.624 371	2.209 585	0.452 574	3.538 900	0.282 574	1.979 255	3
4	1.873 887	0.533 650	2.743 235	0.364 533	5.140 513	0.194 533	3.580 205	4
5	2.192 448	0.456 111	3.199 346	0.312 564	7.014 400	0.142 564	5.404 649	5
6	2.565 164	0.389 839	3.589 185	0.278 615	9.206 848	0.108 615	7.353 842	6
7	3.001 242	0.333 195	3.922 380	0.254 947	11.772 012	0.084 947	9.353 015	7
8	3.511 453	0.284 782	4.207 163	0.237 690	14.773 255	0.067 690	11.346 491	8
9	4.108 400	0.243 404	4.450 566	0.224 691	18.284 708	0.054 691	13.293 721	9
10	4.806 828	0.208 037	4.658 604	0.214 657	22.393 108	0.044 657	15.166 058	10
11	5.623 989	0.177 810	4.836 413	0.206 765	27.199 937	0.036 765	16.944 155	11
12	6.580 067	0.151 974	4.988 387	0.200 466	32.823 926	0.030 466	18.615 870	12
13	7.698 679	0.129 892	5.118 280	0.195 378	39.403 993	0.025 378	20.174 579	13
14	9.007 454	0.111 019	5.229 299	0.191 230	47.102 672	0.021 230	21.617 828	14
15	10.538 721	0.094 888	5.324 187	0.187 822	56.110 126	0.017 822	22.946 263	15
16	12.330 304	0.081 101	5.405 288	0.185 004	66.648 848	0.015 004	24.162 778	16
17	14.426 456	0.069 317	5.474 605	0.182 662	78.979 152	0.012 662	25.271 851	17
18	16.878 953	0.059 245	5.533 851	0.180 706	93.405 608	0.010 706	26.279 023	18
19	19.748 375	0.050 637	5.584 488	0.179 067	110.284 561	0.009 067	27.190 490	19
20	23.105 599	0.043 280	5.627 767	0.177 690	130.032 936	0.007 690	28.012 802	20
21	27.033 551	0.036 991	5.664 758	0.176 530	153.138 535	0.006 530	28.752 623	21
22	31.629 255	0.031 616	5.696 375	0.175 550	180.172 086	0.005 550	29.416 565	22
23	37.006 228	0.027 022	5.723 397	0.174 721	211.801 341	0.004 721	30.011 060	23
24	43.297 287	0.023 096	5.746 493	0.174 019	248.807 569	0.004 019	30.542 271	24
25	50.657 826	0.019 740	5.766 234	0.173 423	292.104 856	0.003 423	31.016 038	25
26	59.269 656	0.016 872	5.783 106	0.172 917	342.762 681	0.002 917	31.437 839	26
27	69.345 497	0.014 421	5.797 526	0.172 487	402.032 337	0.002 487	31.812 773	27
28	81.134 232	0.012 325	5.809 851	0.172 121	471.377 835	0.002 121	32.145 555	28
29	94.927 051	0.010 534	5.820 386	0.171 810	552.512 066	0.001 810	32.440 518	29
30	111.064 650	0.009 004	5.829 390	0.171 545	647.439 118	0.001 545	32.701 627	30
31	129.945 641	0.007 696	5.837 085	0.171 318	758.503 768	0.001 318	32.932 493	31
32	152.036 399	0.006 577	5.843 663	0.171 126	888.449 408	0.001 126	33.136 392	32
33	177.882 587	0.005 622	5.849 284	0.170 961	1 040.485 808	0.000 961	33.316 286	33
34	208.122 627	0.004 805	5.854 089	0.170 821	1 218.368 395	0.000 821	33.474 846	34
35	243.503 474	0.004 107	5.858 196	0.170 701	1 426.491 022	0.000 701	33.614 474	35

(cont.)

(cont.)

17%	Single Payment		Uniform Series				Arithmetic Gradient Series	17%
	Compound Amount Factor (CAF)	Present Worth Factor (PWF)	Present Worth Factor (PWF)	Capital Recovery Factor (CRF)	Compound Amount Factor (CAF)	Sinking Fund Factor (SFF)	Present Worth Factor (PWF)	
n	*F/P*	*P/F*	*P/A*	*A/P*	*F/A*	*A/F*	*P/G*	*n*
36	284.899 064	0.003 510	5.861 706	0.170 599	1 669.994 496	0.000 599	33.737 325	36
37	333.331 905	0.003 000	5.864 706	0.170 512	1 954.893 560	0.000 512	33.845 325	37
38	389.998 329	0.002 564	5.867 270	0.170 437	2 288.225 465	0.000 437	33.940 198	38
39	456.298 045	0.002 192	5.869 461	0.170 373	2 678.223 794	0.000 373	34.023 477	39
40	533.868 713	0.001 873	5.871 335	0.170 319	3 134.521 839	0.000 319	34.096 528	40
41	624.626 394	0.001 601	5.872 936	0.170 273	3 668.390 552	0.000 273	34.160 567	41
42	730.812 881	0.001 368	5.874 304	0.170 233	4 293.016 946	0.000 233	34.216 668	42
43	855.051 071	0.001 170	5.875 473	0.170 199	5 023.829 827	0.000 199	34.265 788	43
44	1 000.409 753	0.001 000	5.876 473	0.170 170	5 878.880 897	0.000 170	34.308 771	44
45	1 170.479 411	0.000 854	5.877 327	0.170 145	6 879.290 650	0.000 145	34.346 362	45
46	1 369.460 910	0.000 730	5.878 058	0.170 124	8 049.770 061	0.000 124	34.379 222	46
47	1 602.269 265	0.000 624	5.878 682	0.170 106	9 419.230 971	0.000 106	34.407 931	47
48	1 874.655 040	0.000 533	5.879 215	0.170 091	11 021.500 236	0.000 091	34.433 002	48
49	2 193.346 397	0.000 456	5.879 671	0.170 078	12 896.155 276	0.000 078	34.454 887	49
50	2 566.215 284	0.000 390	5.880 061	0.170 066	15 089.501 673	0.000 066	34.473 981	50
60	12 335.356 482	0.000 081	5.881 876	0.170 014	72 555.038 129	0.000 014	34.570 659	60
70	59 293.941 729	0.000 017	5.882 254	0.170 003	348 782.010 169	0.000 003	34.594 548	70
72	81 167.476 832	0.000 012	5.882 280	0.170 002	477 449.863 720	0.000 002	34.596 432	72
∞	∞	0	5.882 353	0.170 000	∞	0	34.602 076	∞

18%	Single Payment		Uniform Series				Arithmetic Gradient Series	18%
	Compound Amount Factor (CAF)	Present Worth Factor (PWF)	Present Worth Factor (PWF)	Capital Recovery Factor (CRF)	Compound Amount Factor (CAF)	Sinking Fund Factor (SFF)	Present Worth Factor (PWF)	
n	*F/P*	*P/F*	*P/A*	*A/P*	*F/A*	*A/F*	*P/G*	*n*
1	1.180 000	0.847 458	0.847 458	1.180 000	1.000 000	1.000 000	0.000 000	1
2	1.392 400	0.718 184	1.565 642	0.638 716	2.180 000	0.458 716	0.718 184	2
3	1.643 032	0.608 631	2.174 273	0.459 924	3.572 400	0.279 924	1.935 446	3
4	1.938 778	0.515 789	2.690 062	0.371 739	5.215 432	0.191 739	3.482 813	4
5	2.287 758	0.437 109	3.127 171	0.319 778	7.154 210	0.139 778	5.231 250	5
6	2.699 554	0.370 432	3.497 603	0.285 910	9.441 968	0.105 910	7.083 407	6
7	3.185 474	0.313 925	3.811 528	0.262 362	12.141 522	0.082 362	8.966 958	7
8	3.758 859	0.266 038	4.077 566	0.245 244	15.326 996	0.065 244	10.829 225	8
9	4.435 454	0.225 456	4.303 022	0.232 395	19.085 855	0.052 395	12.632 873	9
10	5.233 836	0.191 064	4.494 086	0.222 515	23.521 309	0.042 515	14.352 453	10
11	6.175 926	0.161 919	4.656 005	0.214 776	28.755 144	0.034 776	15.971 644	11
12	7.287 593	0.137 220	4.793 225	0.208 628	34.931 070	0.028 628	17.481 059	12
13	8.599 359	0.116 288	4.909 513	0.203 686	42.218 663	0.023 686	18.876 511	13
14	10.147 244	0.098 549	5.008 062	0.199 678	50.818 022	0.019 678	20.157 647	14
15	11.973 748	0.083 516	5.091 578	0.196 403	60.965 266	0.016 403	21.326 872	15
16	14.129 023	0.070 776	5.162 354	0.193 710	72.939 014	0.013 710	22.388 517	16
17	16.672 247	0.059 980	5.222 334	0.191 485	87.068 036	0.011 485	23.348 195	17
18	19.673 251	0.050 830	5.273 164	0.189 639	103.740 283	0.009 639	24.212 313	18
19	23.214 436	0.043 077	5.316 241	0.188 103	123.413 534	0.008 103	24.987 692	19
20	27.393 035	0.036 506	5.352 746	0.186 820	146.627 970	0.006 820	25.681 299	20
21	32.323 781	0.030 937	5.383 683	0.185 746	174.021 005	0.005 746	26.300 039	21
22	38.142 061	0.026 218	5.409 901	0.184 846	206.344 785	0.004 846	26.850 612	22
23	45.007 632	0.022 218	5.432 120	0.184 090	244.486 847	0.004 090	27.339 418	23
24	53.109 006	0.018 829	5.450 949	0.183 454	289.494 479	0.003 454	27.772 490	24
25	62.668 627	0.015 957	5.466 906	0.182 919	342.603 486	0.002 919	28.155 456	25
26	73.948 980	0.013 523	5.480 429	0.182 467	405.272 113	0.002 467	28.493 527	26
27	87.259 797	0.011 460	5.491 889	0.182 087	479.221 093	0.002 087	28.791 488	27
28	102.966 560	0.009 712	5.501 601	0.181 765	566.480 890	0.001 765	29.053 709	28
29	121.500 541	0.008 230	5.509 831	0.181 494	669.447 450	0.001 494	29.284 161	29
30	143.370 638	0.006 975	5.516 806	0.181 264	790.947 991	0.001 264	29.486 434	30
31	169.177 353	0.005 911	5.522 717	0.181 070	934.318 630	0.001 070	29.663 763	31
32	199.629 277	0.005 009	5.527 726	0.180 906	1 103.495 983	0.000 906	29.819 050	32
33	235.562 547	0.004 245	5.531 971	0.180 767	1 303.125 260	0.000 767	29.954 895	33
34	277.963 805	0.003 598	5.535 569	0.180 650	1 538.687 807	0.000 650	30.073 616	34
35	327.997 290	0.003 049	5.538 618	0.180 550	1 816.651 612	0.000 550	30.177 275	35

(cont.)

(cont.)

18%	Single Payment		Uniform Series				Arithmetic Gradient Series	18%
	Compound Amount Factor (CAF)	Present Worth Factor (PWF)	Present Worth Factor (PWF)	Capital Recovery Factor (CRF)	Compound Amount Factor (CAF)	Sinking Fund Factor (SFF)	Present Worth Factor (PWF)	
n	*F/P*	*P/F*	*P/A*	*A/P*	*F/A*	*A/F*	*P/G*	*n*
36	387.036 802	0.002 584	5.541 201	0.180 466	2 144.648 902	0.000 466	30.267 706	36
37	456.703 427	0.002 190	5.543 391	0.180 395	2 531.685 705	0.000 395	30.346 532	37
38	538.910 044	0.001 856	5.545 247	0.180 335	2 988.389 132	0.000 335	30.415 189	38
39	635.913 852	0.001 573	5.546 819	0.180 284	3 527.299 175	0.000 284	30.474 945	39
40	750.378 345	0.001 333	5.548 152	0.180 240	4 163.213 027	0.000 240	30.526 919	40
41	885.446 447	0.001 129	5.549 281	0.180 204	4 913.591 372	0.000 204	30.572 094	41
42	1 044.826 807	0.000 957	5.550 238	0.180 172	5 799.037 819	0.000 172	30.611 335	42
43	1 232.895 633	0.000 811	5.551 049	0.180 146	6 843.864 626	0.000 146	30.645 401	43
44	1 454.816 847	0.000 687	5.551 737	0.180 124	8 076.760 259	0.000 124	30.674 958	44
45	1 716.683 879	0.000 583	5.552 319	0.180 105	9 531.577 105	0.000 105	30.700 589	45
46	2 025.686 977	0.000 494	5.552 813	0.180 089	11 248.260 984	0.000 089	30.722 804	46
47	2 390.310 633	0.000 418	5.553 231	0.180 075	13 273.947 961	0.000 075	30.742 048	47
48	2 820.566 547	0.000 355	5.553 586	0.180 064	15 664.258 594	0.000 064	30.758 711	48
49	3 328.268 525	0.000 300	5.553 886	0.180 054	18 484.825 141	0.000 054	30.773 133	49
50	3 927.356 860	0.000 255	5.554 141	0.180 046	21 813.093 666	0.000 046	30.785 610	50
60	20 555.139 966	0.000 049	5.555 285	0.180 009	114 189.666 478	0.000 009	30.846 479	60
70	107 582.222 368	0.000 009	5.555 504	0.180 002	597 673.457 599	0.000 002	30.860 296	70
72	149 797.486 425	0.000 007	5.555 518	0.180 001	832 202.702 361	0.000 001	30.861 321	72
∞	∞	0	5.555 556	0.180 000	∞	0	30.864 198	∞

19%	Single Payment		Uniform Series				Arithmetic Gradient Series	19%
	Compound Amount Factor (CAF)	Present Worth Factor (PWF)	Present Worth Factor (PWF)	Capital Recovery Factor (CRF)	Compound Amount Factor (CAF)	Sinking Fund Factor (SFF)	Present Worth Factor (PWF)	
n	*F/P*	*P/F*	*P/A*	*A/P*	*F/A*	*A/F*	*P/G*	*n*
1	1.190 000	0.840 336	0.840 336	1.190 000	1.000 000	1.000 000	0.000 000	1
2	1.416 100	0.706 165	1.546 501	0.646 621	2.190 000	0.456 621	0.706 165	2
3	1.685 159	0.593 416	2.139 917	0.467 308	3.606 100	0.277 308	1.892 996	3
4	2.005 339	0.498 669	2.638 586	0.378 991	5.291 259	0.188 991	3.389 003	4
5	2.386 354	0.419 049	3.057 635	0.327 050	7.296 598	0.137 050	5.065 200	5
6	2.839 761	0.352 142	3.409 777	0.293 274	9.682 952	0.103 274	6.825 912	6
7	3.379 315	0.295 918	3.705 695	0.269 855	12.522 713	0.079 855	8.601 419	7
8	4.021 385	0.248 671	3.954 366	0.252 885	15.902 028	0.062 885	10.342 113	8
9	4.785 449	0.208 967	4.163 332	0.240 192	19.923 413	0.050 192	12.013 848	9
10	5.694 684	0.175 602	4.338 935	0.230 471	24.708 862	0.040 471	13.594 269	10
11	6.776 674	0.147 565	4.486 500	0.222 891	30.403 546	0.032 891	15.069 919	11
12	8.064 242	0.124 004	4.610 504	0.216 896	37.180 220	0.026 896	16.433 966	12
13	9.596 448	0.104 205	4.714 709	0.212 102	45.244 461	0.022 102	17.684 428	13
14	11.419 773	0.087 567	4.802 277	0.208 235	54.840 909	0.018 235	18.822 805	14
15	13.589 530	0.073 586	4.875 863	0.205 092	66.260 682	0.015 092	19.853 010	15
16	16.171 540	0.061 837	4.937 700	0.202 523	79.850 211	0.012 523	20.780 565	16
17	19.244 133	0.051 964	4.989 664	0.200 414	96.021 751	0.010 414	21.611 987	17
18	22.900 518	0.043 667	5.033 331	0.198 676	115.265 884	0.008 676	22.354 329	18
19	27.251 616	0.036 695	5.070 026	0.197 238	138.166 402	0.007 238	23.014 840	19
20	32.429 423	0.030 836	5.100 862	0.196 045	165.418 018	0.006 045	23.600 728	20
21	38.591 014	0.025 913	5.126 775	0.195 054	197.847 442	0.005 054	24.118 983	21
22	45.923 307	0.021 775	5.148 550	0.194 229	236.438 456	0.004 229	24.576 267	22
23	54.648 735	0.018 299	5.166 849	0.193 542	282.361 762	0.003 542	24.978 838	23
24	65.031 994	0.015 377	5.182 226	0.192 967	337.010 497	0.002 967	25.332 510	24
25	77.388 073	0.012 922	5.195 148	0.192 487	402.042 491	0.002 487	25.642 636	25
26	92.091 807	0.010 859	5.206 007	0.192 086	479.430 565	0.002 086	25.914 104	26
27	109.589 251	0.009 125	5.215 132	0.191 750	571.522 372	0.001 750	26.151 353	27
28	130.411 208	0.007 668	5.222 800	0.191 468	681.111 623	0.001 468	26.358 391	28
29	155.189 338	0.006 444	5.229 243	0.191 232	811.522 831	0.001 232	26.538 816	29
30	184.675 312	0.005 415	5.234 658	0.191 034	966.712 169	0.001 034	26.695 848	30
31	219.763 621	0.004 550	5.239 209	0.190 869	1 151.387 481	0.000 869	26.832 358	31
32	261.518 710	0.003 824	5.243 033	0.190 729	1 371.151 103	0.000 729	26.950 897	32
33	311.207 264	0.003 213	5.246 246	0.190 612	1 632.669 812	0.000 612	27.053 722	33
34	370.336 645	0.002 700	5.248 946	0.190 514	1 943.877 077	0.000 514	27.142 830	34
35	440.700 607	0.002 269	5.251 215	0.190 432	2 314.213 721	0.000 432	27.219 980	35

(cont.)

(cont.)

19%	Single Payment		Uniform Series				Arithmetic Gradient Series	19%
	Compound Amount Factor (CAF)	Present Worth Factor (PWF)	Present Worth Factor (PWF)	Capital Recovery Factor (CRF)	Compound Amount Factor (CAF)	Sinking Fund Factor (SFF)	Present Worth Factor (PWF)	
n	*F/P*	*P/F*	*P/A*	*A/P*	*F/A*	*A/F*	*P/G*	*n*
36	524.433 722	0.001 907	5.253 122	0.190 363	2 754.914 328	0.000 363	27.286 719	36
37	624.076 130	0.001 602	5.254 724	0.190 305	3 279.348 051	0.000 305	27.344 404	37
38	742.650 594	0.001 347	5.256 071	0.190 256	3 903.424 180	0.000 256	27.394 225	38
39	883.754 207	0.001 132	5.257 202	0.190 215	4 646.074 775	0.000 215	27.437 224	39
40	1 051.667 507	0.000 951	5.258 153	0.190 181	5 529.828 982	0.000 181	27.474 308	40
41	1 251.484 333	0.000 799	5.258 952	0.190 152	6 581.496 488	0.000 152	27.506 270	41
42	1 489.266 356	0.000 671	5.259 624	0.190 128	7 832.980 821	0.000 128	27.533 800	42
43	1 772.226 964	0.000 564	5.260 188	0.190 107	9 322.247 177	0.000 107	27.557 499	43
44	2 108.950 087	0.000 474	5.260 662	0.190 090	11 094.474 141	0.000 090	27.577 888	44
45	2 509.650 603	0.000 398	5.261 061	0.190 076	13 203.424 228	0.000 076	27.595 421	45
46	2 986.484 218	0.000 335	5.261 396	0.190 064	15 713.074 831	0.000 064	27.610 489	46
47	3 553.916 219	0.000 281	5.261 677	0.190 053	18 699.559 049	0.000 053	27.623 432	47
48	4 229.160 301	0.000 236	5.261 913	0.190 045	22 253.475 268	0.000 045	27.634 545	48
49	5 032.700 758	0.000 199	5.262 112	0.190 038	26 482.635 569	0.000 038	27.644 083	49
50	5 988.913 902	0.000 167	5.262 279	0.190 032	31 515.336 327	0.000 032	27.652 265	50
60	34 104.970 919	0.000 029	5.263 004	0.190 006	179 494.583 786	0.000 006	27.690 759	60
∞	∞	0	5.263 158	0.190 000	∞	0	27.700 831	∞

20%	Single Payment		Uniform Series				Arithmetic Gradient Series	20%
	Compound Amount Factor (CAF)	Present Worth Factor (PWF)	Present Worth Factor (PWF)	Capital Recovery Factor (CRF)	Compound Amount Factor (CAF)	Sinking Fund Factor (SFF)	Present Worth Factor (PWF)	
n	*F/P*	*P/F*	*P/A*	*A/P*	*F/A*	*A/F*	*P/G*	*n*
1	1.200 000	0.833 333	0.833 333	1.200 000	1.000 000	1.000 000	0.000 000	1
2	1.440 000	0.694 444	1.527 778	0.654 545	2.200 000	0.454 545	0.694 444	2
3	1.728 000	0.578 704	2.106 481	0.474 725	3.640 000	0.274 725	1.851 852	3
4	2.073 600	0.482 253	2.588 735	0.386 289	5.368 000	0.186 289	3.298 611	4
5	2.488 320	0.401 878	2.990 612	0.334 380	7.441 600	0.134 380	4.906 121	5
6	2.985 984	0.334 898	3.325 510	0.300 706	9.929 920	0.100 706	6.580 611	6
7	3.583 181	0.279 082	3.604 592	0.277 424	12.915 904	0.077 424	8.255 101	7
8	4.299 817	0.232 568	3.837 160	0.260 609	16.499 085	0.060 609	9.883 077	8
9	5.159 780	0.193 807	4.030 967	0.248 079	20.798 902	0.048 079	11.433 531	9
10	6.191 736	0.161 506	4.192 472	0.238 523	25.958 682	0.038 523	12.887 081	10
11	7.430 084	0.134 588	4.327 060	0.231 104	32.150 419	0.031 104	14.232 961	11
12	8.916 100	0.112 157	4.439 217	0.225 265	39.580 502	0.025 265	15.466 684	12
13	10.699 321	0.093 464	4.532 681	0.220 620	48.496 603	0.020 620	16.588 251	13
14	12.839 185	0.077 887	4.610 567	0.216 893	59.195 923	0.016 893	17.600 776	14
15	15.407 022	0.064 905	4.675 473	0.213 882	72.035 108	0.013 882	18.509 453	15
16	18.488 426	0.054 088	4.729 561	0.211 436	87.442 129	0.011 436	19.320 771	16
17	22.186 111	0.045 073	4.774 634	0.209 440	105.930 555	0.009 440	20.041 943	17
18	26.623 333	0.037 561	4.812 195	0.207 805	128.116 666	0.007 805	20.680 481	18
19	31.948 000	0.031 301	4.843 496	0.206 462	154.740 000	0.006 462	21.243 896	19
20	38.337 600	0.026 084	4.869 580	0.205 357	186.688 000	0.005 357	21.739 493	20
21	46.005 120	0.021 737	4.891 316	0.204 444	225.025 600	0.004 444	22.174 228	21
22	55.206 144	0.018 114	4.909 430	0.203 690	271.030 719	0.003 690	22.554 620	22
23	66.247 373	0.015 095	4.924 525	0.203 065	326.236 863	0.003 065	22.886 709	23
24	79.496 847	0.012 579	4.937 104	0.202 548	392.484 236	0.002 548	23.176 028	24
25	95.396 217	0.010 483	4.947 587	0.202 119	471.981 083	0.002 119	23.427 611	25
26	114.475 460	0.008 735	4.956 323	0.201 762	567.377 300	0.001 762	23.645 998	26
27	137.370 552	0.007 280	4.963 602	0.201 467	681.852 760	0.001 467	23.835 267	27
28	164.844 662	0.006 066	4.969 668	0.201 221	819.223 312	0.001 221	23.999 058	28
29	197.813 595	0.005 055	4.974 724	0.201 016	984.067 974	0.001 016	24.140 605	29
30	237.376 314	0.004 213	4.978 936	0.200 846	1 181.881 569	0.000 846	24.262 774	30
31	284.851 577	0.003 511	4.982 447	0.200 705	1 419.257 883	0.000 705	24.368 092	31
32	341.821 892	0.002 926	4.985 372	0.200 587	1 704.109 459	0.000 587	24.458 782	32
33	410.186 270	0.002 438	4.987 810	0.200 489	2 045.931 351	0.000 489	24.536 796	33
34	492.223 524	0.002 032	4.989 842	0.200 407	2 456.117 621	0.000 407	24.603 839	34
35	590.668 229	0.001 693	4.991 535	0.200 339	2 948.341 146	0.000 339	24.661 400	35

(cont.)

(cont.)

20%	Single Payment		Uniform Series				Arithmetic Gradient Series	20%
	Compound Amount Factor (CAF)	Present Worth Factor (PWF)	Present Worth Factor (PWF)	Capital Recovery Factor (CRF)	Compound Amount Factor (CAF)	Sinking Fund Factor (SFF)	Present Worth Factor (PWF)	
n	*F/P*	*P/F*	*P/A*	*A/P*	*F/A*	*A/F*	*P/G*	*n*
36	708.801 875	0.001 411	4.992 946	0.200 283	3 539.009 375	0.000 283	24.710 780	36
37	850.562 250	0.001 176	4.994 122	0.200 235	4 247.811 250	0.000 235	24.753 104	37
38	1 020.674 700	0.000 980	4.995 101	0.200 196	5 098.373 500	0.000 196	24.789 355	38
39	1 224.809 640	0.000 816	4.995 918	0.200 163	6 119.048 200	0.000 163	24.820 380	39
40	1 469.771 568	0.000 680	4.996 598	0.200 136	7 343.857 840	0.000 136	24.846 915	40
41	1 763.725 882	0.000 567	4.997 165	0.200 113	8 813.629 408	0.000 113	24.869 594	41
42	2 116.471 058	0.000 472	4.997 638	0.200 095	10 577.355 289	0.000 095	24.888 966	42
43	2 539.765 269	0.000 394	4.998 031	0.200 079	12 693.826 347	0.000 079	24.905 503	43
44	3 047.718 323	0.000 328	4.998 359	0.200 066	15 233.591 617	0.000 066	24.919 612	44
45	3 657.261 988	0.000 273	4.998 633	0.200 055	18 281.309 940	0.000 055	24.931 643	45
46	4 388.714 386	0.000 228	4.998 861	0.200 046	21 938.571 928	0.000 046	24.941 896	46
47	5 266.457 263	0.000 190	4.999 051	0.200 038	26 327.286 314	0.000 038	24.950 631	47
48	6 319.748 715	0.000 158	4.999 209	0.200 032	31 593.743 576	0.000 032	24.958 068	48
49	7 583.698 458	0.000 132	4.999 341	0.200 026	37 913.492 292	0.000 026	24.964 397	49
50	9 100.438 150	0.000 110	4.999 451	0.200 022	45 497.190 750	0.000 022	24.969 782	50
60	56 347.514 353	0.000 018	4.999 911	0.200 004	281 732.571 766	0.000 004	24.994 232	60
∞	∞	0	5.000 000	0.200 000	∞	0	25.000 000	∞

25%	Single Payment		Uniform Series				Arithmetic Gradient Series	25%
	Compound Amount Factor (CAF)	Present Worth Factor (PWF)	Present Worth Factor (PWF)	Capital Recovery Factor (CRF)	Compound Amount Factor (CAF)	Sinking Fund Factor (SFF)	Present Worth Factor (PWF)	
n	*F/P*	*P/F*	*P/A*	*A/P*	*F/A*	*A/F*	*P/G*	*n*
1	1.250 000	0.800 000	0.800 000	1.250 000	1.000 000	1.000 000	0.000 000	1
2	1.562 500	0.640 000	1.440 000	0.694 444	2.250 000	0.444 444	0.640 000	2
3	1.953 125	0.512 000	1.952 000	0.512 295	3.812 500	0.262 295	1.664 000	3
4	2.441 406	0.409 600	2.361 600	0.423 442	5.765 625	0.173 442	2.892 800	4
5	3.051 758	0.327 680	2.689 280	0.371 847	8.207 031	0.121 847	4.203 520	5
6	3.814 697	0.262 144	2.951 424	0.338 819	11.258 789	0.088 819	5.514 240	6
7	4.768 372	0.209 715	3.161 139	0.316 342	15.073 486	0.066 342	6.772 531	7
8	5.960 464	0.167 772	3.328 911	0.300 399	19.841 858	0.050 399	7.946 936	8
9	7.450 581	0.134 218	3.463 129	0.288 756	25.802 322	0.038 756	9.020 678	9
10	9.313 226	0.107 374	3.570 503	0.280 073	33.252 903	0.030 073	9.987 046	10
11	11.641 532	0.085 899	3.656 403	0.273 493	42.566 129	0.023 493	10.846 039	11
12	14.551 915	0.068 719	3.725 122	0.268 448	54.207 661	0.018 448	11.601 953	12
13	18.189 894	0.054 976	3.780 098	0.264 543	68.759 576	0.014 543	12.261 660	13
14	22.737 368	0.043 980	3.824 078	0.261 501	86.949 470	0.011 501	12.833 407	14
15	28.421 709	0.035 184	3.859 263	0.259 117	109.686 838	0.009 117	13.325 988	15
16	35.527 137	0.028 147	3.887 410	0.257 241	138.108 547	0.007 241	13.748 200	16
17	44.408 921	0.022 518	3.909 928	0.255 759	173.635 684	0.005 759	14.108 488	17
18	55.511 151	0.018 014	3.927 942	0.254 586	218.044 605	0.004 586	14.414 733	18
19	69.388 939	0.014 412	3.942 354	0.253 656	273.555 756	0.003 656	14.674 140	19
20	86.736 174	0.011 529	3.953 883	0.252 916	342.944 695	0.002 916	14.893 195	20
21	108.420 217	0.009 223	3.963 107	0.252 327	429.680 869	0.002 327	15.077 663	21
22	135.525 272	0.007 379	3.970 485	0.251 858	538.101 086	0.001 858	15.232 615	22
23	169.406 589	0.005 903	3.976 388	0.251 485	673.626 358	0.001 485	15.362 481	23
24	211.758 237	0.004 722	3.981 111	0.251 186	843.032 947	0.001 186	15.471 095	24
25	264.697 796	0.003 778	3.984 888	0.250 948	1 054.791 184	0.000 948	15.561 764	25
26	330.872 245	0.003 022	3.987 911	0.250 758	1 319.488 980	0.000 758	15.637 322	26
27	413.590 306	0.002 418	3.990 329	0.250 606	1 650.361 225	0.000 606	15.700 186	27
28	516.987 883	0.001 934	3.992 263	0.250 485	2 063.951 531	0.000 485	15.752 412	28
29	646.234 854	0.001 547	3.993 810	0.250 387	2 580.939 414	0.000 387	15.795 740	29
30	807.793 567	0.001 238	3.995 048	0.250 310	3 227.174 268	0.000 310	15.831 640	30
31	1 009.741 959	0.000 990	3.996 039	0.250 248	4 034.967 835	0.000 248	15.861 351	31
32	1 262.177 448	0.000 792	3.996 831	0.250 198	5 044.709 793	0.000 198	15.885 911	32
33	1 577.721 810	0.000 634	3.997 465	0.250 159	6 306.887 242	0.000 159	15.906 194	33
34	1 972.152 263	0.000 507	3.997 972	0.250 127	7 884.609 052	0.000 127	15.922 927	34
35	2 465.190 329	0.000 406	3.998 377	0.250 101	9 856.761 315	0.000 101	15.936 719	35

(cont.)

(cont.)

25%	Single Payment		Uniform Series				Arithmetic Gradient Series	25%
	Compound Amount Factor (CAF)	Present Worth Factor (PWF)	Present Worth Factor (PWF)	Capital Recovery Factor (CRF)	Compound Amount Factor (CAF)	Sinking Fund Factor (SFF)	Present Worth Factor (PWF)	
n	*F/P*	*P/F*	*P/A*	*A/P*	*F/A*	*A/F*	*P/G*	*n*
36	3 081.487 911	0.000 325	3.998 702	0.250 081	12 321.951 644	0.000 081	15.948 077	36
37	3 851.859 889	0.000 260	3.998 962	0.250 065	15 403.439 555	0.000 065	15.957 423	37
38	4 814.824 861	0.000 208	3.999 169	0.250 052	19 255.299 444	0.000 052	15.965 108	38
39	6 018.531 076	0.000 166	3.999 335	0.250 042	24 070.124 305	0.000 042	15.971 422	39
40	7 523.163 845	0.000 133	3.999 468	0.250 033	30 088.655 381	0.000 033	15.976 606	40
41	9 403.954 807	0.000 106	3.999 575	0.250 027	37 611.819 226	0.000 027	15.980 859	41
42	11 754.943 508	0.000 085	3.999 660	0.250 021	47 015.774 033	0.000 021	15.984 347	42
43	14 693.679 385	0.000 068	3.999 728	0.250 017	58 770.717 541	0.000 017	15.987 205	43
44	18 367.099 232	0.000 054	3.999 782	0.250 014	73 464.396 926	0.000 014	15.989 547	44
45	22 958.874 039	0.000 044	3.999 826	0.250 011	91 831.496 158	0.000 011	15.991 463	45
46	28 698.592 549	0.000 035	3.999 861	0.250 009	114 790.370 197	0.000 009	15.993 031	46
47	35 873.240 687	0.000 028	3.999 888	0.250 007	143 488.962 747	0.000 007	15.994 313	47
48	44 841.550 858	0.000 022	3.999 911	0.250 006	179 362.203 434	0.000 006	15.995 361	48
49	56 051.938 573	0.000 018	3.999 929	0.250 004	224 203.754 292	0.000 004	15.996 218	49
50	70 064.923 216	0.000 014	3.999 943	0.250 004	280 255.692 865	0.000 004	15.996 917	50
∞	∞	0	4.000 000	0.250 000	∞	0	16.000 000	∞

30%	Single Payment		Uniform Series				Arithmetic Gradient Series	30%
	Compound Amount Factor (CAF)	Present Worth Factor (PWF)	Present Worth Factor (PWF)	Capital Recovery Factor (CRF)	Compound Amount Factor (CAF)	Sinking Fund Factor (SFF)	Present Worth Factor (PWF)	
n	*F/P*	*P/F*	*P/A*	*A/P*	*F/A*	*A/F*	*P/G*	*n*
1	1.300 000	0.769 231	0.769 231	1.300 000	1.000 000	1.000 000	0.000 000	1
2	1.690 000	0.591 716	1.360 947	0.734 783	2.300 000	0.434 783	0.591 716	2
3	2.197 000	0.455 166	1.816 113	0.550 627	3.990 000	0.250 627	1.502 048	3
4	2.856 100	0.350 128	2.166 241	0.461 629	6.187 000	0.161 629	2.552 432	4
5	3.712 930	0.269 329	2.435 570	0.410 582	9.043 100	0.110 582	3.629 748	5
6	4.826 809	0.207 176	2.642 746	0.378 394	12.756 030	0.078 394	4.665 629	6
7	6.274 852	0.159 366	2.802 112	0.356 874	17.582 839	0.056 874	5.621 827	7
8	8.157 307	0.122 589	2.924 702	0.341 915	23.857 691	0.041 915	6.479 953	8
9	10.604 499	0.094 300	3.019 001	0.331 235	32.014 998	0.031 235	7.234 350	9
10	13.785 849	0.072 538	3.091 539	0.323 463	42.619 497	0.023 463	7.887 193	10
11	17.921 604	0.055 799	3.147 338	0.317 729	56.405 346	0.017 729	8.445 179	11
12	23.298 085	0.042 922	3.190 260	0.313 454	74.326 950	0.013 454	8.917 321	12
13	30.287 511	0.033 017	3.223 277	0.310 243	97.625 036	0.010 243	9.313 524	13
14	39.373 764	0.025 398	3.248 675	0.307 818	127.912 546	0.007 818	9.643 693	14
15	51.185 893	0.019 537	3.268 211	0.305 978	167.286 310	0.005 978	9.917 206	15
16	66.541 661	0.015 028	3.283 239	0.304 577	218.472 203	0.004 577	10.142 628	16
17	86.504 159	0.011 560	3.294 800	0.303 509	285.013 864	0.003 509	10.327 591	17
18	112.455 407	0.008 892	3.303 692	0.302 692	371.518 023	0.002 692	10.478 762	18
19	146.192 029	0.006 840	3.310 532	0.302 066	483.973 430	0.002 066	10.601 887	19
20	190.049 638	0.005 262	3.315 794	0.301 587	630.165 459	0.001 587	10.701 861	20
21	247.064 529	0.004 048	3.319 842	0.301 219	820.215 097	0.001 219	10.782 812	21
22	321.183 888	0.003 113	3.322 955	0.300 937	1 067.279 626	0.000 937	10.848 195	22
23	417.539 054	0.002 395	3.325 350	0.300 720	1 388.463 514	0.000 720	10.900 885	23
24	542.800 770	0.001 842	3.327 192	0.300 554	1 806.002 568	0.000 554	10.943 257	24
25	705.641 001	0.001 417	3.328 609	0.300 426	2 348.803 338	0.000 426	10.977 269	25
26	917.333 302	0.001 090	3.329 700	0.300 327	3 054.444 340	0.000 327	11.004 522	26
27	1 192.533 293	0.000 839	3.330 538	0.300 252	3 971.777 642	0.000 252	11.026 324	27
28	1 550.293 280	0.000 645	3.331 183	0.300 194	5 164.310 934	0.000 194	11.043 740	28
29	2 015.381 264	0.000 496	3.331 679	0.300 149	6 714.604 214	0.000 149	11.057 633	29
30	2 619.995 644	0.000 382	3.332 061	0.300 115	8 729.985 479	0.000 115	11.068 702	30
31	3 405.994 337	0.000 294	3.332 355	0.300 088	11 349.981 122	0.000 088	11.077 510	31
32	4 427.792 638	0.000 226	3.332 581	0.300 068	14 755.975 459	0.000 068	11.084 511	32
33	5 756.130 429	0.000 174	3.332 754	0.300 052	19 183.768 097	0.000 052	11.090 071	33
34	7 482.969 558	0.000 134	3.332 888	0.300 040	24 939.898 526	0.000 040	11.094 481	34
35	9 727.860 425	0.000 103	3.332 991	0.300 031	32 422.868 084	0.000 031	11.097 976	35

(cont.)

(cont.)

30%	Single Payment		Uniform Series				Arithmetic Gradient Series	30%
	Compound Amount Factor (CAF)	Present Worth Factor (PWF)	Present Worth Factor (PWF)	Capital Recovery Factor (CRF)	Compound Amount Factor (CAF)	Sinking Fund Factor (SFF)	Present Worth Factor (PWF)	
n	*F/P*	*P/F*	*P/A*	*A/P*	*F/A*	*A/F*	*P/G*	*n*
36	12 646.218 553	0.000 079	3.333 070	0.300 024	42 150.728 509	0.000 024	11.100 743	36
37	16 440.084 119	0.000 061	3.333 131	0.300 018	54 796.947 062	0.000 018	11.102 933	37
38	21 372.109 354	0.000 047	3.333 177	0.300 014	71 237.031 180	0.000 014	11.104 664	38
39	27 783.742 160	0.000 036	3.333 213	0.300 011	92 609.140 534	0.000 011	11.106 032	39
40	36 118.864 808	0.000 028	3.333 241	0.300 008	120 392.882 695	0.000 008	11.107 112	40
41	46 954.524 251	0.000 021	3.333 262	0.300 006	156 511.747 503	0.000 006	11.107 964	41
42	61 040.881 526	0.000 016	3.333 279	0.300 005	203 466.271 754	0.000 005	11.108 636	42
43	79 353.145 984	0.000 013	3.333 291	0.300 004	264 507.153 281	0.000 004	11.109 165	43
44	103 159.089 779	0.000 010	3.333 301	0.300 003	343 860.299 265	0.000 003	11.109 582	44
45	134 106.816 713	0.000 007	3.333 308	0.300 002	447 019.389 044	0.000 002	11.109 910	45
∞	∞	0	3.333 333	0.300 000	∞	0	11.111 111	∞

35%	Single Payment		Uniform Series				Arithmetic Gradient Series	35%
	Compound Amount Factor (CAF)	Present Worth Factor (PWF)	Present Worth Factor (PWF)	Capital Recovery Factor (CRF)	Compound Amount Factor (CAF)	Sinking Fund Factor (SFF)	Present Worth Factor (PWF)	
n	*F/P*	*P/F*	*P/A*	*A/P*	*F/A*	*A/F*	*P/G*	*n*
1	1.350 000	0.740 741	0.740 741	1.350 000	1.000 000	1.000 000	0.000 000	1
2	1.822 500	0.548 697	1.289 438	0.775 532	2.350 000	0.425 532	0.548 697	2
3	2.460 375	0.406 442	1.695 880	0.589 664	4.172 500	0.239 664	1.361 581	3
4	3.321 506	0.301 068	1.996 948	0.500 764	6.632 875	0.150 764	2.264 786	4
5	4.484 033	0.223 014	2.219 961	0.450 458	9.954 381	0.100 458	3.156 840	5
6	6.053 445	0.165 195	2.385 157	0.419 260	14.438 415	0.069 260	3.982 816	6
7	8.172 151	0.122 367	2.507 523	0.398 800	20.491 860	0.048 800	4.717 017	7
8	11.032 404	0.090 642	2.598 165	0.384 887	28.664 011	0.034 887	5.351 511	8
9	14.893 745	0.067 142	2.665 308	0.375 191	39.696 415	0.025 191	5.888 649	9
10	20.106 556	0.049 735	2.715 043	0.368 318	54.590 160	0.018 318	6.336 264	10
11	27.143 850	0.036 841	2.751 884	0.363 387	74.696 715	0.013 387	6.704 672	11
12	36.644 198	0.027 289	2.779 173	0.359 819	101.840 566	0.009 819	7.004 856	12
13	49.469 667	0.020 214	2.799 387	0.357 221	138.484 764	0.007 221	7.247 429	13
14	66.784 051	0.014 974	2.814 361	0.355 320	187.954 431	0.005 320	7.442 086	14
15	90.158 469	0.011 092	2.825 453	0.353 926	254.738 482	0.003 926	7.597 368	15
16	121.713 933	0.008 216	2.833 669	0.352 899	344.896 951	0.002 899	7.720 608	16
17	164.313 809	0.006 086	2.839 755	0.352 143	466.610 884	0.002 143	7.817 983	17
18	221.823 643	0.004 508	2.844 263	0.351 585	630.924 694	0.001 585	7.894 620	18
19	299.461 918	0.003 339	2.847 602	0.351 173	852.748 336	0.001 173	7.954 728	19
20	404.273 589	0.002 474	2.850 076	0.350 868	1 152.210 254	0.000 868	8.001 726	20
21	545.769 345	0.001 832	2.851 908	0.350 642	1 556.483 843	0.000 642	8.038 371	21
22	736.788 616	0.001 357	2.853 265	0.350 476	2 102.253 188	0.000 476	8.066 873	22
23	994.664 631	0.001 005	2.854 270	0.350 352	2 839.041 804	0.000 352	8.088 991	23
24	1 342.797 252	0.000 745	2.855 015	0.350 261	3 833.706 435	0.000 261	8.106 120	24
25	1 812.776 291	0.000 552	2.855 567	0.350 193	5 176.503 687	0.000 193	8.119 359	25
26	2 447.247 992	0.000 409	2.855 975	0.350 143	6 989.279 978	0.000 143	8.129 575	26
27	3 303.784 789	0.000 303	2.856 278	0.350 106	9 436.527 970	0.000 106	8.137 445	27
28	4 460.109 466	0.000 224	2.856 502	0.350 078	12 740.312 759	0.000 078	8.143 498	28
29	6 021.147 779	0.000 166	2.856 668	0.350 058	17 200.422 225	0.000 058	8.148 149	29
30	8 128.549 501	0.000 123	2.856 791	0.350 043	23 221.570 004	0.000 043	8.151 716	30
31	10 973.541 827	0.000 091	2.856 882	0.350 032	31 350.119 505	0.000 032	8.154 450	31
32	14 814.281 466	0.000 068	2.856 950	0.350 024	42 323.661 332	0.000 024	8.156 543	32
33	19 999.279 979	0.000 050	2.857 000	0.350 018	57 137.942 798	0.000 018	8.158 143	33
34	26 999.027 972	0.000 037	2.857 037	0.350 013	77 137.222 778	0.000 013	8.159 365	34
35	36 448.687 763	0.000 027	2.857 064	0.350 010	104 136.250 750	0.000 010	8.160 298	35
36	49 205.728 479	0.000 020	2.857 085	0.350 007	140 584.938 513	0.000 007	8.161 009	36
37	66 427.733 447	0.000 015	2.857 100	0.350 005	189 790.666 992	0.000 005	8.161 551	37
38	89 677.440 154	0.000 011	2.857 111	0.350 004	256 218.400 440	0.000 004	8.161 964	38
39	121 064.544 208	0.000 008	2.857 119	0.350 003	345 895.840 593	0.000 003	8.162 277	39
40	163 437.134 680	0.000 006	2.857 125	0.350 002	466 960.384 801	0.000 002	8.162 516	40
∞	∞	0	2.857.143	0.350 000	∞	0	8.163 265	∞

40%	Single Payment		Uniform Series				Arithmetic Gradient Series	40%
	Compound Amount Factor (CAF)	Present Worth Factor (PWF)	Present Worth Factor (PWF)	Capital Recovery Factor (CRF)	Compound Amount Factor (CAF)	Sinking Fund Factor (SFF)	Present Worth Factor (PWF)	
n	*F/P*	*P/F*	*P/A*	*A/P*	*F/A*	*A/F*	*P/G*	*n*
1	1.400 000	0.714 286	0.714 286	1.400 000	1.000 000	1.000 000	0.000 000	1
2	1.960 000	0.510 204	1.224 490	0.816 667	2.400 000	0.416 667	0.510 204	2
3	2.744 000	0.364 431	1.588 921	0.629 358	4.360 000	0.229 358	1.239 067	3
4	3.841 600	0.260 308	1.849 229	0.540 766	7.104 000	0.140 766	2.019 992	4
5	5.378 240	0.185 934	2.035 164	0.491 361	10.945 600	0.091 361	2.763 729	5
6	7.529 536	0.132 810	2.167 974	0.461 260	16.323 840	0.061 260	3.427 781	6
7	10.541 350	0.094 865	2.262 839	0.441 923	23.853 376	0.041 923	3.996 968	7
8	14.757 891	0.067 760	2.330 599	0.429 074	34.394 726	0.029 074	4.471 291	8
9	20.661 047	0.048 400	2.378 999	0.420 345	49.152 617	0.020 345	4.858 493	9
10	28.925 465	0.034 572	2.413 571	0.414 324	69.813 664	0.014 324	5.169 637	10
11	40.495 652	0.024 694	2.438 265	0.410 128	98.739 129	0.010 128	5.416 577	11
12	56.693 912	0.017 639	2.455 904	0.407 182	139.234 781	0.007 182	5.610 602	12
13	79.371 477	0.012 599	2.468 503	0.405 104	195.928 693	0.005 104	5.761 789	13
14	111.120 068	0.008 999	2.477 502	0.403 632	275.300 171	0.003 632	5.878 780	14
15	155.568 096	0.006 428	2.483 930	0.402 588	386.420 239	0.002 588	5.968 773	15
16	217.795 334	0.004 591	2.488 521	0.401 845	541.988 334	0.001 845	6.037 645	16
17	304.913 467	0.003 280	2.491 801	0.401 316	759.783 668	0.001 316	6.090 119	17
18	426.878 854	0.002 343	2.494 144	0.400 939	1 064.697 136	0.000 939	6.129 943	18
19	597.630 396	0.001 673	2.495 817	0.400 670	1 491.575 990	0.000 670	6.160 061	19
20	836.682 554	0.001 195	2.497 012	0.400 479	2 089.206 386	0.000 479	6.182 770	20
21	1 171.355 576	0.000 854	2.497 866	0.400 342	2 925.888 940	0.000 342	6.199 844	21
22	1 639.897 806	0.000 610	2.498 476	0.400 244	4 097.244 516	0.000 244	6.212 650	22
23	2 295.856 929	0.000 436	2.498 911	0.400 174	5 737.142 322	0.000 174	6.222 233	23
24	3 214.199 700	0.000 311	2.499 222	0.400 124	8 032.999 251	0.000 124	6.229 388	24
25	4 499.879 581	0.000 222	2.499 444	0.400 089	11 247.198 951	0.000 089	6.234 722	25
26	6 299.831 413	0.000 159	2.499 603	0.400 064	15 747.078 532	0.000 064	6.238 690	26
27	8 819.763 978	0.000 113	2.499 717	0.400 045	22 046.909 945	0.000 045	6.241 638	27
28	12 347.669 569	0.000 081	2.499 798	0.400 032	30 866.673 923	0.000 032	6.243 825	28
29	17 286.737 397	0.000 058	2.499 855	0.400 023	43 214.343 492	0.000 023	6.245 444	29
30	24 201.432 355	0.000 041	2.499 897	0.400 017	60 501.080 889	0.000 017	6.246 643	30
31	33 882.005 298	0.000 030	2.499 926	0.400 012	84 702.513 244	0.000 012	6.247 528	31
32	47 434.807 417	0.000 021	2.499 947	0.400 008	118 584.518 542	0.000 008	6.248 182	32
33	66 408.730 383	0.000 015	2.499 962	0.400 006	166 019.325 959	0.000 006	6.248 664	33
34	92 972.222 537	0.000 011	2.499 973	0.400 004	232 428.056 342	0.000 004	6.249 019	34
35	130 161.111 552	0.000 008	2.499 981	0.400 003	325 400.278 879	0.000 003	6.249 280	35
∞	∞	0	2.500 000	0.400 000	∞	0	6.250 000	∞

45%	Single Payment		Uniform Series				Arithmetic Gradient Series	45%
	Compound Amount Factor (CAF)	Present Worth Factor (PWF)	Present Worth Factor (PWF)	Capital Recovery Factor (CRF)	Compound Amount Factor (CAF)	Sinking Fund Factor (SFF)	Present Worth Factor (PWF)	
n	*F/P*	*P/F*	*P/A*	*A/P*	*F/A*	*A/F*	*P/G*	*n*
1	1.450 000	0.689 655	0.689 655	1.450 000	1.000 000	1.000 000	0.000 000	1
2	2.102 500	0.475 624	1.165 279	0.858 163	2.450 000	0.408 163	0.475 624	2
3	3.048 625	0.328 017	1.493 296	0.669 660	4.552 500	0.219 660	1.131 658	3
4	4.420 506	0.226 218	1.719 515	0.581 559	7.601 125	0.131 559	1.810 313	4
5	6.409 734	0.156 013	1.875 527	0.533 183	12.021 631	0.083 183	2.434 364	5
6	9.294 114	0.107 595	1.983 122	0.504 255	18.431 365	0.054 255	2.972 339	6
7	13.476 466	0.074 203	2.057 326	0.486 068	27.725 480	0.036 068	3.417 559	7
8	19.540 876	0.051 175	2.108 500	0.474 271	41.201 946	0.024 271	3.775 783	8
9	28.334 269	0.035 293	2.143 793	0.466 463	60.742 821	0.016 463	4.058 126	9
10	41.084 691	0.024 340	2.168 133	0.461 226	89.077 091	0.011 226	4.277 186	10
11	59.572 802	0.016 786	2.184 920	0.457 683	130.161 781	0.007 683	4.445 048	11
12	86.380 562	0.011 577	2.196 496	0.455 271	189.734 583	0.005 271	4.572 391	12
13	125.251 815	0.007 984	2.204 480	0.453 622	276.115 145	0.003 622	4.668 198	13
14	181.615 132	0.005 506	2.209 986	0.452 491	401.366 961	0.002 491	4.739 778	14
15	263.341 942	0.003 797	2.213 784	0.451 715	582.982 093	0.001 715	4.792 941	15
16	381.845 816	0.002 619	2.216 403	0.451 182	846.324 035	0.001 182	4.832 224	16
17	553.676 433	0.001 806	2.218 209	0.450 814	1 228.169 850	0.000 814	4.861 122	17
18	802.830 827	0.001 246	2.219 454	0.450 561	1 781.846 283	0.000 561	4.882 297	18
19	1 164.104 699	0.000 859	2.220 313	0.450 387	2 584.677 110	0.000 387	4.897 759	19
20	1 687.951 814	0.000 592	2.220 906	0.450 267	3 748.781 809	0.000 267	4.909 016	20
21	2 447.530 131	0.000 409	2.221 314	0.450 184	5 436.733 623	0.000 184	4.917 187	21
22	3 548.918 689	0.000 282	2.221 596	0.450 127	7 884.263 754	0.000 127	4.923 104	22
23	5 145.932 100	0.000 194	2.221 790	0.450 087	11 433.182 443	0.000 087	4.927 380	23
24	7 461.601 544	0.000 134	2.221 924	0.450 060	16 579.114 543	0.000 060	4.930 462	24
25	10 819.322 239	0.000 092	2.222 017	0.450 042	24 040.716 087	0.000 042	4.932 680	25
26	15 688.017 247	0.000 064	2.222 081	0.450 029	34 860.038 326	0.000 029	4.934 274	26
27	22 747.625 008	0.000 044	2.222 125	0.450 020	50 548.055 573	0.000 020	4.935 417	27
28	32 984.056 262	0.000 030	2.222 155	0.450 014	73 295.680 581	0.000 014	4.936 235	28
29	47 826.881 579	0.000 021	2.222 176	0.450 009	106 279.736 843	0.000 009	4.936 821	29
30	69 348.978 290	0.000 014	2.222 190	0.450 006	154 106.618 422	0.000 006	4.937 239	30
31	100 556.018 521	0.000 010	2.222 200	0.450 004	223 455.596 712	0.000 004	4.937 537	31
32	145 806.226 855	0.000 007	2.222 207	0.450 003	324 011.615 233	0.000 003	4.937 750	32
33	211 419.028 939	0.000 005	2.222 212	0.450 002	469 817.842 088	0.000 002	4.937 901	33
34	306 557.591 962	0.000 003	2.222 215	0.450 001	681 236.871 027	0.000 001	4.938 009	34
35	444 508.508 345	0.000 002	2.222 217	0.450 001	987 794.462 989	0.000 001	4.938 086	35
∞	∞	0	2.222 222	0.450 000	∞	0	4.938 272	∞

50%	Single Payment		Uniform Series				Arithmetic Gradient Series	50%
	Compound Amount Factor (CAF)	Present Worth Factor (PWF)	Present Worth Factor (PWF)	Capital Recovery Factor (CRF)	Compound Amount Factor (CAF)	Sinking Fund Factor (SFF)	Present Worth Factor (PWF)	
n	*F/P*	*P/F*	*P/A*	*A/P*	*F/A*	*A/F*	*P/G*	*n*
1	1.500 000	0.666 667	0.666 667	1.500 000	1.000 000	1.000 000	0.000 000	1
2	2.250 000	0.444 444	1.111 111	0.900 000	2.500 000	0.400 000	0.444 444	2
3	3.375 000	0.296 296	1.407 407	0.710 526	4.750 000	0.210 526	1.037 037	3
4	5.062 500	0.197 531	1.604 938	0.623 077	8.125 000	0.123 077	1.629 630	4
5	7.593 750	0.131 687	1.736 626	0.575 829	13.187 500	0.075 829	2.156 379	5
6	11.390 625	0.087 791	1.824 417	0.548 120	20.781 250	0.048 120	2.595 336	6
7	17.085 938	0.058 528	1.882 945	0.531 083	32.171 875	0.031 083	2.946 502	7
8	25.628 906	0.039 018	1.921 963	0.520 301	49.257 813	0.020 301	3.219 631	8
9	38.443 359	0.026 012	1.947 975	0.513 354	74.886 719	0.013 354	3.427 730	9
10	57.665 039	0.017 342	1.965 317	0.508 824	113.330 078	0.008 824	3.583 803	10
11	86.497 559	0.011 561	1.976 878	0.505 848	170.995 117	0.005 848	3.699 413	11
12	129.746 338	0.007 707	1.984 585	0.503 884	257.492 676	0.003 884	3.784 194	12
13	194.619 507	0.005 138	1.989 724	0.502 582	387.239 014	0.002 582	3.845 853	13
14	291.929 260	0.003 425	1.993 149	0.501 719	581.858 521	0.001 719	3.890 384	14
15	437.893 890	0.002 284	1.995 433	0.501 144	873.787 781	0.001 144	3.922 356	15
16	656.840 836	0.001 522	1.996 955	0.500 762	1 311.681 671	0.000 762	3.945 192	16
17	985.261 253	0.001 015	1.997 970	0.500 508	1 968.522 507	0.000 508	3.961 432	17
18	1 477.891 880	0.000 677	1.998 647	0.500 339	2 953.783 760	0.000 339	3.972 934	18
19	2 216.837 820	0.000 451	1.999 098	0.500 226	4 431.675 640	0.000 226	3.981 054	19
20	3 325.256 730	0.000 301	1.999 399	0.500 150	6 648.513 460	0.000 150	3.986 768	20
21	4 987.885 095	0.000 200	1.999 599	0.500 100	9 973.770 190	0.000 100	3.990 778	21
22	7 481.827 643	0.000 134	1.999 733	0.500 067	14 961.655 285	0.000 067	3.993 584	22
23	11 222.741 464	0.000 089	1.999 822	0.500 045	22 443.482 928	0.000 045	3.995 545	23
24	16 834.112 196	0.000 059	1.999 881	0.500 030	33 666.224 392	0.000 030	3.996 911	24
25	25 251.168 294	0.000 040	1.999 921	0.500 020	50 500.336 588	0.000 020	3.997 861	25
26	37 876.752 441	0.000 026	1.999 947	0.500 013	75 751.504 882	0.000 013	3.998 522	26
27	56 815.128 662	0.000 018	1.999 965	0.500 009	113 628.257 323	0.000 009	3.998 979	27
28	85 222.692 992	0.000 012	1.999 977	0.500 006	170 443.385 985	0.000 006	3.999 296	28
29	127 834.039 489	0.000 008	1.999 984	0.500 004	255 666.078 977	0.000 004	3.999 515	29
30	191 751.059 233	0.000 005	1.999 990	0.500 003	383 500.118 466	0.000 003	3.999 666	30
∞	∞	0	2.000 000	0.500 000	∞	0	4.000 000	∞

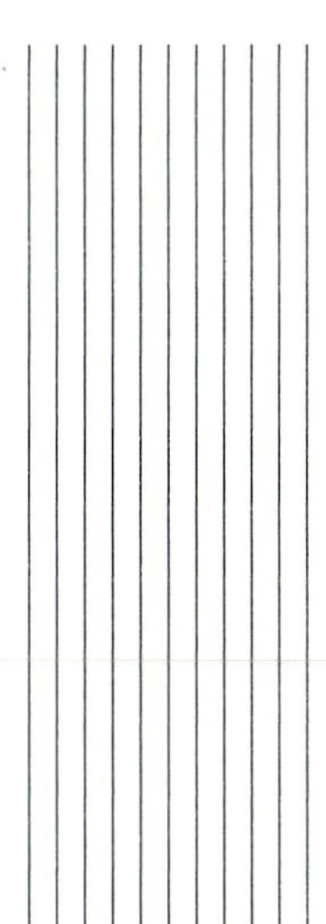

Index

C

D

E

F

G

H

I

K

L

M

N

O

P

R

S

T

U

V

W